Computational Chemical Reaction Engineering

# Computational Chemical Reaction Engineering

## Modeling, Simulation, and Design with MATLAB

*Kamal I. M. Al-Malah*
Process Engineering Department
German University of Technology
Muscat
Oman

WILEY

Published by John Wiley & Sons, Inc., Hoboken, New Jersey.



*Library of Congress Cataloging-in-Publication Data applied for:*

Hardback ISBN: 9781394429196

Cover Design: Wiley
Cover Image: © Image Source/Getty Images

Set in 9.5/12.5pt STIXTwoText by Straive, Chennai, India

# Contents

## About This Book

*MATLAB-Based Computations of Chemical Reaction Engineering* is written to modernize the learning and practice of chemical reaction engineering (CRE) by integrating rigorous scientific foundations with AI-assisted computational tools. The book bridges classical CRE theory, numerical analysis, MATLAB programming, and the emerging use of large language models (LLMs) as engineering copilots. It provides a unified pedagogical framework that reflects the realities of $21^{st}$-century chemical engineering education, where analytical reasoning, computational literacy, and responsible AI usage converge.

The textbook is designed for senior undergraduate and graduate chemical engineering students, practicing engineers, researchers, and instructors who seek a comprehensive and application-driven approach to CRE. Each chapter blends theory, modeling, problem-solving, and MATLAB-based implementations, enabling readers to move seamlessly from fundamental concepts to full computational workflows.

A distinguishing feature of this book is its emphasis on AI-enhanced computation anchored in a scientific discipline. This includes the use of CURVD-compliant prompts (Concise, Unambiguous, Right, Verifiable, and Derived) and a structured methodology for translating chemical algorithms into robust MATLAB code with full physical validation. Through this methodology, the book demonstrates a unique educational paradigm: Exploring AI Mining Using Human Intelligence Minding.

### Chapter Summaries

Chapter 1 introduces the foundations of chemical reaction engineering, including reaction kinetics, thermodynamic relationships, rate laws, and Arrhenius modeling. It then integrates modern AI-assisted MATLAB workflows, addressing prompt engineering, CURVD principles, code verification, and responsible AI usage.

Chapter 2 develops core reactor design models under isothermal conditions for batch, CSTR, PFR, PBR, and FBR systems. It includes MATLAB implementations, cascade reactor configurations, optimization strategies, and real industrial examples.

Chapter 3 addresses chemical equilibrium, including thermodynamic foundations, $\Delta G^\circ$, $\Delta H^\circ$, $K_{eq}$, van't Hoff equation, extent of reaction, and MATLAB-based equilibrium modeling for ideal and nonideal systems such as ammonia synthesis and WGS reaction.

Chapter 4 examines multiple reactions, selectivity, yield, and optimization of desired products. MATLAB codes generate selectivity surfaces, solve multi-reaction networks, perform root-finding operations, and analyze reactor configurations including recycle loops.

Chapter 5 focuses on non-isothermal reaction engineering, incorporating energy balances, temperature coupling, dynamic modeling, multi-steady states, stability analysis, and MATLAB solutions for adiabatic and nonadiabatic BR, CSTR, and PFR systems.

Chapter 6 covers homogeneous and heterogeneous catalysis, catalytic reactor design, mass transfer limitations, Thiele modulus, effectiveness factors, PBR/FBR hydrodynamics, and MATLAB models for pore diffusion, catalytic kinetics, and industrial catalytic processes.

## Method of Approach

The development of this book followed a structured and reproducible methodology grounded in scientific rigor, human expertise, and modern AI-assisted computation. This approach reflects the concept of Exploring AI Mining Using Human Intelligence Minding, where human intelligence directs and validates AI-generated computational output.

1) **Role of the Author:**
   The author developed all scientific content throughout the book, including the formulation of reactor models, derivation of algebraic and differential equations, and construction of physically consistent mathematical representations. He designed CURVD-compliant prompts to communicate these algorithms to the AI system, executed and validated all MATLAB code, and ensured that every solution was numerically stable and physically meaningful. In addition, the author derived analytical solutions for many examples, where applicable, and prepared all end-of-chapter problems to reinforce both theoretical understanding and computational proficiency.
2) **Role of the AI System:**
   The AI system translated human-language algorithms into MATLAB code structures, recommended debugging strategies, and provided alternative formulations. It served as an assistant, not a replacement, to engineering judgment.
3) **Role of MATLAB:**
   MATLAB executed the collaboratively generated code, identified errors, tested numerical stability, generated plots and results, and provided the final verification layer, ensuring scientific correctness.
   Together, this human–AI–MATLAB triad created a modern, efficient, and verifiable workflow for CRE education and research.

## Notice to Readers:

Important Note on Copying and Pasting Code:

When copying text directly from the PDF into the MATLAB editor, formatting mismatches (such as unrecognized non-ASCII characters substituting the power operator ^) may occasionally cause syntax errors. If you encounter a syntax bug, please check the original text in the book and manually correct any character mismatches in your editor.

Alternatively, you can download the original, ready-to-run source code. The corresponding *.m file for every figure in Chapters 1 through 6, as well as the four appendices, is available on the book's companion website, organized by chapter and appendix.

## Acknowledgments

I would like to express my sincere gratitude to the academic and professional communities whose support made this book possible. I extend thanks to my students across Jordan, Bahrain, Saudi Arabia, the UAE, and Oman, whose curiosity and dedication enriched every stage of my teaching and research journey.

My appreciation goes to John Wiley & Sons for their continued confidence in my work and for supporting innovative approaches in engineering education. I also thank the developers of MATLAB and the creators of modern AI tools that enabled a new dimension of computational pedagogy.

Finally, I thank my God, above all, colleagues, friends, and family (Mom, Dad, Fadia, Anwar, Lama, Abdallah, and Mohammad) for their unwavering encouragement throughout the writing of this textbook.

*Dr. Kamal I. M. Al-Malah*

# About the Companion Website

This book is accompanied by a companion website:

**www.wiley.com/go/al-malah/chemicalreaction**

To access the website materials, click on the link under Related Resources on the book's page on wiley.com. You can also find this by searching www.wiley.com for the ISBN 9781394429196.

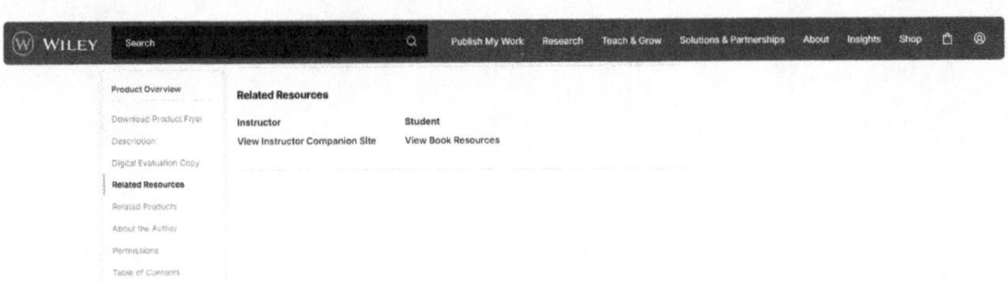


The website includes:

- Answers
- MATLAB Codes

# 1
# Introduction + AI Copilot for AI-Generated MATLAB Code in Chemical Reaction Engineering (CRE)

## 1.1 Introduction to CRE

Chemical reaction engineering (CRE) is a cornerstone course in chemical engineering, providing the fundamental knowledge and tools for designing, analyzing, and optimizing chemical processes. It underpins the development of countless products and technologies that impact our daily lives.

### 1.1.1 Importance and Applications of CRE

1) **Process Design and Optimization:** CRE is central to the design and optimization of chemical reactors, which are the heart of most chemical processes. It enables engineers to select the appropriate reactor type, determine optimal operating conditions, and maximize product yield and selectivity.
2) **Product Development:** CRE plays a crucial role in the development of new products and materials. By understanding reaction kinetics and mechanisms, engineers can design processes to synthesize novel compounds with desired properties.
3) **Environmental Protection:** CRE is essential for developing clean and sustainable technologies. It helps with designing processes that minimize waste generation, reduce energy consumption, and mitigate environmental impact. Developing processes for wastewater treatment, air pollution control, and resource recovery.
4) **Energy Production:** CRE is involved in the production of fuels, chemicals, and materials used in various energy applications, including renewable energy technologies.
5) **Biotechnology and Pharmaceuticals:** CRE is vital to produce biopharmaceuticals, enzymes, and other bio-based products. It encompasses bioreactor design, fermentation processes, and downstream processing.
6) **Petroleum Refining:** Designing catalytic reactors for cracking, reforming, and hydroprocessing of crude oil.
7) **Petrochemicals:** Producing olefins, aromatics, and polymers from hydrocarbon feedstocks.
8) **Food Processing:** Designing reactors for food sterilization, fermentation, and processing.
9) **Materials Science:** Synthesizing advanced materials, such as catalysts, polymers, and ceramics.

### 1.1.2 Chemical Identity

Chemical identity refers to the distinct characteristics that define a chemical species. These characteristics include:

1) **Kind of Atoms:** The specific elements that make up the species, determined by the atomic number (number of protons in the nucleus) of each element.
2) **Number of Atoms:** The number of each type of atom present in the chemical species, which contributes to its molecular formula.
3) **Configuration of Atoms:** The specific arrangement of these atoms within the molecule, including how they are bonded to each other. This includes the geometric structure, bond angles, and the spatial orientation of atoms.

The chemical identity of a species is crucial because it dictates the chemical properties and reactivity of the substance. When a chemical reaction occurs, the original species lose their chemical identity as their atoms are rearranged to form new substances with different identities. For example, when hydrogen gas ($H_2$) reacts with oxygen gas ($O_2$) to form water ($H_2O$), the identity of the hydrogen and oxygen molecules is lost, and a new chemical species, water, with its own unique chemical identity, is born (i.e., formed).

In general, there are three ways a chemical species can lose its chemical identity: decomposition, polymerization (or combination), and isomerization:

1) **Decomposition:** Decomposition is a chemical reaction where a single compound breaks down into two or simpler substances. For example, consider the decomposition of calcium carbonate ($CaCO_3$). When heated, calcium carbonate decomposes into calcium oxide (CaO) and carbon dioxide ($CO_2$):

$$CaCO_3 \rightarrow CaO + CO_2 \tag{1.1}$$

In this process, the original compound, calcium carbonate, loses its chemical identity as it breaks down into two different substances with distinct chemical identities. Another example is the decomposition of ethane into hydrogen and ethene (or ethylene).

$$CH_3CH_3 \rightarrow H_2 + H_2C = CH_2 \tag{1.2}$$

Notice that the molecular mass of individual products is less than that of the reactant. The reaction simply splits a big cake into slices of a cake.

2) **Polymerization (or Combination):** Polymerization is a chemical reaction where small molecules called monomers combine to form a large, chain-like molecule called a polymer. For example, the polymerization of ethylene ($C_2H_4$) to form polyethylene ($-(C_2H_4)n-$):

$$n\,C_2H_4 \rightarrow -(H_2C = CH_2)_n- \tag{1.3}$$

In this process, the individual ethylene monomers lose their separate chemical identities as they bond together to form a much larger molecule with a new chemical identity, called polyethylene. Another example is the combination of nitrogen ($N_2$) and oxygen ($O_2$) to form nitric oxide (NO):

$$2N_2 + 2O_2 \rightarrow 2NO \tag{1.4}$$

Notice that the molecular mass of individual reactants is less than that of the product. The reaction simply binds two pieces of Lego into one big Lego.

3) **Isomerization:** Isomerization is a chemical reaction where a molecule is transformed into another molecule with the same molecular formula but a different structural arrangement. For example, the conversion of n-butane ($C_4H_{10}$) to isobutane ($C_4H_{10}$):

$$CH_3CH_2CH_2CH_3 \rightarrow CH_3CH(CH_3)CH_3 \tag{1.5}$$

In this process, although both *n*-butane and isobutane have the same molecular formula ($C_4H_{10}$), they have different structures and, therefore, different chemical identities. Through isomerization, *n*-butane loses its chemical identity and becomes isobutane, a distinct species with different physical and chemical properties. Another example is the transformation of 1-butene into 2-methylpropene (also known as isobutene):

$$CH_3 - CH_2 - CH = CH_2 \rightarrow CH_2 = C(CH_3)_2 \tag{1.6}$$

Notice that the molecular mass of the reactant is the same as that of the product.

**NOTE**

**A Reversible Versus Irreversible Reaction**

Unlike irreversible reactions, which proceed only in one direction, reversible reactions can go both forward and backward. Whether a reversible reaction moves forward or backward depends on factors like temperature and the amounts of reactants and products present relative to their equilibrium levels. While technically all reactions can be reversed, many reactions have such a strong tendency to move forward that they are considered irreversible for practical purposes.

### 1.1.3 Reaction Kinetics

The reaction rate is the rate at which a species loses its chemical identity per unit volume. The rate of a reaction can be expressed as the rate of disappearance of a reactant or as the rate of appearance of a product. Consider species A:

$$A \xrightarrow{k} B \tag{1.7}$$

$r_A$ = the rate of formation of species $A$ per unit volume
$-r_A$ = the rate of disappearance of species $A$ per unit volume
$r_B$ = the rate of formation of species $B$ per unit volume
$k$ = the rate constant $f$(temperature)

If B is being created at 0.1 moles per decimeter cubed (or, L) per second, then

$$r_B = 0.1\frac{\text{mol}}{\text{dm}^3 \cdot \text{s}} = 0.1\frac{\text{mol}}{\text{L} \cdot \text{s}}$$

Then $A$ is disappearing at the same rate:

$$-r_A = 0.1\frac{\text{mol}}{\text{dm}^3 \cdot \text{s}} = 0.1\frac{\text{mol}}{\text{L} \cdot \text{s}}$$

For a catalytic reaction, we refer to $-r'_A$, which is the rate of disappearance of species A per mass of catalyst as a basis, instead of per volume of a solution as a basis. In general, the volume basis is used for homogeneous reactions, whereas the catalyst mass (or catalyst surface area) is used for heterogeneous reactions.

Consider a reacting species *j*; the following points regard the reaction rate of formation/disappearance of *j*:

- $r_j$ is the rate of formation of species *j* per unit volume $\left[\text{e.g.,} \frac{\text{mol}}{\text{dm}^3 \cdot \text{s}}\right]$.
- $r'_j$ is the rate of formation of species *j* per unit mass of a catalyst $\left[\text{e.g.,} \frac{\text{mol}}{\text{kg catalyst} \cdot \text{s}}\right]$.
- $r_j$ *or* $r'_j$ is a function of concentration, temperature, pressure, and the type of catalyst (if any).
- $r_j$ *or* $r'_j$ is independent of the type of reaction system (batch, plug flow, etc.).

We use an algebraic equation to relate the rate of reaction, $-r_A$, to the concentration of reacting species and to the temperature at which the reaction occurs $\left[\text{e.g.,} -r_A = kC_A^2\right]$. Yet, to find the concentration profile of species *j* as f(time), we need to either analytically or numerically solve the counterpart ordinary differential equation. See upcoming examples.

Examples of other rate laws are:

$$-r_A = \frac{k_1 C_A}{1 + k_2 C_A} - r_A = kC_A^0 = k - r_A = kC_A^n \tag{1.8a–c}$$

**Example 1.1** ***Expressions of*** $-r_A$, $-r_B$, ***and*** $r_C$

$$\text{Consider the reaction : } A + 2B \rightarrow 3C \tag{1.9}$$

The rate of disappearance of *A* is: 3 moles of *A* per dm$^3$ (or, L) per second at the start of the reaction. At the start of the reaction:

a) What is $-r_A$ (rate of disappearance of *A*)?

$$-r_A = 3\frac{\text{mol}}{\text{dm}^3 \cdot \text{s}} = 3\frac{\text{mol}}{\text{L} \cdot \text{s}}$$

b) What is the rate of formation of *A* ($r_A$)? $r_A = -3\frac{\text{mol}}{\text{dm}^3 \cdot \text{s}} = -3\frac{\text{mol}}{\text{L} \cdot \text{s}}$

c) What is the rate of disappearance of *B* ($-r_B$)?

For every one mole of *A* that disappears/forms, two moles of *B* disappear/form. Reactant *B* is disappearing twice as fast as reactant *A*. $-r_B = 2 \times -r_A = 2 \times 3\frac{\text{mol}}{\text{dm}^3 \cdot \text{s}} = 6\frac{\text{mol}}{\text{dm}^3 \cdot \text{s}} = 6\frac{\text{mol}}{\text{L} \cdot \text{s}}$

Thus, the rate of formation of *B* ($r_B$) $= -6\frac{\text{mol}}{\text{dm}^3 \cdot \text{s}} = -6\frac{\text{mol}}{\text{L} \cdot \text{s}}$

d) What is the rate of formation of *C*?

*C* is a product that is being formed three times as fast as *A* is disappearing:

$$r_C = 3 \times (-r_A) = 3 \times 3\frac{\text{mol}}{\text{dm}^3 \cdot \text{s}} = 9\frac{\text{mol}}{\text{L} \cdot \text{s}}$$

Because *C* is a product being formed, its rate of formation is positive.

e) What is the rate of disappearance of *C* ($-r_C$)?

$$-r_C = -9\frac{\text{mol}}{\text{dm}^3 \cdot \text{s}} = -9\frac{\text{mol}}{\text{L} \cdot \text{s}}$$

Because *C* is a product, its rate of disappearance, $-r_C$, is a negative number.

**NOTE**

For *reactants*, the rate of *disappearance* is a positive (+) number.
For *reactants*, the rate of *formation* is a negative (−) number because they are disappearing and not being formed.
For *products*, the rate of *formation* is a positive (+) number.
For *products*, the rate of *disappearance* is a negative number (−) because they are being formed and not disappearing.

Rate laws and reaction order are fundamental concepts in chemical kinetics, which is the study of the rates at which chemical reactions occur.

#### 1.1.3.1 Rate Laws

Rate laws are mathematical expressions that describe the reaction rate as a function of the concentrations of reactants and temperature for a specific reaction. It is an empirical equation that must be determined experimentally. Rate laws are determined through experimental observation. Consequently, rate laws depend on the reaction chemistry itself, not on the type of reactor in which the reactions take place.

For a general reaction:

$$aA + bB \rightarrow cC + dD \tag{1.10}$$

The rate law can be written as:

$$-r_A = -\frac{a}{b}r_B = k[A]^m[B]^n = kC_A^m C_B^n \tag{1.11}$$

Here:

- $-r_A$ is the reaction rate of disappearance of $A$.
- $-r_B$ is the reaction rate of disappearance of $B$.
- $k$ is the rate constant, a proportionality constant specific to the reaction at a given temperature.
- $[A]$ and $[B]$ are the concentrations of the reactants $A$ and $B$.
- $m$ and $n$ are the reaction orders with respect to reactants $A$ and $B$, respectively.

Relative rates indicate how quickly one species is consumed or produced compared with another species in a given reaction. As shown above, we can relate the rate of disappearance of $B$ relative to that of $A$. The limiting reactant, consumed first, is usually chosen as our basis for calculation.

If we have more than one chemical reaction taking place, then the net rate of formation of a species (e.g., $A$) is the sum of its reaction rates across all reactions in which it participates as either a reactant or a product.

#### 1.1.3.2 Elementary Reactions

A reaction *follows* an *elementary rate law* if and only if *the* stoichiometric coefficients are the same as the individual reaction order of each species. For example, for the following reaction:

$$A + B \rightarrow C + D$$

The rate law will be: $-r_A = kC_AC_B$, if it is an elementary reaction.

Another example:

$$2NO + O_2 \rightarrow 2NO_2$$

The reaction rate would be: $-r_{NO} = k_{NO}\,(C_{NO})^2\,C_{O2}$, if the reaction were elementary. However, the observed reaction rate does not conform to the elementary rate law definition. See next section.

#### 1.1.3.3 Nonelementary Rate Laws

A nonelementary reaction is a chemical reaction that occurs through a series of simpler, elementary reaction steps. Unlike elementary reactions, the rate law of a nonelementary reaction cannot be directly determined from its stoichiometric equation. This is because the overall reaction mechanism involves multiple steps, often with intermediate species that are not explicitly shown in the overall equation.

**Key characteristics of nonelementary reactions:**

a) **Multistep Mechanism:** They occur through a sequence of elementary steps.
b) **Intermediate Species:** These are species formed in one step and consumed in a subsequent step, not appearing in the overall reaction equation.
c) **Rate Law is Not Directly Proportional to Stoichiometry:** The rate law is determined by the slowest step (rate-determining step) in the mechanism, which may not directly reflect the overall stoichiometry.

Consider the following reaction:

$$2NO(g) + O_2(g) \rightarrow 2NO_2(g)$$

This reaction might proceed through a two-step mechanism:

1) **Fast Equilibrium Step:** $NO + O_2 \rightleftharpoons NO_3$ (fast)
2) **Slow Rate-determining Step:** $NO_3 + NO \rightarrow 2NO_2$ (slow)

The overall reaction is the sum of these two steps, but the rate law is determined by the slow step, which involves the collision of NO and $NO_3$. This rate law will not directly correspond to the stoichiometry of the overall reaction.

Example: If the rate law for the nonelementary reaction $A + B \rightarrow C + D$ is found to be $-r_A = k\,C_A^2 C_B$, then the reaction will be second order in $A$, first order in $B$, and third order overall.

#### 1.1.3.4 Reaction Order

The reaction order is the sum of the powers of the concentration terms in the rate law. For the example above, the overall reaction order is $m + n$.

**Types of Reaction Orders:**

Consider a reaction:

$$aA \rightarrow Products \tag{1.12}$$

**0. Zero-order reactions:**

The rate is independent of the concentration of the reactants.

Rate law: $-r_A = -\frac{dC_A}{dt} = k$

$$\rightarrow \int_{CA_0}^{C_A} dC_A = -k\int_0^t dt \rightarrow C_A - C_{A0} = -kt \rightarrow C_A = C_{A0} - kt \tag{1.13}$$

**Example 1.2** ***Decomposition Rate of Ammonia on Platinum Surface***
The decomposition of $NH_3$ on the platinum surface is a zero-order reaction. If the rate constant ($k$) is $4\times10^{-3}$ $Ms^{-1}$ $\left(\frac{\text{mol}}{\text{L.s}}\right)$, how long will it take to reduce the initial concentration of $NH_3$ from 0.15 to 0.08 M?

**Solution**

$$-r_{NH3} = -\frac{d[NH_3]}{dt} = k\left[\frac{\text{mol}}{\text{L.s}}\right] \rightarrow d[NH_3] = -kdt$$

$$\rightarrow \int_{NH_{3,0}}^{NH_{3f}} d[NH_3] = -k\int_0^{t_f} dt$$

Integrate from $t_i = 0$ to $t = t_f$, we reach at:

$$[NH_{3f}] - [NH_{3,0}] = -k(t_f - 0) \rightarrow t_f = \frac{([NH_{3f}] - [NH_{3,0}])}{-k} = \frac{([NH_{3,0}] - [NH_{3f}])}{k}$$
$$= \frac{(0.15 - 0.08)\text{M}}{4\times10^{-3}\frac{\text{M}}{\text{s}}} = 17.5\ \text{s}$$

Figure 1.1 shows MATLAB code for symbolically (i.e., analytically) solving a zero-order reaction rate for ammonia decomposition:

The answer is copied from the MATLAB Command Window, where C1 represents the constant of integration, which can be determined by utilizing a given initial condition for y = (x) in general.

```
C1 - k*t
symbolic function inputs: t
The time needed to reach the final concentration of NH3 is 17.5 seconds
```

On the other hand, Figure 1.2 shows the code for the numerical solution of the zero-order ordinary differential equation of the reaction rate.

Figure 1.3 shows the generated plot showing $[NH_3]$ as a function of time. From the plot, one can see that at $t = 17.5$ s, $[NH_3] = 0.08$ M, which coincides with the analytical solution.

Notice that the decomposition of ammonia on a platinum surface can exhibit a reaction order between zero and one. On the one hand, studies have shown that at high ammonia pressures, the reaction becomes zero order. This means the rate of decomposition is constant and independent of the ammonia concentration. This is because all the available catalytic sites on the platinum surface are occupied by ammonia molecules, so further increases in concentration don't affect the reaction rate. On the other hand, at lower ammonia pressures, the reaction typically follows first-order or even higher-order kinetics. This means the reaction rate is directly proportional to (or even more dependent on) the ammonia concentration.

It is reported by Imbihl et al. [1] that values are reached when either the reaction is inhibited by a high $NH_3$ coverage blocking $O_2$ adsorption or when oxygen is strongly in excess so that the rate becomes independent of $p_{O2}$.

```
clear variables;
% Define the symbolic variables
syms NH3(t) k t
% Define the ODE for a zero-order reaction
ode = diff(NH3, t) == -k;
NH3Sol(t) = dsolve(ode);
% Display the solution
disp(NH3Sol)
% Define the initial condition
NH3_0 = 0.15; % Initial concentration of NH3
% Solve the ODE with the initial condition
NH3Sol(t) = dsolve(ode, NH3(0) == NH3_0);
% Define the final concentration of NH3
NH3_final = 0.08;
% Calculate the time required to reach the final concentration symbolically
time_to_reach_final = solve(NH3Sol(t) == NH3_final, t);
% Assign a numerical value to the rate constant k
k_value = 4.00e-3; % Example rate constant in mol/L*s
% Numerically evaluate the time needed
time_numerical = double(subs(time_to_reach_final, k, k_value));
% Display the numerical solution
disp(['The time needed to reach the final concentration of NH3 is ', num2str(time_numerical), ' seconds']);
```

**Figure 1.1** MATLAB code for symbolically solving a simple zero-order rate equation.

```
clear variables;
% Define the parameters
k = 4.0E-3; % example rate constant
NH3_0 = 0.15; % initial concentration of NH3
% Define the time span
tspan = [0 20]; % from t = 0 to t = 20
% Define the ODE function
odefun = @(t, NH3) -k;
% Solve the ODE numerically using ode45
[t, NH3] = ode45(odefun, tspan, NH3_0);
% Plot the solution
figure;
plot(t, NH3, '-o');
xlabel('Time (t)');
ylabel('[NH_3]');
title('Numerical Solution of the ODE d[NH_3]/dt = -k');
grid on;
```

**Figure 1.2** MATLAB code for numerically solving a zero-order reaction rate.

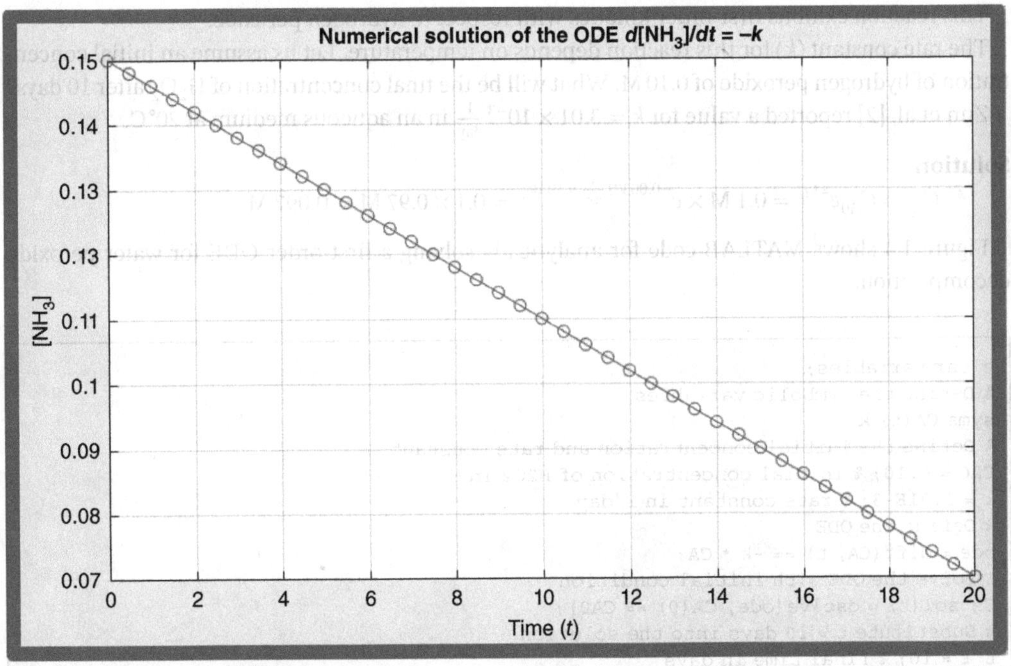


**Figure 1.3** Numerical solution of ODE rate equation showing ammonia concentration as $f(t)$.

**NOTE**

For gas systems, the concentration $C_A$ can be calculated from the inlet mole fraction $y_{A0}$, temperature $T_0$, and pressure $P_0$ using the ideal gas law or another applicable gas law. For an ideal gas:

$$C_{A0} = \frac{P_{A0}}{RT_0} = \frac{Py_{A0}}{RT_0} \tag{1.14}$$

1) **First-order reactions:**

The rate is directly proportional to the concentration of one reactant.

Rate law: $-r_A = -\frac{dC_A}{dt} = kC_A$

Rearrange and carry out integration for t from $t_i = 0$ to $t = t_f$, and for $C_A$ from $C_{A0}$ to $C_{Af}$, we reach at:

$$\frac{dC_A}{C_A} = -kdt \quad \rightarrow \quad \int_{C_{A0}}^{C_{Af}} \frac{dC_A}{C_A} = -k\int_0^{t_f} dt \quad \rightarrow \quad \ln\left(\frac{C_{Af}}{C_{A0}}\right) = -kt_f \tag{1.15}$$

$$C_{Af} = C_{A0}e^{-kt_f} \tag{1.16}$$

**Example 1.3** ***Decomposition Rate of Hydrogen Peroxide***

Consider the first-order reaction of the decomposition of hydrogen peroxide ($H_2O_2$) into water ($H_2O$) and oxygen ($O_2$) at higher temperatures:

$$2H_2O_2(aq) \rightarrow 2H_2O\,(l) + O_2(g) \tag{1.17}$$

This reaction exhibits first-order kinetics with respect to hydrogen peroxide.

The rate constant ($k$) for this reaction depends on temperature. Let us assume an initial concentration of hydrogen peroxide of 0.10 M. What will be the final concentration of $H_2O_2$ after 10 days?

Zun et al. [2] reported a value for $k = 3.01 \times 10^{-3} \frac{1}{\text{day}}$ in an aqueous medium at 20°C.

**Solution**

$$C_{Af} = C_{A0}e^{-kt_f} = 0.1\text{ M} \times e^{-0.00301\frac{1}{\text{day}} \times 10\text{ day}} = 0.1 \times 0.97\text{ M} = 0.097\text{ M}$$

Figure 1.4 shows MATLAB code for analytically solving a first-order ODE for water peroxide decomposition:

```
clear variables;
% Define the symbolic variables
syms CA(t) k
% Define the initial concentration and rate constant
CA0 = 0.10; % initial concentration of H2O2 in M
k = 3.01E-3; % rate constant in 1/day
% Define the ODE
ode = diff(CA, t) == -k * CA;
% Solve the ODE with initial condition
CA_sol(t) = dsolve(ode, CA(0) == CA0);
% Substitute t = 10 days into the solution
t_f = 10; % final time in days
CA_final = subs(CA_sol, t, t_f);
% Display the symbolic solution and final concentration
disp('Symbolic Solution of CA(t):')
disp(CA_sol)
disp('Final concentration after 10 days:')
disp(vpa(CA_final, 4)) % display with 4 significant figures
```

**Figure 1.4** MATLAB code for symbolically solving a first-order ODE reaction rate.

The answer is copied from the Command Window:

```
Symbolic Solution of CA(t):
exp(-(301*t)/100000)/10
symbolic function inputs: t
Final concentration after 10 days: 0.09703
symbolic function inputs: t
```

Figure 1.5 displays the MATLAB implementation of the numerical integration method used to solve the water peroxide decomposition ordinary differential equation.

```
clear variables;
% Define the parameters
CA0 = 0.10; % initial concentration of H2O2 in M
k = 3.01E-3; % rate constant in 1/day
tspan = [0 10]; % time span from 0 to 10 days
% Define the ODE function
odefun = @(t, CA) -k * CA;
% Solve the ODE numerically using ode45
[t, CA] = ode45(odefun, tspan, CA0);
% Plot the solution
figure;
plot(t, CA, '-o');
xlabel('Time (days)');
ylabel('[H_2O_2] (M)');
title('Numerical Solution of the ODE d[H_2O_2]/dt = -k[H_2O_2]');
grid on;
```

**Figure 1.5** MATLAB code of numerical integration of a first-order reaction rate ODE.

Figure 1.6 shows the generated plot showing $[H_2O_2]$ as a function of time. From the plot, one can see that at $t = 10$ days, $[H_2O_2] = 0.097$ M, which coincides with the analytical solution.

**NOTE**

For a first-order reaction, the relationship between the half-life time and the rate constant can be derived:

$$\ln\left(\frac{C_{Af}}{C_{A0}}\right) = -kt_f \quad \rightarrow \quad \ln\left(\frac{\frac{1}{2} * C_{A0}}{C_{A0}}\right) = -kt_f \quad \rightarrow \quad \ln\left(\frac{1}{2}\right) = -kt_{1/2}$$

$$t_{1/2} = \frac{\ln\left(\frac{1}{2}\right)}{-k} = \frac{-0.69315}{-k} = \frac{0.69315}{k} \tag{1.18}$$

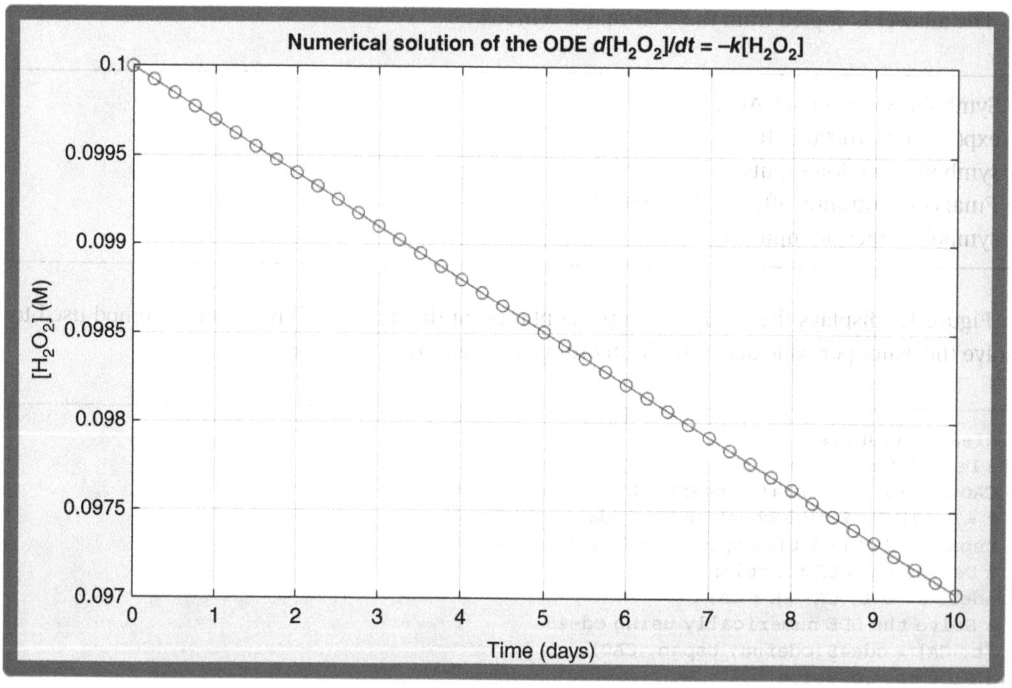


**Figure 1.6** The plot showing numerical integration of the first-order reaction rate ODE for water peroxide decomposition as $f$(time).

2) **Second-order Reactions:**

The rate is proportional to the square of the concentration of one reactant or to the product of the concentrations of two reactants.

$$\text{Rate law: } -r_A = kC_A^2 \qquad \text{for: } aA \rightarrow Products \tag{1.19}$$

$$\text{Rate law: } -r_A = kC_AC_B \qquad \text{for: } aA + bB \rightarrow Products \tag{1.20}$$

**Example 1.4** ***Decomposition Rate of Nitrogen Dioxide into NO and $O_2$***

$$2NO_2\,(g) \text{ ------> } 2NO\,(g) + O_2\,(g) \tag{1.21}$$

This reaction represents the decomposition of nitrogen dioxide ($NO_2$) into nitric oxide (NO) and oxygen ($O_2$). It is considered practically irreversible because the reverse reaction (formation of $NO_2$ from NO and $O_2$) is very slow under typical conditions. The rate constant ($k$) for this reaction depends on temperature. At $T = 300°C$, the value for $k$ is estimated to be 0.54 L/mol/s. Calculate the time needed to drop the initial concentration of $NO_2$ from 0.2 down to 0.05 M.

Reference:
https://chem.libretexts.org/Bookshelves/General_Chemistry/Map%3A_General_Chemistry_(Petrucci_et_al.)/14%3A_Chemical_Kinetics/14.06%3A_Second-Order_Reactions.

$$-r_A = -\frac{dC_A}{dt} = kC_A^2 \rightarrow \frac{dC_A}{C_A^2} = -kdt$$

Rearrange and carry out integration for t from $t_i = 0$ to $t = t_f$, and for $C_A$ from $C_{A0}$ to $C_{Af}$, we reach at:

$$\int_{C_{A0}}^{C_{Af}} \frac{dC_A}{C_A^2} = -k\int_0^{t_f} dt \rightarrow \frac{1}{C_{A0}} - \frac{1}{C_{Af}} = -kt_f \rightarrow \frac{1}{C_{Af}} = \frac{1}{C_{A0}} + kt_f \tag{1.22}$$

$$\frac{1}{C_{Af}} - \frac{1}{C_{A0}} = kt_f \tag{1.23}$$

$$\frac{1}{0.05\ \text{M}} - \frac{1}{0.2\ \text{M}} = 0.54\frac{1}{\text{M.s}} t_f[\text{s}]$$

Solve for $t_f$. $t_f = 27.78$ s.

Figure 1.7 shows MATLAB code for analytically solving the second-order ODE reaction rate for $NO_2$ decomposition. Notice that for $C_{A0} = 0.2$ M, the analytic solution can be written as:

$$C_{Af} = \frac{1}{(kt_f + 5)} \tag{1.24}$$

```
clear variables;
% Define the symbolic variables
syms NO2(t) k
% Define the initial concentration and rate constant
NO2_0 = 0.20; % initial concentration of NO2 in M
NO2_final = 0.05; % final concentration of NO2 in M
k_value = 0.54; % rate constant in L*mol^(-1)*s^(-1)
% Define the ODE
ode = diff(NO2, t) == -k * NO2^2;
% Solve the ODE symbolically with the initial condition
NO2_sol(t) = dsolve(ode, NO2(0) == NO2_0);
% Substitute the rate constant value
NO2_sol(t) = subs(NO2_sol, k, k_value);
% Solve for the time when NO2 drops to NO2_final
time_needed = solve(NO2_sol == NO2_final, t);
% Display the symbolic solution and the time needed
disp('Symbolic Solution of NO2(t):')
disp(NO2_sol)
disp('Time needed to drop from 0.2 M to 0.05 M:')
disp(vpa(time_needed, 4)) % display with 4 significant figures
```

**Figure 1.7** MATLAB code for analytic solution of second-order ODE of reaction rate of $NO_2$ decomposition.

The result is copied from the Command Window of MATLAB:

**Symbolic Solution of NO2(t):**
**1/((27*t)/50 + 5)**
**symbolic function inputs: t**
**Time needed to drop from 0.2 M to 0.05 M:**
**27.78**

Figure 1.8 shows the MATLAB implementation of the numerical integration method used to solve the second-order ordinary differential equation of $NO_2$ decomposition.

```
clear variables;
% Define the parameters
NO2_0 = 0.20; % initial concentration of NO2 in M
NO2_final = 0.05; % final concentration of NO2 in M
k = 0.54; % rate constant in L*mol^(-1)*s^(-1)
% Define the ODE function
odefun = @(t, NO2) -k * NO2^2;
% Define the time span for integration (guess an initial range)
tspan = [0 100]; % time span in seconds, adjust as necessary
% Solve the ODE numerically using ode45
[t, NO2] = ode45(odefun, tspan, NO2_0);
% Find the time when NO2 drops to NO2_final
time_needed = interp1(NO2, t, NO2_final);
% Plot the solution
figure;
plot(t, NO2, '-o');
xlabel('Time (s)');
ylabel('[NO_2] (M)');
title('Numerical Solution of the ODE d[NO_2]/dt = -k[NO_2]^2');
grid on;
% Display the time needed to drop from 0.2 M to 0.05 M
disp('Time needed to drop from 0.2 M to 0.05 M:')
disp(time_needed)
```

**Figure 1.8** MATLAB code for numerical integration of the second-order ODE of $NO_2$ decomposition.

The result is copied from the Command Window.

```
Time needed to drop from 0.2 M to 0.05 M: 27.8255
```

Figure 1.9 shows the generated plot for $[NO_2]$ as $f(t)$ using numerical integration.

For the second case, the rate law: $-r_A = kC_AC_B$ for: $aA + bB \rightarrow cC + dD$, we need to express $C_B$ as $f(C_A)$.

$$(N_{A0} - N_A) = \frac{a}{b} \times (N_{B0} - N_B) \tag{1.25}$$

Rearrange, we have:

$$N_B = N_{B0} - \left[\left(\frac{b}{a}\right)(N_{A0} - N_A)\right] \tag{1.26}$$

For products $C$ and $D$, we have:

$$N_C = N_{C0} + \left[\left(\frac{c}{a}\right)(N_{A0} - N_A)\right] \tag{1.27a}$$

$$N_D = N_{D0} + \left[\left(\frac{d}{a}\right)(N_{A0} - N_A)\right] \tag{1.27b}$$

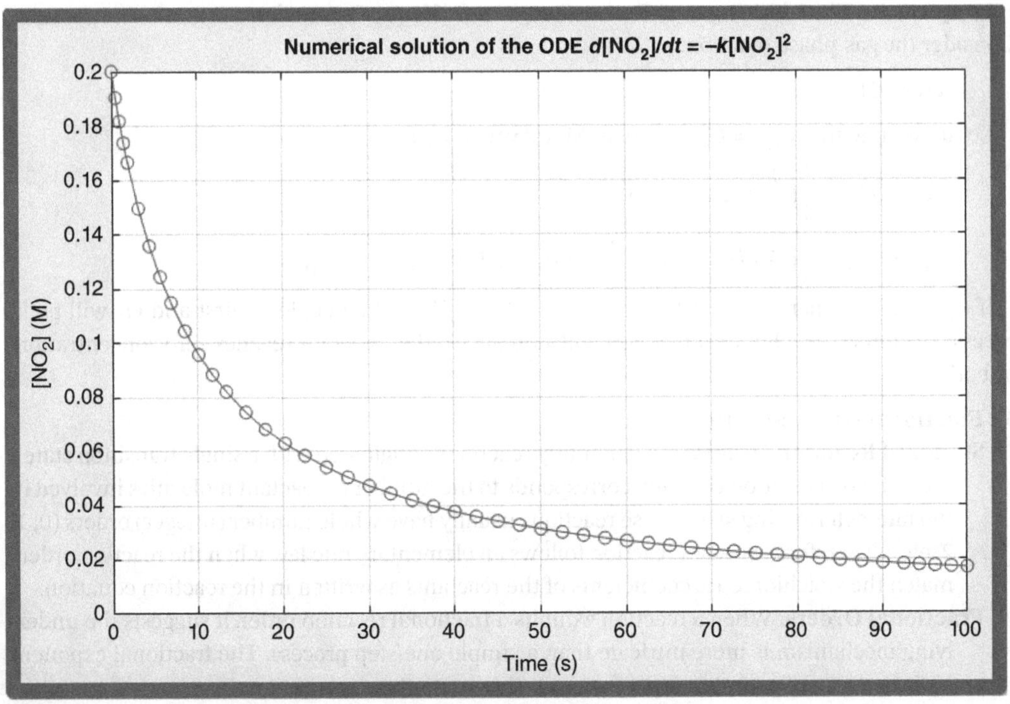


**Figure 1.9** Numerical integration of second-order ODE of $NO_2$ decomposition as $f(t)$.

For a constant-volume reaction medium, we have:

$$C_B = C_{B0} - \left[\left(\frac{b}{a}\right)(C_{A0} - C_A)\right] = C_{A0}\left[\left(\theta_{B0} - \frac{b}{a}X\right)\right] \tag{1.28a}$$

$$C_C = C_{C0} + \left[\left(\frac{c}{a}\right)(C_{A0} - C_A)\right] = C_{A0}\left[\left(\theta_{C0} + \frac{c}{a}X\right)\right] \tag{1.28b}$$

$$C_D = C_{D0} + \left[\left(\frac{d}{a}\right)(C_{A0} - C_A)\right] = C_{A0}\left[\left(\theta_{D0} + \frac{d}{a}X\right)\right] \tag{1.28c}$$

where $\theta_{i0}$ is the initial molar ratio or initial molar concentration with respect to $A$.

**Example 1.5** ***Interrelationship Among Reacting Species***

Consider the gas-phase reaction:

$$H_2 + Cl_2 \rightarrow 2HCl \tag{1.29}$$

Let us assume that $C^0_{H2} = C^0_{Cl2} = 0.5$ M. Moreover, $b = a = 1$.

$$C_B = C_{B0} - \left[\left(\frac{b}{a}\right)(C_{A0} - C_A)\right]$$

$$C_{Cl2} = C^0_{Cl2} - \left[\left(\frac{1}{1}\right)(C^0_{H2} - C_{H2})\right] = 0.5 - [(0.5 - C_{H2})] = C_{H2}$$

The concentration of $Cl_2$, $C_{Cl2}$, can be directly expressed in terms of that of $H_2$, $C_{H2}$.

**Example 1.6** ***The Limiting and Excess Reactant in Water Formation***
Consider the gas-phase reaction:

$$O_2 + 2H_2 \rightarrow 2H_2O$$

Let us assume that $C^0_{O2} = C^0_{H2} = 1.0$ M. Moreover, $b = 2a = 2$.

$$C_B = C_{B0} - \left[\left(\frac{b}{a}\right)(C_{A0} - C_A)\right]$$

$$C_{H2} = C^0_{H2} - \left[\left(\frac{2}{1}\right)(C^0_{O2} - C_{O2})\right] = 1.0 - [2(1.0 - C_{O2})] = 2C_{O2} - 1.0$$

If $C_{O2}$ is 0.5 M, then $C_{H2} = 2*0.5 - 1.0 = 0.0$. Thus, $H_2$ will be depleted first and $O_2$ will be in excess. Of course, the chemical reaction will stop once either reactant reaches zero concentration, or both.

3) **Fractional Order Reactions:**
**Standard Reaction Orders:** In elementary reactions (single step with a single transition state), the order of a reaction typically corresponds to the number of reactant molecules involved in the rate-determining step. These reactions usually have whole number (integer) orders (0, 1, 2, etc.). Therefore, we say a reaction follows an elementary rate law when the reaction orders match the stoichiometric coefficients of the reactants as written in the reaction equation.
**Fractional Orders:** When a reaction exhibits a fractional reaction order, it suggests the underlying mechanism is more intricate than a simple one-step process. The fractional exponent indicates a more complex dependence on the reactant concentrations.

The order can be a non-integer (fractional value). One characteristic of many fractional order reactions, especially with non-integer orders, is that they often proceed through chain mechanisms or other complex pathways. In a chain reaction, intermediates or radicals may be formed, which can significantly affect the overall reaction order.

Chain reactions are multistep processes where the product of one step acts as a reactant in a subsequent step, leading to a domino effect. Intermediates or radicals are often formed and participate in these reactions. The presence of intermediates and radicals can affect the overall rate law expression and lead to fractional exponents. The specific mechanism dictates how the concentrations of reactants and intermediates contribute to the rate.

In general, for $n \neq 1$, we have:

$$\int_{C_{A0}}^{C_{Af}} \frac{dC_A}{C_A^n} = -k\int_0^{t_f} dt \rightarrow \frac{-\frac{1}{n-1}}{C_{Af}^{n-1}} - \frac{-\frac{1}{n-1}}{C_{A0}^{n-1}} = -kt_f \quad \rightarrow \frac{\left(\frac{1}{n-1}\right)}{C_{Af}^{n-1}} - \frac{\left(\frac{1}{n-1}\right)}{C_{A0}^{n-1}} = kt_f \qquad n \neq 1 \quad (1.30)$$

**Example 1.7** ***The Decomposition of Acetaldehyde Reaction Order***
Eusuf and Laidler [3] reported the reaction order and rate constant of uninhibited thermal decomposition of acetaldehyde into $CH_4$ and CO:

$$CH_3C(O)H_{(g)} \rightarrow CH_{4(g)} + CO$$

$$-r_{CH_3C(O)H} = -r_A = -\frac{dC_A}{dt} = kC_A^{1.5}$$

The decomposition was studied from 480 to 525°C and over the pressure range of 30 to 579 mmHg.

Notice that if the relevant literature is searched, there will be different values of reported reaction order and rate constant, depending on the reaction pressure, temperature, and type of catalyst.

The rate constant, $k$, at 500°C = 773 K will be: $0.01275\frac{(L)^{1/2}}{mol^{1/2}.s}$. See next section.

Find the time needed to drop the concentration from 0.5 down to 0.1 M.

$$-r_{CH_3C(O)H} = -r_A = -\frac{dC_A}{dt} = kC_A^{1.5}$$

Making use of Eq. (1.30) and substituting for integration limits for $t$ between 0 and $t_f$ and for $C_A$ between $C_{A0}$ and $C_{Af}$, we reach at:

$$\int_{C_{A0}}^{C_{Af}} \frac{dC_A}{C_A^{1.5}} = -k\int_0^{t_f} dt \rightarrow \frac{-2}{C_{Af}^{0.5}} - \frac{-2}{C_{A0}^{0.5}} = -kt_f \rightarrow \frac{2}{C_{Af}^{0.5}} - \frac{2}{C_{A0}^{0.5}} = kt_f$$

$$\left[\frac{2}{0.1^{0.5}} - \frac{2}{0.5^{0.5}}\right]\frac{(\mathrm{L})^{\frac{1}{2}}}{\mathrm{mol}^{\frac{1}{2}}} = 0.01275\frac{(\mathrm{L})^{\frac{1}{2}}}{\mathrm{mol}^{\frac{1}{2}}.\mathrm{s}}t_f[\mathrm{s}]$$

Solve for $t_f$. $t_f = 274.2\,\mathrm{s} = 4.75$ minutes.

Figure 1.10 shows the analytic solution for the $n^{th}$-order ODE reaction rate for aldehyde decomposition. Two analytic solutions are generated by symbolic MATLAB.

```
% Define the symbolic variables
syms CA(t) k
% Define the initial concentration and rate constant
CA0 = 0.50; % initial concentration of CH3C(O)H in M
t_final = 274.2; % final time in seconds
k_value = 0.01275; % rate constant in L^(1/2) / (mol^(1/2) * s)
% Define the ODE
ode = diff(CA, t) == -k * CA^(3/2);
% Solve the ODE symbolically with the initial condition
CA_sol(t) = dsolve(ode, CA(0) == CA0);
% Display the symbolic solution
disp('Symbolic Solution of CA(t):')
disp(CA_sol)
% Manually select the physically meaningful solution
% Since we are dealing with concentrations, we choose the solution that decreases with time.
CA_sol_correct = 1 / ((k*t)/2 + sqrt(2))^2;
% Substitute the final time into the solution to get the final concentration
CA_final = subs(CA_sol_correct, {k, t}, {k_value, t_final});
% Display the final concentration
disp('Final concentration after 274.2 seconds:')
disp(vpa(CA_final, 4)) % display with 4 significant figures.
```

**Figure 1.10** MATLAB code for the analytic solution of the $n^{th}$-order ODE reaction rate for aldehyde decomposition. The decreasing concentration with time is selected.

Here, the final time is given and the final concentration at the given time will be calculated. The solution with decreasing concentration with time is selected.

The result is copied from the Command window:

**Symbolic Solution of CA(t):**
**1/((k*t)/2 - 2^(1/2))^2**
**1/((k*t)/2 + 2^(1/2))^2**
**symbolic function inputs: t**
**Final concentration after 274.2 seconds: 0.1**

Alternatively, Figure 1.11 shows MATLAB code for analytically solving the $n^{th}$-order ODE reaction rate for aldehyde decomposition. Symbolic MATLAB gives two solutions for $CA = f(t)$. The solution with positive time is selected.

Here, $C_{Afinal}$ is given, and the time needed to achieve this value will be calculated.

```
clear variables;
% Define the symbolic variables
syms t k
% Define the initial concentration, final concentration, and rate constant
CA0 = 0.50; % initial concentration of CH3C(O)H in M
CA_final = 0.10; % final concentration of CH3C(O)H in M
k_value = 0.01275; % rate constant in L^(1/2) / (mol^(1/2) * s)
% MATLAB gives two separate solutions for CA=f(t).
% We select the physically meaningful solution of the ODE where CA decreases with t.
CA_sol_correct = 1 / ((k*t)/2 + sqrt(2))^2;
% Solve for time t when CA(t) = CA_final
t_solutions = solve(CA_sol_correct == CA_final, t);
% Substitute the value of k
t_solutions_subs = subs(t_solutions, k, k_value);
% Convert the solutions to numeric values
t_numeric = double(t_solutions_subs);
% Filter out non-physical (negative) time values
t_positive = t_numeric(t_numeric >= 0);
% Display the time needed.
disp('Time needed to drop from 0.5 M to 0.1 M:')
disp(vpa(t_positive, 4)) % display with 4 significant figures
```

**Figure 1.11** MATLAB code for analytically solving the $n^{th}$-order ODE reaction rate for aldehyde decomposition, with the selection of a positive time solution.

The result is copied from the Command Window.

**Time needed to drop from 0.5 M to 0.1 M:**
**274.2**

Figure 1.12 shows MATLAB code for the numerical integration of the $n^{th}$-order of aldehyde decomposition as $f(t)$.

```
clear variables;
% Define the parameters
CA0 = 0.50; % initial concentration of CH3C(O)H in M
CA_final = 0.10; % final concentration of CH3C(O)H in M
k = 0.01275; % rate constant in L^(1/2) / (mol^(1/2) * s)
% Define the ODE function
odefun = @(t, CA) -k * CA^(3/2);
% Define the time span for integration (guess an initial range)
tspan = [0 500]; % time span in seconds, adjust as necessary
% Solve the ODE numerically using ode45
```

**Figure 1.12** MATLAB code for the numerical integration of the $n^{th}$-order of aldehyde decomposition as $f(t)$.

```
[t, CA] = ode45(odefun, tspan, CA0);
% Find the time when CA drops to CA_final
time_needed = interp1(CA, t, CA_final);
% Plot the solution
figure;
plot(t, CA, '-o');
xlabel('Time (s)');
ylabel('[CH3C(O)H] (M)');
title('Numerical Solution of the ODE d[CH3C(O)H]/dt = -k[CH3C(O)H]^{1.5}');
grid on;
% Display the time needed to drop from 0.5 M to 0.1 M
disp('Time needed to drop from 0.5 M to 0.1 M:')
disp(time_needed)
```

**Figure 1.12** (*Continued*)

The result is copied from the Command Window.

```
>> Fig1_12
Time needed to drop from 0.5 M to 0.1 M:
274.2813
```

Figure 1.13 shows the generated plot of $[CH_3C(O)H]$ as f(t), using the numerical integration.

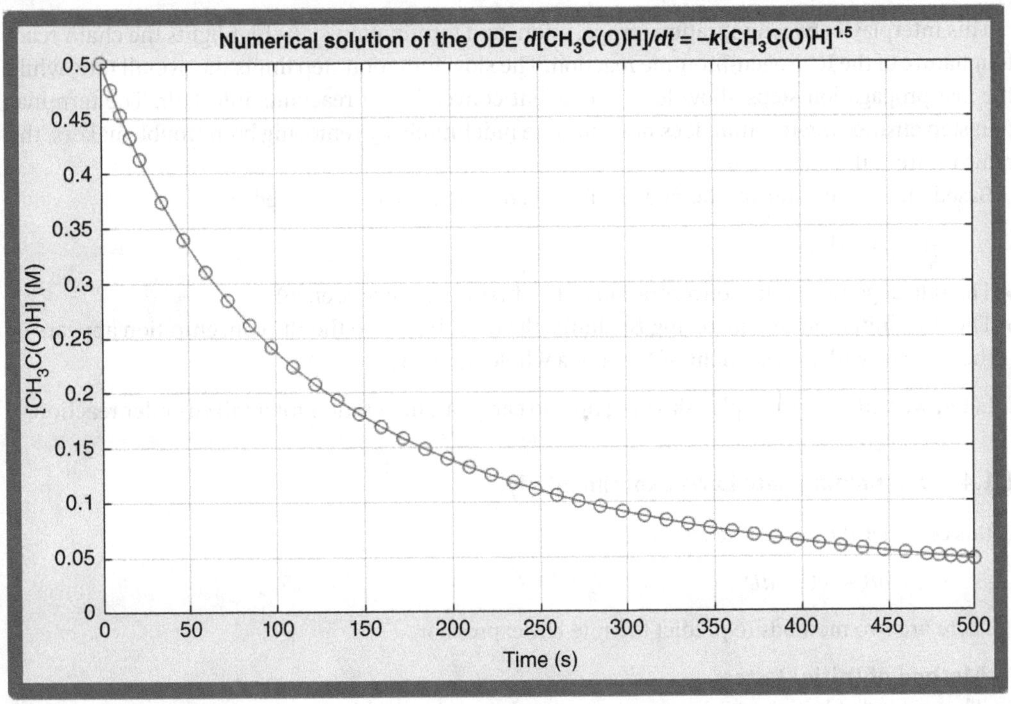


**Figure 1.13** The generated plot of $[CH_3C(O)H]$ as *f(t)*, using the numerical integration.

4) **Mixed-order (or Complex Order) Reactions:**

The rate law involves different orders for different reactants. In mixed-order reactions, the rate of the reaction depends on the concentration of multiple reactants, but not in a simple whole number ratio (like 1:1 or 2:1). The rate law expression for these reactions includes exponents for each reactant that are not integers in general. So, unlike elementary reactions (single-step processes), mixed-order reactions often proceed through multiple steps. The overall rate may be influenced by the concentration of intermediate reactants involved in different steps of the mechanism.

**Example 1.8** ***Hydrogen-bromine multistep reaction***
The reaction between hydrogen ($H_2$) and bromine ($Br_2$) to form hydrogen bromide (HBr) is a well-known example of a mixed-order reaction.

$$H_2 + Br_2 \rightarrow 2HBr\cdot \tag{1.31}$$

**Step 1 (Initiation, Slow):** $Br_2 \rightarrow 2Br\cdot$(rate depends on $Br_2$ concentration)
**Step 2 (Propagation, Fast):** $H_2 + Br\cdot \rightarrow HBr + H\cdot$
**Step 3 (Termination, Fast):** $H\cdot + Br\cdot \rightarrow HBr$
**Sum:** $H_2 + Br_2 \rightarrow 2HBr$

This simplified mechanism effectively captures the essence of the hydrogen-bromine reaction.

1) The slow initiation step ($Br_2 \rightarrow 2Br\cdot$) generates $2Br\cdot$radicals, initiating the chain reaction.
2) The fast propagation step ($H_2 + Br\cdot \rightarrow HBr + H\cdot$) utilizes these $Br\cdot$ radicals to convert $H_2$ into HBr while also creating a new $H\cdot$ radical.
3) The fast termination step ($H\cdot + Br\cdot \rightarrow HBr$) combines both $Br\cdot$ and $H\cdot$ radicals from the cycle and forms another HBr molecule.

This interplay between initiation, propagation, and termination steps highlights the chain reaction nature of the hydrogen-bromine reaction. The slow initiation step limits the overall rate, while the fast propagation steps allow for the efficient conversion of reactants into HBr. The termination step ensures the reaction does not continue indefinitely by removing both troublemakers, the radicals, from the reacting medium.

Based on the mechanism, the rate law for this reaction can be expressed as:

$$-r_{HBr} = k[H_2][Br_2]^{\frac{1}{2}}$$

- The rate depends on the concentration of $H_2$ (first-order dependence).
- The slow initiation step involving $Br_2$ limits the overall rate, so the $Br_2$ concentration appears in the rate law with an exponent of 1/2 (not a whole number).

Later, we will have examples showing how to carry out integration for a mixed-order reaction.

### 1.1.4 Determining Rate Laws Experimentally

Let us consider the general reaction:

$$aA + bB \rightarrow cC + dD$$

There are two methods to predict the rate law expression.

1) **Method of Initial Rates:**
The method of initial rates is a common experimental technique used to determine the order of a reaction with respect to each reactant.

**Steps involved**

1) **Measuring Initial Rates:**
   a. Conduct a series of experiments where the initial concentrations of the reactants are varied.
   b. Measure the initial rate of the reaction (the rate at the very beginning of the reaction) for each set of initial concentrations.
2) **Analyzing Data:**
   a. Compare how changes in the initial concentrations affect the initial rate of the reaction.
   b. Typically, the rate law is assumed to be of the form: $Rate = k[A]^m[B]^n$, where $m$ and $n$ are the orders with respect to reactants $A$ and $B$, respectively, and $k$ is the rate constant.

Before we take an example, let us explain the notion of initial rate. The initial rate of a reaction, typically expressed in units of molarity per second (M/s), is determined by measuring the change in concentration of a reactant or product over a very short period at the beginning of the reaction. Mathematically, it is represented as the derivative of concentration with respect to time, evaluated at $t = 0$. For reactant A, the initial rate of disappearance can be expressed as:

$$Initial\ Rate = -\left.\frac{d[A]}{dt}\right|_{t=0} \tag{1.32}$$

The negative sign indicates that we report the rate of disappearance of reactant $A$, which is a positive quantity. Notice the concentration of $A$ is decreasing over time.
Alternatively, for product $C$, the initial rate of formation can be expressed as:

$$Initial\ Rate = \left.\frac{d[C]}{dt}\right|_{t=0} \tag{1.33}$$

These two expressions highlight that the initial rate is the instantaneous rate of change of concentration at the very start of the reaction. Experimentally, it is often approximated by measuring the concentration change ($\Delta[A]$ or $\Delta[C]$) over a short time interval ($\Delta t$) right after the reaction begins and calculating the ratio:

$$Initial\ Rate \approx -\left.\frac{\Delta[A]}{\Delta t}\right|_{t=0} \tag{1.34}$$

or

$$Initial\ Rate \approx \left.\frac{\Delta[C]}{\Delta t}\right|_{t=0} \tag{1.35}$$

This approximation assumes that $\Delta t$ is sufficiently small so that the rate (or slope) remains approximately constant during this interval.

**Example 1.9** ***Determination of Reaction Order***

Consider the reaction: $A + B \rightarrow C$

Suppose we conduct three experiments with the following initial concentrations and measure the corresponding initial rates:

| Experiment | [A], M | [B], M | Initial Rate (M/s) |
|---|---|---|---|
| 1 | 0.1 | 0.1 | 0.02 |
| 2 | 0.2 | 0.1 | 0.08 |
| 3 | 0.1 | 0.2 | 0.02 |

What is the reaction order with respect to both $A$ and $B$, given the rate constant is $k$?

**Solution**

**Experiments 1 and 2:** $[B]$ is constant, $[A]$ is doubled.
The initial rate increases by a factor of 4. This suggests that the rate is proportional to $[A]^2$, so $m = 2$.

**Experiments 1 and 3:** $[A]$ is constant, $[B]$ is doubled.
The initial rate does not change. This suggests that the rate is independent of [B], so $n = 0$.
Thus, the rate law for this reaction is:

$$Rate = k[A]^m[B]^n = k[A]^2[B]^0 = k[A]^2$$

2) **Integrated Rate Laws**
Integrated rate laws are used to analyze how the concentration of reactants changes over time and to determine the reaction order from these changes.
**Steps Involved**
1) **Collecting Data Over Time**
   a. Measure the concentration of reactants (or products) at various time points as the reaction proceeds.
   b. Use this data to determine how concentration changes over time.
2) **Using Integrated Forms of Rate Laws:**
   a. Depending on the assumed order of the reaction, use the corresponding integrated rate law to fit the data.
   b. Compare the fit of the data to different integrated rate laws to determine the best match, which reveals the reaction order.

Notice that in the previous section, we have found the integral form for each of the following cases:

**Zero-order Reaction:**

$$C_A = C_{Ao} - kt$$

Plot of $[A] = C_A$ versus $t$ is a straight line with slope $-k$.

**First-order Reaction:**

$$\ln\left(\frac{C_{Af}}{C_{A0}}\right) = -kt_f \rightarrow \qquad \ln(C_A) = \ln(C_{A0}) = -kt$$

Plot of $\ln[A] = \ln(C_A)$ versus $t$ is a straight line with slope $-k$.
**Second-order Reaction:**

$$\frac{1}{C_{Af}} - \frac{1}{C_{A0}} = kt_f \rightarrow \frac{1}{C_A} = \frac{1}{C_{A0}} + kt$$

Plot of $1/[A] = 1/(C_A)$ versus $t$ is a straight line with slope $k$.

**$n^{th}$-order Reaction:**

$$\frac{\frac{1}{n-1}}{C_{Af}^{n-1}} - \frac{\frac{1}{n-1}}{C_{A0}^{n-1}} = kt_f \qquad n \neq 1 \qquad \rightarrow \frac{\frac{1}{n-1}}{C_A^{n-1}} = \frac{\frac{1}{n-1}}{C_{A0}^{n-1}} + kt \qquad n \neq 1$$

Here, it is an iterative procedure; one must assume $n$, then make a plot of $\frac{\frac{1}{n-1}}{C_A^{n-1}}$ versus $t$. If the plot is linear, then the slope will be $k$. If the plot is not linear, then the user must try a different value for $n$. The process will be terminated only when the user converges to a linear solution for the previous plot, that is, finding the right $n$. Of course, there are other linear/nonlinear regression techniques to fit the data for such cases where $n$ is not an integer. Here is an example.

**Example 1.10** ***Experimental $-r_A = f(C_A)$ Kinetic Data***

The reaction of stilbene, or 1,2-diphenylethylene, ($A$), $C_{14}H_{12}$ decomposes into naphthalene (B), $C_{12}H_8$, and ethylene (C), that is:

$$C_{14}H_{12}(l) \rightarrow C_{12}H_8(l) + CH_2 = CH_2(g)$$
$$A \rightarrow B + C$$

Table 1.1 shows the isothermal kinetic data for the conversion of $A$ into products:

**Table 1.1** The rate of disappearance of $A$ versus concentration of $A$.

| $C_A$ (mol/L) | 0 | 10 | 20 | 30 | 40 | 50 | 60 | 70 | 80 | 90 | 100 |
|---|---|---|---|---|---|---|---|---|---|---|---|
| $-r_A\left(\frac{\text{mol}}{\text{L}\cdot\text{min}}\right)$ | 0 | 2.54 | 2.913 | 3.159 | 3.35 | 3.50 | 3.63 | 3.74 | 3.84 | 3.93 | 4.02 |

Let us show the MATLAB algorithm to find the best regressed values of the rate constant, $k$, and the reaction order, $n$, for the reaction equation:

$$-r_A = kC_A^n$$

**MATLAB Algorithm:**

1) Read CA

| $C_A$ (mol/L) | 0 | 10 | 20 | 30 | 40 | 50 | 60 | 70 | 80 | 90 | 100 |
|---|---|---|---|---|---|---|---|---|---|---|---|

2) Read $-r_A\left(\frac{\text{mol}}{\text{L}\cdot\text{min}}\right)$:

| $-r_A\left(\frac{\text{mol}}{\text{L}\cdot\text{min}}\right)$ | 0 | 2.54 | 2.913 | 3.159 | 3.35 | 3.50 | 3.63 | 3.74 | 3.84 | 3.93 | 4.02 |
|---|---|---|---|---|---|---|---|---|---|---|---|

3) Use a model of the form:
   $-r_A\left(\frac{\text{mol}}{\text{L}\cdot\text{min}}\right) = kC_A^n$ to curve fit the experimental data and find the best values of $k$ and $n$, which are the rate constant and reaction order, respectively.
4) Initial guesses can be made for both $k$ and $n$.
5) Report the values of both $k$ and $n$.
6) Make a plot showing both the experimental scatter and curve-fitted data on the same plot.
   Figure 1.14 shows MATLAB code for utilizing MATLAB's built-in function called **lsqcurvefit** to curve fit $y$ versus $x$, where y is the reaction rate ($-r_A$) and $x$ is $C_A$. The results will be the best regressed value of $k$ and $n$, and a plot showing the model goodness.

```
clear variables;
% MATLAB Code for Curve Fitting Reaction Rate Model
% 1. Read CA (mol/L)
CA = [0,10,20,30,40,50,60,70,80,90,100]; % Concentration of A (mol/L)
% 2. Read -r_A (mol/(L·min))
rA = -[0, 2.54, 2.913, 3.159, 3.35, 3.50, 3.63, 3.74, 3.84, 3.93, 4.02]; % Reaction rate (mol/(L·min))
% 3. Model: -r_A = k * CA^n
% Define the function to be minimized (model form for curve fitting)
rate_model = @(params, CA) -params(1) * CA.^params(2);
% 4. Initial guesses for k and n
initial_guess = [1, 1]; % [k, n]
% Use lsqcurvefit to fit the model
options = optimset('Display', 'off'); % Turn off display of iterations
[k_n, resnorm] = lsqcurvefit(rate_model, initial_guess, CA, rA, [], [], options);
% Extract the fitted values of k and n
k = k_n(1); % rate constant
n = k_n(2); % reaction order
% 5. Report the values of k and n
fprintf('Best fit value of k: %.4f\n', k);
fprintf('Best fit value of n: %.4f\n', n);
% 6. Plot the experimental data and fitted curve
CA_fine = linspace(min(CA), max(CA), 100); % Fine CA values for plotting fitted curve
rA_fitted = rate_model(k_n, CA_fine); % Fitted reaction rates
figure;
scatter(CA, -rA, 'ro', 'filled'); % Scatter plot of experimental data
hold on;
plot(CA_fine, -rA_fitted, 'b-', 'LineWidth', 2); % Plot fitted curve
xlabel('Concentration of A, C_A (mol/L)');
ylabel('Reaction Rate, -r_A (mol/(L·min))');
title('Curve Fitting: -r_A = k * C_A^n');
legend('Experimental Data', 'Fitted Curve');
grid on;
hold off;
```

**Figure 1.14** MATLAB code for curve-fitting experimental data of y, the reaction rate ($-r_A$), and x, that is, $C_A$.

Results are copied from the MATLAB Command Window.

```
>> Fig1_14
Best fit value of k: 1.6053
Best fit value of n: 0.1992
```

Figure 1.15 shows the generated plot of ($-r_A$) versus $C_A$ for both experimental and curve-fitted values.

### 1.1.5 Arrhenius Law for Temperature Dependence of Rate Constants

Activation energy ($E_a$) is a crucial concept in chemical kinetics and refers to the minimum energy required for reactants to undergo a successful chemical reaction. It represents the energy barrier that must be overcome for reactants to transform into products.

**Key aspects of activation energy:**

1) **Energy Barrier:**
   Activation energy is the energy difference between the reactants and the transition state, the highest energy state during the reaction. The transition state is a short-lived, high-energy configuration where old bonds are breaking and new bonds are forming.

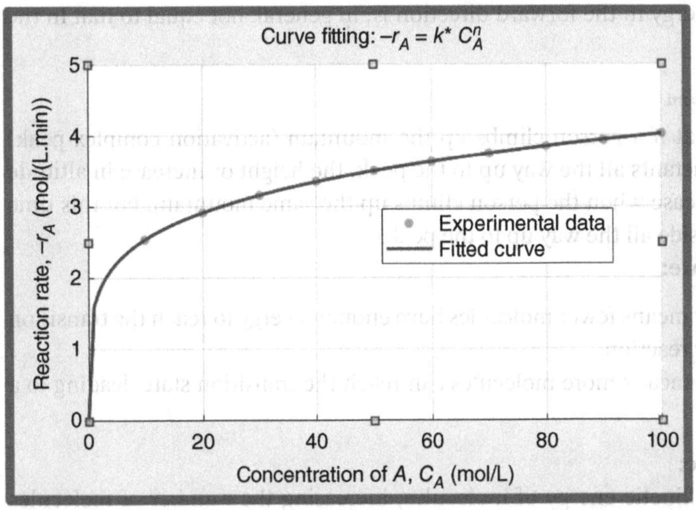


**Figure 1.15** The combined plot of $(-r_A)$ versus $C_A$ for both experimental and curve-fitted values.

2) **Reaction Pathway:**

The reaction pathway is a graphical representation of the energy changes during a reaction. It typically includes:

- **Reactants:** Initial state with a certain energy level.
- **Transition State:** Peak of the energy barrier.
- **Products:** Final state, which can be at a lower or higher energy level compared to reactants.

The activation energy is the difference in energy between the reactants and the transition state, as shown in Figure 1.16. Notice that the activation energy ($E_a$) is always a positive quantity; on the other hand, the heat of reaction can be negative (or exothermic) as demonstrated in Figure 1.16 (left), or it can be positive (or endothermic) as demonstrated in Figure 1.16 (right).

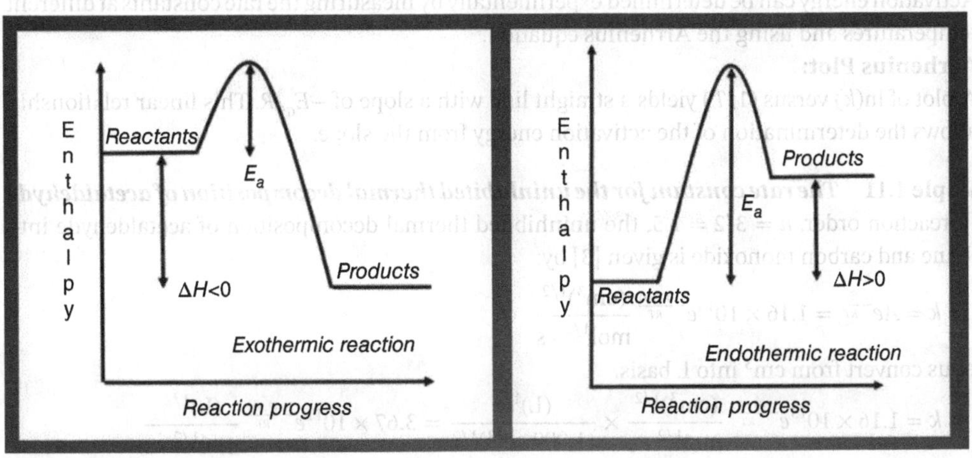


**Figure 1.16** Enthalpy of reacting species as a function of reaction coordinate or progress. A typical exothermic reaction (left) and endothermic reaction (right).

Moreover, the activation energy in the forward direction is, in general, not equal to that in the backward direction.

$$(E_a)_{forward} \neq (E_a)_{backward}$$

This is similar to saying that if a person climbs up the mountain (activation complex peak) starting from the side of reactants all the way up to the peak, the height or increase in altitude will not be the same as the case when the person climbs up the same mountain, but this time starting from the products' side all the way up to the peak.

3) **Influence on Reaction Rate:**

   a) Higher activation energy means fewer molecules have enough energy to reach the transition state, leading to a slower reaction.
   b) Lower activation energy means more molecules can reach the transition state, leading to a faster reaction.

4) **Temperature Dependence:**
   Temperature increases the kinetic energy of molecules, increasing the number of molecules that can overcome the activation energy barrier, thus speeding up the reaction. The Arrhenius equation relates the rate constant $k$ to the activation energy $E_a$ and temperature $T$:

$$k = Ae^{-\frac{E_a}{RT}} \quad \Longrightarrow \ln(k) = \ln(A) - \frac{E_a}{R}\left(\frac{1}{T}\right) = c + m\left(\frac{1}{T}\right) = a0 + a1 * x \tag{1.36}$$

where:

- $A$ is the pre-exponential factor, $\left[M^{(1-n)}\frac{1}{\mathrm{s}}\right] = \left[\left(\frac{\mathrm{mol}}{\mathrm{L}}\right)^{(1-\mathrm{n})}\frac{1}{\mathrm{s}}\right]$.
- $E_a$ is the activation energy, [energy/mol].
- $R$ is the universal gas constant, $\left[\frac{1.987\ \mathrm{cal}}{\mathrm{mol.K}}\right] = \left[\frac{8.314\ \mathrm{J}}{\mathrm{mol.K}}\right]$.
- $T$ is the temperature in Kelvin.
- $c$ is the intercept of the plot of ln($k$) versus (1/T), a straight-line relationship.
- $m$ is the slope of the straight line, equal to $-E_a/R$.

**Measuring Activation Energy:**

1) **Experimental Determination:**
   Activation energy can be determined experimentally by measuring the rate constants at different temperatures and using the Arrhenius equation.
2) **Arrhenius Plot:**
   A plot of ln($k$) versus ($1/T$) yields a straight line with a slope of $-E_a/R$. This linear relationship allows the determination of the activation energy from the slope.

**Example 1.11** ***The rate constant for the uninhibited thermal decomposition of acetaldehyde***
For a reaction order, $n = 3/2 = 1.5$, the uninhibited thermal decomposition of acetaldehyde into methane and carbon monoxide is given [3] by:

$$k = Ae^{-\frac{E_a}{RT}} = 1.16 \times 10^{13} e^{-\frac{47{,}600}{RT}} \frac{(\mathrm{cm}^3)^{1/2}}{\mathrm{mol}^{1/2}\cdot \mathrm{s}}$$

Let us convert from cm$^3$ into L basis.

$$k = 1.16 \times 10^{13} e^{-\frac{47{,}600}{RT}} \frac{(\mathrm{cm}^3)^{1/2}}{\mathrm{mol}^{1/2}\cdot \mathrm{s}} \times \frac{(\mathrm{L})^{1/2}}{(1{,}000\ \mathrm{cm}^3)^{1/2}} = 3.67 \times 10^{11} e^{-\frac{47{,}600}{RT}} \frac{(\mathrm{L})^{1/2}}{\mathrm{mol}^{1/2}\cdot \mathrm{s}}$$

For example, at $T = 500°C \leftrightarrow (500 + 273)$ K = 773 K, we have:

$$k = 3.67 \times 10^{11} e^{-\left(\frac{47{,}600\ \text{cal/mol}}{1.987\frac{\text{cal}}{\text{mol.K}} * 773\ \text{K}}\right)} \frac{(\text{L})^{1/2}}{\text{mol}^{1/2} \cdot \text{s}} = 0.01275 \frac{(\text{L})^{1/2}}{\text{mol}^{1/2} \cdot \text{s}}$$

Notice that the units of either $A$ or $k$ are: $\left[M^{(1-n)}\frac{1}{\text{s}}\right] = \left[\left(\frac{\text{mol}}{\text{L}}\right)^{(1-1.5)}\frac{1}{s}\right] = \left[\left(\frac{\text{mol}}{\text{L}}\right)^{(-0.5)}\frac{1}{\text{s}}\right] = \left[\left(\frac{\text{L}}{\text{mol}}\right)^{(0.5)}\frac{1}{\text{s}}\right] = \left[\frac{(\text{L})^{1/2}}{\text{mol}^{1/2}.\text{s}}\right]$.

**Example 1.12 k = f(T) *Data***

Given the following set of data for the temperature and rate constant, find the activation energy $E_a$ and the pre-exponential factor $A$.

| # | *T* (K) | *k* ($s^{-1}$) × $10^6$ | # | *T* (K) | *k* ($s^{-1}$) × $10^6$ |
|---|---|---|---|---|---|
| 1 | 273.15 | 0.000274 | 9 | 379.8167 | 0.1329 |
| 2 | 286.4833 | 0.000764 | 10 | 393.1500 | 0.2273 |
| 3 | 299.8167 | 0.0019 | 11 | 406.4833 | 0.3755 |
| 4 | 313.1500 | 0.0046 | 12 | 419.8167 | 0.6007 |
| 5 | 326.4833 | 0.0100 | 13 | 433.1500 | 0.9336 |
| 6 | 339.8167 | 0.0206 | 14 | 446.4833 | 1.4133 |
| 7 | 353.1500 | 0.0402 | 15 | 459.8167 | 2.0886 |
| 8 | 366.4833 | 0.0747 | 16 | 473.1500 | 3.0194 |

**Solution**

Figure 1.17 shows the MATLAB code to take in $T$ and $k$ data, transform $T$ into $1/T$ and $k$ into $\ln(k)$, carry out curve-fitting using MATLAB's built-in function, and show both numerical and graphical results.

```
clear variables;
% Given reaction data
T=[ 273.15, 286.4833, 299.8167, 313.15, 326.4833, 339.8167, 353.15, 366.4833, ...
379.8167, 393.15, 406.4833, 419.8167, 433.15, 446.4833, 459.8167, 473.15]; % Temperature in Kelvin
k=1.0E+6*[0.000274, 0.000764, 0.0019, 0.0046, 0.0100, 0.0206, 0.0402, 0.0747, ...
0.1329, 0.2273, 0.3755, 0.6007, 0.9336, 1.4133, 2.0886, 3.0194];% Rate constant in s-1.
R = 8.314; % Gas constant in J/mol·K
% Linearize the Arrhenius equation
ln_k = log(k);
inverse_T = 1 ./ T;
% Perform linear regression to find ln(A) and -Ea/R
coefficients = polyfit(inverse_T, ln_k, 1);
slope = coefficients(1);
intercept = coefficients(2);
% Calculate activation energy (Ea) and pre-exponential factor (A)
Ea = -slope * R;
```

**Figure 1.17** Calculation of Arrhenius kinetic parameters via curve-fitting of $k$ versus $T$ data.

```
A = exp(intercept);
% Display the results
fprintf('Activation energy (Ea): %.4f J/mol\n', Ea);
fprintf('Pre-exponential factor (A): %.4e\n', A);
% Plot the data and the fit
figure;
scatter(inverse_T, ln_k, 'filled');
hold on;
plot(inverse_T, polyval(coefficients, inverse_T), '-r');
xlabel('1/T (1/K)');
ylabel('ln(k)');
title('Arrhenius Plot');
legend('Data', 'Linear Fit', 'Location', 'Best');
grid on;
hold off;
```

**Figure 1.17** (*Continued*)

Results are copied from the Command Window:

**≫ Fig1_17**
**Activation energy (Ea): 50025.6725 J/mol**
**Pre-exponential factor (A): 1.0075e+12**

Figure 1.18 shows the plot of *ln*(*k*) versus (1/*T*), showing a straight-line relationship.

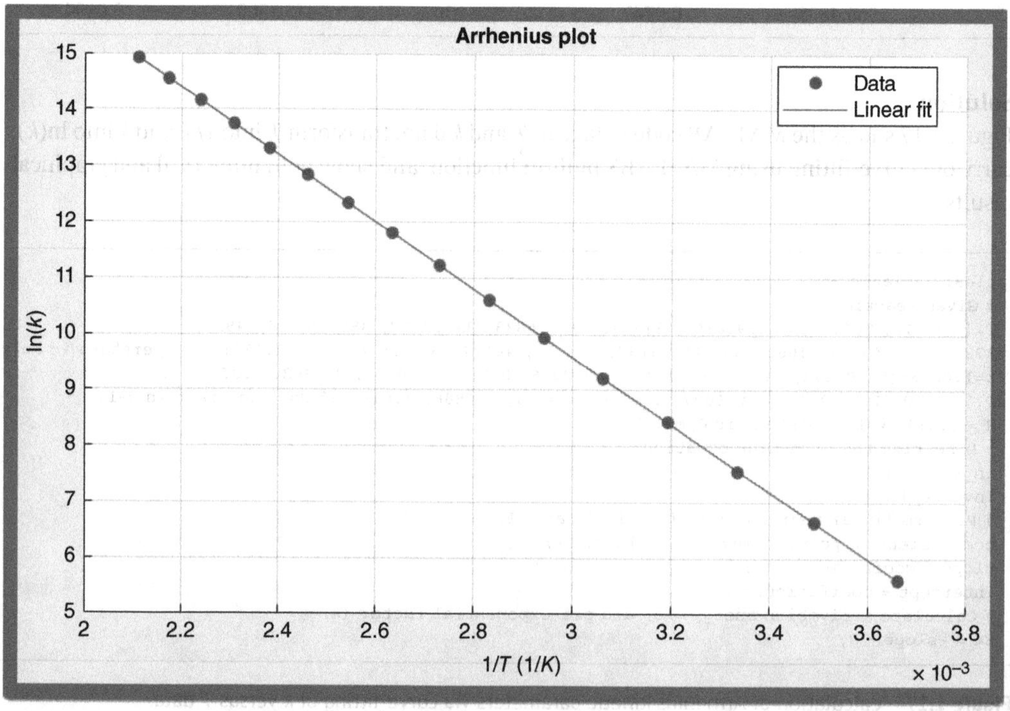


**Figure 1.18** The plot of ln(*k*) versus (1/*T*), showing a straight-line relationship.

**Catalysts and activation energy**

1) **Role of Catalysts:**
   a) Catalysts lower the activation energy of a reaction by providing an alternative reaction pathway with a lower energy barrier. Here, the activation complex peak is visualized as a mountain that separates the world (or village) of reactants from that of the products. Without a catalyst, people must climb up the mountain, reach the peak, and then can easily slide or glide down the hill to reach the village of products. With a catalyst, a tunnel is created underneath where people do not have to travel up to the peak; instead, they can reach the entrance of the tunnel faster than before and have a shortcut pathway to reach the village of products.
   b) They do not change the reactants or products' energy levels, only the transition state energy.
2) **Effect on Reaction Rate:**
   By lowering the activation energy, catalysts increase the rate of reaction without being consumed in the process.

### 1.1.6 Homogeneous Versus Heterogeneous Reactions

The distinction between homogeneous and heterogeneous reactions lies in the phase of the reactants and products involved in the chemical reaction:

1) **Homogeneous Reactions:**
   All reactants and products are in the same phase, whether it be gas or liquid.
   **Characteristics:**
   a) The reactants are uniformly distributed throughout the reaction mixture.
   b) These reactions typically occur faster due to the ease of mixing and interaction between reactants.
   
   Examples include reactions in solutions where all species are dissolved (e.g., acid-base reactions in aqueous solutions) and gas-phase reactions (e.g., the combination of nitrogen and hydrogen gases to form ammonia). Another example is the reaction between hydrogen gas ($H_2$) and chlorine gas ($Cl_2$) to form hydrogen chloride (HCl):

   $$H_2(g)+Cl_2(g)\rightarrow 2HCl(g)$$

2) **Heterogeneous Reactions:**
   Reactants and/or products are in different phases, such as solid, liquid, and gas.
   **Characteristics:**
   a) The reaction occurs at the interface between different phases, such as a solid catalyst in a gas or liquid mixture.
   b) These reactions can be slower due to limitations in the contact area between phases and mass transfer (or diffusion).
   
   Examples include reactions involving solids and gases (e.g., catalytic converters in cars, where gases react on the surface of a solid catalyst) and reactions in multiphase systems like liquid–liquid or liquid–solid mixtures. Another example is the reaction between solid calcium carbonate ($CaCO_3$) and aqueous hydrochloric acid (HCl) to produce calcium chloride ($CaCl_2$), carbon dioxide ($CO_2$), and water ($H_2O$):

   $$CaCO_3(s) + 2HCl(aq) \rightarrow CaCl_2(aq) + CO_2(g) + H_2O(l)$$

In summary, homogeneous reactions involve reactants and products in a single phase with uniform distribution, often leading to faster reaction rates, whereas heterogeneous reactions involve multiple phases with reactions occurring at phase boundaries, often resulting in slower reaction rates due to phase transfer limitations.

### 1.1.7 Reactor Types Used in Chemical Industries

Chemical reactors are vessels designed to contain chemical reactions. The choice of reactor type depends on various factors including the nature of the reaction, production scale, and desired products. Here are some common reactor types used in the chemical engineering industry:

1) **Batch Reactor (BR):**
   **Description:**
   A BR is a closed system where reactants are loaded, the reaction occurs, and then products are removed after the reaction is complete.
   **Advantages:**
   Flexibility in operation, suitable for small-scale production and varied products, and easy to control reaction time and conditions.
   **Disadvantages:**
   Higher labor and time requirements for loading and unloading, and less efficient for large-scale continuous production.
   **Applications:**
   Pharmaceuticals, specialty chemicals, and research and development processes.
2) **Continuous Stirred-tank Reactor (CSTR):**
   **Description**
   In a CSTR, reactants are continuously fed into the reactor, and products are continuously removed. The contents are well mixed, ensuring uniform composition throughout.
   **Advantages**
   Steady-state operation with continuous output and easier to maintain consistent quality and control.
   **Disadvantages**
   More complex control systems requirement and less suitable for reactions with long reaction times.
   **Applications**
   Polymerization processes, fermentation, and wastewater treatment.
3) **Plug Flow Reactor (PFR):**
   **Description**
   In a PFR, reactants flow through a cylindrical pipe and react as they move along the length of the reactor. There is a concentration gradient along the length of the reactor, with no back-mixing.
   **Advantages**
   High conversion efficiency for many reactions and suitable for large-scale production with high throughput.
   **Disadvantages**
   Less flexibility in operation compared to batch reactors and a requirement for careful design to ensure proper flow and heat management.
   **Applications**
   Petrochemical industries, gas-phase reactions, and large-scale chemical manufacturing.

4) **Packed Bed Reactor (PBR):**
   **Description:**
   PBRs are filled with catalyst particles, and reactants flow through the bed, undergoing reaction as they come into contact with the catalyst.
   **Advantages:**
   High surface area for reaction due to the catalyst packing and effective for catalytic processes with high conversion rates.
   **Disadvantages:**
   Pressure drop across the bed can be significant and potential for catalyst deactivation over time.
   **Applications:**
   Catalytic cracking in oil refineries, hydrogenation, and chemical synthesis.
5) **Fluidized Bed Reactor (FBR):**
   **Description:**
   In an FBR, solid catalyst particles are suspended in an upward flow of gas or liquid, creating a fluid-like state.
   **Advantages:**
   Excellent heat and mass transfer, uniform temperature distribution, and high reaction rates.
   **Disadvantages:**
   Complex design and operation, and potential for erosion of reactor internals due to the movement of solid particles.
   **Applications:**
   Gasification, combustion, and catalytic processes.
6) **Tubular Reactor (TR):**
   **Description:**
   A TR is similar to a PFR but can be designed with various configurations (coiled, straight tubes) to enhance heat transfer and accommodate different reaction kinetics.
   **Advantages:**
   Versatile design allows for efficient heat management and is suitable for both endothermic and exothermic reactions.
   **Disadvantages:**
   Requirement of precise control of flow and temperature and potential for fouling in tubes with complex reactions.
   **Applications:**
   Synthesis of chemicals, thermal cracking, and polymerization.
7) **Microreactor (MR):**
   **Description:**
   An MR consists of small channels where reactions take place, allowing for precise control of reaction conditions and rapid mixing.
   **Advantages:**
   High surface-to-volume ratio enhances heat and mass transfer and is suitable for highly exothermic or endothermic reactions.
   **Disadvantages:**
   Limited to small-scale production and the requirement of specialized equipment and design.
   **Applications:**
   Pharmaceutical synthesis, fine chemical production, and laboratory-scale research.
   In brief, these reactor types are selected based on the specific needs of the chemical process, including reaction kinetics, heat and mass transfer requirements, and production scale.

### 1.1.8 MATLAB Code for Fitting $C_A = f(t)$ for Zero-, First-, and Second-Order Reaction

**Example 1.13** $C_A = f(t)$ ***Data 1***

Given time versus concentration data, determine whether the reaction follows zero-, first-, or second-order kinetics by linearizing each rate law, fitting with **polyfit/polyval** MATLAB functions, computing $R^2$, and interpreting slope/intercept to get $k$ and $C_{A0}$.

**Table 1.2** Time versus $C_A$ data.

| $t$ (min) | $C_A$ (mol·L$^{-1}$) | $t$ (min) | $C_A$ (mol·L$^{-1}$) |
|---|---|---|---|
| 0 | 1.035281 | 4 | 0.466112 |
| 1 | 0.825283 | 6 | 0.295307 |
| 2 | 0.683441 | 8 | 0.205733 |
| 3 | 0.573408 | 10 | 0.134926 |

**Solution**

From Eq. (1.13) for a zero-order reaction, we have:

$$\int_{CA_0}^{C_A} dC_A = -k\int_0^t dt \rightarrow C_A - C_{A0} = -kt \rightarrow C_A = C_{A0} - kt$$

**Intercept:** $C_{A0}$ and slope $-k$.

From Eq. (1.15) for a first-order reaction, we have:

$$\frac{dC_A}{C_A} = -kdt \rightarrow \int_{C_{A0}}^{C_{Af}} \frac{dC_A}{C_A} = -k\int_0^{t_f} dt \rightarrow \ln\left(\frac{C_{Af}}{C_{A0}}\right) = -kt_f$$

$$\ln(C_A) = \ln(C_{A0}) - kt$$

**Intercept:** $\ln(C_{A0})$ and slope $-k$.

From Eq. (1.22) for a second-order reaction, we have:

$$\int_{C_{A0}}^{C_{Af}} \frac{dC_A}{C_A^2} = -k\int_0^{t_f} dt \rightarrow \frac{1}{C_{A0}} - \frac{1}{C_{Af}} = -kt_f \rightarrow \frac{1}{C_A} = \frac{1}{C_{A0}} + kt$$

**Intercept:** $\frac{1}{C_{A0}}$ and slope $k$.

Figure 1.19 shows MATLAB code (**Example_1_13.m**) for fitting zero-, first-, and second-order kinetics by linearizing each rate law, fitting with **polyfit/polyval** MATLAB functions, computing $R^2$, and interpreting slope/intercept to get $k$ and $C_{A0}$.

```
clear variables;
% Reaction data (time, concentration)
t = [0 1 2 3 4 6 8 10]'; % min
CA = [1.035281 0.825283 0.683441 0.573408 0.466112 0.295307 0.205733 0.134926]';

% 1) Zero-order fit: C = m0*t + b0 (m0 = -k)
p0 = polyfit(t, CA, 1);
m0 = p0(1); b0 = p0(2);
C_pred0 = polyval(p0, t);
```

**Figure 1.19** MATLAB code (**Example_1_13.m**) for fitting zero-, first-, and second-order kinetics by linearizing each rate law, fitting with **polyfit/polyval** MATLAB functions, computing $R^2$, and interpreting slope/intercept to get $k$ and $C_{A0}$.

```
SSres0 = sum((CA - C_pred0).^2);
SStot0 = sum((CA - mean(CA)).^2);
R20 = 1 - SSres0 / SStot0;
k_zero = -m0;
CA0_zero = b0;

% 2) First-order fit: lnC = m1*t + b1 (m1 = -k, b1 = ln C0)
lnC = log(CA);
p1 = polyfit(t, lnC, 1);
m1 = p1(1); b1 = p1(2);
lnC_pred1 = polyval(p1, t);
SSres1 = sum((lnC - lnC_pred1).^2);
SStot1 = sum((lnC - mean(lnC)).^2);
R21 = 1 - SSres1 / SStot1;
k_first = -m1;
CA0_first = exp(b1);

% 3) Second-order fit: 1/C = m2*t + b2 (m2 = k, b2 = 1/C0)
invC = 1./CA;
p2 = polyfit(t, invC, 1);
m2 = p2(1); b2 = p2(2);
invC_pred2 = polyval(p2, t);
SSres2 = sum((invC - invC_pred2).^2);
SStot2 = sum((invC - mean(invC)).^2);
R22 = 1 - SSres2 / SStot2;
k_second = m2;
CA0_second = 1 / b2;

% Display results
fprintf('Zero-order: k = %.6f (units), CA0 = %.6f, R^2 = %.6f\n', k_zero, CA0_zero, R20);
fprintf('First-order: k = %.6f, CA0 = %.6f, R^2 = %.6f\n', k_first, CA0_first, R21);
fprintf('Second-order: k = %.6f, CA0 = %.6f, R^2 = %.6f\n', k_second, CA0_second, R22);

% Plot raw data and best fit lines
figure;
subplot(3,1,1);
plot(t, CA, 'o'); hold on;
plot(t, C_pred0, '-'); xlabel('t (min)'); ylabel('C_A (mol/L)');
title(sprintf('Zero-order fit, R^2 = %.4f', R20));
set(gca, 'FontSize', 14, 'FontWeight', 'bold');
subplot(3,1,2);
plot(t, lnC, 'o'); hold on;
plot(t, lnC_pred1, '-'); xlabel('t (min)'); ylabel('ln C_A');
title(sprintf('First-order fit (ln C_A vs t), R^2 = %.4f', R21));
set(gca, 'FontSize', 14, 'FontWeight', 'bold');
subplot(3,1,3);
plot(t, invC, 'o'); hold on;
plot(t, invC_pred2, '-'); xlabel('t (min)'); ylabel('1/C_A (L/mol)');
title(sprintf('Second-order fit (1/C_A vs t), R^2 = %.4f', R22));
set(gca, 'FontSize', 14, 'FontWeight', 'bold');
```

**Figure 1.19** *(Continued)*

Results are copied from the Command Window:

```
>> Example_1_13
Zero-order: k = 0.086421 (units), CA0 = 0.894728, R^2 = 0.927789
First-order: k = 0.203067, CA0 = 1.030825, R^2 = 0.999304
Second-order: k = 0.616583, CA0 = 3.595814, R^2 = 0.928835
```

Figure 1.20 shows the generated plot, representing the model fitness for zero-, first-, and second-order kinetics by linearizing each rate law, augmented by $R^2$. The first-order model has the highest $R^2$, indicating that $C_A$ versus $t$ data can be best described by:

$$C_{Af} = C_{A0}e^{-kt_f} = 1.0308e^{-0.20307*t}$$

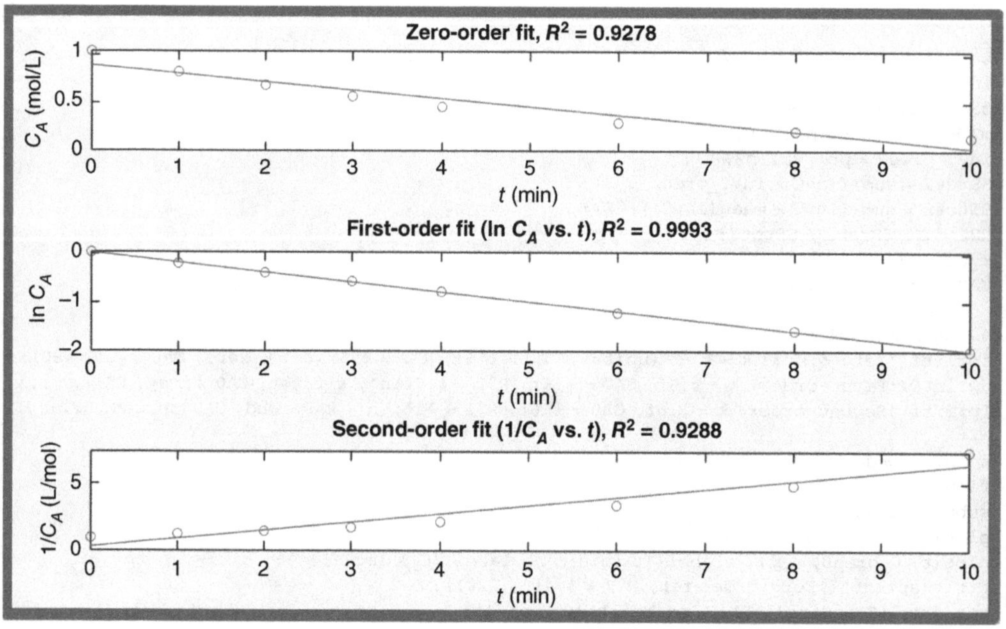


**Figure 1.20** The generated plot, representing the model fitness for zero-, first-, and second-order kinetics by linearizing each rate law, augmented by $R^2$.

### 1.1.9 MATLAB Code for Fitting $C_A = f(t)$ for $n^{th}$-Order Reaction

**Example 1.14** ***$C_A = f(t)$ Data 2***

Given time versus concentration data, determine both the rate constant, $k$, and the reaction order, $n$, augmented by goodness of fit statistical measures, like $R^2$.

**Table 1.3** $C_A = f(t)$ data.

| $t$ (min) | $C_A$ (mol·L$^{-1}$) | $t$ (min) | $C_A$ (mol·L$^{-1}$) | $t$ (min) | $C_A$ (mol·L$^{-1}$) |
|---|---|---|---|---|---|
| 0 | 1.035 | 4 | 0.466 | 8 | 0.206 |
| 1 | 0.825 | 5 | 0.375 | 9 | 0.170 |
| 2 | 0.683 | 6 | 0.295 | 10 | 0.135 |
| 3 | 0.573 | 7 | 0.250 | 12 | 0.090 |

**Solution**

In general, for $n \neq 1$, Eq. (1.30) gives the relationship between $C_A$ and $t$ for a given set of $n$ and $C_{A0}$.

Figure 1.21 shows MATLAB code (**Example_1_14.m**) for fitting $C_A$ versus $t$ data using **lsqcurvefit** (Optimization Toolbox) to fit p = [k, n] simultaneously. If **lsqcurvefit** is not available, then **nlinfit** (Statistics Toolbox) or **fminsearch** (without bounds) can be used.

```
% Simple fit of k and n from CA vs t data
clear;
% – Experimental data (replace with yours) –
t = [0 1 2 3 4 5 6 7 8 9 10 12]';
CA = [1.035 0.825 0.683 0.573 0.466 0.375 0.295 0.250 0.206 0.170 0.135 0.090]';
% – Known initial concentration –
CA0 = CA(1);
% – Model function for arbitrary order –
model = @(p,t) ( CA0^(1-p(2)) - (1-p(2))*p(1).*t ).^(1./(1-p(2)));
% – Initial guess [k, n] and bounds –
p0 = [0.2 1.0]; % guess: k=0.2, n=1.0
lb = [0 0]; % lower bounds
ub = [10 5]; % upper bounds
% – Fit using lsqcurvefit –
opts = optimoptions('lsqcurvefit','Display','off');
[pfit,~,resid] = lsqcurvefit(model,p0,t,CA,lb,ub,opts);
k_fit = pfit(1);
n_fit = pfit(2);
% – Goodness of fit (R2) –
SSres = sum(resid.^2);
SStot = sum((CA - mean(CA)).^2);
R2 = 1 - SSres/SStot;
% – Display results –
fprintf('Best fit: k = %.4f ((L/mol)^(1-n))(min^-1)\n',k_fit);
fprintf(' n = %.4f (reaction order)\n',n_fit);
fprintf('Model R2 = %.4f\n',R2);
% – Plot data and fitted curve –
t_fit = linspace(min(t),max(t),200);
CA_fit = model(pfit,t_fit);
plot(t,CA,'ko','MarkerFaceColor','k'); hold on
plot(t_fit,CA_fit,'b-','LineWidth',1.5);
xlabel('Time (min)'); ylabel('C_A (mol/L)');
legend('Data','Fit','Location','best');
title(sprintf('Fit for n^{th}-Order: k=%.3g, n=%.3g, R^2=%.3f',k_fit,n_fit,R2));
grid on
set(gca, 'FontSize', 14, 'FontWeight', 'bold');
```

**Figure 1.21** MATLAB code (**Example_1_14.m**) for fitting $C_A$ versus $t$ data using **lsqcurvefit** (Optimization Toolbox) to find the best parameters of both $n$ and $k$.

Notice that **lsqcurvefit** is nonlinear in both $n$ and $k$, which is why it uses iterative optimization (usually the trust-region-reflective or Levenberg–Marquardt algorithms) to find the best parameters of both $n$ and $k$.

Results are copied from the Command Window.

```
>> Example_1_14
Best fit: k = 0.2052 ((L/mol)^(1-n))(min^-1)
   n = 1.0149 (reaction order)
Model R2 = 0.9993
```

Figure 1.22 shows the generated plot of $C_A$ versus $t$, fitted to an $n^{th}$-order, where the nonlinear regression method found the best values of $n = 1.0149$ and $k = 0.205$.

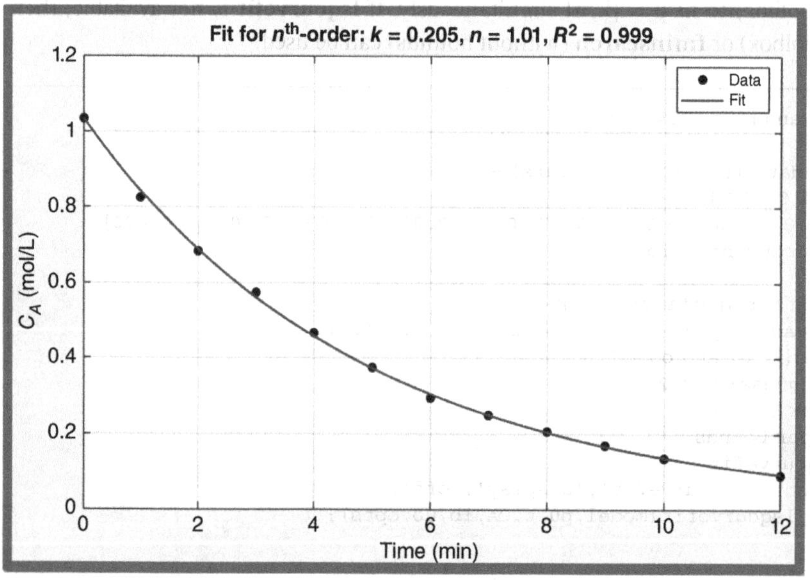


**Figure 1.22** The generated plot of $C_A$ versus $t$, fitted to an $n^{th}$-order, where the nonlinear regression method found the best values of $n = 1.0149$ and $k = 0.205$.

### 1.1.10 AI Copilot for MATLAB-Based Computations in Chemical Reaction Engineering

#### 1.1.10.1 From Numerical Programming to AI Collaboration

For decades, MATLAB has been the engineer's computational laboratory, a language to translate mathematical models into numerical form. Students traditionally learn syntax, operators, and plotting functions before they can express a design algorithm. However, the advent of generative artificial intelligence (AI) has changed this learning order: Now the engineer can describe a problem in natural language and let the AI translate it into executable MATLAB code.

Generative AI models trained on trillions of tokens of text and code (e.g., ChatGPT (GPT-5), Gemini 1.5, DeepSeek-Coder, and Grok-2) can interpret the logic behind a chemical-reaction problem. Instead of starting from syntax, the student begins with conceptual reasoning, the real engineering part, while the AI handles the translation into MATLAB instructions.

#### 1.1.10.2 Why Use AI as a Copilot in CRE Course

Table 1.4 shows the educational benefits gained upon using AI as a copilot in the CRE course.

AI, therefore, becomes a didactic partner, never replacing human reasoning but amplifying it.

#### 1.1.10.3 Challenges and Cautions

Despite its promise, AI-assisted computation has some caveats:

- **Hallucination:** The model may invent constants or equations.
- **Ambiguity Sensitivity:** Poorly phrased prompts yield misleading code.
- **Verification Burden:** Outputs must be checked against chemical principles and units.

**Table 1.4** Educational benefits of using AI in the CRE course.

| Educational Benefit | Description | Example |
|---|---|---|
| Accelerated coding | It converts verbal or written algorithms directly into MATLAB code. | "Simulate a first-order batch reactor with ($k = 0.25\ \text{min}^{-1}$)" → full code with ODE solver. |
| Reinforced understanding | Students must explain governing equations clearly to AI. | Writing the material balance forces articulation of assumptions, like isothermal, steady state, or adiabatic. |
| Immediate feedback | AI highlights missing parameters or inconsistent units. | Alerts that density was not provided for an energy balance. |
| Personalized learning | Different students can ask questions at varying depths. | "Explain the difference between ode45 and ode15s." |

Hence, the core mission is to train students to communicate precisely with AI systems, just as engineers must write unambiguous design specifications.

#### 1.1.10.4 The Large Language Model (LLM) as a Probabilistic Transformer

A large language model (LLM) is a neural network trained to predict the next word or token in a sequence. Its architecture, called a transformer, uses self-attention mechanisms to capture relationships among words, mathematical symbols, and code fragments. When you write:

"Write MATLAB code to solve for the concentration of A in a batch reactor given rate = −k · CA," the model tokenizes every term, compares it to millions of similar examples, and constructs a most-likely continuation: MATLAB syntax with ode45, initial conditions, and plotting commands.

LLMs do not "understand" chemistry; instead, they encode statistical associations between scientific language and programming patterns. The more contextually rich and technically correct your description, the higher the probability of generating accurate, executable MATLAB code.

#### 1.1.10.5 Inside the Prompt–Response Cycle

Table 1.5 illustrates how an LLM processes human language. The model first deconstructs user input, such as text or code, into fundamental units called tokens.

**Table 1.5** Chemical reaction engineering analogy for AI language processing.

| Step | Internal Operation | Chemical Engineering Analogy |
|---|---|---|
| 1. Tokenization | Input is split into discrete symbols (tokens). | Breaking a complex chemical reaction into its elementary steps. |
| 2. Context embedding | Each token is converted into a numerical vector that captures its meaning and context. | Mapping each reactant and product to its unique physicochemical properties (e.g., enthalpy, molecular weight). |
| 3. Attention mechanism | The model weighs the influence of other tokens in the sequence to inform the next prediction. | Identifying how catalysts or reactant concentrations selectively influence the reaction pathway. |
| 4. Decoding | The model assembles the final output by repeatedly selecting the most probable next token. | Predicting the most likely final products and their yields based on reactants and reaction conditions. |

During its training on vast datasets, the model learns the complex statistical relationships between these tokens. This training process allows the model to understand context and nuance by adjusting millions of internal parameters, or weights. When generating a response, the model uses this knowledge to predict the most probable next token in a sequence, given the preceding tokens as context.

By repeating this predictive process token-by-token, the model constructs a full sequence. This sequence of tokens is finally decoded back into the coherent, human-readable language we see as the output.

A helpful analogy is that of an archaeologist deciphering ancient hieroglyphs. The archaeologist learns to map each symbol (a token) to a concept or word in a modern language. The more ancient texts they study (the training data), the more accurately they can translate a sequence of symbols into a meaningful sentence (the output).

#### 1.1.10.6 Strengths and Limitations in Engineering Tasks

**Strengths**

- Handles long, structured prompts with equations.
- Generates complete MATLAB functions and scripts.
- Can explain, refactor, and document code.

**Limitations**

- Lacks true numerical intuition; may mis-scale quantities.
- Sensitive to ambiguous terms ("conversion rate" vs. "reaction rate").
- Does not automatically enforce conservation laws unless prompted.

Thus, human oversight remains the gatekeeper for thermodynamic realism and dimensional correctness.

**Example 1.15** ***AI Translation of a Batch Reactor Algorithm***

**Human-Language Algorithm**

"A liquid-phase, first-order irreversible reaction ($A \xrightarrow{k} B$) occurs in a batch reactor at 25°C with ($k = 0.25\ \text{min}^{-1}$). The initial concentration is ($C_{A0} = 1.0\ \text{mol/L}$). Plot ($C_A(t)$) for 0–10 min."

**Prompt to ChatGPT**

Write MATLAB code to solve $dC_A/dt = -0.25$*CA with $C_A(0) = 1$, integrate for $t = $ 0:0.1:10, and plot $C_A$ versus $t$.

**AI-Generated MATLAB Code**

Figure 1.23 shows the AI-generated MATLAB code (**Example_1_15.m**) upon executing the previous prompt.

```
% Example 1.15: First-order batch reactor
k = 0.25; CA0 = 1; tspan = [0 10];
f = @(t,CA) -k*CA;
[t,CA] = ode45(f,tspan,CA0);
plot(t,CA,'LineWidth',2)
xlabel('Time (min)'), ylabel('C_A (mol/L)')
title('Batch Reactor: First-Order Reaction')
```

**Figure 1.23** The MATLAB code (**Example_1_15.m**) generated upon executing the previous prompt.

It is left as an exercise to try to run the MATLAB code, shown in Figure 1.23, to see the plot of $C_A$ versus time for a first-order reaction.

| NOTE |
|---|
| If the prompt omitted $C_A(0) = 1$, the AI model would invent a default value. Always state initial and boundary conditions explicitly. |

### 1.1.11 What Is Prompt Engineering?

Prompt engineering is the process of designing the input instruction to a generative model so that the output is accurate, relevant, and verifiable. In engineering computation, the prompt replaces the role of program design: it must communicate all governing equations, parameters, and constraints.

### 1.1.12 The CURVD-Compliant Prompt

Table 1.6 explains what the CURVD acronym stands for. The CURVD framework outlines five key principles for writing effective and technically sound prompts, particularly in scientific and engineering contexts. Adhering to these principles ensures that the generated output is accurate, relevant, and useful.

**Table 1.6** CURVD stands for concise, unambiguous, right, verifiable, and derived.

| Letter | Meaning | Practical Implementation |
|---|---|---|
| C: Concise | Eliminate redundant words and focus only on essential information. | Instead of: "I would like you please to kindly write…"<br>Use: "Write MATLAB code for…" |
| U: Unambiguous | Define all variables, symbols, and acronyms explicitly before use. | Example: "…where $C_A$ represents the concentration of species A (mol/L)." |
| R: Right | Ensure all technical information, equations, and constants are factually correct. | Correct: Using the proper differential form, $dC_A/dt = -kC_A$, for a first-order reaction. |
| V: Verifiable | Request an output that can be checked against a known standard or solution. | Example: "Plot the numerical solution for $C_A(t)$ and compare it to the known analytical solution." |
| D: Derived | Ground the request in fundamental principles, laws, or specific data. | Instead of: "Simulate a chemical reaction."<br>Use: "Model the reaction using the provided rate law and stoichiometry." |

### 1.1.13 Poor Versus Good Prompts

**The Impact of Prompt Quality:**

The quality of a prompt directly dictates the quality of the AI's output. A prompt that lacks detail forces the AI to make assumptions, often leading to incorrect or incomplete results. Table 1.7

**Table 1.7** Prompt comparison from vague to verifiable.

| Prompt Level | Example | Likely Outcome and Analysis |
|---|---|---|
| Poor (vague and ambiguous) | "Simulate a reactor where $A$ becomes $B$." | The AI is forced to guess the reaction order, rate constant, reactor type, and units. The resulting code is unreliable and cannot be verified. |
| Improved (specific but incomplete) | "Write MATLAB code for a first-order batch reactor, A→B, $k = 0.25\ \text{min}^{-1}$, $C_A0 = 1\ \text{mol/L}^{-1}$, $t = 0$–$10$ min." | Produces a functionally correct solution (likely using ode45), but it may lack crucial details like axis labels or clear documentation, making it harder to verify or reuse professionally. |
| CURVD-compliant (precise and verifiable) | "Generate MATLAB code that solves $dC_A/dt = -k{\cdot}C_A$, with $k = 0.25\ \text{min}^{-1}$, $C_{A(0)} = 1$ mol/L, over a tspan = [0 10]. Plot $C_{A(t)}$ and label the axes with appropriate units." | Generates accurate, reproducible, and professional-quality code. The output is unambiguous and directly verifiable against the provided equation. |

contrasts a vague prompt with progressively better versions to illustrate how specificity and adherence to the CURVD framework produce superior outcomes.

**Example 1.16** ***Diagnosing an Ambiguous Prompt***

A prompt that is not well-defined forces the AI model to make assumptions. Let's diagnose a common example.

**The Ambiguous Prompt**

"Plot conversion in a reactor as a function of time."

**Likely Outcome and Analysis**

Instead of asking for clarification, a capable AI will often attempt to fill in the blanks on its own. It might make a series of silent, unstated assumptions:

- **It Assumes a Reactor Type:** "This is probably a simple batch reactor."
- **It Assumes Reaction Kinetics:** "I will use a first-order reaction, A → B."
- **It Assumes Parameters:** "I will pick a rate constant, $k = 0.1\ \text{s}^{-1}$."

The AI then generates code that runs perfectly, producing a plausible-looking plot. The danger is that this output appears correct but is scientifically meaningless because it is based on a complete guess, not the user's actual system.

**The Core Problem:**

The prompt fails to define the problem by leaving critical information unspecified:

1) **Reactor Type:** Is it a batch, CSTR, or PFR?
2) **Reaction Kinetics:** What is the reaction order and rate law?
3) **Conditions and Parameters:** What are the initial concentrations and the rate constant?

**Lesson:**

An AI cannot read a user's mind. A prompt must provide all the necessary information, equations, parameters, and constraints to create a uniquely solvable problem. In engineering terms, you must provide enough information to satisfy the system's degrees of freedom. If you don't define the system, the AI will, and its definition will likely be wrong.

### 1.1.14 A Preflight Checklist for CRE Prompts

To ensure prompts for CRE problems are complete and unambiguous, use the following structured template. Think of it as a preflight checklist: by confirming each item is present, you guarantee the AI has all the necessary information to generate accurate and verifiable code.

**The CRE Prompt Template:**

1) **Task:** What is the primary objective? Clearly state the goal using an action verb.
   *Example:* "Solve the material and energy balances for a non-isothermal CSTR to find the steady-state temperature and conversion."
2) **System:** What is the chemical reaction? Define the stoichiometry and kinetics.
   *Example:* "The system involves the irreversible, elementary liquid-phase reaction $A \rightarrow B$."
3) **Governing Equations:** What is the underlying model? Provide specific differential or algebraic equations.
   *Example:* "Model the system using the following material balance: $dC_A/dt = (C_{A0} - C_A)/\tau - k \cdot C_A$."
4) **Parameters and Constants:** What are the fixed values? List all known constants with their corresponding units.
   *Example:* Rate constant $k = 0.05$ L/mol·s, Arrhenius pre-exponential factor $A$, activation energy $E_a$, heat of reaction $\Delta H_{rx}$, specific heats $C_p$, etc.
5) **Conditions:** What is the starting point or boundary? Specify all initial conditions (for dynamic problems) or inlet conditions (for flow reactors).
   *Example:* "Initial conditions are $C_A(0) = 2.0$ mol/L and $T(0) = 298$ K."
6) **Output and verification** : What should the result look like? Describe the desired format for the output. How can the result be verified? Include a step to check the answer's validity.
   *Example:* "Generate a plot of conversion ($X$) versus time ($t$). Also, calculate and display the final steady-state conversion and compare it to the analytical solution."

Using this template transforms prompt engineering from guesswork into a systematic process. It ensures every query is self-contained, verifiable, and aligned with the CURVD principles, leading to the reliable generation of deterministic MATLAB code.

**Exercise 1.1** ***From Theory to Code***

This exercise is designed to put the CURVD framework into practice. You will act as the engineer, using the prompt template to guide an AI in solving a common CRE problem.

**The Scenario**

The goal is to model an isothermal BR performing the liquid-phase isomerization of glucose into fructose. This is a first-order reaction.

Glucose → Fructose

For more information, see Chapter 5 | Section: 5.1.1 Example 5.1: Liquid-phase First-order Isomerization Reaction.

**The Task: A Step-by-Step Guide**

1) **Define the Problem:**
   First, establish a complete and solvable system. For the first-order isomerization of glucose at 60°C, you will need to specify realistic values for:
   a) The first-order rate constant, $k$ (in $s^{-1}$).

b) The initial concentration of glucose, $C_{A0}$ (in mol/L).
c) The total reaction time for the simulation, $t_{final}$ (in seconds).

2) **Construct the Prompt:**
Using the CRE Prompt Template, formulate a detailed, CURVD-compliant prompt. Your prompt must explicitly state the differential mole balance for a BR ($dC_A/dt = -kC_A$) and include all the parameters you defined in step 1.

3) **Generate and Analyze:**
Test your prompt with an LLM. Critically analyze the response:
a) Does the generated code use a proper ODE solver (e.g., MATLAB's ode45) to solve the mole balance?
b) Does the code generate a plot of glucose concentration and conversion as a function of time?
c) Is the output clear and well-documented?

4) **Compare and Document:**
Verify the AI's result. For this first-order reaction, the analytical solution is $C_A = C_{A0} * e^{(-kt)}$. Plot the AI's numerical solution and the analytical solution on the same graph to confirm they match. Document your prompt, the AI's full response, and your verification plot.

**A Note for Instructors**
Encourage students to document and submit both their final prompt and the AI's full, unedited response in their assignments. This practice promotes accountability and provides a valuable opportunity for peer review, shifting the focus from simply getting the right answer to mastering the process of high-quality prompt engineering.

### 1.1.15 The LLM as a Translator of Chemical Algorithms

Traditional programming requires a two-step process: first, the mathematical derivation, and second, the manual translation of those equations into code. With generative AI, the student or engineer provides a clear, descriptive algorithm in natural language, and the AI acts as a "linguistic compiler," translating the scientific description directly into executable syntax.

This process is additive, allowing for the progressive building of complexity, as shown in Table 1.8.

**Table 1.8** Building a prompt: from concept to code.

| Step | Human Input (Building the Prompt) | AI Interpretation (Translating Concepts) | MATLAB Output (Generated Syntax) |
|---|---|---|---|
| 1. The core model | "Simulate an isothermal first-order batch reactor." | Recognizes the governing Ordinary Differential Equation (ODE): $dC_A/dt = -kC_A$. | Creates a function handle for the ODE and implements an ode45 solver call. |
| 2. Add complexity | "Now make it non-isothermal by including the Arrhenius dependence of $k$ on $T$ and the energy balance." | Incorporates the Arrhenius equation $k(T) = A \cdot exp(-E_a/RT)$ and adds the energy balance ODE. | Modifies the function to solve a system of two coupled ODEs for $C_A(t)$ and $T(t)$. |
| 3. Specify output | "Plot the final results for $C_A$ and $T$ versus time on separate, labeled subplots." | Identifies the specific visualization requirements. | Adds subplot, plot, xlabel, ylabel, and legend commands to create a professional figure. |

## 1.1.16 Platforms in Comparison

### 1.1.16.1 Choosing the Right Tool for the Job

While many AI platforms can assist with CRE tasks, they are not interchangeable. Each has distinct strengths and weaknesses, making them suited for different types of work. Understanding these differences allows an engineer to build a versatile "AI toolkit." Table 1.9 shows the strengths and limitations of each AI platform.

**Table 1.9** A comparative look at leading AI platforms.

| Platform | Core Strength ✅ | Limitations ⚠️ | Best Used For... 🎯 |
|---|---|---|---|
| ChatGPT (GPT-4/5) | A strong all-rounder with excellent chemical reasoning and highly accurate MATLAB syntax generation. | May occasionally truncate very long or complex code outputs, requiring follow-up prompts. | Developing and verifying homework problems or generating baseline code for course demonstrations. |
| Gemini 2.5 Pro | Exceptional multimodal reasoning and advanced coding capabilities, ideal for solving complex scientific and mathematical problems and generating complete, robust MATLAB simulations. | While its output is vast (up to 65k tokens), very lengthy, complex code requests may still hit a soft output limit, requiring chunked generation. | Developing and debugging sophisticated process models (e.g., in transport phenomena or reaction kinetics) and generating agentic code applications. |
| DeepSeek-Coder | A code-generation specialist that delivers highly optimized, efficient, and well-commented code blocks. Excels at translating precise technical specifications into clean, functional software. | A specialist, not a generalist. Its performance is contingent on receiving explicit, well-defined requirements. It is not designed for open-ended conceptual reasoning or tasks outside its coding domain. | Building performance-critical software components, algorithm implementation, and generating clean, production-ready code from detailed specifications. Ideal for software engineering tasks where code quality and execution efficiency are paramount. |
| Grok-2 | Excels at conversational debugging, making it feel like a collaborative partner in troubleshooting code. | It can be overly verbose or conversational, sometimes providing more explanation than necessary. | Interactive learning sessions where a student needs to debug their own code with step-by-step guidance. |

While each platform has unique strengths, no one can overcome an imprecise prompt. The effectiveness of every tool on this list is ultimately determined by the quality of the user's input. The **CURVD** principle is the universal key to unlocking their true potential.

### 1.1.16.2 A Note on Process Modeling

This chapter section focuses on applying the final mathematical models for process simulation. The fundamental skill of process modeling, the systematic derivation of the governing algebraic and differential equations, will be covered in detail in subsequent chapters.

For now, the necessary equations will be provided directly. Later in the course, your instructor will guide you through the derivation process, showing how to build these models from scratch based on key factors like: reactor type (e.g., batch, CSTR, and PFR), operating conditions (e.g., isothermal, adiabatic), and reaction phase (gas-phase vs. liquid-phase).

**Example 1.17** ***Batch Reactor (BR) with Temperature-Dependent Rate Constant***
**Human-Language Problem Statement**

"Liquid-phase first-order reaction ($A \xrightarrow{k} B$) is carried out **isothermally** at $T = 350$ K. The rate constant follows Arrhenius behavior with $k_0 = 1.2 \times 10^7$ s$^{-1}$ and $E_a = 75$ kJ/mol. Initial concentration $C_{A0}$=2.0 mol/L. Plot $C_A(t)$ for 0–200 s."

**CURVD-Compliant Prompt**
Generate MATLAB code that:

1) Uses Arrhenius rate law $k = k_0*\exp(-E_a/(R*T))$
2) Solves $dC_A/dt = -k*C_A$ with $C_{A(0)} = 2$
3) Integrates 0–200 s using the ode45 solver
4) Plots $C_{A(t)}$ with labeled axes and units

Given: $k_0 = 1.2e7\,\text{s}^{-1}$, $E_a = 75e3$ J/mol, $T = 350$ K, $R = 8.314$ J/mol·K

**AI-Generated MATLAB Code**
Figure 1.24 shows the AI-generated MATLAB code for the above algorithm, written for a liquid-phase isothermal BR.

```
%% Example 1.17 Arrhenius kinetics for a first-order liquid-phase in BR
k0 = 1.2e7; Ea = 75e3; R = 8.314; T = 350;
k = k0*exp(-Ea/(R*T));
CA0 = 2.0; tspan = [0 200];
f = @(t,CA) -k*CA;
[t,CA] = ode45(f,tspan,CA0);
plot(t,CA,'LineWidth',2)
xlabel('Time (s)'); ylabel('C_A (mol/L)')
title('Isothermal Batch Reactor with Arrhenius Kinetics')
```

**Figure 1.24** The AI-generated MATLAB code for the human language algorithm written for describing an isothermal liquid-phase BR.

It is left as an exercise to run the MATLAB code, shown in Figure 1.24, to see the generated plot of $C_A$ as $f$(time) for the liquid-phase, first-order isothermal BR.

**Example 1.18** ***Steady-State Energy Balance on an Adiabatic CSTR***

The liquid-phase reaction is: $A \xrightarrow{k1}$ B. Since there is no change in the number of moles and it is a constant-density fluid, then $F_0 = F$. We will also assume the heat capacity of the mixture is constant.

**Human Algorithm**

1) Steady-state material balance: $v_0*(C_{A0} - C_A) - V*k_0*\exp(-E_a/(R*T))*C_A = 0;$
2) Steady-state energy balance:

$$v_0*C_{A0}*C_p*(T_0 - T) - (\Delta Hr)*V*k_0*\exp(-E_a/(R*T))*C_A = 0;$$

**Prompt**

Write MATLAB code to compute steady-state $C_A$ and $T$ for a non-isothermal CSTR with parameters:
$C_{A0} = 2$ mol/L, $T_0 = 350$ K, $V = 1{,}000$ L, $v_0 = 1$ L/s,
$k_0 =$ 3e6 s^-1, $E_a = 80{,}000$ J/mol, $\Delta Hr = -60{,}000$ J/mol,
$C_p = 76$ J/mol·K, $R = 8.314$.
Show the numerical values of $C_A$ and $T$ at possible multiple steady states.

**AI-Generated MATLAB Snippet**

Figure 1.25 shows the AI-generated MATLAB code for a steady-state, adiabatic, and liquid-phase CSTR.

```
% Example 1.18 Non-isothermal CSTR steady states - Corrected
% – Parameters
CA0 = 50; % Inlet concentration (mol/L)
T0 = 350; % Inlet temperature (K)
V = 1000; % Reactor volume (L)
k0 = 3e6; % Pre-exponential factor (1/s)
Ea = 8e4; % Activation energy (J/mol)
v0 = 0.1; % Volumetric flow rate (L/s)
dHr = -6e4; % Heat of reaction (J/mol)
Cp = 76.0; % Molar heat capacity (J/mol.K)
R = 8.314; % Ideal gas constant (J/mol.K)
% – Anonymous function with properly scaled equations
% x(1) = CA, x(2) = T
f = @(x) [ ...
  % Eq 1: Full Material Balance (mol/s)
  v0*(CA0 - x(1)) - V*k0*exp(-Ea/(R*x(2)))*x(1);
  % Eq 2: Full Energy Balance (J/s)
  v0*CA0*Cp*(T0 - x(2)) - (dHr)*V*k0*exp(-Ea/(R*x(2)))*x(1) ];
% – Solve using fsolve with a better initial guess
options = optimoptions('fsolve','Display','off'); % Suppress verbose output
% Find the low-temperature steady state
[x_low, ~, ~] = fsolve(f, [45; 360], options);
fprintf('Low-temperature steady state:\n');
fprintf(' CA = %.2f mol/L, T = %.1f K\n\n', x_low(1), x_low(2));
% Find the high-temperature steady state with a different guess
[x_high, ~, ~] = fsolve(f, [5; 440], options);
fprintf('High-temperature steady state:\n');
fprintf(' CA = %.2f mol/L, T = %.1f K\n', x_high(1), x_high(2));
```

**Figure 1.25** The AI-generated MATLAB code for a steady-state, adiabatic, and liquid-phase CSTR.

Results, copied from the Command Window, are shown below:

```
>> Fig1_25
Low-temperature steady state:
  CA = 8.88 mol/L, T = 365.4 K
High-temperature steady state:
  CA = 0.38 mol/L, T = 451.9 K
```

**A Note on the Development Process**

It is important to understand that the code shown in Figure 1.25 is the final product of a collaborative process, not the result of a single AI attempt.

Effective AI-powered development is rarely a one-shot task; it is an iterative conversation. The workflow typically looks like this:

1) The user starts with an initial prompt to generate a basic code structure.
2) The user provides feedback and follow-up requests to refine the output.
3) The AI incorporates the feedback to add features, improve visualizations, or enhance the code's functionality.

### 1.1.17 Verification and Interpretation

Students must always perform the following checks:

1) **Physical Boundaries:** Verify ($C_A \geq 0$), ($T$) within operating range.
2) **Units:** Ensure consistency of $\Delta H$ (J/mol), $v_0$ (L/s), and $C_p$ (J/mol·K).
3) **Sensitivity:** Change $E_a$ or $\Delta Hr$ to see bifurcation behavior.
4) **Cross-model Validation:** Ask another LLM to reproduce the code and compare results.

**Example 1.19** ***Arrhenius Kinetics Data Fitting***

**Prompt**

"Given $T(K) = [300\ 320\ 340\ 360\ 380]$ and $k(s^{-1}) = [0.015\ 0.045\ 0.120\ 0.320\ 0.860]$, generate MATLAB code to fit $\ln(k)$ vs $1/T$ to find $E_a$ and $k_0$ using linear regression."

**AI-Generated Code**

Figure 1.26 shows AI-generated MATLAB code for fitting $k$ versus $T$ to find kinetic parameters: The pre-exponential factor, $k_0$, and the activation energy $E_a$.

```
T = [300 320 340 360 380];
k = [0.015 0.045 0.120 0.320 0.860];
x = 1./T; y = log(k);
p = polyfit(x,y,1);
Ea = -p(1)*8.314; k0 = exp(p(2));
fprintf('Ea = %.2f kJ/mol\n',Ea/1000)
fprintf('k0 = %.2e s^-1\n',k0)
plot(x,y,'o',x,polyval(p,x))
xlabel('1/T (K^{-1})'); ylabel('ln k')
title('Arrhenius Plot by AI-generated MATLAB Code')
```

**Figure 1.26** The AI-generated MATLAB Code for fitting $k$ versus $T$.

Numerical values are copied from the Command Window:

```
>> Fig1_26
Ea = 47.60 kJ/mol
k0 = 2.72e+06 s^-1
```

It is left as an exercise to run MATLAB code, shown in Figure 1.26, to see the generated plot of $\ln(k)$ versus $1/T$.

### 1.1.18 A Note for Instructors: Guiding the Learning Process

The true educational value of the following exercise lies in pushing students beyond simple code generation. Encourage them to engage with the AI as a tool for inquiry.

Challenge your students to:

- **Modify the Data:** Ask them to alter the dataset (e.g., add noise, remove a point) and rerun the analysis to see how it impacts the final result.
- **Prompt for Deeper Meaning:** Instruct them to ask the AI to explain the physical significance of the slope ($-E_a/R$) and intercept ($\ln(A)$) in the context of the Arrhenius plot.

### 1.1.19 Why Guidelines Are Necessary

Although generative LLMs can translate natural-language problems into MATLAB syntax, the correctness of their output depends entirely on how the human defines the system.

Without formal prompting discipline, the model may:

- Confuse rate expressions (e.g., reaction order).
- Use inconsistent units or default constants.
- Mislabel physical quantities.

A structured checklist ensures that AI output remains *scientifically credible* and *numerically verifiable.*

### 1.1.20 State All Modeling Assumptions

Always tell the model what to neglect and what to include:

"Assume ideal gas behavior and constant density. Neglect pressure drop and axial dispersion."

Clear assumptions define the *model domain* and prevent the AI from introducing hidden variables (e.g., mass-transfer terms).

### 1.1.21 Specify Parameters and Units

Provide every symbol's value and unit in one compact list. A simple mis-uniting (J vs. kJ) can yield gross errors. Table 1.10 shows a sample of symbols used in CRE with their proper units.

**Table 1.10** Symbols and units.

| Symbol | Quantity | Unit | Example Value |
|---|---|---|---|
| $k_0$ | Pre-exponential factor | $s^{-1}$ for $n = 1$ $\left[\left(\frac{mol}{L}\right)^{(1-n)}\frac{1}{s}\right]$ for $n \neq 1$ | $2.4 \times 10^6$ |
| $E_a$ | Activation energy | J/mol | 75,000 |
| $\Delta H_r$ | Heat of reaction | J/mol | −60,000 |
| $\rho$ | Density | $kg/m^3$ | 1,000 |
| $C_p$ | Heat capacity | J/kg K | 4,200 |

**Instructor Note**

Always verify dimensional consistency *before* execution. MATLAB will not detect wrong physical units. It is crucial to remember that the engineer is ultimately responsible for validating all outputs from an AI model. An LLM might not always maintain dimensional consistency, and it is the engineer's responsibility to scrutinize the physics and mathematics of the model to ensure it is sound. Think of it as a professional check before accepting any calculation. ✅

### 1.1.22 Include Initial and Boundary Conditions

LLMs cannot infer starting values.

For each differential or algebraic equation, specify boundary conditions explicitly:

At $t = 0$ s: $C_A = 1.0$ mol/L, $T = 300$ K

At $z = 0$ m: $C_A = C_{A0}$, $T = T_0$

### 1.1.23 Request Equation Transparency

Instruct the AI model to display the equations it used:

"Show all governing equations before writing MATLAB code."

This allows students to check *conceptual correctness* prior to coding.

### 1.1.24 Ensure Zero Degrees of Freedom

A solvable system satisfies:

$$\textit{Number of written equations} = \textit{Number of unknown variables}$$

Incomplete prompts (missing parameters or unknown variables) lead to *under-determined* or *ill-posed* MATLAB code.

**Checklist:**
1) Count unknowns.
2) Provide the same number of independent equations.
3) Specify parameters, constants, and initial conditions.

### 1.1.25 Keep Prompts Task-Focused

Instead of giving an AI a single, massive prompt to solve a complex problem from start to finish, you will get far better and more reliable results by breaking the problem down into a series of smaller, single-task prompts. This "divide and conquer" strategy is like building with LEGO bricks; you assemble the final model one well-defined piece at a time🧱.

This approach gives you a checkpoint at each stage to verify the output before proceeding, ensuring the final result is correct.

Table 1.11 shows a workflow example of modeling parallel reactions.

**Table 1.11** A workflow example of modeling parallel reactions.

| Stage | Task | Prompt Example |
|---|---|---|
| 1. Derivation | First, derive the governing equations. Focus only on mathematics without worrying about code. | "For the elementary parallel reactions A → D ($k_1$) and $A \rightarrow U$ ($k_2$), symbolically derive the system of three ordinary differential equations for $dC_A/dt$, $dC_D/dt$, and $dC_U/dt$." |
| 2. Implementation | Next, translate the math into a solvable function. Provide the equations from the previous step. | "Using the derived ODEs, write a MATLAB function ode_system($t$, $C$) that takes the time t and concentration vector $C = [C_A, C_D, C_U]$ as inputs and returns the derivatives." |
| 3. Execution | Now, solve the system numerically. Provide all necessary parameters and initial conditions. | "Use the ode45 solver to solve the ode_system function from $t = 0$ to 50 min, with initial conditions $C_{A0} = 1$ M, $CD_0 = 0$, $CU_0 = 0$, $k_1 = 0.1\ \text{min}^{-1}$, and $k_2 = 0.05\ \text{min}^{-1}$." |
| 4. Visualization | Finally, plot the results. This step is only about presenting the data clearly. | "Plot the concentrations of $A$, $D$, and $U$ versus time on a single graph. Include a title, labeled axes with units, and a legend." |

### 1.1.26 Use Explicit Equation Formats

Write each balance or law in full mathematical form before prompting. For example, in an adiabatic BR for a liquid-phase first-order reaction, we have:

$$\frac{dC_A}{dt} = -kC_A \qquad \frac{dT}{dt} = \frac{(-\Delta H_r)kC_A}{\rho C_p}$$

This prevents AI from guessing hidden dependencies.

### 1.1.27 Specify Preferred Solver

While an LLM can often select an appropriate solver, an expert user guides the AI to the *optimal* choice. Specifying the correct solver prevents errors, improves accuracy, and is essential for numerically challenging problems, such as stiff systems. Think of it as telling a mechanic whether to use a standard wrench or a specialized one. Table 1.12 shows some typical MATLAB ode solvers and their properties. A typical prompt: "... For this non-isothermal CSTR model, you must use the ode15s solver, as the strong coupling between concentration and temperature is expected to cause numerical stiffness." See Table 5.1 for more information on MATLAB ode solvers.

### 1.1.28 Ask for Commented and Readable Code

Request structured output:

"Include section headers, comments, and labeled plots."

Readable AI code promotes pedagogy and easier debugging.

### 1.1.29 Request Unit and Dimension Checks

"Verify that all energy and mass terms have consistent units."

The AI will often echo back dimensional checks, alerting students to hidden errors.

**Table 1.12** MATLAB ode solvers and their applications.

| Solver | Problem Type | Typical CRE Application |
|---|---|---|
| ode45 | Nonstiff ODEs. This is the standard, general-purpose solver. It's the first tool you should try for most dynamic simulations. | An isothermal batch reactor where reaction rates change smoothly and predictably over time. |
| ode15s | Stiff ODEs. Use this for systems with widely different time scales (e.g., a very fast reaction coupled with a slow temperature change). | Mixed reactions with different rates, leading to numerical stiffness. |
| fsolve | Systems of nonlinear algebraic equations. This solver is used when all time derivatives are zero ($dy/dt = 0$). | Finding multiple steady-state operating points of a non-isothermal reactor. |
| dsolve | Symbolic analytical solutions. This is part of the Symbolic Math Toolbox and finds an exact mathematical formula, not a numerical answer. | Validating a simple model, such as a first-order isothermal reaction, by comparing the numerical solution with the exact analytical one. |

**Example 1.20** ***Constructing a Guided Prompt***

**Incomplete Prompt:**

"Write MATLAB code for a CSTR with reaction ($A \xrightarrow{k} B$) and energy balance."

**Likely AI Output:**

Code missing parameters and units.

**Improved Prompt:**

"Write MATLAB code for an adiabatic liquid-phase CSTR with first-order reaction A→B. $k_0$ = 3e10 s^−1, Ea = 80000 J/mol, ΔHr = −70000 J/mol, ρ = 1000 kg/m^3, Cp = 4.2e3 J/kg · K, R = 8.314. Feed: CA0 = 2 mol/L, T0 = 350 K, τ = 30 s. Solve for steady-state CA and T using **fsolve** MATLAB function."

AI now generates a *dimensionally complete* code. Figure 1.27 shows the AI-generated MATLAB code for describing a steady-state, adiabatic liquid-phase CSTR.

```
% ------------------------------------------------------------
% MATLAB script to find the steady-state operating point (CA and T)
% for an adiabatic, liquid-phase Continuous Stirred-Tank Reactor (CSTR).
%
% The script solves the material and energy balance equations simultaneously
% using the built-in 'fsolve' function for nonlinear systems.
% ------------------------------------------------------------
clear; clc;
% – 1. Define Model Parameters and Constants
% Kinetic parameters
k0 = 3e10; % Pre-exponential factor (s^-1)
Ea = 80000; % Activation energy (J/mol)
% Thermodynamic parameters
dHr = -70000; % Heat of reaction (J/mol) (negative for exothermic)
```

**Figure 1.27** AI-generated MATLAB code for a steady-state, adiabatic liquid-phase CSTR.

```
rho = 1000; % Density (kg/m^3)
Cp = 4.2e3; % Specific heat capacity (J/kg.K)
R = 8.314; % Ideal gas constant (J/mol.K)
% Operating Conditions (Feed)
CA0 = 2; % Inlet concentration of A (mol/L)
T0 = 350; % Inlet temperature (K)
tau = 30; % Residence time (s)
% — 2. Calculate Derived Parameters
% Volumetric heat capacity. This is crucial for unit consistency.
% (kg/m^3) * (J/kg.K) * (1 m^3 / 1000 L) -> J/L.K
rhoCp_vol = rho * Cp * (1/1000);
% — 3. Define the System of Equations for fsolve
% The function 'f' must return a vector [F1; F2] where F1 and F2 are the
% residuals of the balance equations that should equal zero at steady state.
% The input 'x' is a vector where x(1) = CA and x(2) = T.
f = @(x) [ ...
  % Equation 1: Material Balance
  % 0 = CA0 - CA - tau * k * CA
  CA0 - x(1) - tau * k0 * exp(-Ea / (R * x(2))) * x(1);
  % Equation 2: Energy Balance (Adiabatic, Q=0)
  % 0 = rho*Cp*(T0 - T) + (-dHr) * tau * k * CA
  rhoCp_vol * (T0 - x(2)) + (-dHr) * tau * k0 * exp(-Ea / (R * x(2))) * x(1)];
% — 4. Solve the System of Equations
% Provide an initial guess for [CA; T]. The feed conditions are a reasonable start.
initial_guess = [0.1*CA0; 380.0];
% Use fsolve to find the solution. 'optimoptions' is used to suppress the
% detailed output from the solver in the command window.
options = optimoptions('fsolve', 'Display', 'off');
[solution, fval, exitflag] = fsolve(f, initial_guess, options);
% — 5. Display the Results
if exitflag > 0
  fprintf('Solution found successfully.\n\n');
  fprintf('Steady-State Concentration (CA): %.4f mol/L\n', solution(1));
  fprintf('Steady-State Temperature (T): %.2f K\n', solution(2));
  % Optional: Calculate and display conversion
  conversion = (CA0 - solution(1)) / CA0 * 100;
  fprintf('Conversion (X):           %.2f %%\n', conversion);
else
  fprintf('Solver did not converge to a solution.\n');
  fprintf('Exit Flag: %d\n', exitflag);
  % Display the last calculated values for debugging
  disp('Last solution attempt:');
  disp(solution);
end
```

**Figure 1.27** (*Continued*)

The results are copied from the Command Window:

**Solution found successfully.**
**Steady-State Concentration (CA): 0.1985 mol/L**
**Steady-State Temperature (T): 380.02 K**
**Conversion (X): 90.07 %**

**Notes on the MATLAB Code (Figure 1.27)**

The MATLAB script shown in Figure 1.27 demonstrates two critical aspects of AI-assisted chemical process simulation: the necessity of unit verification and the challenge of finding multiple steady states.

1) **AI-assisted unit handling:** It is noteworthy that the AI platform successfully handled a deliberate inconsistency in the provided parameters, where units were given on different bases (per unit mass vs. per unit mole). The model correctly derived and applied a conversion factor (rhoCp_vol = rho * Cp * (1/1000)) to ensure the dimensional homogeneity of the energy balance. While this showcases the AI's capability, it also reinforces that the engineer must always verify the physical and mathematical soundness of the final equations.
2) **The Challenge of Multiple Steady States:** For highly nonlinear systems like this non-isothermal CSTR, finding all physically meaningful steady-state solutions is not a matter of luck; it is a matter of strategic exploration. The convergence of a numerical solver like the **fsolve** function is highly dependent on the initial guess. A single, arbitrary guess will often find only the most stable (or trivial) solution. The provided code demonstrates the necessary engineering practice of using multiple, distinct initial guesses to identify all potential operating points (e.g., the low-conversion "extinguished" state and the high-conversion "ignited" state).

### 1.1.30 The Progressive Dialogue

Think of working with an AI not as writing a single command but as having a structured conversation with a brilliant but very literal assistant. A single, complex "mega-prompt" is often too rigid and likely to fail. A better approach is to guide the AI through a progressive dialogue, where each step builds upon the last. This iterative process mirrors the natural workflow of engineering design: You start with the core theory, translate it into a functional model, and then refine and explain it. Table 1.13 shows a workflow example of engineering dialogue.

**Table 1.13** Workflow example of the engineering dialogue.

| Step | Your Role (The Engineer) | AI's Role (The Assistant) |
|---|---|---|
| 1. Establish the physics | **You start with the core concept.** *"Formulate the material and energy balance equations for a non-isothermal PFR."* | **AI provides the foundational model.** It returns the raw differential equations. |
| 2. Translate to code | **Next, you ask for specific implementation.** *"Translate those two ODEs into a functional MATLAB script using the ode45 solver."* | **AI acts as the programmer.** It writes the script to solve the equations. |
| 3. Request explanation | **Finally, you ask for clarification.** *"Now, add comments to that script explaining the physical meaning of each term in the energy balance."* | **AI becomes the annotator.** It adds detailed comments to the code, clarifying the purpose of each line. |

### 1.1.31 Implementing Code Guard Rails and Diagnosing Errors

A numerical solver's only job is to solve the equations you provide. It has no understanding of chemistry or physics and will happily produce results that are mathematically possible but physically absurd, such as negative concentrations or temperatures below absolute zero.

Therefore, the engineer must act as the final check, building "guard rails" into the code to ensure the results are physically feasible. Think of it as installing a safety relief valve on your code; it prevents the simulation from running into a nonsensical state.

After the main solver has finished, always add a block of code to check the validity of the output. The AI may not do this for you, as its primary goal is to solve the mathematical task.

Here is an example of post-simulation sanity checks that you may add toward the end of your main working code.

```
% — POST-SIMULATION SANITY CHECKS —
fprintf('\n— Verifying physical feasibility of results... —\n');
% 1. Check for negative concentrations
if any(Y(:,1) < 0)
  error('FATAL ERROR: Negative concentration detected. Check model or solver tolerances.');
else
  fprintf('Concentration check passed: All values are non-negative.\n');
end
% 2. Check the final temperature (if applicable)
% Example for a non-isothermal case
% if any(Y(:,2) < 0)
%   error('FATAL ERROR: Temperature below absolute zero detected.');
% end
fprintf('Verification complete.\n');
```

Notice that when a simulation produces a physically impossible result, like a negative concentration or temperature, it is a red flag that demands investigation. It is almost never a random error; it is a symptom of a specific problem.

The cause can be traced back to one of two main sources: a model error introduced by the user or a numerical stability issue with the solver itself.

1) **User input and model errors ("garbage in, garbage out")**
   This is the most common cause. The AI, or solver, is performing flawlessly, but it is solving the wrong problem because the instructions were flawed.
   a) **Syntax and Sign Errors:** A simple typo can have dramatic effects. For example, accidentally defining the heat of reaction ($\Delta Hr$) for an exothermic process as a positive number will cause the model to cool down when it should be heating up, potentially leading to nonphysical temperatures.
   b) **Incorrect Governing Equations:** If the material or energy balance provided in the prompt is fundamentally incorrect, the solver will faithfully solve it, even if the results violate physical laws.
2) **Numerical Solver Instability:**
   This is a more subtle issue that arises from the nature of the numerical methods used to solve differential equations. This is a common culprit in chemical kinetics. A system is "stiff" when it involves processes occurring on vastly different time scales. For example, consider a reactor with a very large thermal mass (e.g., a large CSTR or a fixed-bed reactor with a high solid-to-gas ratio):

   - A very fast reaction occurs within the fluid or on a catalyst surface.
   - A much slower change in the bulk reactor temperature, which is dampened by the enormous thermal inertia of the reactor and its contents.

   A standard solver like ode45 might try to take a large time step to efficiently track the slow, gradual change in temperature. However, this large step completely "misses" the rapid chemical reaction (which may have already reached equilibrium or a new steady state), causing the solution to overshoot wildly.

Another classic example is a reaction mechanism with very different rate constants:

- A very fast, initial reaction: $A \xrightarrow{k_1} B$, where $k_1$ is very large, e.g., 103 $s^{-1}$.
- A much slower, subsequent reaction: $B \xrightarrow{k_2} C$, where $k_2$ is very small, e.g., 10-1 $s^{-1}$.

In this system, the concentration of reactant $A$ disappears almost instantly, creating a "pseudo-steady-state" for $B$. A standard solver like ode45 must take femto- or microsecond time steps to accurately capture the rapid disappearance of $A$. However, the solver must continue integrating for many minutes or hours to track the slow formation of $C$. Using microsecond steps for an hour-long process is computationally impossible. But if the solver tries to take a large step (e.g., 1 second) to track $C$, this large step violates the solver's stability limit, which is still dictated by the fast reaction ($k_1$). This causes the numerical solution to become unstable and oscillate, leading to physically impossible negative concentrations and a complete failure of the integration. This requires specialized "stiff" solvers (like ode15s in MATLAB), which have much larger stability regions for these types of problems.

### 1.1.32 Using AI for "Back-of-the-Envelope" Sanity Checks

While an AI can correctly assemble complex equations, it lacks an engineer's intuitive feel for physical reality. It will not recognize if a calculated temperature rise of a million degrees per second is absurd. Therefore, it is a crucial practice to instruct the AI to perform sanity checks on the magnitudes of key physical rates. Think of this as asking the AI to perform a quick, "back-of-the-envelope" calculation to ensure the model's behavior is physically reasonable.

#### 1.1.32.1 Workflow for a Physical Sanity Check

a) **Isolate a Key Rate:** Instruct the AI to calculate the instantaneous rate of a key variable, such as the initial rate of temperature change due to the reaction.
b) **Display the Value with Units:** Command the AI to display this calculated value with its expected units clearly labeled.
c) **Analyze and Diagnose:** As the expert, you assess the result. If the magnitude is physically unrealistic, it points to a potential error in the input parameters.

#### 1.1.32.2 Example Prompt Fragment

"After defining the parameters, calculate the initial rate of adiabatic temperature rise using the expression: rate_$T = (-\Delta H_r) * k * C_A / (\rho * C_p)$. Display this value with the label 'Initial Temperature Rate (K/s)'."

If the simulation returns an absurd value like Initial Temperature Rate $(K/s)$ = 50,000, it will immediately indicate that there is likely an error in the provided values for $\Delta H_r$, $\rho$, or $C_p$, prompting the user to investigate input data before proceeding.

### 1.1.33 Cross-Verification by Multiple Models: The Engineering "Second Opinion"

In critical engineering and scientific analysis, a second, independent verification is not a luxury; it is a necessity. The same principle should be applied to AI-generated code. Never rely on a single model's output as the sole source of truth.

To build robust confidence in your results, perform a cross-verification by presenting the exact same prompt to a different LLM.

#### 1.1.33.1 The Cross-Verification Workflow

1) **Generate:** Develop your primary code using your preferred model (e.g., ChatGPT).
2) **Verify:** Give the identical prompt to a second, distinct model (e.g., Gemini, DeepSeek, or Grok).
3) **Compare:** Critically compare the two outputs.
   - **Agreement:** Do both scripts produce results that agree within a tight numerical tolerance? This provides strong evidence that the solution is robust and not an artifact of a single AI's specific training data.
   - **Discrepancy:** If the results differ, it is a crucial signal to investigate further. The difference may expose a subtle misinterpretation of the prompt by one model or highlight different but equally valid solution strategies.

### 1.1.34 Plot and Result Validation

A simulation is not complete when the code runs; it is complete when the results have been rigorously validated. Just as a manufactured product undergoes a final quality control inspection, every plot and data point generated by your code must be scrutinized for correctness and physical plausibility. Before accepting any figure as correct, perform the following checks:

#### 1.1.34.1 The Plot Validation Checklist

1) **Professional Presentation:**
   - Are all axes clearly labeled with both the variable and its units (e.g., "Temperature (K)," "Concentration (mol/L)")?
   - Does the plot include a clear title and a legend if multiple data series are present?
   - Is the range of the axes appropriate to display the important features of the data without distortion?
2) **Physical and Thermodynamic Sanity:**
   - Do the trends make physical sense? The plot should tell a logical story based on the principles of CRE. For example, reactant concentrations should decrease over time in a reactor.
   - Does the system obey thermodynamic laws? This is a nonnegotiable check.
     - For an exothermic reaction ($\Delta H_r < 0$) in an adiabatic system, does the temperature increase as conversion increases?
     - For an endothermic reaction ($\Delta H_r > 0$), does the temperature decrease?
   - Are the values physically possible? Concentration should never be negative, and conversion percent should never exceed 100%.

This validation step is what separates a coder from an engineer. It ensures your results are not just numbers but a reliable representation of a real-world physical system.

**Example 1.21** ***Post-processing Verification***

A simulation's primary output (like a concentration profile) is often just raw data. The final and most critical step is to post-process this data to calculate key performance indicators, such as conversion, and to verify that these results are physically sound. This is where you translate the raw numbers into engineering insights.

**The Post-processing Workflow**

1) **Prompt Addition:** After the main simulation is complete, instruct the AI to perform calculations on the output data and present it in a clear format.

"Using the results from the PFR simulation, now perform the following post-processing:

1) Calculate the conversion, $X$, as a function of reactor length using the formula $X = 1 - C_A/C_{A0}$ (valid for a constant-density system).
2) Generate a new plot of conversion ($X$) versus reactor length ($z$).
3) Display the final outlet conversion with three decimal places."

2) **The Engineer's Final Check:** The AI will generate the plot and the final number, but it is the engineer's responsibility to perform the final sanity check.

**Verification Checklist**

a) **Does conversion start at zero?** At the reactor inlet ($z = 0$), the conversion must be zero.
b) **Does conversion increase along the reactor?** For a standard reaction, conversion should increase or stay constant, never decrease.
c) **Is the final value physically possible?** The outlet conversion must be within the physical bounds of $\mathbf{0 \leq X \leq 1}$. A value outside this range immediately indicates an error in the model or the code.

### 1.1.35 Use AI for Code Debugging and Optimization

Beyond code generation, one of the most powerful applications of an LLM is as an interactive debugging consultant. When you encounter unexpected behavior in your code, such as numerical instability or oscillations, you can ask the AI to diagnose the problem and suggest a solution, just as you would ask a senior colleague. This transforms the AI from a simple coder into a real-time partner for troubleshooting and optimization. Table 1.14 shows a typical example of debugging dialogue.

**Table 1.14** An example of a debugging dialogue.

| Role | Prompt/Response |
|---|---|
| You (the engineer) | You describe the problem clearly.<br>"I am using ode45 to solve my non-isothermal reactor model. I am seeing strange, non-physical oscillations in the solution when the concentration of A gets close to 1 mol/L. Why is this happening, and what should I do?" |
| AI (the consultant) | The AI provides a diagnosis and a specific, actionable solution.<br>"The oscillations you are seeing are a classic symptom of numerical stiffness. This often occurs in non-isothermal reactors where a small change in temperature causes a very rapid change in the reaction rate. The ode45 solver is not designed for stiff systems.<br>I recommend you switch to the ode15s solver, which is specifically designed for stiff problems. For better accuracy, you should also tighten the relative tolerance. Try this:<br>options = odeset('RelTol', 1e-8);<br>[t, Y] = ode15s(@my_odes, tspan, Y0, options);" |

### 1.1.36 Documentation of AI Usage

In scientific research, the lab notebook is a cornerstone of credibility. It documents every step, assumption, and result, ensuring that the work is transparent, reproducible, and verifiable. When using AI in your academic work, the same principle applies.

Your AI interaction history, the complete record of your prompts and the model's outputs, serves as your digital lab notebook.

#### 1.1.36.1 Submission Requirement

For every assignment that involves the use of an AI model, students must attach the full, unedited transcript of their interaction. This includes every prompt they wrote and the complete response the AI generated.

This policy is not just about avoiding plagiarism; it is about upholding the standards of good engineering practice:

- **Transparency and Reproducibility:** It allows your instructor and peers to follow your workflow and understand how you arrive at your conclusions.
- **Accountability:** It provides a clear record of the information and code you work with.
- **A Tool for Learning:** Reviewing your prompt history is an excellent way to refine your communication and problem-solving skills.

### 1.1.37 Critical Thinking: The Engineer is Always in Command

It is essential to cultivate a mindset of critical oversight when using AI tools. The most effective way to teach this is to emphasize that AI is a powerful but obedient assistant; the human engineer is the expert who must provide direction and critical judgment.

Remember this simple maxim: The AI proposes, but the engineer disposes.

#### 1.1.37.1 Putting Critical Thinking Into Practice

To foster this mindset, instructors should challenge students to actively engage with the AI's output, not just passively accept it.

- **Ask "Why?":** Encourage students to treat the AI's output as the beginning of a conversation. They should always ask *why* the AI chose a particular equation or a specific line of code. Prompting for justification is a powerful learning tool.
- **Take Control and Experiment:** The best way to test comprehension is to move from the virtual to the practical. Instruct students to manually alter key parameters in the AI-generated code and predict the outcome *before* running the new simulation. Does the result match their engineering intuition? If not, why?

### 1.1.38 AI-Assisted Troubleshooting Playbook

When the simulation fails, think of the AI as a diagnostic tool. By describing the symptom accurately, you can get expert-level advice on the root cause and the most likely solution. Table 1.15 shows some common scenarios and how to tackle them.

### 1.1.39 Why Errors Occur in AI-Generated Code

Even a meticulously crafted, CURVD-compliant prompt can produce flawed results. AI is a tool for translation and pattern matching, not a true expert with physical intuition. Understanding the common failure modes is essential for effective verification. Table 1.16 shows common sources of AI-generated errors.

**Table 1.15** Common scenarios and how to solve them.

| Symptom | The Prompt (Describing the Problem) | A Typical AI Response (the Diagnosis and Solution) | The Underlying Principle |
|---|---|---|---|
| Solver fails to converge | "My **fsolve** command is failing to find a steady-state solution for my CSTR model. What is wrong?" | "Convergence failure in **fsolve** often indicates that the initial guess is too far from a valid solution. For systems with multiple steady states, try providing a new initial guess that is physically closer to a different operating point (e.g., a much higher temperature for an 'ignited' state)." | **Numerical Stability:** Nonlinear solvers are "local" and depend heavily on the initial guess. A good guess is critical to finding the desired root. |
| Oscillatory or unstable solution | "My **ode45** solution for concentration versus time has strange, non-physical oscillations." | "Oscillations are a classic symptom of a stiff system, where reaction rates change much faster than other variables. **ode45** is not suitable for these problems. You must switch to a stiff solver like **ode15s** to ensure a stable and accurate solution." | **Stiffness in ODEs:** Standard solvers can become unstable when trying to resolve widely different time scales. Specialized stiff solvers are required. |
| Physically impossible values | "My simulation is producing negative temperatures, which is impossible. What did I do wrong?" | "A negative temperature almost always points to a sign error in the energy balance. For an exothermic reaction, the heat of reaction ($\Delta H_r$) must be negative. Please verify that your $\Delta H_r$ parameter has the correct sign. An incorrect sign would cause the model to remove heat instead of generating it." | **Model Verification:** The physics of the model must be correct. This is a classic "garbage in, garbage out" scenario where the code is executing the user's flawed model perfectly. |

**Table 1.16** Common sources of AI-generated errors.

| Source of Error | Description | Example Symptom |
|---|---|---|
| Subtle prompt omissions | The prompt, while well-structured, unintentionally omits a minor constraint or assumption, forcing the AI to "fill in the gap" with a guess. | Undefined variables appear in the code, or the AI assumes standard atmospheric pressure for a system that is under high pressure. |
| Model "hallucination" | The AI confidently invents plausible but entirely incorrect information, such as physical constants, correlations, or even obscure function names. | The code references a Cp value for a proprietary catalyst, stating, "as found in the literature," when no such value has been published. |
| Numerical instability | The AI selects an inappropriate numerical solver for the problem (e.g., a nonstiff solver for a stiff system) or uses default tolerances that are too loose. | The final plot shows a solution that oscillates wildly or diverges toward infinity, even though the underlying model is physically stable. |
| Physical inconsistency | The code runs without error, but the output violates a fundamental law of conservation or thermodynamics. | For an exothermic adiabatic reaction, the simulation shows the temperature decreasing as the reaction proceeds, directly contradicting the First Law of Thermodynamics. |

### 1.1.40 Categories of MATLAB-Related Errors

Effective troubleshooting begins with an accurate diagnosis. Just as a doctor must distinguish between a virus and a bacterial infection before prescribing treatment, an engineer must identify the specific *type* of error to solve it efficiently. Table 1.17 provides a diagnostic framework, categorizing the most common failure modes encountered when developing MATLAB simulations with AI.

**Table 1.17** A taxonomy of common MATLAB-related errors.

| Error Type | The Core Problem | Example Symptom |
|---|---|---|
| 1. Syntactic errors | The code violates the basic grammar of the MATLAB language. | MATLAB refuses to run the code and provides an immediate, explicit error message, such as "Unbalanced or unexpected parenthesis." |
| 2. Semantic errors | The code is syntactically correct and runs, but its logic is flawed, leading to an incorrect calculation. | The code produces a single numerical answer with no warning, but it is wrong (e.g., using matrix multiplication* instead of element-wise* multiplication, resulting in incorrect vector calculations). |
| 3. Numerical Errors | The error originates from the numerical solver itself, often because its settings are inappropriate for the problem's mathematical nature. | The solution plot oscillates wildly, diverges to infinity, or the solver fails to converge. This is a classic sign of stiffness. |
| 4. Physical errors | The code runs flawlessly, but the output violates fundamental physical laws. This usually means the user's model was flawed. | The simulation produces negative concentrations, conversions greater than 100%, or shows temperature decreasing during an adiabatic exothermic reaction. |
| 5. Contextual errors | The AI correctly solves a problem, but it is the wrong problem. It has fundamentally misinterpreted the engineering intent of the prompt. | The code models a batch reactor when you ask for a CSTR, or it assumes ideal gas behavior for a liquid-phase reaction. |

**Example 1.22** ***Syntactic and Variable-Name Errors***
**Case Study: Diagnosing a Syntactic Error**
This example illustrates one of the most common failure modes: a simple syntax error generated by AI. The key to solving it lies in learning to interpret MATLAB's feedback.

1) **The Faulty AI Output:** An AI, given a prompt to solve an ODE, might generate a line of code that looks plausible but confuses syntax from different programming languages:

```
% Incorrect Code Generated by AI
[t, CA] = ode45(f[t, CA0], [0 10]);
```

2) **MATLAB's Error Message:** When one runs this code, MATLAB will immediately stop and provide a direct error message:
Invalid expression. When calling a function or indexing a variable, use parentheses. Otherwise, check for mismatched delimiters.

3) **The Engineer's Diagnosis and Correction:** This error message is the student's cue. It indicates that the problem is not with logic but with the grammar of the code. The expression f[t, CA0] is incorrect because MATLAB uses parentheses () for function calls and indexing, not brackets [].
   The student's role is to use this clue to correct the syntax according to the standard ode45(odefun, tspan, y0) format:

```
% Corrected Code
[t, CA] = ode45(f, [0 10], CA0);
```

**A Note for Instructors**

This is a critical teaching moment. The learning objective is to train students to read and interpret MATLAB's error messages first, before immediately re-prompting AI. This builds fundamental debugging skills and reinforces their role as the expert who must validate the syntax, not just the scientific output.

**Example 1.23** ***Hallucinated Parameters***

One of the most significant risks in using AI for technical problems is model hallucination, where AI invents plausible sounding but entirely fabricated data to fill gaps in a prompt. Understanding how to prevent this is a critical skill.

1) **The Vague Prompt (The Trap):** An engineer, working quickly, provides a conceptually correct but data-deficient prompt:

   "Simulate a batch reactor for ($A \xrightarrow{k} B$) using the Arrhenius law."

2) **The AI's Hallucinated Response:** The AI, recognizing that it needs parameters to write functional code, does not stop. Instead, it invents a complete set of physically plausible constants:

```
% --- AI Generated Code with HALLUCINATED Parameters ---
Ea = 50000; % Invented Activation Energy (J/mol)
Cp = 75; % Invented Heat Capacity (J/mol.K)
rho = 0.9; % Invented Density (kg/L)
...
```

   The danger is that the code will run perfectly, producing a result that looks correct but is based on completely fabricated physics.

3) **The Corrected CURVD-compliant Prompt (The Solution):** The engineer must provide all necessary data explicitly, leaving no room for invention. The prompt is revised to be self-contained:

   "Using the following parameters: Ea = 75e3 J/mol, Cp = 4.18 J/g·K, and rho = 1,000 g/L, simulate a batch reactor for ($A \xrightarrow{k} B$). Do not assume or introduce any other physical constants."

   **The Reliable AI Output:** AI now acts as a reliable compiler, using only the data provided by the expert user.

   **Lesson:** AI's primary goal is to fulfill the prompt. If you omit the necessary data, it will often "helpfully" invent it. Always provide every required numerical value and explicitly instruct the model not to assume any others.

**Example 1.24** ***Unit Mismatch (Order of Magnitude) Error***

A common and dangerous error in simulation is not a syntax failure but a unit mismatch that leads to an "order of magnitude" error. The code runs perfectly, but the physics it simulates is completely wrong.

1) **The Faulty Code with a Subtle Error:** An engineer provides a value for activation energy without including its proper scientific notation, a common mistake when transcribing from a handbook.

```
% Incorrect parameter definition
Ea = 75; % Missing the 'e3' → it should be 75000 J/mol
R = 8.314;
T = 350; % K
% The Arrhenius equation is syntactically correct
k = k0 * exp(-Ea / (R * T));
```

2) **Unrealistic Reaction Speed:** The code executes without any MATLAB error. However, the calculated rate constant, k, will be enormous. The reaction will be completed almost instantaneously. This is the physical symptom that should alert the engineer that something is fundamentally wrong.
3) **Diagnosis and Remedy:** The engineer's role is to act as an expert and validate physics. The first step is to check the dimensionless group in the exponential, $E_a/RT$.
   a) **Sanity Check:** For most chemical reactions at typical temperatures, the $E_a/RT$ term should be in the range of 20–40.
   b) **Calculation:** 75/(8.314 * 350) ≈ 0.026
   c) **Conclusion:** This value is orders of magnitude too small, indicating the $E_a$ value is incorrect.

   The remedy is to always provide explicit units in the prompt and ensure the code comments reflect those units.

**Improved Prompt Fragment**

"Use an activation energy Ea = 75e3 (in J/mol) and include comments in the generated code that explicitly state the units for each parameter."

This case study highlights the engineer's critical role: The user is not just a prompter but also the validator of the physical model. Always perform a "back-of-the-envelope" check on dimensionless numbers to ensure they are physically reasonable.

**Example 1.25** ***Diagnosing and Curing Numerical Instability***

A common failure mode in dynamic simulations is numerical instability, where the solution appears to "work" but produces physically impossible results, such as oscillations. This is not a coding error but a sign that the wrong numerical tool is being used for the problem.

1) **Unphysical Oscillations:** The engineer runs a simulation using the standard ode45 solver and observes that as the concentration approaches its final value, the solution begins to oscillate wildly, even dipping into negative values.
2) **Diagnosis:** Such oscillations are a classic symptom of a stiff system. This occurs when there are multiple processes happening at vastly different time scales. The ode45 solver, designed for nonstiff problems, takes steps that are too large to accurately capture the fast process, causing it to overshoot and become unstable.

3) **The Prescription: A Targeted Prompt Fix:** The engineer, acting as the expert, must diagnose the stiffness and instruct the AI to use the appropriate specialized tool.

**Improved Prompt Fragment**
"The current ode45 solution is showing non-physical oscillations, which indicates the system is stiff. Please modify the code to use the ode15s solver, which is designed for stiff problems. For high accuracy, set the relative tolerance to 1e−8 and the absolute tolerance to 1e−10. Then, plot the results from both solvers on the same graph for comparison."

4) **The Corrected Code:** The AI, following the above instructions, adjusts the code:

```
% AI adjusts the code to use a stiff solver and tighter tolerances.
options = odeset('RelTol', 1e-8, 'AbsTol', 1e-10);
[t, CA] = ode15s(f, [0 10], CA0, options);
```

The lesson is that the engineer's role is to interpret the physical symptoms of the simulation's output and guide the AI to the correct numerical method. Recognizing the signs of stiffness is a critical diagnostic skill.

**Example 1.26** ***Physical Inconsistency***
A simulation can be mathematically perfect but physically nonsensical. The most dangerous errors are not the ones that crash the code, but the ones that produce a plausible looking but completely wrong answer. This is where the engineer's critical thinking becomes indispensable.

1) **A Physically Implausible Result:** A simulation of a mild exothermic reaction in a CSTR completes without any numerical errors. However, the output shows a final steady-state temperature of 1,200 K, a value that is far too high to be physically reasonable for the given system.
2) **Cross-Examining the AI:** The engineer's next step is not to trust the number, but to use the AI to diagnose the model itself.
   - **The Diagnostic Prompt:**
     "The simulation ran successfully, but the final temperature of 1,200 K is physically unrealistic for this mildly exothermic reaction. Please analyze the energy balance equation used in the model and identify the most likely parameters that could cause such an excessive temperature rise."
   - **A Typical AI Explanation:**
     "An unexpectedly high temperature rise from an exothermic reaction typically points to two potential issues in the model's parameters:
     1. **The Heat of Reaction ($\Delta H_r$):** The magnitude might be too large, or its sign could be incorrect.
     2. **The Heat Capacity ($C_p$):** The value for the reacting mixture might be unrealistically low, causing the same amount of energy release to produce a much larger temperature change."
3) **The Engineer's Corrective Action:** Armed with this diagnosis, the engineer performs the final validation
   - **Verify the Heat of Reaction ($\Delta H_r$):** The expert checks their sources. For an exothermic reaction, is ΔHr correctly entered as a negative number? Is its magnitude correct?
   - **Sanity-Check the Heat Capacity ($C_p$):** The engineer reviews the $C_p$ value. Is it in the correct order of magnitude for the substance (e.g., for a liquid stream, is it reasonably close to the value for water, ~4.18 J/g·K)?

### 1.1.41 When the Code is Right, but Chemistry Is Wrong

A subtle but critical error occurs when the AI generates syntactically perfect code that models the wrong physical reality. This often happens if the prompt is not explicit about the underlying scientific principles, such as the form of the rate law. This case study illustrates why the **"R" (Right)** in the CURVD framework is nonnegotiable.

1) **The Ambiguous Prompt:** An engineer provides a prompt that correctly identifies the reactants but is vague about the reaction kinetics:

   "Write the rate law for the reaction $2A + B \rightarrow$ Products."

2) **The AI's Logical (but Incorrect) Assumption:** The AI, lacking explicit instruction, might assume the reaction order is equal to the stoichiometric coefficients, a common but not universally true assumption. This leads to a fundamentally incorrect rate law.
   - **AI's Incorrect Output:** $r = -k * C_A^2 * C_B$
   - **The Flaw:** The AI has incorrectly assumed the reaction is second-order with respect to reactant $A$.
3) **The Consequence, a Flawed Model:** Any simulation based on this rate law will be mathematically sound but physically meaningless. The results will not reflect the actual behavior of the chemical system.
4) **The Corrected, CURVD-compliant Prompt:** The engineer must provide the correct, explicit scientific model, leaving no room for interpretation.
   - **Revised Prompt:**

"For the non-elementary reaction 2A + B → Products, provide the rate law. The reaction is first-order with respect to each reactant. The final expression must be $r = -k$ * CA * CB."

This example demonstrates that the engineer's primary role is to be an expert on the underlying physics and chemistry. The coder must provide the correct scientific foundation; only then can the AI reliably translate that foundation into correct code.

### 1.1.42 The Power of a Diagnostic Dialogue

The most effective way to troubleshoot is to engage AI in an iterative conversation. By describing one symptom at a time, one can get targeted feedback that guides him/her to a solution. This process mirrors a real-world diagnostic workflow and is a powerful way to learn cause-and-effect reasoning. Table 1.18 shows an example of a troubleshooting conversation.

### 1.1.43 A Systematic Workflow for AI-Assisted Debugging

Debugging is not just about fixing errors; it is about understanding why they occurred. Using an AI as a partner can transform this process from a frustrating task into a structured learning loop. Follow this systematic workflow to diagnose and solve issues effectively.

1) **Observe and Isolate the Symptom:** Run MATLAB code and carefully observe the outcome. Is it a specific error message from MATLAB? Is it an unexpected result in a plot (e.g., oscillations, physically impossible values)? Isolate the problem to a specific part of your code or output.
2) **Query the AI with Context:** Formulate a precise prompt for the AI. Do not just ask "How do I fix this?" Instead, provide the necessary context:
   a) Copy the exact error message.
   b) Provide the relevant block of code that is causing the error.

**Table 1.18** An example of a troubleshooting conversation.

| Iteration | Your Role (The Engineer) | AI's Role (The Consultant) | The Outcome |
|---|---|---|---|
| 1. The initial problem | "My simulation is wrong. The concentration of my reactant, $C_A$, is increasing over time instead of decreasing." | "That indicates a sign error in your rate expression. The rate of consumption should be negative. Please check the line where you define your ODE." | The student finds the error and corrects the rate law to $dC_A/dt = -k \cdot C_A$. The model is now chemically correct. |
| 2. The second symptom | "Okay, the concentration profile is correct now, but the reactor's temperature is dropping much too quickly for this reaction." | "A rapid temperature change often points to an issue with either the heat of reaction ($\Delta H_r$) or the heat capacity ($C_p$) terms. Please confirm that $\Delta H_r$ has the correct sign (negative for exothermic) and that your $C_p$ value has units consistent with the rest of the energy balance." | The student realizes their $C_p$ value was in J/kg·K instead of J/mol·K and corrects it. The model is now physically consistent. |
| 3. The final polish | "The results look much better, but the plot is still a bit noisy and jagged near the end." | "That noise is a sign of numerical instability, likely because the system has become stiff. I recommend you switch from the ode45 solver to ode15s, which is designed for stiff systems." | The student implements the stiff solver. The final simulation is now stable, accurate, and physically correct. |

   c) Describe the symptoms you observed.
   d) **Example Prompt:** "I am running this MATLAB code to solve CSTR dynamics. The solver is reporting 'Matrix dimensions must agree'. Here is my code block. Can you explain the source of this error and suggest a correction?"
3) **Interpret and Hypothesize:** Analyze the AI's response. Don't just copy the corrected code. Understand *why* AI's suggestion is the right fix. Is it a syntax error? A mismatched variable? A fundamental logic flaw? Formulate a hypothesis about the root cause of the problem.
4) **Implement, Re-execute, and Compare:** Modify your code based on AI's advice. Rerun the script and compare the new output to the previous, faulty result. Did the change fix the problem? Did it create a new one?
5) **Document the Correction:** In your code comments or lab notebook, document the fix. Crucially, don't just write *what* you changed; write *why* the change was necessary. This final step solidifies your understanding and turns a simple bug fix into a lasting lesson.

### 1.1.44 Numerical Accuracy and Tolerance Control

A numerical solver is like a high-precision manufacturing tool. The quality of the final product (the solution) depends directly on the tolerances you set. While default settings are fine for simple problems, the highly nonlinear and sensitive nature of CRE systems demands tighter control to ensure the results are both accurate and physically meaningful. Table 1.19 shows key solver tolerances and recommendations for CRE applications.

**Table 1.19** Key Solver Tolerances and Recommendations for CRE applications.

| Parameter | Description and Typical Default | Recommendation for CRE |
|---|---|---|
| RelTol | **Relative Tolerance:** Controls the accuracy relative to the magnitude of the solution (Default: 1e−3). | **1e−6 – 1e−8** |
| AbsTol | **Absolute Tolerance:** Controls the accuracy when the solution approaches zero (Default: 1e−6). | **1e−8 – 1e−10** |
| MaxStep | **Maximum Step Size:** Limits the largest time step the solver can take (Default: automatic). | Enforce a max step relative to a characteristic time, e.g., 0.01 * tau. |

#### 1.1.44.1 Why This Matters in CRE

The Arrhenius equation's exponential dependence on temperature ($k \propto exp(-E_a/RT)$) makes reactor models extremely sensitive. A small, seemingly acceptable error in temperature can be magnified as a massive error in the calculated reaction rate. The default solver tolerances are often too loose to handle this sensitivity, which can lead to oscillations, divergence, or convergence to a nonphysical solution. By tightening **RelTol** and **AbsTol**, the coder instructs the solver to be more careful, taking smaller steps to accurately capture the system's stiff and nonlinear behavior. Here is a portion of MATLAB code that shows how to define an **odeset** statement prior to calling any ode solver, like **ode15s**.

```
% --- Define a characteristic time scale for the simulation ---
tau = 10; % Example: CSTR residence time of 10 seconds
% --- Create the solver options structure ---
% This command bundles all the specified tolerances into a single variable.
options = odeset('RelTol', 1e-8, 'AbsTol', 1e-10, 'MaxStep', 0.01 * tau);
% [t, Y] = ode15s(odefun, tspan, y0, options);
[t, Y] = ode15s(@my_reactor_odes, [0 100], [CA0; T0], options);
```

### 1.1.45 From MATLAB Mastery to AI Literacy

Historically, chemical-engineering curricula trained students first in MATLAB syntax, then in CRE modeling. Generative AI reverses this sequence: it enables students to begin with conceptual reasoning, reaction kinetics, balances, and energy coupling, while delegating syntax generation to the AI.

Thus, the educational focus shifts from "How do I code this?" to "How do I describe this system precisely enough that the code becomes self-evident?"

The AI becomes a computational partner, encouraging verbalization of thought and deeper conceptual articulation.

### 1.1.46 Components of an AI-Enhanced Classroom

Table 1.20 details a framework for transforming AI from a passive calculation tool into an active pedagogical partner within an engineering curriculum. It outlines four key components designed to

**Table 1.20** Framework for an AI-enhanced engineering classroom.

| Component | Pedagogical Role | Example Activity (CRE Focus) |
|---|---|---|
| Live interactive prompting | Bridging conceptual algorithms with executable code and developing foundational prompt engineering skills in real time. | Instructors provide a human-language algorithm (e.g., for a numerical method); students collaboratively refine prompts until a functional MATLAB script is generated. |
| AI-assisted problem-solving | To train students in using AI as a codeveloper for complex modeling tasks and to emphasize documenting the AI-human interaction. | Assignment: Model an adiabatic CSTR using an AI partner. Students submit the final code, the full prompt dialogue, and a critique of the AI's output. |
| AI-enhanced laboratories | To leverage AI for hypothesis generation, complex data analysis, and fitting models to experimental results. | Use an AI tool to suggest and apply appropriate regression routines (e.g., nonlinear regression) to fit experimental kinetic (rate-constant) data. |
| Collaborative prompt refinement | To foster a community of practices where students learn from each other's AI interactions and collaboratively improve querying strategies. | Students post challenging prompts and their AI outputs on a discussion board; peers then review and suggest prompt improvements (e.g., using the CURVD rubric). |

interactively integrate AI into the classroom, laboratory, and homework assignments, emphasizing practical application and collaborative refinement.

These components convert AI from a passive calculator into an interactive partner, fostering a more dynamic and effective pedagogical environment.

**Exercise 1.2** ***An AI-Integrated Class Exercise***
**Objective:** Verify Arrhenius parameters using experimental data.

1) Instructor provides datasets.
2) Student groups formulate CURVD-compliant prompts.
3) Each group uses a different platform (ChatGPT, Gemini, DeepSeek, and Grok).
4) Groups present results: fitted (Ea), (k0), residual plots, and AI dialogues.
5) Class compares the numerical and linguistic performance of each platform.

This exercise transforms the classroom into an AI laboratory of ideas, where both human and AI models are subject to analysis.

### 1.1.47 Roles in the AI-Augmented Ecosystem

The integration of AI transforms the educational landscape from a traditional hierarchy into a dynamic ecosystem. This shift requires a clear redefinition of the roles and responsibilities for all participants. Table 1.21 outlines the new functions of the student, instructor, AI model, and the institution, detailing how each contributes to this collaborative learning environment.

**Table 1.21** A Framework of responsibilities and key actions for participants in an AI-augmented education.

| Participant | Primary Role/Responsibility | Key AI-Related Actions |
|---|---|---|
| Student | To transition from *information consumer* to active problem-solver and critical validator of AI-generated solutions. | Strategic prompt engineering; iterative debugging in a human–AI loop; analyzing AI outputs for correctness, bias, and limitations. |
| Instructor | To shift from *sole knowledge source* to learning architect and expert guide, assessing the *process* of problem solving, not just the final answer. | Designing AI-integrated assignments; demonstrating expert-level prompting and verification; guiding discussions on ethical use and academic integrity. |
| AI model | To serve as an on-demand coprocessor and conceptual tutor, accelerating tasks and offering alternative perspectives. | Generating/refactoring code (e.g., MATLAB, Python); explaining complex algorithms or theories; suggesting solution pathways; summarizing data. |
| Institution | To provide the foundational framework (policy, infrastructure, and support) that enables effective and ethical AI integration. | Establishing clear academic integrity policies, providing secure, equitable access to AI platforms, and offering faculty development and training. |

### 1.1.48 Ethical and Responsible AI Integration

To foster academic integrity and develop critical thinking in an AI-augmented environment, the following four ethical pillars must be strictly adhered to:

1) **Transparency and Documentation**
   **Principle:** Every AI-assisted contribution must be fully auditable and traceable, shifting the focus from the final answer to the process of engagement.
   - **Required Documentation:** For every task where AI is used, the submission must include:
     a. The complete prompt text, including the AI platform (e.g., ChatGPT, Gemini) and version number used.
     b. The generated output (e.g., code, text, or data summary).
     c. A log of all subsequent human modifications and the rationale for those changes.
2) **Appropriate Attribution**
   **Principle:** AI models are defined as sophisticated *tools*, not cognitive partners or coauthors, that aid in the execution of work.
   - **Citation Standard:** Contributions must be explicitly cited in the methodology or notes section. A suggested format: "MATLAB script and initial data fitting generated using [AI Platform Name] (version/date), with subsequent critical verification and refinement by the author."
   - **Academic Integrity:** The student remains the sole author and responsible party for the final submitted work.
3) **Critical Verification and Engineering Rigor**
   **Principle:** Students must always employ their fundamental engineering knowledge to validate AI-derived results, ensuring they satisfy physical laws and professional standards.
   - **Validation Checkpoints:** Students must explicitly demonstrate that AI outputs satisfy:
     1. **Conservation Principles:** Fulfillment of the conservation of mass and energy (e.g., mass balances and heat duty calculations).

2. **Dimensional and Unit Consistency:** Rigorous checking of dimensional homogeneity across all equations and variables.
3. **Physical/Thermodynamic Feasibility:** Confirming that results (e.g., temperatures, pressures, compositions, and yields) are physically plausible and do not violate thermodynamic constraints.

4) **Equity and Access**

**Principle:** The use of AI should not create a technology gap; all students must have fair and equal opportunity to utilize the augmented resources.

- **Infrastructure Commitment:** The institution must provide secure, reliable, and uniform access to approved AI tools for all enrolled students.
- **Accessibility Policy:** Where online or subscription-based tools are required, the institution must provide equivalent, institutionally funded, or offline alternatives to ensure no student is disadvantaged by personal economic or technological constraints.

### 1.1.49 Metrics for Evaluating AI-Assisted Learning

To ensure that the integration of AI enhances, rather than diminishes, academic rigor, a structured set of metrics is required to evaluate student performance. Table 1.22 shows metrics for evaluating student performance with a shift assessment focus from the final answer to the student's critical process skills, including their ability to guide, validate, and ethically document the AI interactions.

Such focused rubrics are essential for maintaining academic rigor while rewarding the analytical and verification skills crucial for modern engineering practice.

**Table 1.22** Metrics for evaluating student proficiency in critical analysis and AI interactions.

| Metric | Description | Example Indicator (CRE Focus) |
|---|---|---|
| Prompt engineering efficacy | Assessment of the student's ability to craft precise, detailed prompts that guide the AI toward an effective and efficient solution path. | CURVD score ≥4/5 and minimal prompt-to-solution iterations. |
| Code and solution validation | The measure of the final output's technical correctness and physical realism requires the student to verify AI-generated code and results. | Code execution error = 0. Final results (e.g., Conversion ($X$), Temperature ($T$), or Yield) satisfy mass/energy balances and lie within thermodynamically feasible limits. |
| Reflective critical analysis | The quality and depth of the student's commentary on the AI-human interactions, focusing on necessary corrections and the rationale for the AI's failures or successes. | Dedicated commentary section detailing where the AI failed, how the student corrected it (e.g., debugging dimensional inconsistencies), and the knowledge gained from the fix. |
| Ethical and professional compliance | Adherence to policies regarding transparency, proper attribution, and responsible use of AI tools in all submitted academic work. | AI-generated contributions are fully documented (prompt, output, modifications) and cited according to established guidelines on the cover page or methodology section. |

### 1.1.50 AI as a Personalized Tutor

Generative AI's greatest potential in engineering education lies in its ability to function as a Personalized Tutor, providing adaptive scaffolding that tailors complex explanations to the individual student's immediate needs, background, and current cognitive level. This level of customization is challenging to replicate in traditional large-class settings.

#### 1.1.50.1 The Mechanism of Personalized Scaffolding

The core of this capability is the AI model's ability to interpret the intent and implied knowledge level within a student's natural-language query. It acts as a dynamic Intelligent Tutoring System (ITS) that adjusts its output in real time.

1) **Adaptation to Knowledge Level:**
   The AI adapts the depth and complexity of its explanation based on the query:
   - **Beginner User (Simple Query):** A query like, "Explain what ode45 does in simple terms," prompts the AI to provide a high-level conceptual analogy. For a chemical engineering student, the AI might describe it as an automated reaction-progress tracking tool that determines how the concentrations and temperature change over time in a reactor. The focus is on function and application.
   - **Advanced User (Complex Query):** A request such as "Derive the local truncation error expression for ode45" signals that the student possesses a strong mathematical and numerical methods background. The AI responds with a rigorous, derivation-based explanation involving Taylor series expansions and Runge-Kutta method theory. The focus shifts to mechanism and mathematical rigor.
2) **Contextualization to Discipline:**
   Beyond just the complexity, the AI can contextualize the explanation to the student's discipline (e.g., chemical engineering, as in your field).
   - If a student is asking about Aspen Plus or MATLAB, the AI can frame its response using concepts like CSTRs, distillation columns, mass balances, or kinetic models.
   - This makes the abstract theory more concrete, bridging the gap between theoretical knowledge and professional engineering practice.

#### 1.1.50.2 Benefits of the AI Personalized Tutor

1) **Fosters the Zone of Proximal Development (ZPD):**
   AI excels at operating within the student's ZPD, the space between what a student can do without help and what they can do with guidance. By adjusting the level of detail, the AI ensures the explanation is challenging enough to promote learning but not so complex as to cause frustration, thereby maintaining student engagement and flow.
2) **Available for Low-Stakes Exploration:**
   Students often hesitate to ask "simple" or foundational questions in front of peers or the instructor. The AI provides an always-available, nonjudgmental partner for self-paced learning and exploring basic concepts, allowing students to solidify foundational knowledge before tackling advanced topics in class.
3) **Supports Advanced Conceptual Exploration:**
   For doctoral students or those engaged in specialized research (like modeling, simulation, and optimization), the AI can serve as a sounding board to test assumptions, troubleshoot complex code, or rapidly retrieve obscure constants and relationships, significantly accelerating the iterative nature of research.

### 1.1.51 Instructor Implementation Model: A Phased Approach

This four-phase model provides a structured, progressive framework for integrating AI tools into an engineering course, moving students from guided exposure to independent innovation. Table 1.23 shows the instructor deployment strategy: progression from AI orientation to capstone innovation.

**Table 1.23** A four-phase deployment model for fostering AI proficiency and critical thinking.

| Phase | Title and Duration | Core Activity and Focus | Deliverable/Assessment |
|---|---|---|---|
| Phase 1 | Orientation and skill acquisition (weeks 1–3) | **Focus:** Establishing ethical boundaries (transparency, attribution) and foundational prompting skills using the CURVD principle. | Instructor-Led Live Prompting Session (e.g., using AI to solve a steady-state mass balance); short ethics quiz on proper citation. |
| Phase 2 | Guided application and verification (weeks 4–9) | **Focus:** Utilizing AI for complex problem solving. Students must use critical verification to validate AI-generated code against engineering principles. | AI-Assisted Homework Assignments (e.g., modeling a non-isothermal CSTR). Submission includes the AI dialogue log and the final, human-verified MATLAB code. |
| Phase 3 | Reflection and troubleshooting (weeks 10–13) | **Focus:** Deepening metacognitive skills by analyzing human–AI interactions. Students must identify, explain, and correct AI errors. | Technical Reflection Essay or Case Study Report evaluating the AI's performance, detailing required corrective actions (e.g., fixing dimensional errors), and assessing the AI's overall reliability. |
| Phase 4 | Innovation and customization (weeks 14+) | **Focus:** Applying advanced skills to create novel solutions. This serves as a high-level capstone experience focused on research and optimization. | **Capstone Project:** Students develop, train, and test a domain-specific AI assistant (e.g., a retrieval-augmented generation (RAG) tool) fine-tuned for a complex topic like nonideal chemical equilibrium systems or heat exchanger network optimization. |

**Example 1.27** ***AI-Assisted Design Challenge***
Table 1.24 presents a high-value example, demonstrating the effective utilization of AI in a core CRE design problem. The example illustrates a collaborative workflow where the AI handles computational tasks while the student applies crucial engineering judgment for final design validation and optimization.

### 1.1.52 The Future of Hybrid AI–Human Computation

The trajectory of AI integration points toward a future where the boundary between human cognition and artificial computation is not just blurred but functionally eliminated, creating a coevolutionary intelligence in the engineering workspace. Table 1.25 shows key future trends driving the coevolution of AI and engineering cognition.

**Table 1.24** Example of an AI-augmented workflow for non-isothermal CSTR network design.

| Component | Description | Learning Objective Focus |
|---|---|---|
| Challenge task | Design a minimum two-reactor non-isothermal Continuous Stirred Tank Reactor (CSTR) network to achieve 90% conversion ($X = 0.9$) for a highly exothermic reaction, subject to specified heat-exchange temperature constraints. | Optimization, energy balance, and safety |
| Augmented workflow | 1. **Conceptualization:** Students define the system boundaries, governing equations (mass and energy balances), and kinetic model.<br>2. **Code Generation:** AI generates the initial MATLAB script for solving the system of ordinary differential equations (ODEs).<br>3. **AI Guidance:** AI suggests and executes automated parameter sweeps (e.g., varying residence time and jacket temperature).<br>4. **Critical Validation:** Students manually check the AI-generated solutions against thermodynamic feasibility and stability criteria. | Modeling, numerical methods, and critical verification |
| Required outcome | Final report must include the optimal design parameters (reactor volumes, heat duties) and a comparative analysis detailing the strengths and weaknesses of human-guided optimization versus AI's parameter search strategy. | Design feasibility, comparative analysis, engineering judgment |
| Core insight | Students learn that while AI drastically accelerates numerical computation and hypothesis generation, ultimate design feasibility, constraint satisfaction, and risk mitigation require profound engineering judgment and validation based on first principles (e.g., conservation laws). | Metacognition, professionalism |

**Table 1.25** Emerging trends in hybrid AI–human computational engineering.

| Trend | Description | Engineering Impact |
|---|---|---|
| Multimodal model integration | Advanced models capable of processing and translating diverse data inputs, such as hand-drawn process flow diagrams (PFDs), complex chemical structures, and scanned mathematical equations, directly into executable code (e.g., MATLAB, Python) or simulation environments. | Eliminates manual data entry and coding for routine tasks, allowing engineers to prototype complex systems based on graphical and textual blueprints instantly. |
| Real-time simulation chat | Generative AI agents are embedded directly within simulation platforms (e.g., Simulink, Aspen Plus) to enable instant parameter tuning, sensitivity analysis, and troubleshooting via natural-language queries. | Accelerates optimization cycles and facilitates "what-if" scenario testing in dynamic systems without halting the simulation or requiring manual script edits. |
| Decentralized AI agents and cross-validation | Utilizing collaborative AI agents (e.g., a "Code Generator" agent interacting with a "Thermodynamic Validator" agent) to cross-check solutions, verify dimensional consistency, and predict potential points of failure. | Increases solution robustness and inherent reliability by building a multilayered verification protocol that reduces reliance on a single model's output. |

The long-term vision is a symbiotic relationship: engineers define the high-level problems and ethical constraints in human language, AI executes the visualization and numerical solution, and both human and machine refine their collective understanding through continuous feedback loops and iteration.

## References


1 Imbihl, R., Scheibe, A., Zeng, Y.F. et al. (2007). Catalytic ammonia oxidation on platinum: mechanism and catalyst restructuring at high and low pressure. *Physical Chemistry Chemical Physics* 9: 3522–3540. https://doi.org/10.1039/b700866j.

2 Zun, M., Dwornicka, D., Wojciechowska, K. et al. Kinetics of the decomposition and the estimation of the stability of 10% aqueous and non-aqueous hydrogen peroxide solutions. *Current Issues in Pharmacy and Medical Sciences* 27 (4): 213–216. https://doi.org/10.1515/cipms-2015-0017.

3 Eusuf, M. and Laidler, K.J. (1964). Kinetics and mechanisms of the thermal decomposition of acetaldehyde. I. The uninhibited reaction. *Canadian Journal of Chemistry* 12: 1851–1860. https://doi.org/10.1139/v64-276.


## End of Chapter Problems

**CHW 1.1 Nitrous Oxide ($N_2O$) Decomposition Into Nitrogen Gas ($N_2$) and Oxygen Gas ($O_2$)**

The balanced chemical equation for this reaction is:

$$2N_2O \rightarrow 2N_2 + O_2$$

This decomposition typically occurs at high temperatures or in the presence of a catalyst. Under very specific conditions, such as high concentrations of $N_2O$ or the presence of certain catalysts, the reaction might exhibit pseudo-zero-order behavior. Given a rate constant = $2.46 \times 10^{-3}$ M/s.

Reference: Zumdahl, S.S. and Zumdahl, S.A. (2013). Study Guide Ninth Edition, Chemistry. Prepared by Paul Kelter Northern Illinois University. Brooks/Cole, Cengage Learning, U.S.A. www.cengage.com/highered. https://vdoc.pub/download/study-guide-for-zumdahl-zumdahls-chemistry-7qb3bq8kqem0.

If the initial concentration $[N_2O]_0$ is 0.45 M, how long will it take to drop the concentration of $[N_2O]$ down to 0.0515 M?

Exploiting MATLAB, calculate the time needed to reach the final concentration of 0.0515 M using both symbolic and numerical integration for the zero-order ODE of $[N_2O]$ decomposition. Generate a plot of $[N_2O]$ versus time.

**CHW 1.2 Sulfuryl Chloride Undergoing First-order Decomposition at 320°C with a Half-life of 8.75 h**

$$SO_2Cl_2(g) \rightarrow SO_2(g) + Cl_2(g)$$

Reference: Zumdahl Chapter 12 Kinetics. https://www.cusd80.com/cms/lib/AZ01001175/Centricity/Domain/2155/Zumdahl%20Chapter%2012%20Kinetics.pdf

Analytically, find the time needed to drop the concentration of $SO_2Cl_2$ from 0.5 down to 0.1 M at 320°C. Use MATLAB to evaluate the time needed using both symbolic and numerical integration methods for a solution of a first-order ODE of sulfuryl chloride ($SO_2Cl_2$) decomposition at 320°C.

**CHW 1.3 Second-order Oxidation of Nitrous Oxide ($N_2O$) Into Nitric Oxide (NO)**

$$N_2O + O \rightarrow 2NO \quad -r_{N2O} = k_1 C_{N2O} C_O$$

With the rate constant:

$$k_1 = 6.23 \times 10^{13} e^{-\left(\frac{24,540}{RT}\right)} \frac{\text{cm}^3}{\text{mol.s}} \times \frac{1\,L}{1,000\ \text{cm}^3}$$

$$= 6.23 \times 10^{10} e^{-\left(\frac{24,540}{RT}\right)} \frac{\text{L}}{\text{mol.s}} \quad [1,815 - 3,365\ \text{K}]$$

Reference: Monat, J.P., Hanson, R.K., and Kruger, C.H. (1977). Kinetics of nitrous oxide decomposition. *Combustion Science and Technology* 16(1–2): 21–28. 10.1080/00102207708946790. https://www.tandfonline.com/doi/abs/10.1080/00102207708946790.

Let us assume that $C^0_{N2O} = C^0_O = 1.0$ M. Assume $T = 1,815$ K.

Analytically, calculate the time needed for $N_2O$ to drop down to 0.001 M. Use MATLAB to evaluate the time needed using both symbolic and numerical integration methods for a solution of a second-order ODE of $N_2O$ decomposition at 1,815 K.

**CHW 1.4 Third-order Gas-Phase Reaction**

$$Cl_2(g) + 2NO(g) \rightarrow 2NOCl(g)$$

$$\text{with} - r_{Cl2} = -r_A = k[Cl_2][NO]^2 = k[A][B]^2$$

$$\text{The rate constant } k = 3.0 \left(\frac{\text{L}}{\text{mol}}\right)^2 \left(\frac{1}{\text{s}}\right)$$

Reference: Flowers, P., Theopold, K., Langley, R., Stephen, F., and Robinson, W.R. *Chemistry 2e*, an OpenStax resource. https://openstax.org/books/chemistry-2e/pages/12-3-rate-laws.

Starting with $[NO]_0 = 2$ M and $[Cl_2]_0 = 1.0$ M, analytically calculate the time needed for $[Cl_2]$ to drop down to 0.1 M. Use MATLAB to evaluate the time needed using both symbolic and numerical integration methods for a solution of a third-order ODE of NOCl formation.

**CHW 1.5 Arrhenius Kinetics**

A set of data for the temperature and rate constant is shown in Table 1.26.
Find the activation energy $E_a$ and the pre-exponential factor, $A$.

**CHW 1.6 $CA = f(t)$ Data 1**

Given time versus concentration data as shown in Table 1.27, determine whether the reaction follows zero-, first-, or second-order kinetics by linearizing each rate law, fitting with **polyfit/polyval** MATLAB functions, computing $R^2$, and interpreting slope/intercept to get $k$ and $CA_0$.

**CHW 1.7 $CA = f(t)$ Data 2**

**Table 1.26** The rate constant, $k$, versus temperature, $T$.

| # | $T$ (K) | $k$ ($s^{-1}$) × $10^6$ | # | $T$ (K) | $k$ ($s^{-1}$) × $10^6$ |
|---|---|---|---|---|---|
| 1 | 273.1500 | 0.000248 | 9 | 379.8167 | 0.064731 |
| 2 | 286.4833 | 0.0006235 | 10 | 393.1500 | 0.10495 |
| 3 | 299.8167 | 0.0014445 | 11 | 406.4833 | 0.16486 |
| 4 | 313.1500 | 0.0031157 | 12 | 419.8167 | 0.25164 |
| 5 | 326.4833 | 0.0063112 | 13 | 433.1500 | 0.37423 |
| 6 | 339.8167 | 0.0121000 | 14 | 446.4833 | 0.54351 |
| 7 | 353.1500 | 0.022069 | 15 | 459.8167 | 0.77244 |
| 8 | 366.4833 | 0.038543 | 16 | 473.1500 | 1.0763 |

**Table 1.27** Time versus $C_A$ data.

| $t$ (min) | $C_A$ (mol/L) | $t$ (min) | $C_A$ (mol/L) |
|---|---|---|---|
| 0 | 1.0360 | 4 | 0.2841 |
| 1 | 0.6234 | 6 | 0.2084 |
| 2 | 0.4459 | 8 | 0.1646 |
| 3 | 0.3470 | 10 | 0.1360 |

Given time versus concentration data as shown in Table 1.28, determine both the rate constant, $k$, and the reaction order, $n$, augmented by goodness of fit statistical measures, like $R^2$.

**Table 1.28** $C_A = f(t)$ data.

| $t$ (min) | $C_A$ (mol/L) | $t$ (min) | $C_A$ (mol/L) | $t$ (min) | $C_A$ (mol/L) |
|---|---|---|---|---|---|
| 0 | 1.05 | 4 | 0.450 | 8 | 0.200 |
| 1 | 0.925 | 5 | 0.360 | 9 | 0.160 |
| 2 | 0.700 | 6 | 0.300 | 10 | 0.130 |
| 3 | 0.600 | 7 | 0.250 | 12 | 0.080 |

**CHW 1.8 A Case Study in Prompt Refinement**

**Learning Outcome:** To demonstrate how to transform a vague initial idea into a high-quality, verifiable prompt for a CSTR simulation.

Follow the three stages to generate the required MATLAB code:

**Stage 1: The Vague Prompt (the Starting Point)**

Let a student begin with a simple, high-level goal, with an initial prompt:

"Write MATLAB code to simulate a CSTR."

**Stage 2: The Improved Prompt (Adding Key Details)**

The student realizes he/she needs to provide more information and adds crucial parameters.

**Stage 3: The CURVD-compliant Prompt (The Final Goal)**

The student now refines the prompt to be fully compliant with the CURVD framework, making it concise, unambiguous, right, verifiable, and derived from a specific model.

**CHW 1.9 Symbolic Versus Numeric Solution for an Isothermal BR**

**Learning Outcome:** Reinforce the equivalence of analytic and numeric approaches.

1) **Problem Definition**

We consider a single, irreversible, first-order liquid-phase reaction ($A \xrightarrow{k} B$) carried out in an isothermal, constant-volume BR.

**Reaction Rate Law:** The rate of disappearance of reactant A is given by:

$$-rA = k\,CA$$

**Design Parameters:** | Parameter | Symbol | Value | Units

Initial concentration of A, CA0 | 2.0 | mol/L

Reaction rate constant, $k$ | 0.5 | min^-1

Final time | $t_{final}$ | 10.0 | min

With the aforementioned reaction rate law and design parameters, provide a CURVD-compliant prompt to show both solutions: Analytic (symbolic) and numeric (using ode solver). On the same plot, show $CA = f(time)$.

**CHW 1.10 Multi-model Validation**

**Learning Outcome:** Develop critical evaluation skills.

Compare outputs from ChatGPT and Gemini using the same following prompt: "Explain the design equation for an isothermal batch reactor for a first-order reaction ($A \xrightarrow{k} B$), including the derivation and final expression for concentration vs. time. Also provide a simple MATLAB code to plot the concentration profile."

**CHW 1.11 MATLAB Debugging Challenge: Flawed OLS Regression**

**Learning Outcome:** Practice systematic error detection and correction.

**The Flawed Code:**

The following MATLAB function, intended to calculate the slope ($m$) and intercept ($c$) of a simple linear regression line ($y = mx + c$) using Ordinary Least Squares (OLS), contains five critical flaws.

```
function [m, c] = calculate_regression_matlab(x_vals, y_vals)
%CALCULATE_REGRESSION_MATLAB Calculates OLS slope (m) and intercept (c).
% Check for array length
 if length(x_vals) ~= length(y_vals)
    disp('Error: Data arrays must have the same length.');
    m = NaN;
```

*(Continued)*

**(Continued)**

```
    c = NaN;
    return;
  end
  n = length(x_vals);
  % Calculate the means
  mean_x = sum(x_vals) / n;
  mean_y = sum(x_vals) / n; % Flaw 1: Incorrect array used for mean_y
  numerator = 0.0;
  denominator = 0.0;
  % Loop to calculate numerator and denominator for slope
  for i = 1 : n-1 % Flaw 2: Off-by-one error, skipping the last element
  term_x = x_vals(i) - mean_x;
  term_y = y_vals(i) - mean_y;
  numerator = numerator + term_x * term_y;
  denominator = denominator + term_x; % Flaw 3: Denominator term is not squared
  end
  % Calculate slope
  m = numerator / denominator;
  % Calculate intercept
  c = mean_y + m * mean_x; % Flaw 4: Incorrect arithmetic sign in intercept formula
end
```

MATLAB Test Script (to run the flawed function)
You can run this script to see the incorrect output:

```
% --- Test Case ---
x_data = [1, 2, 3, 4, 5];
y_data = [2, 4, 5, 4, 5];
% The mathematically correct result should be: m = 0.6, c = 2.2
[slope, intercept] = calculate_regression_matlab(x_data, y_data);
disp(['AI''s Calculated Slope (m): ', num2str(slope)]);
disp(['AI''s Calculated Intercept (c): ', num2str(intercept)]);
disp(' ');
disp('--- Expected Output (Approx) ---');
disp('Slope: 0.6');
disp('Intercept: 2.2');
% Note: You must also address the robustness issue (Flaw 5: Division by Zero)
% which this simple test case does not trigger.
```

**NOTE:** Create an m file where it has both sections, the main (calling) code, and, underneath, the called function named **calculate_regression_matlab.**

**Your Task**

1) Identify and explain all five flaws (Flaws 1-4 are commented; Flaw 5 is the robustness issue).

2) Provide the complete, corrected, and robust MATLAB function **calculate_regression_matlab**.

**CHW 1.12 The "Black Box" Code**

Learning outcome: Develop skills in code interpretation and reinforce the link between the mathematical model and the final code.

Below are two MATLAB files: a main script to define parameters and solve the system and a function file containing the undocumented differential equations.

The two scripts together are fully functional but deliberately undocumented and cryptically-named, MATLAB code for a non-isothermal CSTR with a first-order exothermic reaction ($A \xrightarrow{k} B$).

1) **Main Script**

This script defines all the physical parameters and calls the **ode45** solver. The student must reverse-engineer what each of these variables represents and how they are used.

```
% Main Code
  P = [1.0 1.0 1.987 11843 34930800 -5960 500 150 299];
  I = [10.0 300.0];
  tspan = [0 5];
  X0 = [2.0 373.0];
  opts = odeset('RelTol',1e-4,'AbsTol',1e-6);
  [T,Y] = ode45(@f_ode, tspan, X0, opts, P, I);
  figure(1);
  plot(T, Y(:,1), 'r-', 'LineWidth', 2);
  xlabel('Time');
  ylabel('State Variable 1');
  title('Component Profile');
  grid on;
  figure(2);
  plot(T, Y(:,2), 'b-', 'LineWidth', 2);
  xlabel('Time');
  ylabel('State Variable 2');
  title('Thermal Profile');
  grid on;
```

2) **ODE Function**

This function contains the core governing differential equations for concentration (*C_A*) and temperature (*T*).

```
function dY = f_ode(~, Y, P, I)
  V_R = P(1); F_0 = P(2); R_C = P(3); E_A = P(4); k_0 = P(5);
  H_R = P(6); rCp = P(7); Ua = P(8); T_J = P(9);
  CAf = I(1); Tf = I(2);
```

*(Continued)*

**(Continued)**

```
    C_A = Y(1); T_R = Y(2);
    k = k_0 * exp(-E_A / (R_C * T_R));
    rA = k * C_A;
    dY = zeros(2, 1);
    dY(1) = (F_0/V_R) * (CAf - C_A) - rA;
    Q_G = -H_R * rA;
    Q_R = Ua/V_R * (T_R - T_J);
    dY(2) = (F_0/V_R) * (Tf - T_R) + Q_G/rCp - Q_R/rCp;
end
```

**Your Task:**

1. Use an AI tool to analyze the f_ode.m function. For example, explain the purpose of the two differential equations (dY(1) and dY(2)). Identify the physical meaning of the parameters defined in the P and I vectors, based on their mathematical placement (e.g., P(3)).
2. Map the variables and terms back to the fundamental principles of CRE. Derive the two governing equations, starting from the general balances.
3. Produce a final, fully commented, and professional-grade version of the code. Rewrite the **f_ode** function using the standard chemical engineering notation and add comments that explain the physical meaning of each term (e.g., "This term is the rate of heat generation due to reaction").

**CHW 1.13 Chemical Reaction Selectivity and Yield**

You are a MATLAB code-generation assistant helping a chemical engineering professor, Dr. Kamal I. M. Al-Malah, develop pedagogically clear and physically meaningful code for his textbook, *MATLAB-based Computations of Chemical Reaction Engineering.*

The problem concerns a non-isothermal CSTR in which two parallel first-order liquid-phase reactions occur:

$$A \overset{k1}{\Longrightarrow} D \text{ (desired)}$$

$$A \overset{k2}{\Longrightarrow} U \text{ (undesired)}$$

The system can operate adiabatically or with external heat exchange.

You need to instruct the LLM platform to generate a MATLAB script such that:

1) The user will select one of three kinetic cases:

$$k_1 < k_2,$$

$$k_1 = k_2,$$

$$k_1 > k_2.$$

2) The user will choose adiabatic or nonadiabatic operation.
3) It computes the steady-state conversion $X_A$ and the **selectivity D**, $S_{D/U} = (k_1/(k_1 + k_2)$.
4) It computes the overall **yield**, $Y = S_{D/U} * X_A$ at the exit or final $X_A$.

5) It uses Arrhenius temperature dependence for rate constants.
6) It maximizes selectivity of D with respect to reactor temperature $T$ using the **fminsearch** function.
7) It ensures physical constraints $0 \leq X_A \leq 1$.
8) It provides readable, formatted output showing the optimum temperature, conversion, selectivity, and yield.
9) MATLAB syntax must be fully runnable without additional toolboxes.
10) It properly handles both adiabatic and nonadiabatic energy balances.
11) It optimizes the control variable:
    - If nonadiabatic: varies coolant temperature Tc.
    - If adiabatic: varies feed temperature T0.

**NOTE**: Ask the LLM platform to properly assign values with the proper units for the following physical terms that will appear in describing the non-isothermal behavior of CSTR. Such parameters are $R$, $\Delta H_r$, $\rho$, $C_p$, $U$, $A_{HT}$, $T_c$, $V$, $v_0$, and $C_{A0}$. Else, your instructor will help you with such an assignment.

**CHW 1.14 Inconsistent Units**

"Consider the first-order, exothermic, liquid-phase reaction: $(A \overset{k1}{\Longrightarrow} B)$

**Reaction Rate:**

$r_A = -k\,C_A$, where $k = k_0 \exp(-E_a/RT)$

**Energy Balance:**

$$\frac{dT}{dV} = \frac{-\Delta H_r\, r_A + UA_H(T_c - T)}{F_A C_{p,mix}}$$

Write MATLAB code to model the above liquid-phase, first-order, exothermic reaction carried out in a Plug Flow Reactor (PFR). Use exactly as are the following data:

- Feed concentration, $C_{A0} = 1.5$ mol/L
- Feed temperature, $T_0 = 350$ K
- Activation energy, $E_a = 12.5$ kcal/mol
- Pre-exponential factor, $k_0 = 2.0 \times 10^7\ \text{s}^{-1}$
- Heat of reaction, $\Delta H_r = -45$ kJ/mol
- Heat transfer coefficient × Area, $UA_H = 240$ J/(s · L · K)
- Coolant temperature, $T_c = 300$ K
- Flow rate, $F_{A0} = 1.2$ mol/s
- Heat capacity, $C_p = 4.18$ J/(g · K)
- Average molecular weight = 60 g/mol
- Reactor volume = 1,000 L

Use MATLAB's ode45 to integrate from $V = 0$ to 1,000 L and plot conversion and temperature versus reactor volume."

**Expected Outcome:**

The AI will likely produce a numerically correct-looking code but physically nonsensical results (e.g., temperature spikes of thousands of K or negative conversions) because of unit inconsistencies among:

- $E_a$ in kcal/mol (but $R$ in J/mol·K),
- $\Delta H_r$ in kJ/mol,
- $C_p$ in J/g·K, and
- $C_A$ in mol/L while volume is in L, not $m^3$.

The student should *diagnose* that the issue is not the ODE solver but the unit system inconsistency.

If AI does not correctly account for the inconsistency in the unit system, then the user can either ask LLM to account for the inconsistency, or the user can manually correct the above list of parameters, entering the consistent units for each.

# 2

# Steady-State Chemical Reactors

**Reaction Conversion Overview:**
Reaction conversion is a key concept in chemical engineering that quantifies the extent to which reactants are transformed into products in a chemical reaction. It is typically expressed as the ratio of the amount of a particular reactant that has reacted to the initial amount of that reactant present in the system. Conversion is an important metric for evaluating the efficiency of a chemical process, as it directly impacts the yield of desired products and the economic viability of the reaction. In industrial processes, achieving a high conversion rate often requires optimization of reaction conditions such as temperature, pressure, catalyst presence, and feed reactant concentrations.

**Applications and Importance:**
In practical applications, conversion is used to determine reactor performance, size, and design. For instance, in continuous flow reactors, the conversion rate can indicate how long the reactants need to remain in the reactor to achieve the desired level of product formation. High conversion is generally desirable, but it must be balanced with considerations such as reaction rate, selectivity, and energy efficiency. Additionally, in processes where unreacted feedstock can be recycled, understanding conversion helps in designing appropriate separation and recycle loops. Conversion analysis is thus integral to both the economic and environmental optimization of chemical processes, ensuring minimal waste and maximum resource utilization.

## 2.1 Conversion

Consider the general equation:

$$aA + bB \xrightarrow{k} cC + dD$$

**A will be chosen as our basis of calculation.**

$$A + \frac{b}{a}B \xrightarrow{k} \frac{c}{a}C + \frac{d}{a}D$$

The basis of calculation is the limiting reactant. The conversion ($X$) of species **A** in a reaction is equal to the number of moles of **A** reacted per mole of **A** fed.

**Batch-type Reactor:**

$$X = \frac{N_{A0} - N_A}{N_{A0}} \tag{2.1}$$

$$\text{Or, } N_A = N_{A0}(1 - X) \tag{2.2}$$

NOTICE that $N_A$ represents the number of moles of $A$ at any time $t$ and $N_{A0}$ the initial number of moles of $A$.

For a constant-volume reactor, we can further express

$$X = \frac{C_{A0} - C_A}{C_{A0}} \tag{2.3}$$

$$\text{Or, } C_A = C_{A0}(1 - X) \tag{2.4}$$

NOTICE that $C_A$ represents the concentration $A$ at any time $t$ and $C_{A0}$ the initial concentration of $A$.

**Flow-type Reactor:**

$$\mathrm{X} = \frac{F_{A0} - F_A}{F_{A0}} \tag{2.5}$$

$$\text{Or, } F_A = F_{A0}(1 - X) \tag{2.6}$$

NOTICE that:

$$F_{A0} = C_{A0} \times v_0 = C_{A0} \times Q = C_{A0} \times \dot{V} \tag{2.7}$$

And

$$F_A = C_A \times v = C_A \times Q = C_A \times \dot{V} \tag{2.8}$$

where $v_0 = Q = \dot{V}$ represents the volumetric flow rate (volume/time) of the entering feed, and $F_{A0} = C_{A0} \times v_0 = \left(\frac{\text{mol}}{\text{volume}}\right)\left(\frac{\text{volume}}{\text{time}}\right) = \left(\frac{\text{mol}}{\text{time}}\right)$ represents the feed molar flow rate of A. Notation for volumetric flow rate is not standardized across chemical engineering textbooks.

For *irreversible reactions*, the maximum value of conversion, $X$, is that for complete conversion, that is, $X = 1.0$.

For *reversible reactions*, the maximum value of conversion, $X$, is the equilibrium conversion, that is, $X = X_e$.

Notice that the following important rules of thumb are related to chemical reaction engineering principles:

- In general, the molar flow of each species will be different between the feed and product stream simply as a result of a chemical reaction taking place inside the reactor. The total mass flow rate is conserved in a chemically reactive system; nevertheless, the total molar flow rate is not necessarily conserved. The same notion is applicable to a batch reactor (BR) where the total mass is always conserved, but it is not necessarily true for the total number of moles.
- The chemical reaction stoichiometry will dictate the interrelationship among reacting species, both as a reactant and as a product. The approach will be to monitor the change in the number of moles of A per unit time as a function of reaction kinetics and reactor operating conditions. Later, we will express the molar flow rate of $B$, $C$, and $D$ in terms of that of $A$.
- The energy balance will be carried out later to monitor how the reaction medium temperature changes as a function of reaction progress, reactor length, and/or time for a nonsteady/nonuniform state operating condition. The reactor can be operated under steady-state material

balance, yet the energy balance can be either steady- or unsteady-state condition. Of course, a steady-state material balance is a prerequisite for having a steady-state energy balance, but not the opposite. Needless to say, a batch or a semi-BR is inherently transient; that is, the reactor dynamics violate the notion of a steady-state condition.

- As a quick example, imagine taking a water shower where the cold water enters the water heater (or water sprinkler) while leaving as hot water. We have the inlet mass flow rate of water equal to the outlet water mass flow rate. Thus, a steady-state material balance is maintained, yet a nonsteady-state energy balance is dynamically inherent in this case. Nevertheless, a steady-state energy balance can be perceived if the amount of heat supplied via the electric coil is exactly equal to the sensible heat enthalpy difference between the inlet and outlet streams of water. In the latter case, the water temperature inside the heater, which is also equal to the exit water temperature, will become steady with time, $\left(\frac{\delta T}{\delta t} = 0\right)$.

In the following sections, we will consider different reactor types commonly used in chemical reaction engineering industries.

## 2.2 Isothermal Batch Reactor (BR)

Consider the sketch of a typical BR as shown in Figure 2.1. The BR simply means no mass in and no mass out of the control mass, which is the reactor itself with its reacting medium. The energy balance will be postponed to a later chapter.

**Component Material Balance of *A***

$$\text{In} - \text{Out} + \text{Generation} - \text{Consumption} = \text{Accumulation} \tag{2.9}$$

**Figure 2.1** A schematic for a BR with no input or output stream.

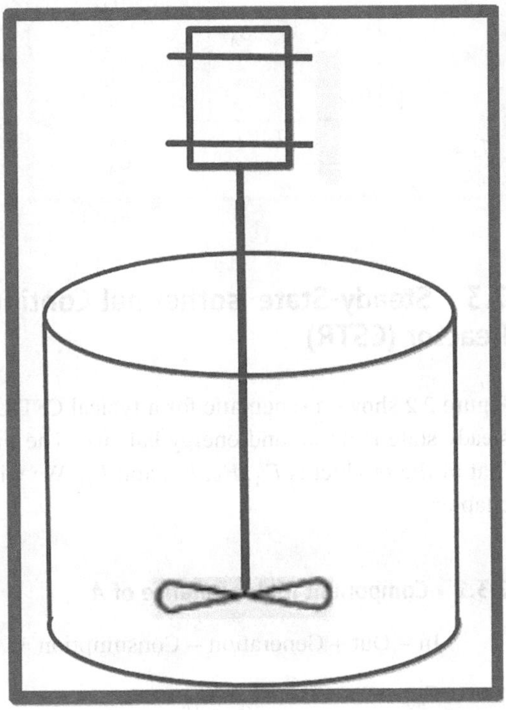

$$0 - 0 + 0 - (-r_A V) = \frac{dN_A}{dt} \tag{2.10}$$

$$\text{From Eq. (2.1): } N_A = N_{Ao} - N_{Ao}X \text{ AND } dN_A = 0 - N_{Ao}dX \tag{2.11}$$

$$\text{Equation (2.10) becomes: } (+r_A V) = \frac{dN_A}{dt} = -N_{Ao}\frac{dX}{dt} \tag{2.12}$$

$$\text{Or, } (-r_A V) = N_{Ao}\frac{dX}{dt} \rightarrow dt = \frac{N_{Ao}}{(-r_A V)}dX \tag{2.13}$$

At $t = 0, X = 0$

$$t_{BR} = N_{Ao}\int_0^X \frac{dX}{(-r_A V)} \tag{2.14}$$

Equation (2.14) is valid for a constant-temperature (i.e., isothermal) BR operation.

Alternatively, for a constant volume BR, we can express the rate of change in the number of moles as:

$$(+r_A V) = \frac{dN_A}{dt} = \frac{d(C_A V)}{dt} = V\frac{dC_A}{dt} \rightarrow (+r_A) = V\frac{dC_A}{dt} \rightarrow -kC_A^n = \frac{dC_A}{dt}$$
$$\frac{dC_A}{C_A^n} = -kdt \rightarrow \quad -\frac{1}{k}\int_{C_{A0}}^{C_A}\frac{dC_A}{C_A^n} = \int_0^t dt = t_{BR} \tag{2.15}$$

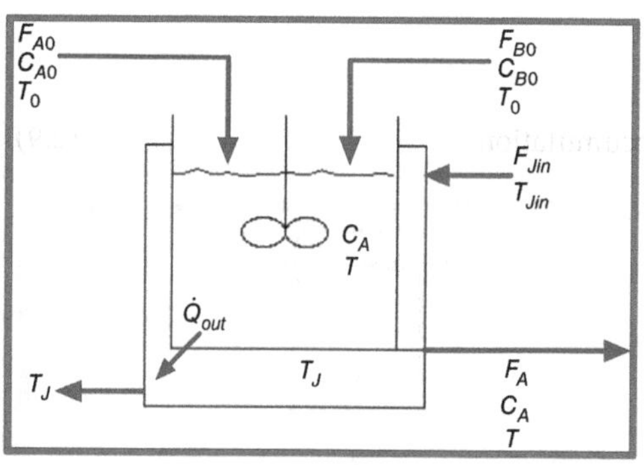


**Figure 2.2** A CSTR for carrying out a chemical reaction, augmented by a cooling/heating jacket. Inlet and outlet mass flow rates are equal.

## 2.3 Steady-State Isothermal Continuous Stirred Tank Reactor (CSTR)

Figure 2.2 shows a schematic for a typical CSTR. We will consider the case where we have both steady-state material and energy balance. The molar flow rate for the feed is $F_{A0}$ and $F_{B0}$, and that of the product is $F_A$, $F_B$, $F_C$, and $F_D$. We will postpone the energy balance to a subsequent chapter.

### 2.3.1 Component Molar Balance of *A*

$$\text{In} - \text{Out} + \text{Generation} - \text{Consumption} = \text{Accumulation}$$

$$F_{A0} - F_A + 0 - (-r_A V_{CSTR}) = 0 \tag{2.16}$$

Rearrangement of Eq. (2.16) yields:

$$V = \frac{F_{A0} - F_A}{-r_A} \tag{2.17}$$

Given that $F_A = F_{A0} - F_{A0}X$

$$V_{CSTR} = \frac{F_{A0} - F_A}{-r_A} = \frac{F_{A0} - (F_{A0} - F_{A0}X)}{-r_A} = \frac{F_{A0}X}{-r_A} \tag{2.18a}$$

Alternatively, the volume can be expressed in terms of $C_A$ as:

$$V_{CSTR} = \frac{F_{A0} - F_A}{-r_A} = \frac{\upsilon_0 C_{A0} - \upsilon C_A}{-r_A} = \frac{\upsilon_0 C_{A0} - \upsilon C_A}{kC_A^n} \tag{2.18b}$$

If we have a constant-volume flow rate, then $\upsilon = \upsilon_0$, then we have:

$$\tau = \frac{V_{CSTR}}{\upsilon_0} = \frac{C_{A0} - C_A}{-r_A} = \frac{C_{A0} - C_A}{kC_A^n} \tag{2.18c}$$

Equations (2.18a–c) are valid for a constant-temperature (i.e., isothermal) steady-state CSTR reactor operation.

## 2.4 Steady-State Plug-Flow Reactor (PFR)

Consider the sketch for a PFR as shown in Figure 2.3. Energy balance will be postponed to a later chapter. We will consider the case where we have both steady-state material and energy balance.

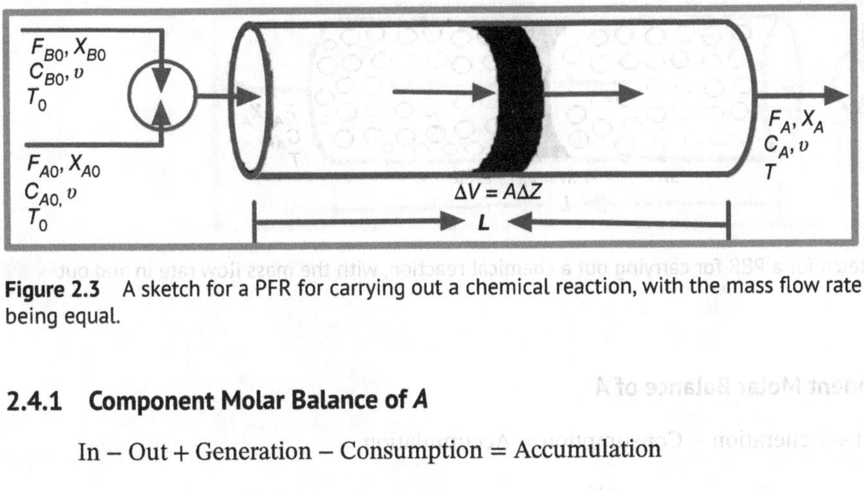


**Figure 2.3** A sketch for a PFR for carrying out a chemical reaction, with the mass flow rate in and out being equal.

### 2.4.1 Component Molar Balance of *A*

In – Out + Generation – Consumption = Accumulation

$$F_{A0} - F_A + 0 - \left(-\int r_A dV\right) = \frac{dN_A}{dt} = 0 \tag{2.19}$$

Differentiate with respect to $V$:

$$0 - \frac{dF_A}{dV} + r_A = 0 \tag{2.20}$$

Rearrangement of Eq. (2.20) yields:

$$\frac{dF_A}{dV} = r_A \tag{2.21}$$

Given that $F_A = F_{A0} - F_{A0}X$,  $dF_A = 0 - F_{A0}dX$

$$-F_{A0}\frac{dX}{dV} = r_A, \qquad F_{A0}\frac{dX}{dV} = -r_A \tag{2.22}$$

Rearrange, we have:

$$F_{A0}\frac{dX}{-r_A} = dV \tag{2.23}$$

At $V = 0, X = 0$. Thus:

$$V_{PFR} = F_{A0}\int_0^X \frac{dX}{-r_A} \tag{2.24a}$$

Alternatively, in terms of $C_A$, we may write:
At $V = 0$, $CA = CA_0$. Thus:

$$V_{PFR} = F_{A0}\int_0^X \frac{dX}{-r_A} = \upsilon_0 C_{A0}\int_{C_{A0}}^{C_A} \frac{-\frac{1}{C_{A0}}dC_A}{-r_A} = \upsilon_0\int_{C_A}^{C_{A0}} \frac{dC_A}{-r_A} \tag{2.24b}$$

## 2.5 Steady-State Packed Bed Reactor (PBR)

Consider the sketch for a PBR as shown in Figure 2.4. Energy balance will be postponed to a later chapter. We will consider the case where we have both steady-state material and energy balance.

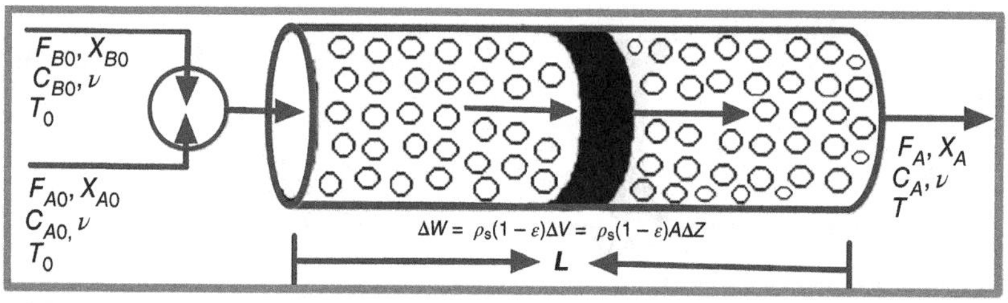


**Figure 2.4** A sketch for a PBR for carrying out a chemical reaction, with the mass flow rate in and out being equal.

### 2.5.1 Component Molar Balance of *A*

In − Out + Generation − Consumption = Accumulation

$$F_{A0} - F_A + 0 - \int (-r'_A)\, dW = \frac{dN_A}{dt} = 0 \tag{2.25}$$

Differentiate with respect to $W$:

$$0 - \frac{dF_A}{dW} + r'_A = 0 \tag{2.26}$$

Rearrangement of Eq. (2.26) yields:

$$\frac{dF_A}{dW} = r'_A \tag{2.27}$$

Given that $F_A = F_{A0} - F_{A0}X, \quad dF_A = 0 - F_{A0}dX$

$$-F_{A0}\frac{dX}{dW} = r'_A, \qquad F_{A0}\frac{dX}{dW} = -r'_A \tag{2.28}$$

$$F_{A0}\frac{dX}{-r'_A} = dW \tag{2.29}$$

At $W = 0$, $X = 0$. Thus:

$$W_{PBR} = F_{A0}\int_0^X \frac{dX}{-r'_A} \tag{2.30}$$

The total volume of the catalyst is given by:

$$V_{PBR} = \frac{W_{PBR}}{\rho_P}$$

**Example 2.1** If we load 50 kg of catalyst with a particle density of 2,500 kg/m$^3$, the solid catalyst volume is: $V_{solid} = \frac{50\ \text{kg}}{2{,}500\ \frac{\text{kg}}{\text{m}^3}} = 0.02\ \text{m}^3 = 20\ \text{L}$

**Example 2.2** If the packed bed volume $V_{bed}$ is 0.05 m$^3$ = 50 L, the bulk density $\rho_b$ is 1,000 kg/m$^3$, and the particle density $\rho_P$ is 2,500 kg/m$^3$. Find the volume of solid particles.

$$\text{Mass of catalyst: } M_{cat} = \rho_b * V_{bed} = 1{,}000\ \frac{\text{kg}}{\text{m}^3} \times 0.05\ \text{m}^3 = 50\ \text{kg}$$

$$\text{Solid catalyst volume, } V_{solid} = \frac{50\ \text{kg}}{2{,}500\ \frac{\text{kg}}{\text{m}^3}} = 0.02\ \text{m}^3 = 20\ \text{L}$$

$$\text{Or: } V_{solid} = \frac{M_{cat}}{\rho_P} = \frac{\rho_b * V_{bed}}{\rho_P} = \frac{1{,}000\ \frac{\text{kg}}{\text{m}^3} \times 0.05\ \text{m}^3}{2{,}500\ \frac{\text{kg}}{\text{m}^3}} = 0.02\ \text{m}^3 = 20\ \text{L}$$

**Example 2.3** If the packed bed volume $V_{bed}$ is 0.05 m$^3$ and the void fraction ε is known to be 0.6 (or 60%), find the volume of the solid catalyst.

Solid catalyst volume: $V_{solid} = V_{bed}(1 - \varepsilon) = 0.05\ \text{m}^3 \times (1 - 0.6) = 0.02\ \text{m}^3$

Notice that we can find the void fraction if we know both densities: $\varepsilon = 1 - \frac{\rho_b}{\rho_P}$

Thus,

$$V_{solid} = V_{bed}(1 - \varepsilon) = V_{bed}\left(1 - \left(1 - \frac{\rho_b}{\rho_P}\right)\right) = V_{bed}\left(\frac{\rho_b}{\rho_P}\right)$$

**NOTE #1**

In design equations: (2.14), (2.18a–c), (2.24a–b), and (2.30), it will be impractical to calculate the time or volume requirement for a given reactor if we let $X$ become identically 1.0. As such, the time or volume requirement will approach infinity (∞).

On the other hand, for reversible reactions (e.g., A ⇄ B), the maximum conversion is the equilibrium conversion $X_{Eq}$. At equilibrium, the reaction rate is zero ($r \equiv 0$). Therefore, since

*(Continued)*

**(Continued)**

$-r_A$ lies in the denominator of such design equations, then it will again blow up the value of time, volume, or both. The designer should avoid this mathematical divergence issue.

$-r'_A$, appearing in Eqs. (2.26)–(2.30), can be expressed either as $\frac{\text{mol}}{\text{unit mass catalyst} \cdot \text{time}}$ or as $\frac{\text{mol}}{\text{unit area ofcatalyst. time}}$. On the other hand, the units of the rate constant, $k'$, will be either

$$\frac{L}{\textit{mass catalyst} \cdot \textit{time}} \times \left(\frac{\text{L}}{\text{mol}}\right)^{n-1} \quad \text{or} \quad \frac{L}{\textit{unit area of catalyst} \cdot \textit{time}} \times \left(\frac{\text{L}}{\text{mol}}\right)^{n-1}.$$ $n$ is the reaction order.

Alternatively, if the partial pressure is used instead of the molar concentration, then $k'$ will be either

$$\frac{\textit{mol}}{\textit{mass catalyst} \cdot \textit{time}} \times \left(\frac{1}{\textit{unit pressure}}\right)^{n} \quad \text{or} \quad \frac{\textit{mol}}{\textit{unit area of catalyst} \cdot \textit{time}} \times \left(\frac{1}{\textit{unit pressure}}\right)^{n}$$

More on PBRs is explained in Chapter 6: "Catalysis in Packed Bed and Fluidized Bed Reactors."

**Example 2.4** ***PBR Catalyst Weight***

A PBR reactor is used to decompose a pollutant ($a$) into harmless products. The reaction is first order with respect to $A$. The initial concentration of $A$ is 10.0 mol/L.

**Given Data:** Reaction rate constant ($k'$) = 0.1 $\frac{\text{L}}{\text{g catalyst} \cdot \text{min}}$ and $v = v_0 = 1.0$ L/min.

Calculate the catalyst weight needed to achieve a conversion of 95%.

**Solution**

$$W_{PBR} = F_{A0}\int_0^X \frac{dX}{-r'_A} = F_{A0}\int_0^X \frac{dX}{k'C_A} = F_{A0}\int_0^X \frac{dX}{k'C_{A0}(1-X)}$$

$$= \frac{F_{Ao}}{k'C_{A0}}\int_0^X \frac{dX}{(1-X)} = \frac{v_0}{k'}\int_0^{0.95} \frac{dX}{(1-X)}$$

$$W_{PBR} = \frac{1\,\frac{\text{L}}{\text{min}}}{0.1\,\frac{\text{L}}{\text{g catalyst} \cdot \text{min}}}\int_0^{0.95} \frac{dX}{(1-X)} = 10\,\text{g}\,\textit{catalyst} * \ln(1-X)\rfloor_{0.95}^{0}$$

$$W_{PBR} = 10\,\text{g}\,\textit{catalyst} * \{\ln(1) - \ln(0.05)\} = 29.96\,\text{g}\,\textit{catalyst}$$

Videos on PBRs/fixed bed reactors are available at:

1) https://www.youtube.com/shorts/waY5L3Nx2cg
2) https://www.youtube.com/watch?v=xjNfrLsMm44
3) https://www.youtube.com/watch?v=QRvndv7xVQ4
4) https://www.youtube.com/watch?v=K4YpnwHaNNs

More on PBRs is covered in Chapter 6.

## 2.6 Fluidized Bed Reactor (FBR)

A FBR is a type of reactor widely used in various industrial applications, including catalytic reactions, combustion, and gas–solid interactions, where maintaining uniform temperature and effective mass and heat transfer are crucial. The FBR operates by suspending solid particles in an upward flow of gas or liquid, achieving fluid-like behavior of the solid phase. Here are more details on the key design elements.

## 2.6.1 Basic Operating Principle

In an FBR, the feed gas or liquid enters from the bottom of the reactor and passes through a distributor plate or grid, designed to evenly spread the fluid. This fluid flow lifts and suspends the solid particles, creating a fluidized bed where solids and fluid are well-mixed. If the reactor operates isothermally, it means the temperature is maintained constant throughout the bed. This is typically achieved via external cooling/heating jackets or internal heat exchangers. FBRs and PBRs will be covered later in Chapter 6: Catalysis in Packed Bed and Fluidized Bed Reactors.

## 2.6.2 Design Components

Figure 2.5 shows a schematic for FBR, the main unit, augmented by a catalyst regenerator and cyclone separator for trapping catalyst solid particles and recycling them to the reactor. It has the following hardware components:

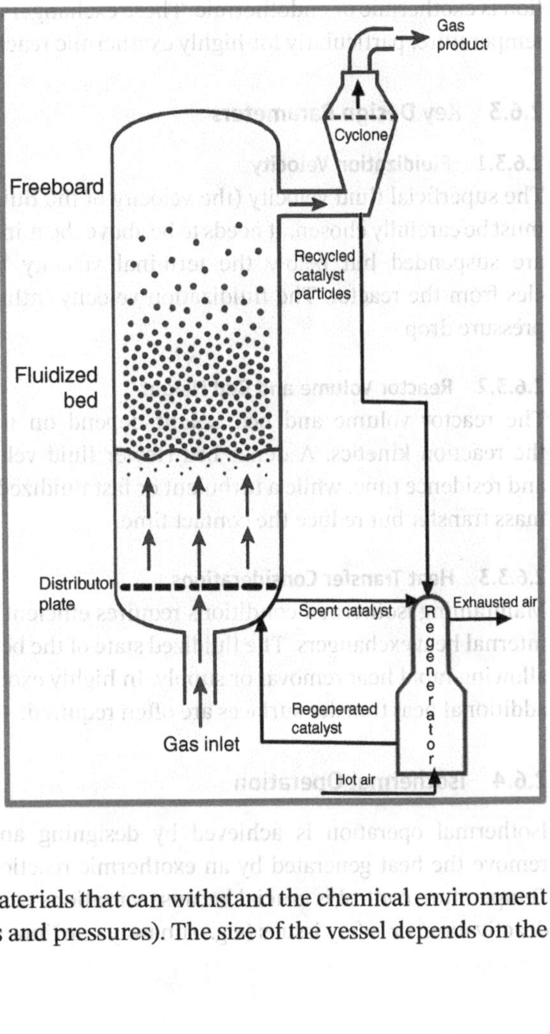


**Figure 2.5** A schematic for FBR, the main unit, augmented by a catalyst regenerator and cyclone separator for trapping catalyst solid particles and recycling them to the reactor.

### 2.6.2.1 Reactor Vessel

The vessel is typically cylindrical, made of materials that can withstand the chemical environment and operating conditions (high temperatures and pressures). The size of the vessel depends on the required reaction rate, throughput, and desired conversion. The vessel height must accommodate different zones, including the gas distributor, the fluidized zone, and possibly a disengagement zone for separating solids from the gas phase.

#### 2.6.2.2 Distributor Plate

A gas distributor plate is critical to achieving uniform fluidization. It can be a perforated plate, sintered metal, or a bubble cap design. It ensures even distribution of the incoming fluid to avoid channeling or stagnant zones in the bed, which could cause uneven reaction rates or poor temperature control.

#### 2.6.2.3 Bed Material (Catalyst or Inert Solids)

The solid particles can be catalytic or inert materials with a specific size and density that allows fluidization at the desired gas velocity. The size distribution of the particles influences fluidization behavior. Particles that are too fine might get entrained out of the bed, while larger particles might cause uneven mixing or poor gas–solid contact.

#### 2.6.2.4 Internal Heat Exchangers

If an FBR operates isothermally, internal heat exchangers (such as cooling coils or heat transfer tubes) can be placed within the bed to either remove or add heat, depending on whether the reaction is exothermic or endothermic. These exchangers enhance heat transfer and maintain uniform temperature, particularly for highly exothermic reactions.

### 2.6.3 Key Design Parameters

#### 2.6.3.1 Fluidization Velocity

The superficial fluid velocity (the velocity of the fluid relative to the empty reactor cross-section) must be carefully chosen. It needs to be above the minimum fluidization velocity to ensure particles are suspended but below the terminal velocity to prevent excessive entrainment of particles from the reactor. The fluidization velocity influences bed expansion, mixing efficiency, and pressure drop.

#### 2.6.3.2 Reactor Volume and Bed Height

The reactor volume and bed height depend on the desired conversion, residence time, and the reaction kinetics. A dense bed (lower fluid velocity) generally has better gas–solid contact and residence time, while a turbulent or fast fluidized bed (higher velocity) may enhance heat and mass transfer but reduce the contact time.

#### 2.6.3.3 Heat Transfer Considerations

Maintaining isothermal conditions requires efficient heat transfer between the bed and external/internal heat exchangers. The fluidized state of the bed leads to excellent heat transfer coefficients, allowing rapid heat removal or supply. In highly exothermic reactions, external cooling jackets or additional heat transfer surfaces are often required.

### 2.6.4 Isothermal Operation

Isothermal operation is achieved by designing an effective heat-exchange system to either remove the heat generated by an exothermic reaction or supply heat in endothermic processes. Temperature control is crucial because variations can lead to poor reaction performance, catalyst deactivation, or safety issues (e.g., runaway reactions in exothermic processes).

### 2.6.5 Advantages of Isothermal FBR Design

a) **High Mass and Heat Transfer Efficiency:** Due to the vigorous mixing of solids and gas, heat and mass transfer are extremely efficient, reducing temperature gradients and improving conversion rates.
b) **Temperature Uniformity:** Isothermal design ensures uniform temperature throughout the reactor, critical for processes that are sensitive to temperature variations.
c) **Ability to Handle High Throughputs:** FBRs are often used for processes with large feed volumes due to their excellent flow properties and scalability.
d) **Catalyst Regeneration and Maintenance:** Fluidized beds allow easy removal and replacement of spent catalyst during operation, or the catalyst can be regenerated in situ.

### 2.6.6 Applications

a) **Catalytic Cracking** (e.g., fluid catalytic cracking in petroleum refining).
b) **Coal Combustion and Gasification:** Fluidized beds are widely used for clean combustion and gasification of coal.
c) **Polymerization Reactors:** Isothermal FBRs are common in processes like polyethylene production, where uniform temperature is crucial for product quality.

### 2.6.7 Challenges/Limitations

a) **Particle Attrition:** The constant movement and collision of particles can lead to wear and tear consequences; that is, the generation of fine particles, potentially leading to operational difficulties.
b) **Entrainment:** At high velocities, fine particles may be entrained and carried out of the reactor with the exit gas, necessitating the use of cyclones or bag filters to recover the particles.
c) **Pressure Drop:** Fluidized beds experience pressure drops, which must be accounted for in the overall design.

**Videos on FBRs are available at:**

1) https://www.youtube.com/shorts/32rcN3QmStI
2) https://www.youtube.com/shorts/rHKzTL7i5gw
3) https://www.youtube.com/shorts/wHkV8q9_sWs
4) https://www.youtube.com/watch?v=3_ILu2Ye8gQ
5) https://www.youtube.com/watch?v=obz3R-Iwiys
6) https://www.youtube.com/watch?v=Oedi6TIcswk

More on FBRs is covered in Chapter 6.

## 2.7 Space Time Calculations for BR, CSTR, and PFR

The **space time**, tau, is obtained by dividing the reactor volume by the volumetric flow rate entering the reactor:

$$\tau = \frac{V}{v_o} = \frac{V}{Q} = \frac{V}{\dot{V}} \tag{2.31}$$

Space time is the time necessary to process one volume of reactor fluid at the entrance conditions. This is the time it takes for the amount of fluid that fills in the entire volume of an empty reactor or the time needed to completely flush the fluid-filled reactor.

**Example 2.4** ***Consider the Following Liquid-phase Reaction:*** $A \xrightarrow{k} B$

The rate constant is 0.2 min$^{-1}$, and it is first order with respect to $A$ ($-r_A = kC_A = 0.2\ \text{min}^{-1} \times C_A$). Calculate the time or volume needed to achieve 99% conversion in a BR:

1) **Batch Reactor:** From Eq. (2.15)

$$t_{BR} = N_{A0}\int_0^X \frac{dX}{(kC_AV)} = N_{A0}\int_0^X \frac{dX}{(kN_A)} = N_{A0}\int_0^X \frac{dX}{(kN_{A0}(1-X))} \tag{2.32}$$

$$t_{BR} = \int_0^X \frac{dX}{(k(1-X))} = \frac{1}{0.2}\int_0^X \frac{dX}{(1-X)}$$

$$= 5\int_0^{0.99} \frac{dX}{(1-X)} = 5\ \text{min} \times \ln\left[\frac{(1-0)}{(1-0.99)}\right] = 5\ \text{min} \times 4.6052 = 23.026\ \text{min}$$

Figure 2.6 shows MATLAB code for carrying out symbolic (i.e., analytic) integration of Eq. (2.32).

```
Clear variables;
% Define the symbolic variables
syms X k
% Define the function to integrate
integrand = 1 / (k * (1 - X));
% Perform the integration
result = int(integrand, X, 0, 0.99);
% Substitute the value of k
k_value = 0.2;
final_result = subs(result, k, k_value);
% Convert the symbolic result to a numerical value
numerical_result = vpa(final_result);
% Display the numerical result
disp('The numerical value of the integral is:');
disp(numerical_result);
```

**Figure 2.6** MATLAB code for symbolically solving Eq. (2.32).

The result is copied from the Command Window:

**The numerical value of the integral is: 23.025850929940456840179914546844**

Figure 2.7 shows MATLAB code for the numerical integration method for solving Eq. (2.32). The results are copied from the Command Window:

**The numerical value of the integral is: 23.0259**

```
clear variables;
% Define the function to integrate
k = 0.2;
integrand = @(X) 1 ./ (k * (1 - X));
% Perform the numerical integration
numerical_result = integral(integrand, 0, 0.99);
% Display the numerical result
disp('The numerical value of the integral is:');
disp(numerical_result);
```

**Figure 2.7** MATLAB code for the numerical integration of Eq. (2.32).

Figure 2.8 shows MATLAB code for making a plot of $X$ versus time for a BR.

```
clear variables;
% Define symbolic variables
syms X(t) k
% Define the differential equation
ode = diff(X, t) == 1 / (k * (1 - X));
% Solve the differential equation symbolically with the initial condition X(0) = 0
sol = dsolve(ode, X(0) == 0);
% Substitute the value of k
k_value = 0.2;
sol = subs(sol, k, k_value);
% Simplify the solution to avoid any complex parts
sol = simplify(sol, 'IgnoreAnalyticConstraints', true);
% Ensure the solution is real-valued by extracting the real part
sol = real(sol);
disp('The analytical solution for X(t) is:');
disp(sol);
% Convert the solution to a function for numerical evaluation
X_func = matlabFunction(sol);
% Define the time vector for plotting
t_values = linspace(0, 20, 100);
% Calculate the corresponding X values using the analytical solution
X_values = X_func(t_values);
% Plot X versus t
figure;
plot(t_values, X_values, 'LineWidth', 2);
xlabel('Time (t)');
ylabel('Conversion (X)');
title('Plot of X versus t using Analytical Solution');
grid on;
```

**Figure 2.8** MATLAB code for creating a plot of $X$ versus time for the BR.

The result is copied from the Command Window:

```
≫ Fig2_8
The analytical solution for X(t) is:
1 - real((1 - 10*t)^(1/2))
```

Figure 2.9 shows the plot of $X$ versus time for a first-order reaction in a BR.

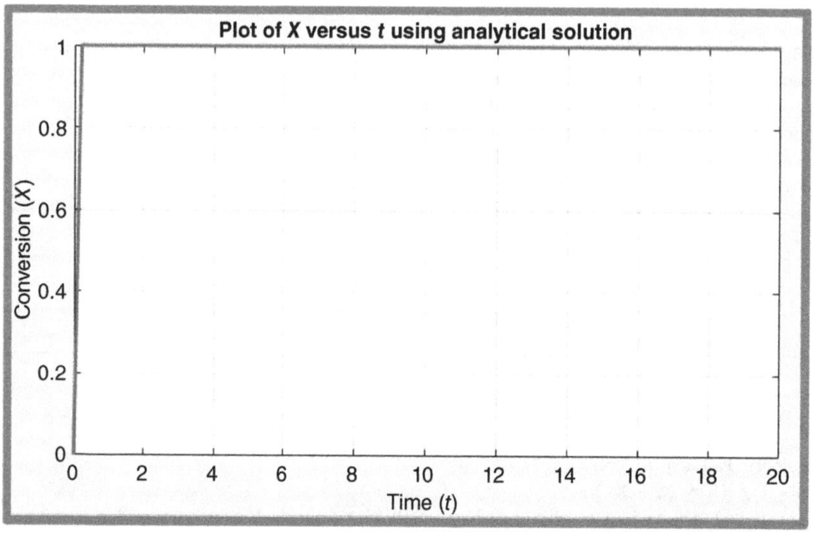


**Figure 2.9** A plot of $X$ versus time for a batch reactor with first-order kinetics.

**Example 2.5** Repeat Example 2.4, but for a CSTR while assuming that $v_0 = v$.

From Eqs. (2.18a–c), we have:

$$V_{CSTR} = \frac{F_{A0}X}{-r_A} = \frac{F_{A0}X}{kC_A} = \frac{F_{A0}X}{k\frac{F_A}{v}} = \frac{vF_{A0}X}{kF_A}$$

$$\tau = \frac{V_{CSTR}}{v} = \frac{F_{A0}X}{kF_A} = \frac{F_{A0}X}{kF_{A0}(1-X)} = \frac{X}{k(1-X)} = \frac{0.99}{0.2\ \text{min}^{-1} \times (1-0.99)} = 495\ \text{min}$$

**Example 2.6** Repeat Example 2.4, but for a PFR while assuming that $v_0 = v$.

From Eqs. (2.24a–b), we have:

$$V_{PFR} = F_{A0}\int_0^X \frac{dX}{-r_A} = F_{A0}\int_0^X \frac{dX}{kC_A} = F_{A0}\int_0^X \frac{dX}{k\frac{F_A}{v}} = vF_{A0}\int_0^X \frac{dX}{kF_A}$$

$$\tau_{PFR} = \frac{V_{PFR}}{v} = F_{A0}\int_0^X \frac{dX}{kF_A} = F_{A0}\int_0^X \frac{dX}{kF_{A0}(1-X)} = \int_0^X \frac{dX}{k(1-X)} = \frac{1}{k}\int_0^X \frac{dX}{(1-X)}$$

$$\tau_{PFR} = \frac{V}{v} = 5\int_0^{0.99} \frac{dX}{(1-X)} = 5 \times \ln\left[\frac{(1-0)}{(1-0.99)}\right] = 5\ \text{min} \times 4.6052 = 23.026\ \text{min}$$

Notice that the residence time of both the BR and PFR is the same.

**Exercise 2.1** Consider the following liquid-phase reaction: $A \xrightarrow{k} B$

The rate constant is 0.2 $\left(\frac{\text{L}}{\text{mol}\cdot\text{min}}\right)$ and it is second order with respect to A ($-r_A = kC_A^2 = 0.2\left(\frac{\text{L}}{\text{mol}\cdot\text{min}}\right) \times C_A^2$.

Calculate the time or volume needed to achieve 99% conversion in the following reactors: Batch, CSTR, and PFR.

## 2.8 Reactor Sizing of $N^{th}$-Order Liquid-Phase Reaction

Let us consider the kinetic data of the liquid-phase reaction:

$$-r_A\left[\frac{\text{mol}}{\text{L}\cdot\text{min}}\right] = kC_A^n = 1.6\left(\frac{\text{mol}^{0.8}}{\text{L}^{0.8}\cdot\text{min}}\right)\times C_A^{0.2}$$

1) Calculate the BR time to achieve 90% conversion when 500 mol of $A$ are charged to a 500-L reactor.
2) What are the CSTR and PFR volumes necessary to achieve 90% conversion? The entering molar flow rate of $A$, $F_{A0} = 100\ \frac{\text{mol}}{\text{min}}$, and $\upsilon_0 = 100\frac{\text{L}}{\text{min}}$
3) Over what range of conversions would the CSTR and PFR reactor volumes be identical?
4) What is the maximum conversion that can be achieved in a 120-L CSTR?
5) What conversion can be achieved if a 50-L PFR is followed in series by a 40-L CSTR?
6) What conversion can be achieved if a 40-L CSTR is followed in a series by a 40-L PFR?
7) Plot the conversion as a function of PFR reactor volume up to a volume of 80 L.

**Solution**

$$C_{A0} = \frac{N_{A0}}{V_0} = \frac{500\text{ mol}}{500\text{ L}} = 1\frac{\text{mol}}{\text{L}}$$

1) From Eq. (2.14), we have:

$$t_{BR} = N_{A0}\int_0^X \frac{dX}{(-r_A V)} = C_{A0}\int_0^{0.9}\frac{dX}{\left(kC_{A0}^{0.2}(1-X)^{0.2}\right)} = \frac{C_{A0}^{0.8}}{k}\int_0^{0.9}\frac{dX}{(1-X)^{0.2}}$$

We need to find the value of the limited integral.
Assume: $u = (1-X)$, $du = -dx$.

$$\int_0^{0.9}\frac{dX}{(1-X)^{0.2}} = \int_1^{0.1}\frac{-du}{(u)^{0.2}} = -\left.\frac{u^{0.8}}{0.8}\right|_1^{0.1} = \left.\frac{u^{0.8}}{0.8}\right|_{0.1}^{1} = \frac{1}{0.8}[1-0.1^{0.8}] = 1.052$$

$$t_{BR} = \frac{C_{A0}^{0.8}}{k}\int_0^{0.9}\frac{dX}{(1-X)^{0.2}} = \left(\frac{C_{A0}^{0.8}}{k}\right)\text{min}\times 1.052 = 0.625\text{ min}\times 1.052 = \frac{1.052}{1.6} = 0.657\text{ min}$$

2) $F_{A0} = 100\frac{\text{mol}}{\text{min}} = C_{A0}\times\upsilon_0 = C_{A0}\times 100\frac{\text{L}}{\text{min}} \rightarrow C_{A0} = 1\frac{\text{mol}}{\text{L}}$

2.a. From Eq. (2.18a), we have:

$$V_{CSTR} = \frac{F_{A0}X}{-r_A} = \frac{F_{A0}X}{\left(kC_{A0}^{0.2}(1-X)^{0.2}\right)} = \frac{100\frac{\text{L}}{\text{min}}\times C_{A0}\times 0.9}{\left(k\times C_{A0}^{0.2}(1-0.9)^{0.2}\right)}$$

$$V_{CSTR} = \frac{100\frac{\text{L}}{\text{min}}\times C_{A0}^{0.8}\times 0.9}{(k\times(1-0.9)^{0.2})} = \left(\frac{C_{A0}^{0.8}}{k}\right)\text{min}\times\frac{100\frac{\text{L}}{\text{min}}\times 0.9}{((1-0.9)^{0.2})}$$

$$= 0.625\text{ min}\times 142.6\frac{\text{L}}{\text{min}} = 89.1\ L$$

2.b. From Eqs. (2.24a–b), we have:

$$V_{PFR} = F_{A0}\int_0^X\frac{dX}{-r_A} = F_{A0}\int_0^{0.9}\frac{dX}{\left(kC_{A0}^{0.2}(1-X)^{0.2}\right)} = \frac{C_{A0}\times\upsilon_0}{k}\int_0^{0.9}\frac{dX}{\left(C_{A0}^{0.2}(1-X)^{0.2}\right)} =$$

$$V_{PFR} = \frac{C_{A0}^{0.8}\times\upsilon_0}{k}\int_0^{0.9}\frac{dX}{((1-X)^{0.2})} = \left(\frac{1^{0.8}}{1.6}\right)\text{min}\times 100\frac{\text{L}}{\text{min}}\times 1.052 = 65.75\text{ L}$$

3) Over what range of conversions would the CSTR and PFR reactor volumes be identical?

$$V_{CSTR} = \frac{F_{A0}X}{\left(kC_{A0}^{0.2}(1-X)^{0.2}\right)} = V_{PFR} = F_{A0}\int_0^X \frac{dX}{\left(kC_{A0}^{0.2}(1-X)^{0.2}\right)}$$

Simplify the above equality, we have:

$$\frac{X}{((1-X)^{0.2})} = \int_0^X \frac{dX}{((1-X)^{0.2})}$$

Assume: $u = (1-X)$, $du = -dx$.

$$\int_0^X \frac{dX}{((1-X)^{0.2})} = \int_0^X \frac{-du}{(u)^{0.2}} = \left.\frac{u^{0.8}}{0.8}\right|_X^0 = 1.25 * \left.(1-X)^{0.8}\right|_X^0 = 1.25(1-(1-X)^{0.8})$$

Equate both sides: $\frac{X}{((1-X)^{0.2})} = 1.25(1-(1-X)^{0.8})$

Which means: $f(X) = \frac{X}{((1-X)^{0.2})} - 1.25(1-(1-X)^{0.8}) = 0$

Write MATLAB code for finding the roots of the given equation.

1) Find the roots for the following $f(X) = \frac{X}{((1-X)^{0.2})} - 1.25(1-(1-X)^{0.8}) = 0$
2) Assume an initial guess for $X$. $X$ varies between 0 and 1.
3) Report the numerical values of the roots that satisfy the above $f(X) = 0$

Figure 2.10 shows MATLAB code for finding the roots (i.e., conversion) for $f(X) = 0$, where we have an equal volume of both CSTR and PFR.

```
clear variables;
% MATLAB code to find the roots of f(X) = X / ((1-X)^0.2) - 1.25 * (1 - (1-X)^0.8)
% Define the function f(X)
f = @(X) X ./ ((1 - X).^0.2) - 1.25 .* (1 - (1-X).^0.8);
% Define the range for X (between 0 and 1)
X_initial_guess = linspace(0, 1, 1000); % Initial guess within the range
% Use 'fzero' to find the roots
% To ensure multiple roots within the range 0 and 1 are found,
% we evaluate in smaller subintervals.
roots = [];
for i = 1:length(X_initial_guess)-1
  try
    % Narrow down the interval and solve for the root
    root = fzero(f, [X_initial_guess(i), X_initial_guess(i+1)]);
    % Store only unique roots (to avoid duplications)
    if isempty(roots) || all(abs(roots - root) > 1e-6)
      roots(end+1) = root;
    end
  catch
    % In case no root is found in the given interval, skip
  end
end
% Display the roots found
disp('The roots that satisfy f(X) = 0 are:');
disp(roots);
```

**Figure 2.10** MATLAB code for finding the roots for $f(X) = 0$, where PFR and CSTR volumes are identical.

Results are copied from MATLAB Command Window:

```
≫ Roots1
The roots that satisfy f(X) = 0 are:   0
```

For this kind of chemical reaction, for all possible $X$ values between 0 and 1, $f(X) \geq 0$, which indicates that $V_{CSTR} > V_{PFR}$. In other words, no root could be found for $X > 0$.

4) What is the maximum conversion that can be achieved in a 120-L CSTR?

$$V_{CSTR} = 120\text{ L} = \frac{F_{A0}X}{\left(kC_{A0}^{0.2}(1-X)^{0.2}\right)} = \frac{100 * X}{(1.6 * 1(1-X)^{0.2})}$$

Find $X$.

$$120 - \frac{100 * X}{(1.6 * 1(1-X)^{0.2})} = 0$$

Rerun the code in Figure 2.10, but with the following modification:

```
f= @(X) 120.0-(100*X./(1.6*(1-X)^0.2));
```

Results are copied from MATLAB Command Window:

```
≫ Roots2
The roots that satisfy f(X) = 0 are: 0.9675
```

For the given $V_{CSTR} = 120$ L, $X = 0.9675$.

5) What conversion can be achieved if a 50-L PFR is followed in series by a 40-L CSTR?

$$V_{PFR} = \frac{C_{A0} \times v_0}{k} \int_0^X \frac{dX}{\left(C_{A0}^{0.2}(1-X)^{0.2}\right)} = \frac{C_{A0}^{0.8} \times v_0}{k} \int_0^X \frac{dX}{((1-X)^{0.2})}$$

$$V_{PFR} = 50\text{ L} = \frac{C_{A0}^{0.8} \times v_0}{k} * 1.25(1-(1-X)^{0.8}) = \frac{1 \times 100}{1.6} * 1.25(1-(1-X)^{0.8})$$

$$\frac{50 * 1.6}{125} = 0.64 = 1-(1-X)^{0.8} \rightarrow (1-X)^{0.8} = 0.36$$

$$(1-X) = (0.36)^{1.25} = 0.2789 \rightarrow X_{PFR} = 0.7211$$

$$V_{CSTR} = 40\text{ L} = \frac{F_{A0}(X_{CSTR} - X_{PFR})}{\left(kC_{A0}^{0.2}(1-X_{CSTR})^{0.2}\right)} = \frac{100 * (X_{CSTR} - 0.7211)}{\left(1.6 * (1-X_{CSTR})^{0.2}\right)}$$

$$f(X_{CSTR}) = 40 - \frac{100 * (X_{CSTR} - 0.7211)}{\left(1.6 * (1-X_{CSTR})^{0.2}\right)} = 0$$

Rerun the code in Figure 2.10, but with the following modification:

```
f= @(X) 40.0-(100*(X-0.7211)./(1.6*(1-X)^0.2));
```

Results are copied from MATLAB Command Window:

```
>> Roots3
The roots that satisfy f(X) = 0 are: 0.9875
```

So, the CSTR exit conversion will be $X = 0.9875$.

6) What conversion can be achieved if a 40-L CSTR is followed in a series by a 40-L PFR?

$$V_{CSTR} = 40 \text{ L} = \frac{F_{Ao}(X_{CSTR} - 0)}{\left(kC_{A0}^{0.2}(1 - X_{CSTR})^{0.2}\right)} = \frac{100 * (X_{CSTR} - 0)}{\left(1.6 * (1 - X_{CSTR})^{0.2}\right)}$$

$$f(X_{CSTR}) = 40 - \frac{100 * (X_{CSTR} - 0.0)}{\left(1.6 * (1 - X_{CSTR})^{0.2}\right)} = 0$$

Rerun the code in Figure 2.10, but with the following modification:

```
f= @(X) 40.0-(100*(X-0.7211)./(1.6*(1-X)^0.2));
```

Results are copied from MATLAB Command Window:

```
>> Roots4
The roots that satisfy f(X) = 0 are: 0.5464
```

So, the CSTR exit conversion will be $X = 0.5464$.

$$V_{PFR} = 40 \text{ L} = \frac{C_{A0} \times v_0}{k} \int_{0.5464}^{X} \frac{dX}{\left(C_{A0}^{0.2}(1 - X)^{0.2}\right)} = \frac{C_{A0}^{0.8} \times v_0}{k} \int_{0.5464}^{X} \frac{dX}{((1 - X)^{0.2})}$$

$$40 \text{ L} = 62.5 * \int_{0.5464}^{X} \frac{dX}{(1 - X)^{0.2}}$$

$$\int_{0.5464}^{X} \frac{-du}{(u)^{0.2}} = -\left.\frac{u^{0.8}}{0.8}\right|_{0.5464}^{X} = -\left.\frac{(1 - X)^{0.8}}{0.8}\right|_{0.5464}^{X} = -\frac{1}{0.8}[(1 - X)^{0.8} - (1 - 0.5464)^{0.8}]$$

$$\int_{0.5464}^{X} \frac{dX}{(1 - X)^{0.2}} = -1.25[(1 - X)^{0.8} - 0.5313] = 0.6641 - 1.25(1 - X)^{0.8}$$

$$40L = 62.5 * \int_{0.5464}^{X} \frac{dX}{(1 - X)^{0.2}} = 62.5 * (0.6641 - 1.25(1 - X)^{0.8})$$

Rerun the code in Figure 2.10, but with the following modification:

```
f= @(X) 40.0-(62.5*(0.6641-(1.25*(1-X)^0.8)));
```

Results are copied from MATLAB Command window:

**≫ Roots5**
**The roots that satisfy f(X) = 0 are: 0.9928**

So, the PFR exit conversion will be $X = 0.9928$.

7) Plot the conversion as a function of PFR reactor volume up to a volume of 80 L.

$$dV = \frac{C_{A0}^{0.8} \times v_0}{k} \frac{dX}{((1 - X)^{0.2})}$$

**MATLAB algorithm:**

1) CA0 = 1 mol/L; $v_0 = 100$ L/min; and $k = 1.6$
2) Carry out symbolic integration of $X$ as $f(V)$, where $X$ goes from 0 to 1 and $V$ from 0 to 80 L. The differential equation is given below:

$$dV = \frac{C_{A0}^{0.8} \times v_0}{k} \frac{dX}{((1 - X)^{0.2})}$$

3) Make a plot for $X$ versus $V$ between 0 and 80 L.
   Figure 2.11 shows MATLAB code for plotting $X$ versus $V$ for a PFR.

```
clear variables;
% Step 1: Define symbolic variables
syms X V C % Declare C as a symbolic constant
% Step 2: Define constants
CA0 = 1;  % mol/L
v0 = 100;  % L/min
k = 1.6;  % given constant
% Step 3: Define the differential equation
dV_dX = (CA0^0.8 * v0) / (k * (1 - X)^0.2);
% Step 4: Perform symbolic integration to find V as a function of X
V_solution = int(dV_dX, X) + C;  % Add integration constant C
% Step 5: Solve for the constant C, knowing that V = 0 when X = 0
C_value = solve(subs(V_solution, X, 0) == 0, C); % Find C
% Step 6: Substitute the constant C into the solution
V_solution = subs(V_solution, C, C_value);
% Step 7: Create a plot for X vs. V
% Generate numeric values for X and evaluate V
X_vals = linspace(0, 1, 100); % X from 0 to 1
V_vals = double(subs(V_solution, X, X_vals)); % Substitute X values in V_solution
% Step 8: Plot the results
figure;
plot(V_vals, X_vals, 'LineWidth', 2);
xlabel('Volume (V) [L]');
ylabel('Conversion (X)');
title('Conversion (X) vs. Volume (V)');
grid on;
```

**Figure 2.11** MATLAB code for plotting $X$ versus $V$ for a PFR.

Figure 2.12 shows the generated plot.

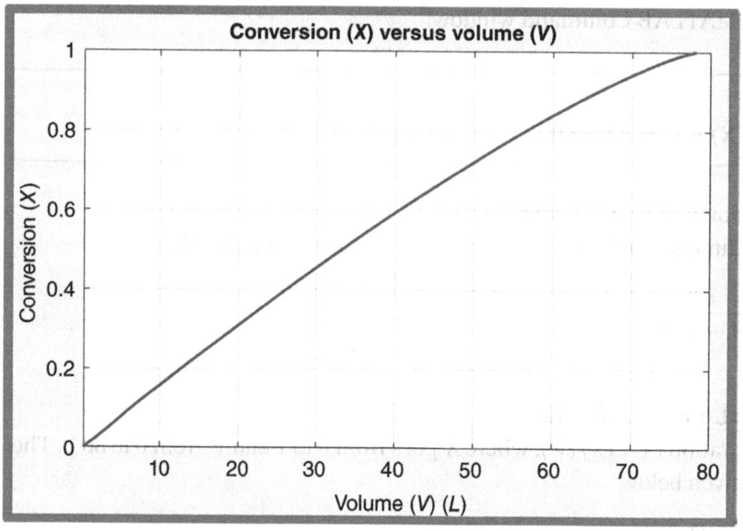


**Figure 2.12** The plot of $X$ versus $V$ for a PFR.

## 2.9 Plot of $\left(\frac{F_{A0}}{-r_A}\right)$ Versus X for a Homogeneous Gas-Phase Reaction

The adiabatic exothermic irreversible gas-phase reaction

$$2A + B \rightarrow 2C$$

is to be carried out in a flow reactor for an equimolar feed of $A$ and $B$. The plot of $\left(\frac{F_{A0}}{-r_A}\right)$, expressed in $m^3$, versus $X$, the conversion for the above reaction, is shown in Figure 2.13.

1) What CSTR volume is necessary to achieve 50% conversion?
2) What PFR volume is necessary to achieve 50% conversion?

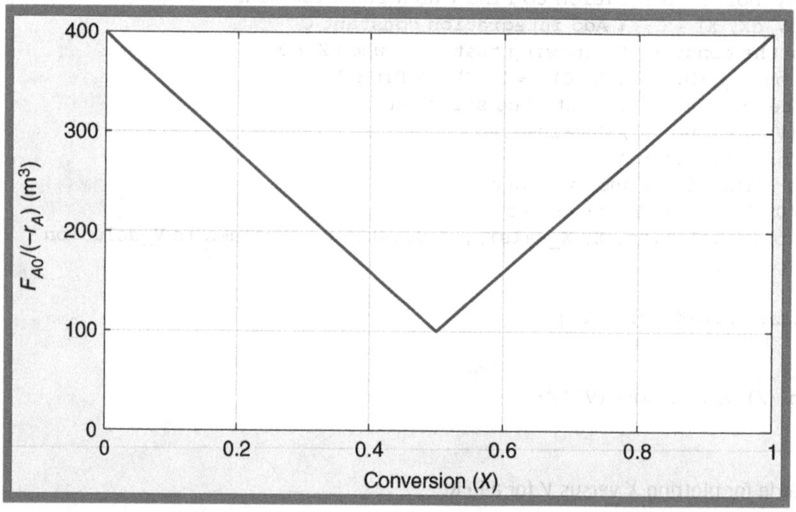


**Figure 2.13** A plot of $\left(\frac{F_{A0}}{-r_A}\right)$ versus X for an adiabatic exothermic irreversible gas-phase reaction.

3) What is the volume of a second CSTR added in series to the first CSTR **(Part 1)** necessary to achieve an overall conversion of 80%?
4) What PFR volume must be added to the first CSTR **(Part 1)** to raise the conversion to 80%?

**Solution**

Notice that the area under the curve (or line in Figure 2.13) gives the volume requirement for a PFR, in general. On the other hand, the volume requirement for a CSTR represents the area of a rectangle drawn with a width $\Delta X$ and a height evaluated at the exit conversion.

1) The CSTR volume necessary to achieve 50% conversion is represented by a rectangular area limited between $X = 0, X = 0.5, V = 0$, and $V = 100\ \text{m}^3$. Thus,

$$V_{CSTR1} = \frac{F_{A0}\,(X-0)}{-r_A} = (100-0)\ \text{m}^3 * (0.5-0) = 50\ \text{m}^3$$

2) The PFR volume necessary to achieve 50% conversion is represented by the area below the line between 0 and $X = 0.5$, which is basically the area of the left-side triangle and the bottom rectangle area, below the triangle. So, it is a trapezoidal area made of two segments, the top triangle and the bottom rectangle.
From Figure 2.13, we have:

$$V_{PFR1} = F_{Ao}\int_0^{0.5}\frac{dX}{-r_A} = Area_{Triangle} + Area_{Rectangle}$$

$$V_{PFR1} = \frac{(400-100) * (0.5-0)}{2} + (100-0) * (0.5-0) = 75 + 50 = 125\ \text{m}^3$$

3) The volume of a second CSTR added in series to the first CSTR **(Part 1)** necessary to achieve an overall conversion of 80% will be to move the conversion forward from $X = 0.5$ up to $X = 0.8$. So, it will be an area of a rectangle with $\Delta X = 0.8–0.5 = 0.3$ and a height evaluated at $X = 0.8$.

$$V_{CSTR2} = \frac{F_{A0}\,(X_2 - X_1)}{-r_A} = (278-0)\ \text{m}^3 * (0.8-0.5) = 83.4\ \text{m}^3$$

The total volume of 2 reactors in series $= V_{CSTR1} + V_{CSTR2} = 50 + 83.4 = 133.4\ \text{m}^3$

4) The PFR volume that must be added to the first CSTR (Part 1) to raise the conversion to 80% will be equal to the area of the right-side triangle and the bottom rectangle area, below the triangle. So, it is a trapezoidal area made of two segments, the top triangle and the bottom rectangle.

$$V_{PFR2} = F_{Ao}\int_{0.5}^{0.8}\frac{dX}{-r_A} = Area_{Triangle} + Area_{Rectangle}$$

$$V_{PFR2} = \frac{(278-100) * (0.8-0.5)}{2} + (100-0) * (0.8-0.5) = 26.7 + 30 = 56.7\ \text{m}^3$$

The total volume of 2 reactors in series $= V_{PFR1} + V_{PFR2} = 125 + 56.7 = 181.7\ \text{m}^3$

## 2.10 Plot of $\left(\frac{F_{A0}}{-r'_A}\right)$ Versus X for a Heterogeneous Gas-Phase Reaction

Figure 2.14 shows a typical gas–solid catalytic exothermic reaction carried out adiabatically for: $A \rightarrow B + C$.

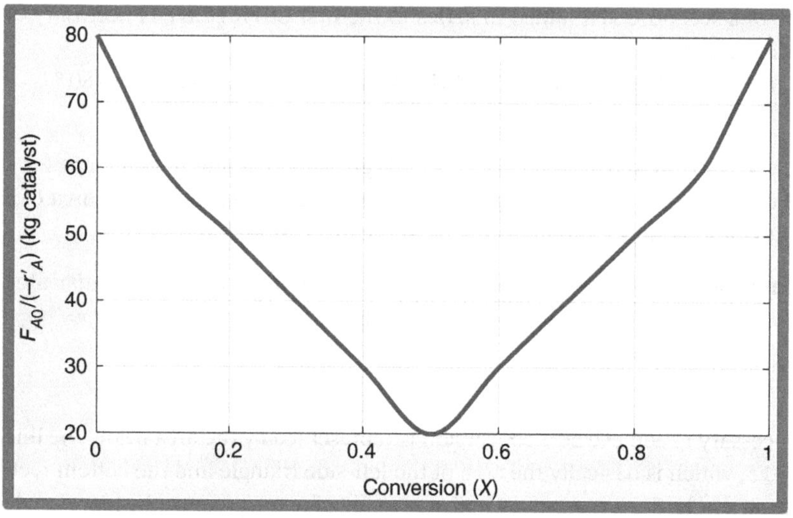


**Figure 2.14** A plot of the catalyst weight, kg, versus conversion for a heterogeneous gas-phase reaction.

1) Assuming that we have a fluidized CSTR and a PBR, how should they be arranged for this adiabatic reaction? Use the smallest amount of catalyst weight to achieve 80% conversion of $A$. The best configuration that will minimize the catalyst weight requirement will be installing a fluidized CSTR from $X = 0$ up to $X = 0.5$, followed by a PFR from $X = 0.5$ up to 0.8.

$$W_{CSTR1} = \frac{F_{Ao}(X_f - X_0)}{-r'_A} = Area_{Rectangle} = (20 - 0) * (0.5 - 0) = 10 \text{ kg cat.}$$

$$W_{PFR2} = F_{Ao}\int_0^X \frac{dX}{-r'_A} \cong Area_{Trapezoid} \cong Area_{Triangle} + Area_{Rectangle}$$

$$W_{PFR2} \cong \frac{(50 - 20) * (0.8 - 0.5)}{2} + (20 - 0) * (0.8 - 0.5) = 4.5 + 6 = 10.5 \text{ kg cat.}$$

2) What is the catalyst weight necessary to achieve 80% conversion in a fluidized CSTR?

$$W_{CSTR} = \frac{F_{Ao}(X_f - X_0)}{-r'_A} = Area_{Rectangle} = (50 - 0) * (0.8 - 0) = 40 \text{ kg cat.}$$

3) What fluidized CSTR weight is necessary to achieve 40% conversion?

$$W_{CSTR1} = \frac{F_{Ao}(X_f - X_0)}{-r'_A} = Area_{Rectangle} = (30 - 0) * (0.4 - 0) = 12 \text{ kg cat.}$$

4) What PBR weight is necessary to achieve 80% conversion?

$$W_{PFR} = F_{Ao}\int_0^X \frac{dX}{-r'_A} \cong Area_{Left-Triangle} + Area_{Right-Triangle} + Area_{Bottom\ Rectangle}$$

$$W_{PFR} \cong \frac{(80 - 20) * (0.5 - 0.0)}{2} + \frac{(50 - 20) * (0.8 - 0.5)}{2} + (20 - 0) * (0.8 - 0.0)$$
$$= 15 + 4.5 + 16 = 35.5 \text{ kg cat.}$$

5) What PBR weight is necessary to achieve 40% conversion?

$$W_{PFR} = F_{Ao} \int_0^X \frac{dX}{-r'_A} \cong Area_{Trapezoid} \cong Area_{Triangle} + Area_{Rectangle}$$

$$W_{PFR} \cong \frac{(80-30) * (0.4-0.0)}{2} + (30-0) * (0.4-0.0) = 10 + 12 = 22 \text{ kg cat}$$

## 2.11 MATLAB-Based Gas-Phase BR, CSTR, and PBR

Consider the following solid platinum-catalyzed gas-phase reaction:

$$SO_2 + ½\, O_2 \rightarrow SO_3 \qquad A + ½\, B \rightarrow C$$

A 100-mol mixture of 29.58% $SO_2$ and 70.42% air is charged to a BR in which it is oxidized. The total initial pressure is 1,500 kPa (15 bar or 14.804 atm), and the initial temperature is 300 K. However, the reactor temperature will later be kept constant at 400 °C by an external cooling medium to offset the exothermic heat of reaction.

The reaction rate is given by [1]:

$$-r'_{SO2}\left[\frac{\text{mol SO}_2}{\text{h}\cdot\text{g}_{\text{cat}}}\right] = \frac{k\left[P_{SO2}\sqrt{P_{O2}} - \frac{P_{SO3}}{K_P}\right]}{[1+\sqrt{P_{O2}K_{O2}} + P_{SO3}K_{SO3}]^2} \tag{2.33}$$

$$k = 9.7 \frac{\text{mol SO2}}{\text{atm}^{1.5}\cdot\text{h}\cdot\text{g}_{\text{cat}}} \tag{2.34}$$

$$K_{O2} = 38.5\ \text{atm}^{-1} \qquad K_{SO3} = 42.5\ \text{atm}^{-1} \qquad K_P = 930\ \text{atm}^{-1/2} \tag{2.35a–c}$$

Calculate the batch time per 100 g catalyst weight to achieve 99% conversion.

**Solution**

$$(-r'_A W) = \frac{dN_A}{dt} = N_{Ao}\frac{dX}{dt}$$

Or:

$$dt = \frac{N_{A0}}{(-r'_A W)} dX \tag{2.36}$$

At $t = 0, X = 0$

$$t = N_{A0}\int_0^X \frac{dX}{(-r'_A W)} \tag{2.37}$$

To understand why such a basis of composition is selected is simply to let $O_2$ enter at the stoichiometric ratio with respect to $SO_2$. Hence, $2x$ stands for initial moles of $SO_2$ (or $x$ for initial moles of $O_2$) and $y$ for air.

$$2x = 100 - y \tag{a}$$

$$x = y^* 0.21 \quad \text{or} \quad 2x = 0.42^* y \tag{b}$$

Substitute (b) in (a)

$0.42y = 100 - y$. Solve for $y$. $\rightarrow y = 70.42$

Assume that we have 100 mol of the feed mixture. So, we have:

$$N^0_{SO2} = 29.58 \text{ moles} \rightarrow N^0_{Air} = 70.42 \text{ moles}$$

$$N^0_{Air} = 70.42 \text{ moles} \rightarrow 14.79 \text{ moles } N^0_{O2} \text{ and } 55.63\ N^0_{N2}$$

For simplicity, we will assume an ideal gas mixture.

$$P_i = C_i RT$$

Let us start with $T_0 = 300$ K and $P_0 = 14.804$ atm. As a result of initial heating, $T$ will increase to 400 °C = 673 K. The oxidation reaction, by nature, is exothermic, which can be initially utilized, and later cooling is a must to keep the temperature constant.

$$\varepsilon = \left(\frac{c}{a} - \frac{b}{a} - 1\right) y^0_{SO2} = \left(1 - \frac{1}{2} - 1\right) 0.2958 = -0.1479$$

Let us apply the ideal gas law at the initial and final conditions:

$$P_0 V_0 = n_0 R T_0$$

$$PV = nRT$$

Keep in mind that $n_0 = 100$ moles. As a result of the operative chemical reaction and assuming 100% conversion, then 29.58 mol of $SO_2$ and 14.79 mol of $O_2$ will be consumed. In return, 29.58 mol of $SO_3$ will be generated. The net will be −14.79 mol. Hence, $n = 100 - 14.79 = 85.21$ moles.

The initial and final volumes will be the same. Rearranging the two equations for the benefit of $P$, we have:

$$P = P_0 \times \left(\frac{n}{n_0}\right)\left(\frac{T}{T_0}\right) = 14.804 \text{ atm} \times \left(\frac{85.21 \text{ mol}}{100 \text{ mol}}\right)\left(\frac{673 \text{ K}}{300 \text{ K}}\right) = 28.2985 \text{ atm}$$

Initial concentration of $SO_2$: $C^0_{SO2} = \left(\frac{P_0 y^0_{SO2}}{0.08206 * T_0}\right) = \frac{14.804 \text{ atm} \times 0.2958}{0.08206 \frac{\text{atm.L}}{\text{mol.K}} \times 300 \text{ K}} = 0.1776 \frac{\text{mol}}{\text{L}}$

**MATLAB Algorithm:**

1) Prompt the user to enter the initial temperature, $T_0$.
2) Prompt the user to enter the reactor temperature, T.
3) Prompt the user to enter the initial pressure, $P_0$.
4) Prompt the user to enter the initial moles, $n_0$.
5) Prompt the user to enter the initial mole fraction of SO2, ySO2_0.
6) Prompt the user to enter the initial mole fraction of O2, yO2_0.
7) Prompt the user to enter the initial mole fraction of SO3, ySO3_0.
8) Prompt the user to enter the stoichiometric coefficient of SO2, a
9) Prompt the user to enter the stoichiometric coefficient of O2, b
10) Prompt the user to enter the stoichiometric coefficient of SO3, c
11) Calculate epsilon, $\varepsilon = \left(\frac{c}{a} - \frac{b}{a} - 1\right) * ySO2\_0$
12) Calculate 100% number of moles, $n = n0 * (1 + \varepsilon)$
13) Calculate the reactor pressure: $P = P_0 \times \left(\frac{n}{n_0}\right)\left(\frac{T}{T_0}\right)$
14) Calculate the initial concentration of SO2, CSO2_0 = $\left(\frac{P_0 * ySO2\_0}{0.08206 * T_0}\right)$
15) Calculate O2 molar ratio, theta O2 = yO2_0/ySO2_0
16) Calculate SO3 molar ratio, theta SO3 = ySO3_0/ySO2_0

17) Express the partial pressure of $SO_2$ as f(X), $PSO2 = CSO2\_0 * \left[\frac{(1-X)}{(1+\varepsilon X)} \frac{P}{P_0} \frac{T_0}{T}\right] * 0.08206 * T$, where X stands for the conversion.
18) Express the partial pressure of $O_2$ as f(X), $PO2 = CSO2\_0 * \left[\frac{\left(\text{thetaO2} - \left(\frac{b}{a}\right) * X\right)}{(1+\varepsilon X)} \frac{P}{P_0} \frac{T_0}{T}\right] * 0.08206 * T$
19) Express the partial pressure of $SO_3$ as f(X), $PSO3 = CSO2\_0 * \left[\frac{\left(\text{thetaSO3} + \left(\frac{c}{a}\right) * X\right)}{(1+\varepsilon X)} \frac{P}{P_0} \frac{T_0}{T}\right] * 0.08206 * T$
20) $k = 9.7$; $K_{O2} = 38.5$; $K_{SO3} = 42.5$; $K_P = 930.0$
21) Prompt the user for the final conversion, Xf.
22) Express reaction rate, $-rSO2 = \frac{k\left[P_{SO2}\sqrt{P_{O2}} - \frac{P_{SO3}}{K_P}\right]}{[1+\sqrt{P_{O2}K_{O2}} + P_{SO3}K_{SO3}]^2}$. Notice that PSO2, PO2, and PSO3 are f(X) as expressed in steps #17, #18, and #19, respectively.
23) Prompt the user to select one of three cases:

**Case 1:** BR

24) Prompt the user to enter the weight of batch catalyst, W in gram.

$$tau = (n0 * ySO_2\_0) \int_0^{Xf} \frac{dX}{(-rSO_2 * W)}$$

Print or display the residence time, tau, expressed in hours
Print or display the weight of the catalyst, W, expressed in grams
Print or display the final conversion, Xf

**Case 2:** CSTR

Display for the user that the initial number of moles of A will be converted in molar flow rate of A per unit time

$$F_{A0} = n0^* ySO_2\_0$$

$$W_{CSTR} = \frac{F_{A0} Xf}{-rSO_2}$$

Print or display the weight of the catalyst, $W_{CSTR}$, expressed in grams
Print or display the final conversion, Xf

**Case 3:** PBR

Display for the user that the initial number of moles of A will be converted in molar flow rate of A per unit time

$$F_{A0} = n0^* ySO_2\_0$$

$$W_{PBR} = F_{A0} \int_0^{Xf} \frac{dX}{-rSO_2}$$

Print or display the weight of the catalyst, $W_{PBR}$, expressed in grams
Print or display the final conversion, Xf
Figure 2.15 shows MATLAB code (**SO2_Cat_Oxid.m**) for calculating the residence time or the weight of catalyst needed for a given conversion.

```
% MATLAB Algorithm for Reactor Design
clear variables;
% 1. Prompt the user to enter initial temperature, T0
T0 = input('Enter the initial temperature (K): ');
% 2. Prompt the user to enter reactor temperature, T
T = input('Enter the reactor temperature (K): ');
% 3. Prompt the user to enter the initial pressure, P0
P0 = input('Enter the initial pressure (atm): ');
% 4. Prompt the user to enter the initial moles, n0
n0 = input('Enter the initial moles: ');
% 5. Prompt the user to enter the initial mole fraction of SO2, ySO2_0
ySO2_0 = input('Enter the initial mole fraction of SO2: ');
% 6. Prompt the user to enter the initial mole fraction of O2, yO2_0
yO2_0 = input('Enter the initial mole fraction of O2: ');
% 7. Prompt the user to enter the initial mole fraction of SO3, ySO3_0
ySO3_0 = input('Enter the initial mole fraction of SO3: ');
% 8-10. Prompt the user to enter the stoichiometric coefficients of SO2 (a), O2 (b), and SO3 (c)
a = input('Enter the stoichiometric coefficient of SO2 (a): ');
b = input('Enter the stoichiometric coefficient of O2 (b): ');
c = input('Enter the stoichiometric coefficient of SO3 (c): ');
% 11. Calculate epsilon, ε=(c/a - b/a - 1) * ySO2_0
epsilon = (c/a - b/a - 1) * ySO2_0;
% 12. Calculate 100% number of moles, n = n0 * (1 + ε)
n = n0 * (1 + epsilon);
% 13. Calculate the reactor pressure, P = P0 * (n/n0) * (T/T0)
P = P0 * (n/n0) * (T/T0);
% 14. Calculate the initial concentration of SO2, CSO2_0 = ((P0 * ySO2_0)/(0.08206 * T0))
CSO2_0 = (P0 * ySO2_0) / (0.08206 * T0);
% 15. Calculate O2 molar ratio, thetaO2 = yO2_0 / ySO2_0
thetaO2 = yO2_0 / ySO2_0;
% 16. Calculate SO3 molar ratio, thetaSO3 = ySO3_0 / ySO2_0
thetaSO3 = ySO3_0 / ySO2_0;
% Define symbolic variable X for conversion
syms X
% 17. Express the partial pressure of SO2 as f(X)
PSO2_sym = CSO2_0 * ((1 - X) / (1 + epsilon * X)) * (P/P0) * (T0/T) * 0.08206 * T;
% 18. Express the partial pressure of O2 as f(X)
PO2_sym = CSO2_0 * ((thetaO2 - (b/a) * X) / (1 + epsilon * X)) * (P/P0) * (T0/T) * 0.08206 * T;
% 19. Express the partial pressure of SO3 as f(X)
PSO3_sym = CSO2_0 * ((thetaSO3 + (c/a) * X) / (1 + epsilon * X)) * (P/P0) * (T0/T) * 0.08206 * T;
% Convert to numeric functions for numerical calculations
PSO2 = matlabFunction(PSO2_sym);
PO2 = matlabFunction(PO2_sym);
PSO3 = matlabFunction(PSO3_sym);
% 20. Given constants
k = 9.7;
K_O2 = 38.5;
K_SO3 = 42.5;
K_P = 930.0;
% 21. Prompt the user for the final conversion, Xf
Xf = input('Enter the final conversion Xf: ');
% 22. Express reaction rate, -rSO2
rSO2 = @(X) -k * (PSO2(X) .* sqrt(PO2(X)) - PSO3(X)/K_P) ./ (1 + sqrt(PO2(X)/K_O2) + PSO3(X)/K_SO3).^2;
```

**Figure 2.15** MATLAB code (**$SO_2$_Cat_Oxid.m**) for calculating the residence time or the weight of catalyst needed for a given conversion.

```
%% 23. Prompt the user to select a case for the reactor
disp('Select reactor type: 1 - Batch, 2 - CSTR, 3 - PBR');
case_option = input('Enter your choice: ');
switch case_option
    case 1 % Batch Reactor
% 24. Prompt the user to enter the weight of the batch catalyst, W
        W = input('Enter the weight of batch catalyst (g): ');
        tau = (n0 * ySO2_0) * integral(@(X) 1 ./ (-rSO2(X)), 0, Xf) / W; % Residence time in hours
        fprintf('Residence time (tau): %.4f hr\n', tau);
        fprintf('Residence time (tau): %.4f sec\n', tau*3600.0);
        fprintf('Weight of catalyst: %.4f grams\n', W);
        fprintf('Final conversion (Xf): %.4f\n', Xf);
    case 2 % CSTR
        disp ('The initial number of moles of A will be converted in molar flow rate of A per unit time ');
        FA0 = n0 * ySO2_0;
        W_CSTR = FA0 * Xf / (-rSO2(Xf));

        fprintf('Weight of catalyst (CSTR): %.4f grams\n', W_CSTR);
        fprintf('Final conversion (Xf): %.4f\n', Xf);
    case 3 % PBR
        disp ('The initial number of moles of A will be converted in molar flow rate of A per unit time ');
        FA0 = n0 * ySO2_0;
        W_PBR = FA0 * integral(@(X) 1 ./ (-rSO2(X)), 0, Xf);
        fprintf('Weight of catalyst (PBR): %.4f grams\n', W_PBR);
        fprintf('Final conversion (Xf): %.4f\n', Xf);
    otherwise
        disp('Invalid selection');
end
```

**Figure 2.15** *(Continued)*

A demonstration is shown below showing the user-input parameters and the results for three types of reactors.

```
>> SO2_Cat_Oxid
Enter the initial temperature (K): 300
Enter the reactor temperature (K): 673
Enter the initial pressure (atm): 14.804
Enter the initial moles: 1000
Enter the initial mole fraction of SO2: 0.2958
Enter the initial mole fraction of O2: 0.1479
Enter the initial mole fraction of SO3: 0
Enter the stoichiometric coefficient of SO2 (a): 1
Enter the stoichiometric coefficient of O2 (b): 0.5
Enter the stoichiometric coefficient of SO3 (c): 1
Enter the final conversion Xf: 0.99
```

*(Continued)*

**(Continued)**

```
Select reactor type: 1 - Batch, 2 - CSTR, 3 - PBR
Enter your choice: 1
Enter the weight of batch catalyst (g): 100
Residence time (tau): 0.5122 hr
Residence time (tau): 1844.0364 sec
Weight of catalyst: 100.0000 grams
Final conversion (Xf): 0.9900

Select reactor type: 1 - Batch, 2 - CSTR, 3 - PBR
Enter your choice: 2
The initial number of moles of A will be converted in molar flow rate of A per unit time
Weight of catalyst (CSTR): 4278.5028 grams
Final conversion (Xf): 0.9900

Select reactor type: 1 - Batch, 2 - CSTR, 3 - PBR
Enter your choice: 3
The initial number of moles of A will be converted in molar flow rate of A per unit time
Weight of catalyst (PBR): 51.2232 grams
Final conversion (Xf): 0.9900
```

Notice that in terms of performance, expressed in catalyst weight requirement, PBR ranks number one, followed by BR and then by CSTR.

## 2.12 CSTRs in Series Configuration

When designing chemical reactor systems, using multiple reactors in series is often more efficient than a single reactor. The two most common reactor types used in series are CSTR and PFR. Configurations with different numbers and types of these reactors can significantly impact conversion, selectivity, and overall process efficiency.

Multiple CSTRs can be connected in series, where the outlet of one reactor feeds directly into the next. The conversion increases progressively through each reactor. The series of a large number of CSTRs mimics the behavior of a PFR, with higher conversion than a single CSTR for the same total volume.

**Configuration Advantages:** Easy control and operation since each reactor operates at steady-state, and flexibility in adjusting residence time and reaction conditions in each reactor.

**Configuration Drawbacks:** It requires more space and more capital investment compared with a single large CSTR. It may need intermediate cooling or heating to maintain optimal reaction conditions.

In one CSTR, conversion may be low. Using N CSTRs in series increases conversion and approximates PFR behavior as $N \to \infty$:

$$X_N = 1 - \left(\frac{1}{1 + k\tau/N}\right)^N \tag{A}$$

As $N \to \infty, X_\infty \to 1 - e^{-k\tau}$.

So, in the context of N equal-volume CSTRs in series:

- Each individual CSTR has a volume of $V/N$
- Each has a residence time of $\tau/N$
- The sum of residence times across all reactors equals the total $\tau$

$$\tau = \tau_1 + \tau_2 + \cdots + \tau_N = N\left(\frac{\tau}{N}\right) \quad \text{(B)}$$

## 2.13 Examples of Reactions Using CSTRs in Series

The use of multiple CSTRs in series is common in industrial reaction engineering when a single stirred tank cannot achieve the desired conversion or control process conditions effectively. By dividing the total reactor volume into a series of smaller CSTRs, the system provides a gradually decreasing reactant concentration profile, improving overall conversion and approaching the performance of a PFR as the number of reactors increases. This configuration is particularly beneficial for first-order irreversible reactions, highly exothermic systems requiring staged temperature control, and multiphase or complex reactions such as hydrogenation, polymerization, fermentation, and neutralization processes. In many practical applications, such as hydrogenation of vegetable oils, wastewater treatment, and biochemical reactors, multiple CSTRs allow better mixing, safer heat management, greater operational flexibility, and higher product quality compared with a single tank or a pure plug-flow configuration.

### 2.13.1 First-Order Irreversible Reaction ($A \to B$)

Typical examples:

- Hydrolysis reactions
- Thermal decomposition reactions

### 2.13.2 Neutralization Reactions

Neutralization of sulfuric acid using sodium hydroxide:

$$H_2SO_4 + 2NaOH \to Na_2SO_4 + 2H_2O$$

Used in wastewater treatment:

- Multiple CSTRs ensure smooth pH control
- Each reactor reduces large pH gradients and prevents overshooting

### 2.13.3 Polymerization Reactions

Free-radical polymerization of styrene:

Styrene → Polystyrene

Why CSTRs in series?

- **Reactor 1:** low viscosity, rapid initiation
- **Reactor 2–3:** propagation as viscosity increases
- Allows better temperature control and prevents auto-acceleration (gel effect)

### 2.13.4 Gas–Liquid Reactions

Hydrogenation of vegetable oils:

Vegetable oil + $H_2$ → Partially hydrogenated oil

Reasons for series CSTRs:

- Better mixing improves mass transfer
- Control of saturation level (butter versus margarine consistency)
- Allows catalyst addition/removal between stages

### 2.13.5 Biological/Biochemical Processes

#### 2.13.5.1 Fermentation

Substrate → Biomass + Product

Reasons for using multiple CSTR bioreactors:

- More stable cell growth phases
- Prevents oxygen starvation
- Used for ethanol, citric acid, antibiotics, and enzymes

### 2.13.6 Highly Exothermic Reactions

#### 2.13.6.1 Nitric Acid Absorption/Oxidation

$NO \rightarrow NO_2 \rightarrow HNO_3$

Multiple reactors are used to:

- Maintain safe temperatures
- Avoid thermal runaway
- Optimize conversion at each stage

## 2.14 PFRs in Series Configuration

Multiple PFRs can be placed in series, although this is less common since a single PFR can often achieve the desired conversion. Each PFR operates with a concentration gradient along its length, maintaining a high driving force for the reaction.

**Configuration Advantages:**
High conversion per unit volume, and it is suitable for reactions with favorable kinetics over the reactor's entire length.

**Configuration Drawbacks:**
It has less operational flexibility compared to CSTRs. It is challenging to maintain temperature control along the length of the reactor if the reaction is highly exothermic or endothermic.

## 2.15 Examples of Reactions Using PFRs in Series

### 2.15.1 First-Order Irreversible Reaction ($A \rightarrow B$)

Used frequently in:

- Thermal cracking of hydrocarbons
- Hydrogen peroxide decomposition
- Polymerization initiator decay

If conversion in a single long PFR causes undesirable temperature profiles or hot spots, multiple smaller PFRs in series are used, with cooling/heating between stages:

$$-r_A = kC_A$$

Advantages:

- Temperature can be adjusted between stages
- Better control of runaway kinetics (exothermic reactions)

### 2.15.2 Consecutive Reactions ($A \rightarrow B \rightarrow C$)

Nitration of toluene:

$$C_6H_5CH_3 \rightarrow \text{o-/p-nitrotoluene} \rightarrow \text{dinitrotoluene}$$

**Goal:** Maximize intermediate *B* (used in polyurethane intermediates, dyes, pharmaceuticals, and agrochemicals), minimize by-product *C*.

Using 2 or 3 PFRs:

- Stop the reaction early in the first stage
- Separate product *B*
- Continue reaction if needed

### 2.15.3 Parallel Reactions

$$A \rightarrow B(\text{desired})$$

$$A \rightarrow C(\text{undesired})$$

Oxidation of ethylene into ethylene oxide:

$$A\text{: } CH_2 = CH_2 + \frac{1}{2}O_2 \rightarrow B\text{: } CH_2\text{-O-}CH_2 \text{ Ethylene Oxide (desired)}$$

$$A\text{: } CH_2 = CH_2 + 3O_2 \rightarrow C\text{: } 2CO_2 + 2H_2O \text{ (undesired, complete combustion)}$$

The selectivity toward product *B* versus *C* is defined as:

$$S_{B/C} = \frac{r_B}{r_C} = \frac{k_B C_A^{\alpha}}{k_C C_A^{\beta}} = \frac{k_B}{k_C} C_A^{(\alpha-\beta)} \quad \alpha > \beta \tag{C}$$

Notice that ethylene oxide is a high-value intermediate used for ethylene glycol, PET plastics, and detergent production.

PFRs maintain a high concentration of reactant $A$ at the reactor inlet along the reactor length, improving selectivity toward the desired product. When used in series, conversion can be increased stepwise while keeping $A$ high in each stage, maximizing selectivity. PFRs also enhance safety and thermal stability in strongly exothermic systems (e.g., oxidation).

### 2.15.4 Reversible Reactions Approaching Equilibrium

Esterification or hydrogenation:

$$A + B \rightleftharpoons C$$

As conversion approaches equilibrium, the reaction rate slows down as $X_{eq}$ is approached gradually along the reactor length. On the other hand, using 2–3 PFRs in series with:

- Product removal (using membranes)
- Interstage heating/cooling
- Recycle reactant

will improve overall conversion significantly without a single very long reactor.

### 2.15.5 Gas-Phase Oxidation Reactions

$SO_2$ oxidation:

$$2SO_2 + O_2 \rightarrow 2SO_3$$

is highly exothermic; hence, the catalyst deactivates if overheated.

In industry:

- 4 catalytic PFRs in series
- Interstage heat removal
- Temperature optimization for high conversion and catalyst protection

## 2.16 Hybrid Configurations (CSTRs and PFRs in Series)

CSTRs and PFRs can be combined in various sequences to optimize reactor performance, particularly when balancing conversion, selectivity, heat management, and safety considerations.

A CSTR is placed first when:

- Heat removal and safety control are the primary concerns
- The reaction is highly exothermic and prone to thermal runaway
- The rate of an undesired side reaction increases strongly at high reactant concentration
- Early dilution of reactant A is needed to suppress undesired pathways

A CSTR maintains a relatively low concentration of reactant $A$ because of complete mixing. This minimizes the formation of undesired products, especially in parallel reactions $A$ the undesired pathway has a higher reaction order (i.e., $\beta > \alpha$; see Eq. (C) in Section 2.15C). It also enables uniform temperature control and safe dissipation of heat.

On the other hand, a PFR is placed first when:

- Rapid initial conversion is desirable

- The desired reaction pathway benefits from high reactant concentration
- Selectivity improves when the concentration of $A$ is initially high
- It is advantageous to shift the equilibrium quickly in the forward direction before slower finishing steps

A PFR maintains a high concentration of $A$ along the reactor inlet region, maximizing the desired reaction rate and improving selectivity toward the primary product. A CSTR may be placed second when the remaining conversion required is modest, and the objective is to provide additional residence time under controlled and uniform operating conditions, or when stabilization of composition, equilibrium approach, or temperature management outweighs volume efficiency. However, if the reaction becomes extremely slow at low reactant concentration, using a second PFR (or multiple PFRs in series) is typically more economical and effective than employing a large CSTR, which would require excessive volume to achieve the same incremental conversion.

## 2.17 Examples of Hybrid Reactor Configurations

### 2.17.1 CSTR Followed by PFR (CSTR → PFR)

Liquid-phase esterification:

$$\text{Acetic acid} + \text{ethanol} \rightleftharpoons \text{ethyl acetate} + \text{water}$$

- Reaction is reversible and equilibrium-limited
- CSTR provides good mixing and temperature control
- PFR pushes the reaction further toward equilibrium due to a higher concentration gradient and residence-time distribution

  **CSTR role:** Quickly consumes reactants from a high initial concentration and provides good heat removal.

  **PFR Role:** Higher conversion beyond equilibrium of the first tank and uses the remaining driving force.

### 2.17.2 CSTRs Followed by a PFR in Polymerization

Free-radical polymerization of styrene:

$$\text{Styrene} \rightarrow \text{Polystyrene}$$

- First two reactors are CSTRs: control temperature during initiation
- Final reactor is PFR: handles high viscosity and allows completion

#### 2.17.2.1 Reason

- CSTR prevents gel effect and hot spots.
- PFR provides uniform residence time for high-conversion finishing.

### 2.17.3 PFR Followed by CSTR (PFR → CSTR)

Consecutive Reaction $A \rightarrow B \rightarrow C$:

$$\text{Toluene} \rightarrow \text{Mononitrotoluene} \rightarrow \text{Dinitrotoluene}$$

**Goal:** Maximize intermediate *B*.
**PFR Role:** Maximizes selectivity by rapid conversion of $A \rightarrow B$, without back-mixing.
**CSTR Role:** Prevents long tail time distribution; stops overreaction to *C*.

### 2.17.4 Hydrogenation of Vegetable Oils

$$\text{Unsaturated oil} + H_2 \rightarrow \text{Partially hydrogenated fat}$$

**Configuration:**

- CSTR for mixing with catalyst slurry
- PFR polishing reactor after catalyst removal

**Benefits:**

- Good contacting + adjustable selectivity
- PFR finishing step to reach the precise iodine value

### 2.17.5 Biological Fermentation + PFR Bioreactor

$$\text{Sugar} \rightarrow \text{Ethanol} + CO_2$$

- CSTR fermentation tanks maintain microbial growth control
- PFR maturation reactor allows completion of the slow conversion

Table 2.1 shows the general guidelines of hybrid reactor configuration for some selected scenarios.

**Table 2.1** Hybrid reactor configuration guidelines for some selected scenarios.

| Scenario | Conditions | Recommended Configuration |
|---|---|---|
| 1 | Highly exothermic reaction, strong temperature sensitivity | **CSTR–PFR:** CSTR for heat control + PFR for conversion |
| 2 | Exothermic with high conversion required | CSTR–PFR |
| 3 | Reversible reaction with equilibrium limitations | **PFR–CSTR:** PFR for rapid conversion + CSTR for equilibrium approach |
| 4 | Equilibrium-limited | PFR–Multiple CSTRs |
| 5 | Complex network, intermediate desired | Multiple CSTRs |
| 6 | Reaction order changes with conversion | High order → PFR, Low order → CSTR (arranged by reaction order profile) |
| 7 | Series reactions with competing side reactions | PFR–CSTR |
| 8 | High viscosity mixture | Multiple CSTRs (+optional PFR) |

### 2.17.6 Optimizing Reactor Configurations

The goal is to maximize conversion, selectivity, and yield while minimizing production cost, energy consumption, and reactor volume. Key optimization considerations include:

1) **Reaction Kinetics:** Fast reactions that are not severely exothermic may benefit from a PFR due to a high driving force at the reactor inlet, whereas reactions prone to thermal runaway

often require an initial CSTR for safe temperature control. For reversible, equilibrium-limited reactions, a PFR is commonly placed first to rapidly achieve high conversion before equilibrium slows the rate.

2) **Temperature Control:** Highly exothermic reactions frequently require multiple reactors in series (PFRs or CSTRs) with intermediate cooling stages to maintain safe operation and optimal reaction rate.
3) **Selectivity:** Reactor configuration strongly influences product selectivity in complex reaction networks. A PFR may favor desired products in series reactions by quickly forming intermediates, while a CSTR can suppress undesired parallel reactions when the undesired pathway exhibits a higher reaction order.
4) **Capital and Operating Costs:** The optimal configuration balances the cost of reactor volume, heat-exchange utilities, mixing requirements, and control complexity with the desired conversion and selectivity.

### 2.17.7 General Guidelines for Optimization

1) **CSTRs** are ideal when precise temperature control, efficient heat removal, and rapid mixing are required. They are particularly useful for highly exothermic, hazardous, or complex multiphase reactions, where preventing temperature gradients and hot spots is essential. CSTRs are also advantageous when maintaining uniform composition improves process stability or catalyst performance.
2) **PFRs** are preferred when taking advantage of high reactant concentrations, and maintaining concentration gradients drives higher reaction rates and conversion. They are especially well-suited for fast reactions, reactions with simple kinetics, or cases where approaching equilibrium along the reactor length is beneficial. PFRs typically achieve higher conversion per unit volume than CSTRs under the same operating conditions.
3) **Hybrid Configurations** (CSTRs and PFRs in series) become valuable when dealing with multiple intermediate products, varying reaction orders, competing pathways, or reactions requiring staged temperature control. Combining reactor types enables engineers to balance high initial reaction rates, selectivity control, and final conversion efficiency in ways that a single reactor type cannot achieve alone.
4) **Economic Analysis** is essential to determine the most cost-effective configuration. Optimization requires comparing capital costs (reactor volume, materials, heat exchangers, and instrumentation) and operational costs (energy, maintenance, and utilities) against expected performance improvements in conversion, selectivity, quality, and productivity. The optimal design minimizes total cost while meeting safety and performance targets.

## 2.18 Design Equations for Multiple Reactors

### 2.18.1 CSTRs in Series

We can modify Eq. (2.18a) to account for having more than one CSTR in series. Let us say we have three CSTRs where the first reactor operates from $X = 0$ to $X_1$, the second from $X = X_1$ to $X = X_2$, and the third from $X = X_2$ to $X = X_3$. We have the following three design equations:

$$V_1 = \frac{F_{A0} - F_{A1}}{-r_{A1}} = \frac{F_{A0} - (F_{A0} - F_{A0}X_1)}{-r_{A1}} = \frac{F_{A0}X_1}{(-r_A)|_{X=X_1}} \tag{2.38}$$

$$V_2 = \frac{F_{A1} - F_{A2}}{-r_{A2}} = \frac{(F_{A0} - F_{A0}X_1) - (F_{A0} - F_{A0}X_2)}{-r_{A2}} = \frac{F_{A0}(X_2 - X_1)}{(-r_A)|_{X=X_2}} \tag{2.39}$$

$$V_3 = \frac{F_{A2} - F_{A3}}{-r_{A3}} = \frac{(F_{A0} - F_{A0}X_2) - (F_{A0} - F_{A0}X_3)}{-r_{A3}} = \frac{F_{A0}(X_3 - X_2)}{(-r_A)|_{X=X_3}} \tag{2.40}$$

### 2.18.2 PFRs in Series

We can modify Eq. (2.24a) to account for having more than one PFR in series. Let us say we have three PFRs, where the first reactor operates from $X = 0$ to $X_1$, the second from $X = X_1$ to $X = X_2$, and the third from $X = X_2$ to $X = X_3$. We have the following three design equations:

$$V_1 = F_{A0} \int_0^{X_1} \frac{dX}{-r_A} \tag{2.41}$$

$$V_2 = F_{A0} \int_{X_1}^{X_2} \frac{dX}{-r_A} \tag{2.42}$$

$$V_3 = F_{A0} \int_{X_2}^{X_3} \frac{dX}{-r_A} \tag{2.43}$$

### 2.18.3 Hybrid Reactors

#### 2.18.3.1 PFR + CSTR + PFR

Let us say we have three reactors: PFR + CSTR + PFR, where the first reactor operates from $X = 0$ to $X_1$, the second from $X = X_1$ to $X = X_2$, and the third from $X = X_2$ to $X = X_3$. We have the following three design equations:

$$V_1 = F_{A0} \int_0^{X_1} \frac{dX}{-r_A} \tag{2.44}$$

$$V_2 = \frac{F_{A1} - F_{A2}}{-r_{A2}} = \frac{(F_{A0} - F_{A0}X_1) - (F_{A0} - F_{A0}X_2)}{-r_{A2}} = \frac{F_{A0}(X_2 - X_1)}{(-r_A)|_{X=X_2}} \tag{2.45}$$

$$V_3 = F_{A0} \int_{X_2}^{X_3} \frac{dX}{-r_A} \tag{2.46}$$

#### 2.18.3.2 CSTR + PFR + CSTR

Let us say we have three reactors: CSTR + PFR + CSTR, where the first reactor operates from $X = 0$ to $X_1$, the second from $X = X_1$ to $X = X_2$, and the third from $X = X_2$ to $X = X_3$. We have the following three design equations:

$$V_1 = \frac{F_{A0} - F_{A1}}{-r_{A1}} = \frac{F_{A0} - (F_{A0} - F_{A0}X_1)}{-r_{A1}} = \frac{F_{A0}X_1}{(-r_A)|_{X=X_1}} \tag{2.47}$$

$$V_2 = F_{A0} \int_{X_1}^{X_2} \frac{dX}{-r_A} \tag{2.48}$$

$$V_3 = \frac{F_{A2} - F_{A3}}{-r_{A3}} = \frac{(F_{A0} - F_{A0}X_2) - (F_{A0} - F_{A0}X_3)}{-r_{A3}} = \frac{F_{A0}(X_3 - X_2)}{(-r_A)|_{X=X_3}} \tag{2.49}$$

Notice that Eqs. (2.38)–(2.49) can be used to find the volume requirement for each reactor in series for a given $X$, or vice versa, the conversion can be calculated for a given volume of the reactor in series. The optimization strategy is to minimize the total volume/residence-time requirement for a given final $X$ (i.e., constraint).

## 2.19 MATLAB-Based Liquid-Phase First-Order Reaction Using Different Reactor Cascades

Consider the following liquid-phase reaction: $A \xrightarrow{k} B$

The rate constant is 0.2 min$^{-1}$, and it is first-order with respect to $A$ ($-r_A = kC_A = 0.2\ \text{min}^{-1} \times C_A$).

Calculate the time needed to achieve $X_3 = 99\%$ conversion using the following scheme of reactors:

1) Three CSTRs with $X_1 = 0.33$ and $X_2 = 0.67$.
2) Three PFRs with $X_1 = 0.33$ and $X_2 = 0.67$.
3) PFR + CSTR + PFR with $X_1 = 0.33$ and $X_2 = 0.67$.
4) CSTR + PFR + CSTR with $X_1 = 0.33$ and $X_2 = 0.67$.

**Solution**

We will use MATLAB to carry out calculations for each step.

**MATLAB Algorithm:**

1) Prompt the user to select one out of four scenarios:

**Case 1:** CSTR1 + CSTR2 + CSTR3 with $X_1 = 0.33$ and $X_2 = 0.67$.

$$\tau_{CSTR1} = \frac{X_1}{k(1 - X_1)}$$

$$\tau_{CSTR2} = \frac{(X_2 - X_1)}{k(1 - X_2)}$$

$$\tau_{CSTR3} = \frac{(X_3 - X_2)}{k(1 - X_3)}$$

$$\tau_{total1} = \tau_{CSTR1} + \tau_{CSTR2} + \tau_{CSTR3}$$

**Case 2:** PFR1 + PFR2 + PFR3 with $X_1 = 0.33$ and $X_2 = 0.67$.

$$\tau_{PFR1} = \frac{1}{k}\int_0^{X_1} \frac{dX}{(1 - X)}$$

$$\tau_{PFR2} = \frac{1}{k}\int_{X_1}^{X_2} \frac{dX}{(1 - X)}$$

$$\tau_{PFR3} = \frac{1}{k}\int_{X_2}^{X_3} \frac{dX}{(1 - X)}$$

$$\tau_{total2} = \tau_{PFR1} + \tau_{PFR2} + \tau_{PFR3}$$

**Case 3:** PFR1 + CSTR2 + PFR3 with $X_1 = 0.33$ and $X_2 = 0.67$.

$$\tau_{PFR1} = \frac{1}{k}\int_0^{X_1} \frac{dX}{(1 - X)}$$

$$\tau_{CSTR2} = \frac{(X_2 - X_1)}{k(1 - X_2)}$$

$$\tau_{PFR3} = \frac{1}{k}\int_{X_2}^{X_3} \frac{dX}{(1 - X)}$$

$$\tau_{total3} = \tau_{PFR1} + \tau_{CSTR2} + \tau_{PFR3}$$

**Case 4:** CSTR1 + PFR2 + CSTR3 with $X_1 = 0.33$ and $X_2 = 0.67$.

$$\tau_{CSTR1} = \frac{X_1}{k(1 - X_1)}$$

$$\tau_{PFR2} = \frac{1}{k}\int_{X_1}^{X_2} \frac{dX}{(1 - X)}$$

$$\tau_{CSTR3} = \frac{(X_3 - X_2)}{k(1 - X_3)}$$

$$\tau_{total4} = \tau_{CSTR1} + \tau_{PFR2} + \tau_{CSTR3}$$

2) Carry out either numerical or symbolic integration for any integration appearing in previous cases.
3) Report the numerical value for $\tau_{total1}$, $\tau_{total2}$, $\tau_{total3}$, and $\tau_{total4}$ for Case 1, Case 2, Case 3, and Case 4, respectively.
   Figure 2.16 shows MATLAB code (**SeriesReactors.m**) for calculating the total residence-time requirement for each of the four cases. The user may, however, modify the parameters: $k$, $X_1$, $X_2$, and $X_3$ prior to finding the total time requirement of the three serial reactors.

```
clear variables;
% MATLAB Code for Calculating τ_total based on the selected scenario
% Prompt the user to select a case
disp('Select one of the following scenarios:');
disp('1: CSTR1 + CSTR2 + CSTR3');
disp('2: PFR1 + PFR2 + PFR3');
disp('3: PFR1 + CSTR2 + PFR3');
disp('4: CSTR1 + PFR2 + CSTR3');
choice = input('Enter the number corresponding to your choice: ');
% Define constants
X1 = 0.33; %Modify as needed.
X2 = 0.67; %Modify as needed.
X3 = 0.99; %Modify as needed.
k = 0.1; % min^-1. Modify as needed.
% Initialize t_total
t_total = 0;
switch choice
   case 1 % Case 1: CSTR1 + CSTR2 + CSTR3
      t_CSTR1 = X1 / (k * (1 - X1));
      t_CSTR2 = (X2 - X1) / (k * (1 - X2));
      t_CSTR3 = (X3 - X2) / (k * (1 - X3));
      t_total = t_CSTR1 + t_CSTR2 + t_CSTR3;
```

**Figure 2.16** MATLAB code (**SeriesReactors.m**) for calculating the time requirement for the three cascade reactors made of CSTR and/or PFR.

```
  case 2 % Case 2: PFR1 + PFR2 + PFR3
    t_PFR1 = integral(@(X) 1./(k*(1-X)), 0, X1);
    t_PFR2 = integral(@(X) 1./(k*(1-X)), X1, X2);
    t_PFR3 = integral(@(X) 1./(k*(1-X)), X2, X3);
    t_total = t_PFR1 + t_PFR2 + t_PFR3;

  case 3 % Case 3: PFR1 + CSTR2 + PFR3
    t_PFR1 = integral(@(X) 1./(k*(1-X)), 0, X1);
    t_CSTR2 = (X2 - X1) / (k * (1 - X2));
    t_PFR3 = integral(@(X) 1./(k*(1-X)), X2, X3);
    t_total = t_PFR1 + t_CSTR2 + t_PFR3;

  case 4 % Case 4: CSTR1 + PFR2 + CSTR3
    t_CSTR1 = X1 / (k * (1 - X1));
    t_PFR2 = integral(@(X) 1./(k*(1-X)), X1, X2);
    t_CSTR3 = (X3 - X2) / (k * (1 - X3));
    t_total = t_CSTR1 + t_PFR2 + t_CSTR3;

  otherwise
    disp('Invalid selection. Please choose a number between 1 and 4.');
end
% Display the result
if t_total > 0
fprintf('t_total for Case %d is: %.4f minutes\n', choice, t_total);
end
```

**Figure 2.16** *(Continued)*

Results are copied from the Command Window.

```
>> SeriesReactors
Select one of the following scenarios:
1: CSTR1 + CSTR2 + CSTR3
2: PFR1 + PFR2 + PFR3
3: PFR1 + CSTR2 + PFR3
4: CSTR1 + PFR2 + CSTR3
Enter the number corresponding to your choice: 1
t_total for Case 1 is: 335.2284 minutes
Enter the number corresponding to your choice: 2
t_total for Case 2 is: 46.0517 minutes
Enter the number corresponding to your choice: 3
t_total for Case 3 is: 49.2729 minutes
Enter the number corresponding to your choice: 4
t_total for Case 4 is: 332.0072 minutes
```

Notice that the best scenario of the reactors scheme, in terms of minimum total residence-time requirement, is Case 2, where we have 3 PFRs in series, followed by Case 3, which is quite comparable with Case 2. Cases 1 and 4 are the worst.

## 2.20 MATLAB-Based Liquid-Phase $n^{th}$-Order Reaction Using Different Reactor Cascades

Consider the following liquid-phase reaction:

$$A + \frac{b}{a}B \xrightarrow{k} \frac{c}{a}C + \frac{d}{a}D \rightarrow A + B \xrightarrow{k} C + D \tag{2.50}$$

$$\text{The reaction rate is: } \left(-r_A = kC_A^{\alpha}C_B^{\beta} = 0.01\left(s^{-1} \times \left(\frac{\text{L}}{\text{mol}}\right)^{\alpha+\beta-1}\right) C_A^{1.1}C_B^{1.2}\right) \tag{2.51}$$

Calculate the total volume requirement needed to achieve the final conversion for a given set of parameters: $k$, $\alpha$, $\beta$, $b$, $a$, $X_1, X_2, X_3, F_{A0}$, $C_{A0}$, and $\theta_{B0}$, using the following scheme of reactors:

1) Three CSTRs
2) Three PFRs
3) PFR + CSTR + PFR
4) CSTR + PFR + CSTR

**Solution**

From Eq. (1.28a), we have:

$$C_B = C_{B0} - \left[\left(\frac{b}{a}\right)(C_{A0} - C_A)\right] = C_{A0}\left[\theta_{B0} - \left[\left(\frac{b}{a}\right)\left(\frac{C_{A0} - C_A}{C_{A0}}\right)\right]\right] = C_{A0}\left[\theta_{B0} - \left[\left(\frac{b}{a}\right)(X_A)\right]\right] \tag{2.52}$$

where $\theta_{B0} = \frac{F_{B0}}{F_{A0}}$

Or for a constant volumetric flow rate:

$$\theta_{B0} = \frac{C_{B0}}{C_{A0}}$$

We will use MATLAB to carry out calculations for each step.

**MATLAB Algorithm:**

1) Prompt the user to enter the values of $k$, $\alpha$, $\beta$, $b$, $a$, $X_1, X_2, X_3, F_{A0}$, $C_{A0}$, and $\theta_{B0}$.
2) Prompt the user to select one out of four scenarios:

   **Case 1:** CSTR1 + CSTR2 + CSTR3

$$V_{CSTR1} = \frac{F_{A0}X_1}{k[C_{A0}(1 - X_1)]^{\alpha}\left[C_{A0}\left[\theta_{B0} - \left[\left(\frac{b}{a}\right)(X_1)\right]\right]\right]^{\beta}} \tag{2.53}$$

$$V_{CSTR2} = \frac{F_{A0}(X_2 - X_1)}{k[C_{A0}(1 - X_2)]^{\alpha}\left[C_{A0}\left[\theta_{B0} - \left[\left(\frac{b}{a}\right)(X_2)\right]\right]\right]^{\beta}} \tag{2.54}$$

$$V_{CSTR3} = \frac{F_{A0}(X_3 - X_2)}{k[C_{A0}(1 - X_3)]^{\alpha}\left[C_{A0}\left[\theta_{B0} - \left[\left(\frac{b}{a}\right)(X_3)\right]\right]\right]^{\beta}} \tag{2.55}$$

$$V_{total} = V_{CSTR1} + V_{CSTR2} + V_{CSTR3} \tag{2.56}$$

**Case 2:** PFR1 + PFR2 + PFR3

$$V_{PFR1} = F_{A0}\int_{0}^{X_1} \frac{dX}{k[C_{A0}(1-X)]^{\alpha}\left[C_{A0}\left[\theta_{B0} - \left[\left(\frac{b}{a}\right)(X)\right]\right]\right]^{\beta}} \tag{2.57}$$

$$V_{PFR2} = F_{A0}\int_{X_1}^{X_2} \frac{dX}{k[C_{A0}(1-X)]^{\alpha}\left[C_{A0}\left[\theta_{B0} - \left[\left(\frac{b}{a}\right)(X)\right]\right]\right]^{\beta}} \tag{2.58}$$

$$V_{PFR3} = F_{A0}\int_{X_2}^{X_3} \frac{dX}{k[C_{A0}(1-X)]^{\alpha}\left[C_{A0}\left[\theta_{B0} - \left[\left(\frac{b}{a}\right)(X)\right]\right]\right]^{\beta}} \tag{2.59}$$

$$V_{total} = V_{PFR1} + V_{PFR2} + V_{PFR3} \tag{2.60}$$

**Case 3:** PFR + CSTR + PFR

$$V_{PFR1} = F_{A0}\int_{0}^{X_1} \frac{dX}{k[C_{A0}(1-X)]^{\alpha}\left[C_{A0}\left[\theta_{B0} - \left[\left(\frac{b}{a}\right)(X)\right]\right]\right]^{\beta}} \tag{2.61}$$

$$V_{CSTR2} = \frac{F_{A0}(X_2 - X_1)}{k[C_{A0}(1-X_2)]^{\alpha}\left[C_{A0}\left[\theta_{B0} - \left[\left(\frac{b}{a}\right)(X_2)\right]\right]\right]^{\beta}} \tag{2.62}$$

$$V_{PFR3} = F_{A0}\int_{X_2}^{X_3} \frac{dX}{k[C_{A0}(1-X)]^{\alpha}\left[C_{A0}\left[\theta_{B0} - \left[\left(\frac{b}{a}\right)(X)\right]\right]\right]^{\beta}} \tag{2.63}$$

$$V_{total} = V_{PFR1} + V_{CSTR2} + V_{PFR3} \tag{2.64}$$

**Case 4:** CSTR + PFR + CSTR

$$V_{CSTR1} = \frac{F_{A0}X_1}{k[C_{A0}(1-X_1)]^{\alpha}\left[C_{A0}\left[\theta_{B0} - \left[\left(\frac{b}{a}\right)(X_1)\right]\right]\right]^{\beta}} \tag{2.65}$$

$$V_{PFR2} = F_{A0}\int_{X_1}^{X_2} \frac{dX}{k[C_{A0}(1-X)]^{\alpha}\left[C_{A0}\left[\theta_{B0} - \left[\left(\frac{b}{a}\right)(X)\right]\right]\right]^{\beta}} \tag{2.66}$$

$$V_{CSTR3} = \frac{F_{A0}(X_3 - X_2)}{k[C_{A0}(1-X_3)]^{\alpha}\left[C_{A0}\left[\theta_{B0} - \left[\left(\frac{b}{a}\right)(X_3)\right]\right]\right]^{\beta}} \tag{2.67}$$

$$V_{total} = V_{CSTR1} + V_{PFR2} + V_{CSTR3} \tag{2.68}$$

3) Carry out either numerical or symbolic integration for any integration appearing in previous cases.
4) Report the numerical value for $V_{total}$ for the selected case out of Case 1, Case 2, Case 3, or Case 4.

Figure 2.17 shows MATLAB code (**SeriesReactors_N_Order.m**) for calculating the total volume requirement for each of the four cases. The user may, however, modify the parameter values of $k$, $\alpha$, $\beta$, $b$, $a$, $X_1$, $X_2$, $X_3$, $F_{A0}$, $C_{A0}$, and $\theta_{B0}$, prior to finding the total volume of the three serial reactors.

```
clear variables;
% MATLAB Code for Calculating V_total based on the selected scenario
% Step 1: Prompt the user to enter the required input values
k = input('Enter the value of the rate constant k:, s^(-1)×(L/mol)^(α+β-1) ');
alpha = input('Enter the value of α, reaction order with respect to A: ');
beta = input('Enter the value of β, reaction order with respect to B: ');
a = input('Enter the stoichiometric coefficient a: ');
b = input('Enter the stoichiometric coefficient b: ');
X1 = input('Enter the conversion X1: ');
X2 = input('Enter the conversion X2: ');
X3 = input('Enter the conversion X3: ');
FA0 = input('Enter the molar flow rate of A (FA0), mol/s: ');
CA0 = input('Enter the initial concentration of A (C_A0): ');
theta_B0 = input('Enter the initial mole ratio θ_B0, CB0/CA0: ');
% Step 2: Prompt the user to select a scenario
disp('Select one of the following scenarios:');
disp('1: CSTR1 + CSTR2 + CSTR3');
disp('2: PFR1 + PFR2 + PFR3');
disp('3: PFR1 + CSTR2 + PFR3');
disp('4: CSTR1 + PFR2 + CSTR3');
choice = input('Enter the number corresponding to your choice: ');
% Initialize V_total
V_total = 0;
% Step 3: Calculate V_total based on the selected case
switch choice
   case 1  % Case 1: CSTR1 + CSTR2 + CSTR3
   V_CSTR1 = (FA0 * X1) / (k * (CA0 * (1 - X1))^alpha * (CA0 * (theta_B0 - (b/a) * X1))^beta);
   V_CSTR2 = (FA0 * (X2 - X1)) / (k * (CA0 * (1 - X2))^alpha * (CA0 * (theta_B0 - (b/a) * X2))^beta);
   V_CSTR3 = (FA0 * (X3 - X2)) / (k * (CA0 * (1 - X3))^alpha * (CA0 * (theta_B0 - (b/a) * X3))^beta);
   V_total = V_CSTR1 + V_CSTR2 + V_CSTR3;
   Tau=V_total*CA0/FA0;

   case 2  % Case 2: PFR1 + PFR2 + PFR3
   V_PFR1 = FA0 * integral(@(X) 1./(k*(CA0*(1-X)).^alpha .* (CA0 * (theta_B0 - (b/a) * X)).^beta), 0, X1);
   V_PFR2 = FA0 * integral(@(X) 1./(k*(CA0*(1-X)).^alpha .* (CA0 * (theta_B0 - (b/a) * X)).^beta), X1, X2);
   V_PFR3 = FA0 * integral(@(X) 1./(k*(CA0*(1-X)).^alpha .* (CA0 * (theta_B0 - (b/a) * X)).^beta), X2, X3);
   V_total = V_PFR1 + V_PFR2 + V_PFR3;
   Tau=V_total*CA0/FA0;

   case 3  % Case 3: PFR1 + CSTR2 + PFR3
   V_PFR1 = FA0 * integral(@(X) 1./(k*(CA0* (1-X)).^alpha .* (CA0*(theta_B0 - (b/a) * X)).^beta), 0, X1);
   V_CSTR2 = (FA0 * (X2 - X1)) / (k *(CA0 * (1 - X2))^alpha * (CA0 *(theta_B0 - (b/a) * X2))^beta);
   V_PFR3 = FA0 * integral(@(X) 1./(k*(CA0* (1-X)).^alpha .* (CA0*(theta_B0 - (b/a) * X)).^beta), X2, X3);
   V_total = V_PFR1 + V_CSTR2 + V_PFR3;
   Tau=V_total*CA0/FA0;

   case 4  % Case 4: CSTR1 + PFR2 + CSTR3
   V_CSTR1 = (FA0 * X1) / (k * (CA0 * (1 - X1))^alpha * (CA0 *(theta_B0 - (b/a) * X1))^beta);
   V_PFR2 = FA0 * integral(@(X) 1./(k* (CA0 * (1-X)).^alpha .*(CA0 *(theta_B0 - (b/a) * X)).^beta), X1, X2);
   V_CSTR3 = (FA0 * (X3 - X2)) / (k * (CA0 * (1 - X3))^alpha *(CA0 *(theta_B0 - (b/a) * X3))^beta);
   V_total = V_CSTR1 + V_PFR2 + V_CSTR3;
   Tau=V_total*CA0/FA0;

   otherwise
      disp('Invalid selection. Please choose a number between 1 and 4.');
end
% Step 4: Display the result
if V_total > 0
   fprintf('V_total for Case %d is: %.4f Liter\n', choice, V_total);
   fprintf('Tau for Case %d is: %.4f Seconds\n', choice, Tau);
   fprintf('τ_total for Case %d is: %.4f Hours\n', choice, Tau/3600.0);
end
```

**Figure 2.17** MATLAB (**SeriesReactors_N_Order.m**) code for calculating the volume requirement for the three cascade reactors made of CSTR and/or PFR for a liquid-phase $n^{th}$-order reaction.

Results are copied from the Command Window.

```
>> SeriesReactors_N_Order
Enter the value of α, reaction order with respect to A: 1.1
Enter the value of β, reaction order with respect to B: 1.2
Enter the value of the rate constant k,s^(-1)×(L/mol)^(α+β-1): 0.01
Enter the conversion X1: 0.33
Enter the conversion X2: 0.67
Enter the conversion X3: 0.99
Enter the molar flow rate of A (FA0), mol/s: 1
Enter the initial concentration of A (C_A0): 10
Enter the initial mole ratio of B to A, θ_B0: 1
Enter the stoichiometric coefficient a: 1
Enter the stoichiometric coefficient b: 1
Select one of the following scenarios:
1: CSTR1 + CSTR2 + CSTR3
2: PFR1 + PFR2 + PFR3
3: PFR1 + CSTR2 + PFR3
4: CSTR1 + PFR2 + CSTR3
Enter the number corresponding to your choice: 1
V_total for Case 1 is: 6387.4371 Liter
τ_total for Case 1 is: 63874.3709 Seconds
τ_total for Case 1 is: 17.7429 Hours
Enter the number corresponding to your choice: 2
V_total for Case 2 is: 153.0962 Liter
τ_total for Case 2 is: 1530.9619 Seconds
τ_total for Case 2 is: 0.4253 Hours
Enter the number corresponding to your choice: 3
V_total for Case 3 is: 154.2980 Liter
τ_total for Case 3 is: 1542.9802 Seconds
τ_total for Case 3 is: 0.4286 Hours
Enter the number corresponding to your choice: 4
V_total for Case 4 is: 6386.2353 Liter
τ_total for Case 4 is: 63862.3526 Seconds
τ_total for Case 4 is: 17.7395 Hours
```

Notice that the best scenario of reactors scheme, in terms of minimum total volume/time requirement, is Case 2, where we have 3 PFRs in series, followed immediately by Case 3, which is quite comparable with Case 2. Cases 1 and 4 are the worst.

**NOTE #2**

The best scenario case may change based on reaction order ($\alpha$ and $\beta$) and stoichiometry ($b$ and $a$), initial molar ratios and concentration ($F_{A0}$, $C_{A0}$, and $\theta_{B0}$), and selection of intermediate and final conversion values ($X_1$, $X_2$, and $X_3$). However, Case #2 will not be affected by the selection of intermediate and final conversion values ($X_1$, $X_2$, and $X_3$). See CHW 2.23 problem.

Let us show that the MATLAB code of Figure 2.17 can also be used for a chemical reaction that is expressed in terms of reactant A only.

$$aA \rightarrow Products \tag{2.69}$$

$$\text{Let us say that} -r_A = kC_A^2 \tag{2.70}$$

From Eq. (2.24a), we have:
At $V = 0$, $X = 0$. Thus:

$$V = F_{A0}\int_0^X \frac{dX}{-r_A} = F_{A0}\int_0^X \frac{dX}{kC_A^2} = \int_0^X \frac{dX}{k[C_{A0}(1-X)]^2} = F_{A0}\int_0^X \frac{dX}{k[C_{A0}(1-X)]^2} \tag{2.71}$$

Let us take $F_{A0} = 1\frac{\text{mol}}{\text{s}}$, $C_{A0} = 10\,\text{mol/L}$, $k = 0.01\,\text{L/(mol.s)}$, and $X = 0.99$.

Figure 2.18 shows MATLAB code (**Second_Order_A_PFR.m**) for finding the required volume of PFR for a second-order reaction in $A$.

```
% Given parameters
FA0 = 1;      % mol/s
CA0 = 10;     % mol/L
k = 0.01;     % L/(mol.s)
X = 0.99;     % Conversion
% Define the integrand as an anonymous function
integrand = @(X) 1./(k * (CA0 * (1-X)).^2);
% Perform the numerical integration using MATLAB's integral function
V = FA0 * integral(integrand, 0, X);
Tau=V*CA0/FA0;
% Display the result
fprintf('The value of V is: %.4f L\n', V);
fprintf('The value of τ_total is: %.4f sec\n', Tau);
fprintf('The value of τ_total is: %.4f hour\n', Tau/3600.0);
```

**Figure 2.18** MATLAB code for finding the PFR volume of a liquid-phase, second-order reaction in $A$.

The result is copied from the Command Window:

```
>> Second_Order_A_PFR
The value of V is: 99.0000 L
The value of τ_total is: 990.0000 sec
The value of τ_total is: 0.2750 hour
```

We can run the code shown in Figure 2.17 and use the second case, that is, three PFRs in series with a final conversion equal to 0.99. For such a case, the intermediate values $X_1$ and $X_2$ do not affect the final result, as the integration, overall, will be carried out from $X = 0$ to $X = 0.99$.

Copied from the Command Window, here are the results of running the code while showing the input parameters:

```
>> SeriesReactors_N_Order
Enter the value of the rate constant k: 0.01
Enter the value of α: 1
```

```
Enter the value of β: 1
Enter the conversion X1: 0.33
Enter the conversion X2: 0.67
Enter the conversion X3: 0.99
Enter the molar flow rate of A (FA0): 1
Enter the initial concentration of A (C_A0): 10
Enter the initial mole ratio θ_B0: 1
Enter the stoichiometric coefficient b: 1
Enter the stoichiometric coefficient a: 1
Select one of the following scenarios:
1: CSTR1 + CSTR2 + CSTR3
2: PFR1 + PFR2 + PFR3
3: PFR1 + CSTR2 + PFR3
4: CSTR1 + PFR2 + CSTR3
Enter the number corresponding to your choice: 2
V_total for Case 2 is: 99.0000 Liter
τ_total for Case 2 is: 990.0000 Seconds
τ_total for Case 2 is: 0.2750 Hours
```

So, the code of Figure 2.17 gives the same volume and residence time as those given by the code of Figure 2.18. This simply says that we can replace *B* by *A* and have only *A*. In effect, as far as the volume/time requirements are concerned, a reaction with a second-order in *A* is equivalent to having a reaction which is first-order with respect to *A* and also first-order with respect to *B*.

The same thing can be said: As far as the volume/time requirements are concerned, a reaction with 1.5th-order in *A* is equivalent to having a reaction which is first-order with respect to *A* and also half-order with respect to *B* for using the code in Figure 2.17. Of course, in such cases, the following parameters will be assumed to be unity: The initial mole ratio *θ_B0*, the stoichiometric coefficient *b*, and the stoichiometric coefficient *a*.

In conclusion, the MATLAB code in Figure 2.17 can be used for

$$\left(-r_A = kC_A^{\alpha}C_B^{\beta} = k\left(\text{min}^{-1} \times \left(\frac{\text{L}}{\text{mol}}\right)^{\alpha+\beta-1}\right)C_A^{\alpha}C_B^{\beta}\right) \tag{2.72}$$

and for

$$\left(-r_A = kC_A^{n} = k\left(\text{min}^{-1} \times \left(\frac{\text{L}}{\text{mol}}\right)^{n-1}\right)C_A^{n}\right) \tag{2.73}$$

With $n = \alpha + \beta$.

## 2.21 MATLAB-Based Gas-Phase $n^{th}$-Order Reaction Using Different Reactor Cascades

### 2.21.1 Gas-Phase Volume Expansion Effect

For the gas-phase reactions, we need to express $C_A$ in terms of conversion and consider the volume expansion effects due to changes in pressure, temperature, and change in number of moles. In this regard, Fogler's approach [1] will be adopted.

For a general gas-phase reaction:

$$A + \frac{b}{a}B \rightarrow \frac{c}{a}C + \frac{d}{a}D$$

$C_A$ becomes:

$$C_A = \frac{F_A}{v} = \frac{F_{A0}(1-X)}{v_0(1+\varepsilon X)}\frac{P}{P_0}\frac{T_0}{T} = C_{A0}\frac{(1-X)}{(1+\varepsilon X)}\frac{P}{P_0}\frac{T_0}{T} \tag{2.74}$$

and

$$C_j = C_{A0}\frac{(\theta_{j0} + v_j X)}{(1+\varepsilon X)}\frac{P}{P_0}\frac{T_0}{T} \tag{2.75}$$

where:

$$\varepsilon = \delta \times \frac{F_{A0}}{F_0} = \delta \times y_{A0} = \left(\frac{c}{a} + \frac{d}{a} - \frac{b}{a} - 1\right) y_{A0} \tag{2.76}$$

$v_j$: The stoichiometric coefficient associated with each reacting species.

For example, $C_B$, $C_C$, and $C_D$ become:

$$C_B = C_{A0}\frac{\left(\theta_{B0} - \frac{b}{a}X\right)}{(1+\varepsilon X)}\frac{P}{P_0}\frac{T_0}{T} \quad C_C = C_{A0}\frac{\left(\theta_{C0} + \frac{c}{a}X\right)}{(1+\varepsilon X)}\frac{P}{P_0}\frac{T_0}{T} \quad C_D = C_{A0}\frac{\left(\theta_{D0} + \frac{d}{a}X\right)}{(1+\varepsilon X)}\frac{P}{P_0}\frac{T_0}{T} \tag{2.77}$$

Consider the following gas-phase reaction: $A + B \xrightarrow{k} C + 2D$ (2.78)

The reaction rate is given by $\left(-r_A = kC_A^{\alpha}C_B^{\beta} = k\left(\mathrm{s}^{-1} \times \left(\frac{\mathrm{L}}{\mathrm{mol}}\right)^{\alpha+\beta-1}\right)C_A^{\alpha}C_B^{\beta}\right)$ (2.79)

Calculate the total volume/time requirement needed to achieve the final conversion for a given set of kinetic parameters, intermediate and final conversions, initial molar flow rates, fractions, and ratios; the stoichiometric coefficients: *a*, *b*, *c*, and *d*; and the feed and reactor temperature and pressure: $T_0$, $T$, $P_0$, and $P$.

Use the following scheme of reactors:

1) Three CSTRs
2) Three PFRs
3) PFR + CSTR + PFR
4) CSTR + PFR + CSTR

**Solution**

From Eq. (1.28a), we have: $C_B = C_{A0}\frac{\left(\theta_{B0} - \frac{b}{a}X\right)}{(1+\varepsilon X)}\frac{P}{P_0}\frac{T_0}{T}$; where $\theta_{B0} = \frac{F_{B0}}{F_{A0}}$

We will use MATLAB to carry out calculations for each step.

**MATLAB Algorithm:**

1) Prompt the user to enter the values of kinetic parameters: $k$, $\alpha$, and $\beta$.
2) Prompt the user to enter the intermediate and final conversions: $X_1$, $X_2$, and $X_3$.
3a) Prompt the user to enter the initial molar flow rate of A: $F_{A0}$.
3b) Prompt the user to enter the initial molar ratio of B over A: $\theta_{B0}$.
4a) Prompt the user to enter the initial mole fraction of A: *yA0*.
4b) Prompt the user to enter the initial mole fraction of B: *yB0*.
5) Prompt the user to enter the values of stoichiometric coefficients: *a*, *b*, *c*, and *d*.

6) Prompt the user to enter the values of feed and reactor temperature and pressure: *T0, T, P0, P,*
8) Calculate epsilon: $\varepsilon = \left(\frac{c}{a} + \frac{d}{a} - \frac{b}{a} - 1\right) y_{A0}$
9) Calculate the initial concentration of A: $C_{A0} = \left(\frac{P_0 y_{A0}}{0.08206 * T_0}\right)$

Prompt the user to select one out of four scenarios:

**Case 1:** CSTR1 + CSTR2 + CSTR3

$$V_{CSTR1} = \frac{F_{A0}X_1}{k\left[C_{A0}\frac{(1-X_1)}{(1+\varepsilon X_1)}\frac{P}{P_0}\frac{T_0}{T}\right]^{\alpha}\left[C_{A0}\frac{\left(\theta_{B0}-\frac{b}{a}X_1\right)}{(1+\varepsilon X_1)}\frac{P}{P_0}\frac{T_0}{T}\right]^{\beta}} \tag{2.80}$$

$$V_{CSTR2} = \frac{F_{A0}(X_2 - X_1)}{k\left[C_{A0}\frac{(1-X_2)}{(1+\varepsilon X_2)}\frac{P}{P_0}\frac{T_0}{T}\right]^{\alpha}\left[C_{A0}\frac{\left(\theta_{B0}-\frac{b}{a}X_2\right)}{(1+\varepsilon X_2)}\frac{P}{P_0}\frac{T_0}{T}\right]^{\beta}} \tag{2.81}$$

$$V_{CSTR3} = \frac{F_{A0}(X_3 - X_2)}{k\left[C_{A0}\frac{(1-X_3)}{(1+\varepsilon X_3)}\frac{P}{P_0}\frac{T_0}{T}\right]^{\alpha}\left[C_{A0}\frac{\left(\theta_{B0}-\frac{b}{a}X_3\right)}{(1+\varepsilon X_3)}\frac{P}{P_0}\frac{T_0}{T}\right]^{\beta}} \tag{2.82}$$

$$V_{total} = V_{CSTR1} + V_{CSTR2} + V_{CSTR3} \tag{2.83}$$

**Case 2:** PFR1 + PFR2 + PFR3

$$V_{PFR1} = F_{A0}\int_0^{X_1} \frac{dX}{k\left[C_{A0}\frac{(1-X)}{(1+\varepsilon X)}\frac{P}{P_0}\frac{T_0}{T}\right]^{\alpha}\left[C_{A0}\frac{\left(\theta_{B0}-\frac{b}{a}X\right)}{(1+\varepsilon X)}\frac{P}{P_0}\frac{T_0}{T}\right]^{\beta}} \tag{2.84}$$

$$V_{PFR2} = F_{A0}\int_{X_1}^{X_2} \frac{dX}{k\left[C_{A0}\frac{(1-X)}{(1+\varepsilon X)}\frac{P}{P_0}\frac{T_0}{T}\right]^{\alpha}\left[C_{A0}\frac{\left(\theta_{B0}-\frac{b}{a}X\right)}{(1+\varepsilon X)}\frac{P}{P_0}\frac{T_0}{T}\right]^{\beta}} \tag{2.85}$$

$$V_{PFR3} = F_{A0}\int_{X_2}^{X_3} \frac{dX}{k\left[C_{A0}\frac{(1-X)}{(1+\varepsilon X)}\frac{P}{P_0}\frac{T_0}{T}\right]^{\alpha}\left[C_{A0}\frac{\left(\theta_{B0}-\frac{b}{a}X\right)}{(1+\varepsilon X)}\frac{P}{P_0}\frac{T_0}{T}\right]^{\beta}} \tag{2.86}$$

$$V_{total} = V_{PFR1} + V_{PFR2} + V_{PFR3} \tag{2.87}$$

**Case 3:** PFR + CSTR + PFR

$$V_{PFR1} = F_{A0}\int_0^{X_1} \frac{dX}{k\left[C_{A0}\frac{(1-X)}{(1+\varepsilon X)}\frac{P}{P_0}\frac{T_0}{T}\right]^{\alpha}\left[C_{A0}\frac{\left(\theta_{B0}-\frac{b}{a}X\right)}{(1+\varepsilon X)}\frac{P}{P_0}\frac{T_0}{T}\right]^{\beta}} \tag{2.88}$$

$$V_{CSTR2} = \frac{F_{A0}(X_2 - X_1)}{k\left[C_{A0}\frac{(1-X_2)}{(1+\varepsilon X_2)}\frac{P}{P_0}\frac{T_0}{T}\right]^{\alpha}\left[C_{A0}\frac{\left(\theta_{B0}-\frac{b}{a}X_2\right)}{(1+\varepsilon X_2)}\frac{P}{P_0}\frac{T_0}{T}\right]^{\beta}} \tag{2.89}$$

$$V_{PFR3} = F_{A0}\int_{X_2}^{X_3} \frac{dX}{k\left[C_{A0}\frac{(1-X)}{(1+\varepsilon X)}\frac{P}{P_0}\frac{T_0}{T}\right]^{\alpha}\left[C_{A0}\frac{\left(\theta_{B0}-\frac{b}{a}X\right)}{(1+\varepsilon X)}\frac{P}{P_0}\frac{T_0}{T}\right]^{\beta}} \tag{2.90}$$

$$V_{total} = V_{PFR1} + V_{CSTR2} + V_{PFR3} \tag{2.91}$$

**Case 4:** CSTR + PFR + CSTR

$$V_{CSTR1} = \frac{F_{A0}X_1}{k\left[C_{A0}\frac{(1-X_1)}{(1+\varepsilon X_1)}\frac{P}{P_0}\frac{T_0}{T}\right]^{\alpha}\left[C_{A0}\frac{\left(\theta_{B0}-\frac{b}{a}X_1\right)}{(1+\varepsilon X_1)}\frac{P}{P_0}\frac{T_0}{T}\right]^{\beta}} \tag{2.92}$$

$$V_{PFR2} = F_{A0}\int_{X_1}^{X_2}\frac{dX}{k\left[C_{A0}\frac{(1-X)}{(1+\varepsilon X)}\frac{P}{P_0}\frac{T_0}{T}\right]^{\alpha}\left[C_{A0}\frac{\left(\theta_{B0}-\frac{b}{a}X\right)}{(1+\varepsilon X)}\frac{P}{P_0}\frac{T_0}{T}\right]^{\beta}} \tag{2.93}$$

$$V_{CSTR3} = \frac{F_{A0}(X_3 - X_2)}{k\left[C_{A0}\frac{(1-X_3)}{(1+\varepsilon X_3)}\frac{P}{P_0}\frac{T_0}{T}\right]^{\alpha}\left[C_{A0}\frac{\left(\theta_{B0}-\frac{b}{a}X_3\right)}{(1+\varepsilon X_3)}\frac{P}{P_0}\frac{T_0}{T}\right]^{\beta}} \tag{2.94}$$

$$V_{total} = V_{CSTR1} + V_{PFR2} + V_{CSTR3} \tag{2.95}$$

10) Carry out either numerical or symbolic integration for any integration appearing in previous cases.
11) Report the numerical value for $V_{total}$ for the selected case out of Case 1, Case 2, Case 3, or Case 4.

Figure 2.19 shows MATLAB code (**SeriesReactors_Gas.m**) for calculating the total volume/ total residence-time requirement for each of the four cases. The user may, however, modify the list of parameter values, indicated in steps #1 up to #6, prior to finding the total volume and total residence-time requirement of the three serial reactors.

```
clear variables;
% Step 1: Prompt the user to enter the kinetic parameters
alpha = input('Enter the value of α, reaction order with respect to A: ');
beta = input('Enter the value of β, reaction order with respect to B: ');
k = input('Enter the value of the rate constant k,s^(-1)×(L/mol)^(α+β-1): ');
% Step 2: Prompt the user to enter the intermediate and final conversions
X1 = input('Enter the 1st intermediate conversion X1: ');
X2 = input('Enter the 2nd intermediate conversion X2: ');
X3 = input('Enter the final conversion X3: ');

% Step 3: Prompt the user to enter the initial molar flow rate of A and molar ratio of B
FA0 = input('Enter the initial molar flow rate of A, FA0 (mol/s): ');
theta_B0 = input('Enter the molar ratio of B to A, θ_B0: ');

% Step 4: Prompt the user to enter the initial mole-fraction of A and B
yA0 = input('Enter the initial mole fraction of A, yA0: ');
yB0 = input('Enter the initial mole fraction of B, yB0: ');

% Step 5: Prompt the user to enter the stoichiometric coefficients
a = input('Enter the stoichiometric coefficient for A, a: ');
b = input('Enter the stoichiometric coefficient for B, b: ');
c = input('Enter the stoichiometric coefficient for C, c: ');
d = input('Enter the stoichiometric coefficient for D, d: ');
```

**Figure 2.19** MATLAB code (**SeriesReactors_Gas.m**) for calculating the volume requirement for the three cascade reactors made of CSTR and/or PFR for a gas-phase $n^{th}$-order reaction.

```
% Step 6: Prompt the user to enter feed and reactor temperature and pressure
T0 = input('Enter the feed temperature T0 (K): ');
T = input('Enter the reactor temperature T (K): ');
P0 = input('Enter the feed pressure P0 (atm): ');
P = input('Enter the reactor pressure P (atm): ');

% Step 7: Calculate epsilon
epsilon = (c/a + d/a - b/a - 1) * yA0;

% Step 8: Calculate initial concentration of A
CA0 = (P0 * yA0) / (0.08206 * T0); % mol/L

%% Step 9: Prompt the user to select a scenario
disp('Select one scenario:');
disp('1: CSTR1 + CSTR2 + CSTR3');
disp('2: PFR1 + PFR2 + PFR3');
disp('3: PFR + CSTR + PFR');
disp('4: CSTR + PFR + CSTR');
scenario = input('Enter the number corresponding to the scenario: ');

% Initialize V_total
V_total = 0;

% Step 10: Calculate V_total based on the selected scenario
switch scenario
case 1 % CSTR1 + CSTR2 + CSTR3
     V_CSTR1 = (FA0 * X1) / (k * CA0 * (((1 - X1) / (1 + epsilon * X1)) * P / P0 * T0 / T)^alpha ...
              * ((theta_B0 - (b/a) * X1) / (1 + epsilon * X1) * P / P0 * T0 / T)^beta);
     V_CSTR2 = (FA0 * (X2 - X1)) / (k * CA0 * (((1 - X2) / (1 + epsilon * X2)) * P / P0 * T0 / T)^alpha ...
              * ((theta_B0 - (b/a) * X2) / (1 + epsilon * X2) * P / P0 * T0 / T)^beta);
     V_CSTR3 = (FA0 * (X3 - X2)) / (k * CA0 * (((1 - X3) / (1 + epsilon * X3)) * P / P0 * T0 / T)^alpha ...
              * ((theta_B0 - (b/a) * X3) / (1 + epsilon * X3) * P / P0 * T0 / T)^beta);
     V_total = V_CSTR1 + V_CSTR2 + V_CSTR3;
     Tau=V_total*CA0/FA0;

case 2 % PFR1 + PFR2 + PFR3
     integrand = @(X) 1 ./ (k * CA0 * (((1 - X) ./ (1 + epsilon * X)) * P ./ P0 * T0 ./ T).^alpha ...
              .* ((theta_B0 - (b/a) * X) ./ (1 + epsilon * X) * P ./ P0 * T0 / T).^beta);
     V_PFR1 = FA0 * integral(integrand, 0, X1);
     V_PFR2 = FA0 * integral(integrand, X1, X2);
     V_PFR3 = FA0 * integral(integrand, X2, X3);
     V_total = V_PFR1 + V_PFR2 + V_PFR3;
     Tau=V_total*CA0/FA0;

case 3 % PFR + CSTR + PFR
     integrand1 = @(X) 1 ./ (k * CA0 * (((1 - X) ./ (1 + epsilon * X)) .* P ./ P0 * T0 / T).^alpha ...
              .* ((theta_B0 - (b/a) * X) ./ (1 + epsilon * X) * P ./ P0 * T0 / T).^beta);
     V_PFR1 = FA0 * integral(integrand1, 0, X1);
     V_CSTR2 = (FA0 * (X2 - X1)) / (k * CA0 * (((1 - X2) / (1 + epsilon * X2)) * P / P0 * T0 / T).^alpha ...
              * ((theta_B0 - b/a * X2) / (1 + epsilon * X2) * P / P0 * T0 / T).^beta);
       integrand3 = @(X) 1 ./ (k * CA0 * (((1 - X) ./ (1 + epsilon .* X)) * P ./ P0 * T0 / T).^alpha ...
              .* ((theta_B0 - (b/a) * X) ./ (1 + epsilon * X) * P ./ P0 * T0 / T).^beta);
     V_PFR3 = FA0 * integral(integrand3, X2, X3);
     V_total = V_PFR1 + V_CSTR2 + V_PFR3;
     Tau=V_total*CA0/FA0;
```

**Figure 2.19** (*Continued*)

```
case 4 % CSTR + PFR + CSTR
    V_CSTR1 = (FA0 * X1) / (k * CA0 * (((1 - X1) / (1 + epsilon * X1)) * P / P0 * T0 / T).^alpha ...
               * ((theta_B0 - b/a * X1) / (1 + epsilon * X1) * P / P0 * T0 / T).^beta);
    integrand2 = @(X) 1 ./ (k * CA0 .* (((1 - X) ./ (1 + epsilon .* X)) * P ./ P0 * T0 / T).^alpha ...
               .* ((theta_B0 - (b/a) .* X) ./ (1 + epsilon .* X) * P / P0 * T0 / T).^beta);
    V_PFR2 = FA0 * integral(integrand2, X1, X2);
    V_CSTR3 = (FA0 * (X3 - X2)) / (k * CA0 * (((1 - X3) / (1 + epsilon * X3)) * P / P0 * T0 / T).^alpha ...
               * ((theta_B0 - b/a * X3) / (1 + epsilon * X3) * P / P0 * T0 / T).^beta);
    V_total = V_CSTR1 + V_PFR2 + V_CSTR3;
    Tau=V_total*CA0/FA0;
  otherwise
    disp('Invalid scenario selected');
end
% Step 11: Report the numerical value for V_total
fprintf('The total reactor volume V_total is: %.4f L\n', V_total);
fprintf('τ_total for Case %d is: %.4f Seconds\n', scenario, Tau);
fprintf('τ_total for Case %d is: %.4f Hours\n', scenario, Tau/3600.0);
```

**Figure 2.19** *(Continued)*

Copied from the Command Window, here are the results of running the code while showing the input parameters:

```
≫ SeriesReactors_Gas
Enter the reaction order with respect to A, α: 1
Enter the reaction order with respect to B, β: 1
Enter the rate constant k): 0.01
Enter the intermediate conversion X1: 0.33
Enter the intermediate conversion X2: 0.67
Enter the final conversion X3: 0.99
Enter the initial molar flow rate of A, FA0 (mol/s): 0.1
Enter the molar ratio of B to A, θ_B0: 1
Enter the initial mole fraction of A, yA0: 0.5
Enter the initial mole fraction of B, yB0: 0.5
Enter the stoichiometric coefficient for A, a: 1
Enter the stoichiometric coefficient for B, b: 1
Enter the stoichiometric coefficient for C, c: 1
Enter the stoichiometric coefficient for D, d: 2
Enter the feed temperature T0 (K): 300
Enter the reactor temperature T (K): 300
Enter the feed pressure P0 (atm): 10
Enter the reactor pressure P (atm): 10
Select one scenario:
1: CSTR1 + CSTR2 + CSTR3
2: PFR1 + PFR2 + PFR3
3: PFR + CSTR + PFR
4: CSTR + PFR + CSTR
Enter the number corresponding to the scenario: 1
```

```
The total reactor volume V_total is: 352462.9013 L
τ_total for Case 1 is: 715864.2076 Seconds
τ_total for Case 1 is: 198.8512 Hours
Enter the number corresponding to the scenario: 2
The total reactor volume V_total is: 10639.3947 L
τ_total for Case 2 is: 21608.9745 Seconds
τ_total for Case 2 is: 6.0025 Hours
Enter the number corresponding to the scenario: 3
The total reactor volume V_total is: 10791.1225 L
τ_total for Case 3 is: 21917.1388 Seconds
τ_total for Case 3 is: 6.0881 Hours
Enter the number corresponding to the scenario: 4
The total reactor volume V_total is: 352311.1735 L
τ_total for Case 4 is: 715556.0433 Seconds
τ_total for Case 4 is: 198.7656 Hours
```

**NOTES on MATLAB Code (SeriesReactors_Gas.m) of** Figure 2.19

1) The code is valid for an isothermal and isobaric gas-phase reaction. However, if either $T$ or $P$ changes from those of the feed condition, then for a PFR, the average value of $P$ and $T$ between the feed and outlet conditions can be taken. Later, we will incorporate the energy balance to account for temperature change as a variable, not a parameter, as it stands out now. The same issue can be said about CSTR, although the variation is restricted to a transient condition. Notice that a PFR can have a steady-state temperature value, yet it may vary with both the axial and radial directions.
2) The configuration of having more than three reactors in series can be extended to include additional units of PFR, CSTR, or both. The design equation of each type is given in equations within the appropriate section.
3) The code assumes an ideal gas mixture behavior. So, a $P$ value up to 100 bar is acceptable for an ideal gas behavior. Yet, for relatively elevated $P$ values, one may introduce the compressibility factor, $Z$, which accounts for deviation from ideality. See Step #8 in the previous algorithm.
4) The treatment of $\left(-r_A = kC_A^n = k\left(\text{min}^{-1} \times \left(\frac{\text{L}}{\text{mol}}\right)^{n-1}\right) C_A^n\right)$, with $n = \alpha + \beta$, can be handled as we did earlier for the liquid-phase $n^{\text{th}}$-order reaction case.
5) The best configuration of serial reactors will depend on the aforementioned parameters listed in steps #1 up to #6 in the previous algorithm. Thus, for a given set of reaction kinetic parameters (i.e., step #1) and stoichiometric coefficients (i.e., step #5), the user may attempt to vary the numerical values of parameters appearing in other steps to come up with the minimum volume/residence-time requirements.
6) The rate constant $k$ will be later explicitly expressed as a function of temperature.

## Reference

1 H.S. Fogler. "Elements of Chemical Reaction Engineering". Chapter 4: pp. 270–337. Pearson Education, Inc.; 6th edition (July 24, 2020). ISBN-10: 013548622X. ISBN-13: 978-0135486221.

## End of Chapter Problems

**CHW 2.1** Which of the following equations represents the mole balance for a batch reactor where $A \rightarrow$ Products?

**A** $(-r_A V) = \frac{dN_A}{dt}$
**B** $(r_A V) = \frac{dN_A}{dt}$
**C** $(-r_A F_{A0}) = \frac{dN_A}{dt}$
**D** $(r_A F_{A0}) = \frac{dN_A}{dt}$

**CHW 2.2** In a flow-type reactor, where $A \rightarrow$ Products, the molar flow rate of species $A$ at any point in the reactor is given by:

**A** $F_A = F_{A0} + F_{A0}X$
**B** $F_A = F_{A0}X$
**C** $F_A = F_{A0} - F_{A0}X$
**D** $F_A = F_{A0} - X$

**CHW 2.3** For a reversible reaction, the maximum conversion achievable in a BR is:

**A** 1.0
**B** $X_e$
**C** 0.5
**D** Infinite

**CHW 2.4** Batch reactor time
Consider a first-order irreversible reaction $A \rightarrow B$ taking place in a constant-volume BR.
**Given Data:**
- Initial concentration $C_{A0} = 2.0$ mol/L
- Rate constant $k = 0.1\ \text{min}^{-1}$
- Reactor volume $V = 10$ L

Calculate the time ($t_{BR}$) required to achieve a conversion $X = 0.98$.

**CHW 2.5** The volume of CSTR
Consider an irreversible, second-order liquid-phase reaction: $A \rightarrow B$ occurring in a CSTR. The reaction rate is given by: $-r_A = kC_A^2$.
**Given Data:** Inlet molar flow rate of $A$, $F_{A0} = 5$ mol/min, $C_{A0} = 2$ mol/L, desired conversion, $X = 0.85$, and rate constant, $k = 0.2\ \frac{\text{L}}{\text{mol} \cdot \text{min}}$. Calculate the volume requirement, expressed in L.

**CHW 2.6** PBR catalyst weight
A PBR reactor is used to decompose a pollutant ($A$) into harmless products. The reaction is second order with respect to $A$. The initial concentration of $A$ is 10.0 mol/L.
**Given Data:** Reaction rate constant $(k') = 0.1\ \frac{\text{L}^2}{\text{g catalyst} \cdot \text{mol} \cdot \text{min}}$ and $v = v_0 = 1.0$ L/min.
Calculate the catalyst weight needed to achieve a conversion of 95%.

**CHW 2.7** What is the primary purpose of the distributor plate in a FBR?

**A** To prevent the bed from collapsing

**B** To ensure even distribution of the fluid
**C** To increase the reaction rate
**D** To remove heat from the reactor

**CHW 2.8** If a FBR is operating isothermally, how is the temperature maintained constant?
**A** By increasing the gas velocity
**B** By reducing the solid particle size
**C** By using external/internal cooling or heating
**D** By increasing the reactor pressure

**CHW 2.9** What is the primary purpose of maintaining isothermal conditions in a FBR?
**A** To increase the reaction rate
**B** To prevent catalyst deactivation
**C** To improve mass transfer
**D** All the above

**CHW 2.10** If a FBR is operating at a very high fluid velocity, what might be the potential consequences?
**A** Increased gas–solid contact
**B** Decreased residence time
**C** Reduced heat transfer
**D** Lower pressure drop

**CHW 2.11** Which of the following is a key advantage of isothermal operation in a FBR?
**A** Reduced temperature gradients
**B** Increased pressure drop
**C** Decreased conversion rate
**D** Lower catalyst activity

**CHW 2.12** Which of the following is NOT an advantage of an isothermal FBR design?
**A** High mass and heat transfer efficiency
**B** Catalyst regeneration and maintenance
**C** Inconsistent temperature control
**D** Ability to handle high throughputs

**CHW 2.13** Why is isothermal operation important in exothermic or endothermic reactions?
**A** It prevents particle attrition in the reactor
**B** It maintains uniform temperature to avoid safety issues like runaway reactions
**C** It increases the pressure drop across the reactor
**D** It ensures high velocities for better particle entrainment

**CHW 2.14** BR time for 1.5th-order reaction
Consider the following liquid-phase reaction: $A \xrightarrow{k} B$
It has a reaction order of 1.5 with respect to A: $(-r_A = kC_A^{1.5} = 0.2\left(\frac{\text{L}^{0.5}}{\text{mol}^{0.5}\cdot\text{min}}\right) \times C_A^{1.5}$.
Calculate the time needed to achieve 99% conversion in BR, when 500 mol of A are charged to a 500-L reactor.

**CHW 2.15** Volume of CSTR for 1.5th-order reaction

Consider the following liquid-phase reaction: $A \xrightarrow{k} B$

It has a reaction order of 1.5 with respect to A: $(-r_A = kC_A^{1.5} = 0.2\left(\frac{L^{0.5}}{mol^{0.5}\cdot min}\right) \times C_A^{1.5}$.

Calculate the volume needed to achieve 99% conversion in CSTR, when the entering molar flow rate of $A$, $F_{A0} = 100$ mol/min, and $\upsilon_0 = 100\frac{L}{min}$

**CHW 2.16** Volume of PFR for 1.5th-order reaction

Consider the following liquid-phase reaction: $A \xrightarrow{k} B$

It has a reaction order of 1.5 with respect to $A$: $(-r_A = kC_A^{1.5} = 0.2\left(\frac{L^{0.5}}{mol^{0.5}\cdot min}\right) \times C_A^{1.5}$.

Calculate the volume needed to achieve 99% conversion in PFR, when the entering molar flow rate of A, $F_{A0} = 100$ mol/min, and $\upsilon_0 = 100\frac{L}{min}$

**CHW 2.17** Maximum conversion for a given CSTR volume

Consider the following liquid-phase reaction: $A \xrightarrow{k} B$

It has a reaction order of 1.5 with respect to $A$: $(-r_A = kC_A^{1.5} = 1.8\left(\frac{L^{0.5}}{mol^{0.5}\cdot min}\right) \times C_A^{1.5}$.

What is the maximum conversion that can be achieved in a 500-L CSTR?

The entering molar flow rate of $A$, $F_{A0} = 100$ mol/min, and $\upsilon_0 = 100\frac{L}{min}$

**CHW 2.18** Maximum conversion for a given PFR volume

Consider the following liquid-phase reaction: $A \xrightarrow{k} B$

It has a reaction order of 1.8 with respect to A: $(-r_A = kC_A^{1.8} = 1.8\left(\frac{L^{0.8}}{mol^{0.8}. min}\right) \times C_A^{1.8}$.

The entering molar flow rate of $A$, $F_{A0} = 100$ mol/min, and $\upsilon_0 = 100\frac{L}{min}$

What conversion can be achieved if a 100-L PFR is followed in series by a 150-L PFR?

**CHW 2.19** Conversion of CSTR + PFR

Consider the following liquid-phase reaction: $A \xrightarrow{k} B$

It has a reaction order of 1.9 with respect to A: $(-r_A = kC_A^{1.9} = 1.9\left(\frac{L^{0.9}}{mol^{0.9}. min}\right) \times C_A^{1.9}$.

The entering molar flow rate of $A$, $F_{A0} = 100$ mol/min, and $\upsilon_0 = 100\frac{L}{min}$

What conversion can be achieved if a 60-L CSTR is followed in a series by a 100-L PFR?

**CHW 2.20** Volume requirement for $\left(\frac{F_{A0}}{-r_A}\right)$ versus conversion plot

The adiabatic exothermic irreversible gas-phase reaction

$$A + 2B \rightarrow 2C$$

is to be carried out in a flow reactor for an equimolar feed of $A$ and $B$. The plot of $\left(\frac{F_{A0}}{-r_A}\right)$, expressed in m$^3$, versus $X$, the conversion for the above reaction is shown in Figure CHW 2.20.

1) What CSTR volume is necessary to achieve 50% conversion?
2) What PFR volume is necessary to achieve 50% conversion?
3) What is the volume of a second CSTR added in series to the first CSTR (Part 1) necessary to achieve an overall conversion of 80%?
4) What PFR volume must be added to the first CSTR (Part 1) to raise the conversion to 80%?

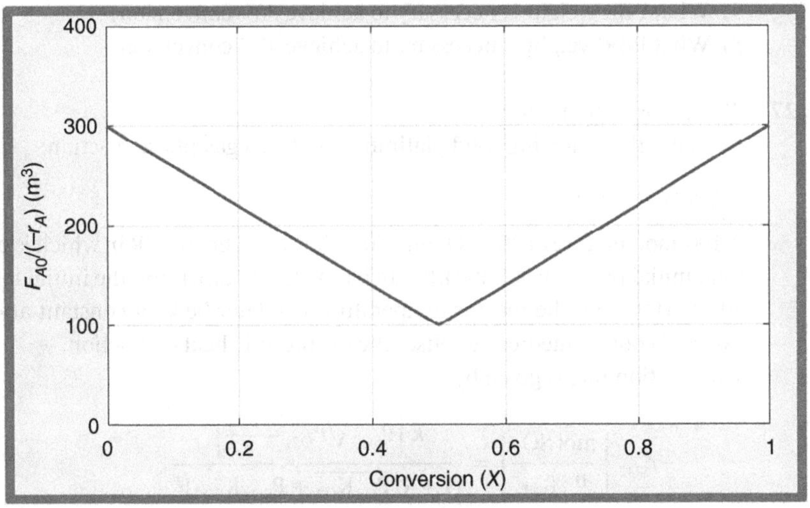


**CHW 2.20** A plot of $\left(\frac{F_{A0}}{-r_A}\right)$ versus X for an adiabatic exothermic irreversible gas-phase reaction.

**CHW 2.21** Volume requirement for $\left(\frac{F_{A0}}{-r'_A}\right)$ versus conversion plot

Figure CHW 2.21 shows a typical gas–solid catalytic exothermic reaction carried out adiabatically for: $A \rightarrow B + C$.

1) If we have a fluidized CSTR and a PBR, how should they be arranged for this adiabatic reaction? Use the smallest amount of catalyst weight to achieve 80% conversion of $A$.
2) What is the catalyst weight necessary to achieve 80% conversion in a fluidized CSTR?
3) What fluidized CSTR weight is necessary to achieve 40% conversion?

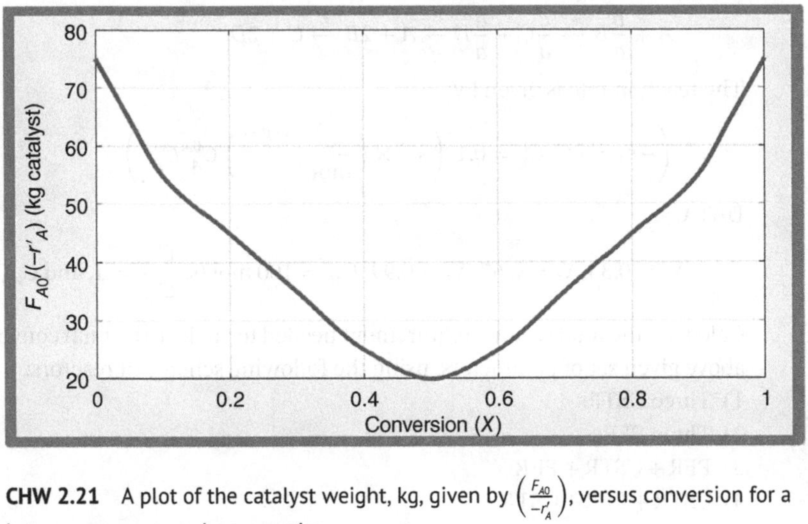


**CHW 2.21** A plot of the catalyst weight, kg, given by $\left(\frac{F_{A0}}{-r'_A}\right)$, versus conversion for a heterogeneous gas-phase reaction.

4) What PBR weight is necessary to achieve 80% conversion?
5) What PBR weight is necessary to achieve 40% conversion?

**CHW 2.22** $SO_2$ gas oxidation into $SO_3$

Consider the following solid platinum-catalyzed gas-phase reaction:

$$SO_2 + \tfrac{1}{2}O_2 \rightarrow SO_3$$

A 100-mol mixture of 30% $SO_2$ and 70% air is charged to a BR in which is oxidized. The total initial pressure is 1,400 kPa (14 bar or 13.817 atm), and the initial temperature is 300 K. However, the reactor temperature will later be kept constant at 423 °C by an external cooling medium to offset the exothermic heat of reaction.

The reaction rate is given by:

$$-r'_{SO2}\left[\frac{\text{mol SO}_2}{\text{h}\cdot\text{g}_{\text{cat}}}\right] = \frac{k\left[P_{SO2}\sqrt{P_{O2}} - \frac{P_{SO3}}{K_P}\right]}{[1+\sqrt{P_{O2}K_{O2}} + P_{SO3}K_{SO3}]^2}$$

$$k = 10.5\,\frac{\text{mol SO}_2}{\text{atm}^{1.5}\cdot\text{h}\cdot\text{g}_{\text{cat}}}$$

$$K_{O2} = 38.0\ \text{atm}^{-1} \qquad K_{SO3} = 42.0\ \text{atm}^{-1} \qquad K_P = 929\ \text{atm}^{-1/2}$$

While modifying the code to account for the new $k$, $K_{O2}$, $K_{SO3}$, and $K_P$,use MATLAB code (SO_2_Cat_Oxid.m) to do the following three tasks:

1) Calculate the batch time per 150 g catalyst weight to achieve 97% conversion.
2) Repeat the process for a CSTR with the same initial and operating conditions and find the amount of catalyst needed to achieve 97% conversion.
3) Repeat the process for a PBR with the same initial and operating conditions and find the amount of catalyst needed to achieve 97% conversion.

**CHW 2.23** Reactors in series for liquid phase

Consider the following liquid-phase reaction:

$$A + \frac{b}{a}B \xrightarrow{k} \frac{c}{a}C + \frac{d}{a}D \rightarrow A + 2B \xrightarrow{k} C + 2D$$

The reaction rate is given by:

$$\left(-r_A = kC_A^{\alpha}C_B^{\beta} = 0.1\left(\text{s}^{-1}\times\left(\frac{\text{L}}{\text{mol}}\right)^{\alpha+\beta-1}\right)C_A^{0.4}C_B^{0.5}\right)$$

DATA:

$$X_1 = 0.33, X_2 = 0.67, X_3 = 0.99, F_{A0} = 100\ \text{mol/s}, \frac{C_{B0}}{C_{A0}} = 1, \text{and } C_{A0} = 10\ \text{mol/L}$$

Calculate the total volume requirement needed to achieve the final conversion for the above given set of parameters, using the following scheme of reactors:

1) Three CSTRs
2) Three PFRs
3) PFR + CSTR + PFR
4) CSTR + PFR + CSTR

Which scheme or cascade gives the minimum total volume requirement?

**CHW 2.24** Reactors in series for gas phase

Consider the following gas-phase reaction: $A + 2B \xrightarrow{k} C + 2D$

The reaction rate is given by $\left(-r_A = kC_A^\alpha C_B^\beta = 0.1\left(\mathrm{s}^{-1} \times \left(\frac{\mathrm{L}}{\mathrm{mol}}\right)^{\alpha+\beta-1}\right) C_A^{1.2} C_B^{1.1}\right)$

1) Calculate the total volume/time requirement needed to achieve the final conversion for a given set of parameters:
2) The intermediate and final conversions: $X_1 = 0.33$, $X_2 = 0.67$, and $X_3 = 0.99$
3) The initial molar flow rate of $A$ and the molar ratio of $B$: $F_{A0} = 1$ mol/s and $\theta_{B0} = 2$
4) The initial mole fraction of $A$ and $B$: $y_{A0} = 0.33$ and $y_{B0} = 0.67$
5) The feed and reactor temperature and pressure: $T_0 = 300$ K, $T = 500$ K, $P_0 = 10$ atm, and $P = 15$ atm.

In terms of minimization of total time/volume requirement, find the best of the following scheme of reactors:

1) Three CSTRs
2) Three PFRs
3) PFR + CSTR + PFR
4) CSTR + PFR + CSTR

# 3

# Equilibrium Reactions

## 3.1 Introduction

Equilibrium and nonequilibrium reactions represent two fundamental categories of chemical processes that exhibit distinct behaviors based on thermodynamics and kinetics. An equilibrium reaction occurs when the rate of the forward reaction (reactants converting to products) equals the rate of the reverse reaction (products converting back to reactants). In such reactions, the system reaches a dynamic balance, where the concentrations of both reactants and products remain constant over time. The composition of an equilibrium mixture remains the same, regardless of the direction from which the equilibrium is reached.

A YouTube video titled "Introduction to Dynamic Equilibrium" can be seen at:
https://youtu.be/4AJbFuzW2cs.

This state of dynamic equilibrium is dictated by the equilibrium constant $K_{eq}$, which is a function of temperature and the relative stabilities of the reactants and products. The concept of chemical equilibrium is central to understanding reversible reactions, as it provides insights into how conditions, such as pressure, temperature, and concentration, affect the yield of products. A classic example of an equilibrium reaction is the synthesis of ammonia via the Haber process:

$$N_2(g) + 3H_2(g) \rightleftharpoons 2NH_3(g)$$

Here, the reaction does not proceed to completion but rather establishes a balance between the reactants and products, which can be shifted by altering the reaction conditions (e.g., through Le Chatelier's principle).

In contrast, nonequilibrium reactions are characterized by the absence of such a balance. These reactions proceed in one direction, often until the reactants are fully converted into products. The key distinction is that these reactions do not reverse, and the system does not achieve a steady state of constant concentrations of reactants and products. Nonequilibrium processes are typically seen in systems that are far from thermodynamic equilibrium, and their rates are often determined by external driving forces such as temperature gradients, pressure differences, or the continuous supply of reactants. A familiar example of a nonequilibrium reaction is combustion, where a fuel reacts with oxygen to form combustion products (e.g., $CO_2$ and $H_2O$) irreversibly.

Nonequilibrium reactions occur in systems that are not at thermodynamic equilibrium, meaning they do not achieve a balance between forward and reverse reaction rates. These reactions are typically time-dependent and are often influenced by external driving forces such as energy input, concentration gradients, or environmental conditions. In biological systems, for example, metabolic pathways consist of a series of enzyme-catalyzed reactions that are inherently nonequilibrium. A well-known example is glycolysis, where glucose is broken down to produce pyruvate, ATP, and NADH. This process is irreversible under physiological conditions because cells continuously consume the products (e.g., ATP) and regenerate reactants (e.g., glucose), driving the pathway forward. The system is kept away from equilibrium by the constant input of energy (from ATP hydrolysis) and the regulation of enzymes.

In materials science, nonequilibrium reactions are also prevalent, particularly in polymerization and phase transitions. For instance, in free-radical polymerization, monomers react to form long-chain polymers in a process that is irreversible. This reaction proceeds until the monomer supply is exhausted or the reaction is stopped, making it a nonequilibrium process. Similarly, phase transitions such as the crystallization of metals or the glass transition in polymers are examples of nonequilibrium processes. During these transitions, the system moves from one phase to another, often irreversibly, without reaching equilibrium. For example, when a liquid cools rapidly, it may solidify into a glassy state without forming an ordered crystal lattice, resulting in a nonequilibrium glassy structure.

In the context of chemical industries, nonequilibrium reactions are foundational to numerous processes. For example, the production of ammonia via the Haber-Bosch process, though technically a reversible reaction, operates under nonequilibrium conditions in practice. High pressures and temperatures are used to continuously push the reaction toward product formation, while the removal of ammonia shifts the equilibrium further toward completion. Similarly, petrochemical refining processes, such as catalytic cracking, rely on nonequilibrium conditions to break down large hydrocarbon molecules into more useful smaller fractions. These processes operate under conditions far from equilibrium to achieve high conversion rates and economic viability, highlighting the importance of understanding and manipulating nonequilibrium reactions for industrial efficiency.

In energy storage and transfer, nonequilibrium reactions are at the heart of batteries, fuel cells, and energy conversion technologies. In batteries, electrochemical reactions occur at the electrodes, with electrons and ions moving between them. The system is kept away from equilibrium by the applied electrical potential, enabling the continuous generation or storage of energy. Fuel cells, which convert chemical energy into electrical energy, also operate under nonequilibrium conditions, where reactants such as hydrogen and oxygen are continuously consumed, and the products are removed to sustain the energy conversion. These technologies rely on nonequilibrium principles to ensure that energy can be efficiently stored or transferred, making them essential for modern energy infrastructure and sustainable technologies.

In the field of engineering materials, nonequilibrium processes are critical for the synthesis and design of advanced materials. For instance, nonequilibrium solidification processes, such as rapid quenching in metallurgy, allow the formation of amorphous metals or metallic glasses, which possess unique mechanical properties that are not achievable under equilibrium conditions. Such materials are known for their high strength and corrosion resistance.

Similarly, nonequilibrium phase transitions are exploited in the production of nanomaterials and metamaterials, where the controlled manipulation of reaction pathways can lead to the formation of structures with tailored optical, electrical, or mechanical properties. These materials have broad applications in electronics, photonics, and structural engineering.

### 3.1.1 Practical Irreversibility in Chemical Reactions: Balancing Theoretical Reversibility and Industrial Realities

The statement "Strictly speaking, all chemical reactions are reversible" is rooted in fundamental thermodynamics, where reactions can, in principle, proceed both forward (reactants to products) and backward (products to reactants) under the right conditions. However, in practice, many reactions appear irreversible because the rate of the reverse reaction is negligibly small compared to the forward reaction. This practical irreversibility occurs when the forward reaction is highly favorable, either due to thermodynamic driving forces (such as a large negative Gibbs free energy change) or kinetic factors (such as high activation energy for the reverse reaction).

From a thermodynamic perspective, the concept of reversibility is based on the fact that if a system reaches equilibrium, both the forward and reverse reactions continue to occur, but at equal rates, so the concentrations of reactants and products remain constant. However, when the forward reaction dominates to such an extent that the concentration of products becomes overwhelmingly larger than that of reactants, the backward reaction may proceed so slowly that it is practically negligible over the timescale of interest. In such cases, the reaction is often treated as irreversible for practical purposes.

An example of this occurs in combustion reactions, such as the burning of methane:

$$CH_4 + 2O_2 \rightarrow CO_2 + 2H_2O$$

In theory, the reverse reaction (recombining $CO_2$ and $H_2O$ to form methane and oxygen) can occur. However, the conditions required to drive this backward reaction, such as extreme temperatures or very specific catalysts, are so far removed from ordinary combustion conditions that the reverse reaction rate is insignificant. Thus, for practical purposes, combustion is considered an irreversible process, even though the reverse reaction is theoretically possible.

The same principle applies to many other chemical processes. For example, in industrial synthesis or biological systems, certain reactions are driven so strongly in the forward direction by the removal of products or continuous input of energy that the reverse reaction becomes practically irrelevant. A classic case in industry is the large-scale synthesis of sulfuric acid via the contact process:

$$2SO_2(g) + O_2(g) \rightarrow 2SO_3(g)$$

While technically reversible, under the high temperatures and catalytic conditions employed in the process, the forward reaction is heavily favored, and the amount of $SO_2$ present at equilibrium is negligible. As a result, the reaction is treated as practically irreversible.

From a kinetic perspective, the apparent irreversibility of a reaction can also be explained by the relative rates of the forward and reverse reactions. If the activation energy for the reverse reaction is significantly higher than for the forward reaction, then the backward reaction will occur extremely slowly, making the system appear to proceed only in the forward direction. In such cases, the products accumulate, and the reaction behaves as though it is one-way, despite the theoretical possibility of reversal.

This practical irreversibility is particularly important in the design of industrial chemical processes, where engineers often assume that reactions go to completion to maximize product yield. However, recognizing the theoretical reversibility is still critical because it underpins our understanding of how equilibrium shifts can be controlled. For example, in reversible reactions that approach equilibrium under certain conditions, Le Chatelier's principle can be used to manipulate conditions (such as temperature or pressure) to favor either the forward or reverse reaction, optimizing yields based on process requirements.

The following YouTube videos demonstrate the concept of a reversible reaction, an irreversible reaction, and a reversible chemical equilibrium:

1) https://www.youtube.com/watch?v=ty9TczsW5ew.
2) https://www.youtube.com/watch?v=br8lKynV1Hc.
3) https://www.youtube.com/watch?v=I1XovwvQmUM.
4) https://www.youtube.com/watch?v=WE6TbCytMBk.
5) https://www.youtube.com/watch?v=pIP3QIVajzI.

## 3.2 Thermodynamic Foundations of Chemical Equilibrium

The concept of chemical equilibrium is deeply rooted in thermodynamics, which governs how energy and matter behave in chemical systems. At equilibrium, a system has reached a state where the rates of the forward and reverse reactions are equal, and no net change occurs in the concentrations of reactants and products. This balance is driven by the system's tendency to minimize its Gibbs free energy ($G$), which combines the effects of enthalpy ($H$) and entropy ($S$).

The Gibbs free energy change ($\Delta G$) of a reaction provides the key to understanding chemical equilibrium. For any reaction at constant temperature and pressure:

$$aA + bB \Longrightarrow cC + dD$$

$\Delta G$ is related to the reaction quotient $Q$ by the equation:

$$\Delta G = \Delta G^0 + RT \ln Q = \Delta G^0 + RT \ln \frac{[D]^d[C]^c}{[A]^a[B]^b} \tag{3.1}$$

where

- $\Delta G^0$ is the standard molar Gibbs free energy change,
- $R$ is the universal gas constant,
- $T$ is the absolute temperature, and
- $Q$ is the reaction quotient, a ratio of the concentrations of products to reactants at any given moment, not necessarily at equilibrium.

Thus, at equilibrium, Eq. (3.1) becomes:

$$\Delta G = \Delta G^0 + RT \ln Q = \Delta G^0 + RT \ln \left( \frac{[D]^d[C]^c}{[A]^a[B]^b} \right)_{eq} = 0$$

$$\Delta G^0 = -RT \ln K_{eq} = -RT \ln \left( \frac{[D]^d[C]^c}{[A]^a[B]^b} \right)_{eq} \tag{3.2}$$

Alternatively, $K_{eq}$ can be expressed in terms of $\Delta G^0$ as:

$$K_{eq} = e^{-\frac{\Delta G^0}{RT}} \tag{3.3}$$

Equation (3.3) illustrates the thermodynamic driving force behind chemical reactions. If $\Delta G^0$ is negative, the forward reaction will be favored, and equilibrium will result in a larger concentration of products relative to reactants (i.e., $K_{eq} > 1$). Conversely, if $\Delta G^0$ is positive, the reverse reaction will be favored (i.e., $K_{eq} < 1$).

**Example 3.1** ***$K_{eq}$ as f($C_{eq}$)***
Express $K_{eq}$ for the following reaction:

$$2CO(g) + O_2(g) \rightleftharpoons 2CO_2(g)$$

**Solution**

$$K_{eq} = \frac{[CO_2]^2}{[CO]^2[O_2]}$$

### 3.2.1 Comparing *Q* and $K_{eq}$

The comparison between $Q$ and $K_{eq}$ reveals whether the reaction mixture needs to shift toward products or reactants to reach equilibrium:

a) $Q < K_{eq}$: The concentration of products is too low relative to the reactants, meaning the reaction will shift to the right (toward products) to reach equilibrium. This indicates that more products will form as the system approaches equilibrium.
b) $Q > K_{eq}$: The concentration of products is too high relative to the reactants, so the reaction will shift to the left (toward reactants) to reach equilibrium. In this case, some of the products will convert back to reactants until equilibrium is achieved.
c) $Q = K_{eq}$: The system is already at equilibrium, so there is no net shift in either direction. The reaction rates of the forward and reverse processes are equal, and the concentrations of reactants and products remain constant.

This dynamic behavior explains why reactions do not instantaneously reach equilibrium but rather evolve toward it over time, with the rate of approach influenced by factors such as temperature, pressure, and the presence of catalysts.

In industrial and laboratory settings, the reaction quotient is a valuable tool for adjusting reaction conditions to maximize yield. By understanding how far $Q$ is from $K_{eq}$, engineers can decide whether to add reactants, remove products, or change the temperature or pressure to achieve the desired equilibrium position efficiently.

For instance, in the production of ammonia via the Haber process:

$$N_2(g) + 3H_2(g) \rightleftharpoons 2NH_3(g)$$

If the reaction starts with only nitrogen and hydrogen, $Q = 0$, and the reaction strongly favors the formation of ammonia to approach equilibrium. As ammonia forms, $Q$ increases until it eventually equals $K_{eq}$, at which point the reaction rates in both directions become equal.

**Example 3.2** ***The Reaction Quotient for $NH_3$ Synthesis***
A 4-L flask contains 4.0 g ammonia ($NH_3$), 8 g nitrogen ($N_2$), and 7 g hydrogen ($H_2$) at a certain point of time during the gas phase reaction: $N_2(g) + 3H_2(g) \leftrightarrow 2NH_3(g)$, with the equilibrium constant, $K_C = 0.05$ at the given temperature and pressure. Evaluate the direction of the reaction under the given conditions.

$$[N_2]_0 = \frac{N_2 \text{ moles}}{\text{Vol}} = \frac{8.0\,\text{g} \times \frac{\text{mol}}{28\,\text{g}}}{4\,\text{L}} = 0.0714\frac{\text{mol}}{\text{L}} = 0.0714\,\text{M}$$

$$[H_2]_0 = \frac{H_2 \text{ moles}}{\text{Vol}} = \frac{7.0\,\text{g} \times \frac{\text{mol}}{2\,\text{g}}}{4\,\text{L}} = 0.875\frac{\text{mol}}{\text{L}} = 0.875\,\text{M}$$

$$[NH_3]_0 = \frac{NH_3 \text{ moles}}{Vol} = \frac{4.0\,g \times \frac{mol}{17\,g}}{4\,L} = 0.0588\frac{mol}{L} = 0.0588\,M$$

$$Q = \frac{[NH_3]_0^2}{[N_2]_0^1[H_2]_0^3} = \frac{[0.0588]^2}{[0.0714]^1[0.875]^3} = 0.0723$$

Since $Q > K_C$, the reaction will proceed in a backward direction, resulting in the formation of more reactants, which will lead to a higher value of the denominator as well as to a lower value of the numerator on the right-hand side of the equation for $Q$, thus pushing the value of $Q$ down to the equilibrium constant, $K_C$.

### 3.2.2 $K_{eq}$ as $f(\Delta G^0)$

$\Delta G^0$ is either tabulated in literature for some common chemical reactions or can be deduced from the standard molar Gibbs free energy of formation for products and reactants:

$$\Delta G^0 = \sum_{i=1}^{N} \nu_i \left(\Delta G_f^0\right)_{Products} + \sum_{i=1}^{N} \nu_i \left(\Delta G_f^0\right)_{Reactants} \tag{3.4}$$

where $\nu_i$ is positive for products and negative for reactants.

The standard molar Gibbs free energy of formation, $\Delta G_f^0$, can be found for some organic and inorganic compounds on the following web pages:

1) https://www.engineeringtoolbox.com/standard-enthalpy-formation-value-Gibbs-free-energy-entropy-heat-capacity-organic-d_1979.html
2) https://www.engineeringtoolbox.com/standard-state-enthalpy-formation-definition-value-Gibbs-free-energy-entropy-molar-heat-capacity-d_1978.html
3) https://en.wikipedia.org/wiki/Standard_Gibbs_free_energy_of_formation
4) https://www.wiredchemist.com/chemistry/data/entropies-inorganic
5) https://www.wiredchemist.com/chemistry/data/entropies-organic
6) https://www.studocu.com/da/document/danmarks-tekniske-universitet/fundamental-chemistry/free-gibbs-energy-enthalpy-and-enthropy/39897479

**NOTE**

The presence of a catalyst affects the rate at which equilibrium is reached but does not change the equilibrium position itself or the extent of equilibrium manifested via $X_{eq}$.

In general, if we have a chemical reaction in the form of: $aA + bB \rightleftharpoons cC + dD$, then the standard molar Gibbs free energy change for the reaction can be expressed as:

$$\Delta G^0 = d\left(\Delta G_f^0\right)_D + c\left(\Delta G_f^0\right)_C - b\left(\Delta G_f^0\right)_B - a\left(\Delta G_f^0\right)_A \tag{3.5}$$

So does the molar heat of reaction is expressed as:

$$\Delta H^0 = d\left(\Delta H_f^0\right)_D + c\left(\Delta H_f^0\right)_C - b\left(\Delta H_f^0\right)_B - a\left(\Delta H_f^0\right)_A \tag{3.6}$$

**Example 3.3** ***Ammonia Formation***

Let us consider the synthesis of ammonia from nitrogen and hydrogen:

$$N_{2(g)} + 3H_{2(g)} \rightleftharpoons 2NH_{3(g)}$$

**DATA:**

$\left(\Delta G_f^0\right)_{NH3}$ : −16.3 kJ/mol (https://www.wiredchemist.com/chemistry/data/entropies-inorganic).

$\left(\Delta G_f^0\right)_{N2}$ : 0.0 (pure element found in nature)

$\left(\Delta G_f^0\right)_{H2}$ : 0.0 (pure element found in nature)

Calculate $K_{eq}$ for this reaction.

**Solution**

$$\Delta G^0 = 2\left(\Delta G_f^0\right)_{NH_3} - 1\left(\Delta G_f^0\right)_{N_2} - 3\left(\Delta G_f^0\right)_{H_2} = 2(-16.3) - 1(0) - 3(0) = -32.6\ \text{kJ/mol}$$

From Eq. (3.3), we have:

$$K_{eq} = e^{-\frac{\Delta G^0}{RT}} = e^{-\frac{-32,600\ \frac{J}{mol}}{8.314\ \frac{J}{mol.K} * 298\ K}} = e^{\frac{32,600}{8.314*298}} = e^{13.158} = 5.182 \times 10^5$$

This relatively large value of $K_{eq}$ indicates that, at equilibrium, the reaction strongly favors the formation of ammonia ($NH_3$), as expected given the large negative value of $\Delta G^0$.

**Example 3.4** ***Calculation of $K_{eq}$ for Formation of Methyl Formate***

Consider the reaction between formic acid and methanol to produce methyl formate and water:

$$CHCOOH_{(l)} + CH_3OH_{(l)} \rightleftharpoons CH_3OOCH_{(l)} + H_2O_{(l)}$$

In symbolic form:

$$FA + MOH \rightleftharpoons MF + H_2O$$

We have initially one mole of FA and one mole of MOH placed in $V$ (L) solution.

Calculate $K_C$ for this reaction.

**DATA:**

$\left(\Delta G_f^0\right)_{MF} = -297.4\frac{kJ}{mol}$ (https://www.chemeo.com/cid/41-465-0/Methyl-formate)

$\left(\Delta G_f^0\right)_{H_2O(l)} = -237.14\frac{kJ}{mol}$

(https://en.wikipedia.org/wiki/Standard_Gibbs_free_energy_of_formation)

$\left(\Delta G_f^0\right)_{MOH} = -162.6\frac{kJ}{mol}$ (https://www.chemeo.com/cid/69-136-5/Methyl-Alcohol)

$\left(\Delta G_f^0\right)_{FA} = -351.2\frac{kJ}{mol}$ (https://www.chemeo.com/cid/39-377-1/Formic-acid)

**Solution**

Refer to Eq. (3.5):

$$\Delta G^0 = d\left(\Delta G_f^0\right)_D + c\left(\Delta G_f^0\right)_C - b\left(\Delta G_f^0\right)_B - a\left(\Delta G_f^0\right)_A$$

$$\Delta G^0 = \left(\Delta G_f^0\right)_{H_2O} + \left(\Delta G_f^0\right)_{MF} - \left(\Delta G_f^0\right)_{MOH} - \left(\Delta G_f^0\right)_{FA}$$

$$\Delta G^0 = -237.14 - 297.4 - (-162.6) - (-351.2) = -20.74\frac{kJ}{mol}$$

From Eq. (3.3), we have:

$$K_{eq} = e^{-\left(\frac{\Delta G^0}{RT}\right)} = e^{-\left(\frac{-20{,}740\,\frac{J}{mol}}{8.314\,\frac{J}{mol\cdot K}(298.15)K}\right)} = 4.3 \times 10^3$$

The thermodynamic foundations of equilibrium extend beyond Gibbs free energy to include the concepts of enthalpy and entropy. Enthalpy represents the heat exchange in the reaction, while entropy reflects the disorder or randomness. These two quantities determine the spontaneity of the reaction. A reaction with a large negative $\Delta H^0_{rxn}$ (exothermic) and a positive $\Delta S^0_{rxn}$ (increase in disorder) will strongly favor product formation at equilibrium.

### 3.2.3 Van't Hoff Equation

The Van't Hoff equation relates the temperature dependence of the equilibrium constant to the enthalpy change:

$$\frac{d\ln(K_{eq})}{dT} = \frac{\Delta H^0_{rxn}}{RT^2} \tag{3.7}$$

where $\Delta H^0_{rxn}$ is the heat of reaction at the given $T$.

If $\Delta H^0_{rxn}$ is assumed constant, independent of $T$, then we have, upon integration:

$$\ln\left(\frac{K_{eq,2}}{K_{eq,1}}\right) = -\frac{\Delta H^0_{rxn}}{R}\left(\frac{1}{T_2} - \frac{1}{T_1}\right) \tag{3.8}$$

Or

$$\ln\left(\frac{K_{eq}}{K_{eq,1}}\right) = -\frac{\Delta H^0_{rxn}}{R}\left(\frac{1}{T} - \frac{1}{T_1}\right) \rightarrow \ln(K_{eq}) = \ln(K_{eq,1}) - \frac{\Delta H^0_{rxn}}{R}\left(\frac{1}{T} - \frac{1}{T_1}\right) \tag{3.9}$$

A plot of $\ln(K_{eq})$ versus $(1/T)$ will give a straight line with a slope equal to $-\frac{\Delta H^0_{rxn}}{R}$. If the reaction is endothermic, then the slope will be negative. This case resembles that of the Arrhenius case for the reaction rate constant, $k$, where $\Delta H^0_{rxn}$ replaces the activation energy term, $E_a$. This means $K_{eq}$ decreases with $1/T$ or increases with $T$. Vice versa, if the reaction is exothermic, then the slope will be positive. This case is an anti-Arrhenius case. This means $K_{eq}$ increases with $1/T$ or decreases with $T$.

More on the Van't Hoff equation can be seen at the following YouTube video: https://www.youtube.com/watch?v=4vk6idAXp_A.

**Example 3.5** ***Effect of T on $K_P$***

A gas-phase reaction, with a standard heat of reaction of $\Delta H^0_{rxn}$ = 324 kJ/mol, has a value of $K_P$ = 0.035 at 33 °C. Determine the value of $K_P$ for this reaction at 47 °CF.

**Solution**

Convert the given temperatures to their absolute values. $T_1$ = 273.15 + 33 °C = 306.15 K. $T_2$ = 273.15 + 47 °C = 320.15 K.

From Eq. (3.8), we have:

$$\ln\left(\frac{K_{eq,2}}{0.035}\right) = -\frac{324{,}000\,\frac{J}{mol}}{8.314\,\frac{J}{mol\cdot K}}\left(\frac{1}{320.15} - \frac{1}{306.15}\right)\frac{1}{K} = 5.5664$$

$$\frac{K_{eq,2}}{0.035} = e^{5.5664} \rightarrow K_{eq,2} = 0.035 * 261.494 = 9.1523$$

**Example 3.6** ***Finding*** $\Delta H^0_{rxn}$

$H_2$ and $N_2$ react to form $NH_3$ according to the following balanced chemical equation:

$$N_2(g) + 3H_2(g) \rightleftharpoons 2NH_3(g)$$

Values of the equilibrium constant at various temperatures were reported as:

$$K_{25°C} = 561724.97 \qquad K_{75°C} = 83838.35$$

Assuming a constant heat of reaction, find $\Delta H^0_{rxn}$.

From Eq. (3.8) we have:

$$\ln\left(\frac{83838.35}{561724.97}\right) = -\frac{\Delta H^0_{rxn}}{R}\left(\frac{1}{348} - \frac{1}{298}\right) \rightarrow \ln(0.1493) = -\frac{\Delta H^0_{rxn}}{R}(-0.00046)\ \text{K}^{-1}$$

$$-\frac{\Delta H^0_{rxn}}{R} = \frac{\ln(0.1493)}{(-0.00046)\ \text{K}^{-1}} = 4.134 \times 10^3\ \text{K}$$

$$\Delta H^0_{rxn} = -8.314\,\frac{\text{J}}{\text{mol}\cdot\text{K}} \times 4.134 \times 10^3\ \text{K}$$

$$\Delta H^0_{rxn} = -3.437 \times 10^4\,\frac{\text{J}}{\text{mol}} = -34.37\,\frac{\text{kJ}}{\text{mol}}$$

Notice that we assume the heat of reaction is: constant, which is valid for a short temperature range, say, less than 50 °C. In general, the heat of reaction is: a function of the heat capacity difference between products and reactants, as it will be shown later.

### 3.2.4 MATLAB Code for $NH_3$ Equilibrium Constant, $K_{eq}$ as $f(T)$

Consider $NH_3$ reaction shown in Example 3.6. Make a plot of $K_{eq}$ as $f(T)$.

Figure 3.1 shows MATLAB code (**KeqNH3.m**) for expressing $K_{eq}$ of $NH_3$ synthesis as a function of $T$, using van't Hoff equation (Eq. 3.8).

Figure 3.2 shows the generated plot for $K_{eq}$ as $f(T)$ for $NH_3$ synthesis.

### 3.2.5 Le Chatelier's Principle: Predicting Equilibrium Shifts

Le Chatelier's principle is a fundamental concept in chemical dynamic equilibrium, stating that when a system at equilibrium is subjected to an external change (such as a change in concentration, pressure, or temperature), the system will respond by adjusting the position of the dynamic equilibrium to counteract the imposed change. This principle allows chemists and engineers to predict how modifying reaction conditions can influence the equilibrium composition, thereby controlling product yields and optimizing reaction processes.

The following YouTube video shows the effect of concentration, pressure, and temperature in light of Le Chatelier's principle:

https://www.youtube.com/watch?v=rwn4UwFZ46o.

#### 3.2.5.1 Concentration Changes

When the concentration of a reactant or product is altered, the system shifts to restore equilibrium by either consuming or producing the substance that was altered. For example, in the reaction:

$$N_2(g) + 3H_2(g) \rightleftharpoons 2NH_3(g)$$

```
function Keq = vanthoff(T, dH_ref, T_ref, Keq_ref)
  % Calculate Keq at temperature T using Van't Hoff equation
  % Inputs:
  %  T      = Temperature [K] (scalar or vector)
  %  dH_ref = ΔH⁰ at T_ref [J/mol]
  %  T_ref  = Reference temperature [K]
  %  Keq_ref= Keq at T_ref
  clear variables;
  R = 8.314;
  Keq = Keq_ref .* exp(-(dH_ref/R) .* (1./T - 1/T_ref));
end
% Ammonia synthesis: N2 + 3H2 ↔ 2NH3 (ΔH⁰ = -33.97 kJ/mol)
T = 300:50:350;          % Temperature range [K]
Keq_298 = 561724.97;       % Keq at 500K (reference)
Keq_vals = vanthoff(T, -33970, 298, Keq_298);
% Plotting
plot(T, Keq_vals, 'b-', 'LineWidth', 2);
grid on;
xlabel('Temperature (K)');
ylabel('K_{eq}');
title('Van''t Hoff Plot: NH_3 Synthesis');
annotation('textbox', [0.4, 0.7, 0.2, 0.1], 'String', ...
  'Exothermic reaction: K_{eq} decreases with T');
ax = gca;
% Set new font size for all axes text (including labels)
ax.FontSize = 16;
ax.FontWeight="bold";
```

**Figure 3.1** MATLAB code (**KeqNH3.m**) for expressing $K_{eq}$ of $NH_3$ synthesis as a function of $T$, using van't Hoff equation.

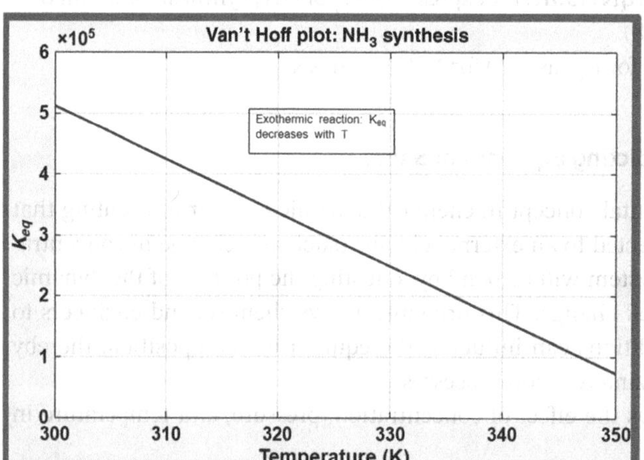


**Figure 3.2** A plot of $K_{eq}$ as $f(T)$ for $NH_3$ synthesis.

If additional $N_2$ is added to the system, the equilibrium will shift to the right, favoring the production of $NH_3$ to reduce the excess $N_2$. Conversely, removing $NH_3$ from the system will also shift the equilibrium to the right, encouraging further production of $NH_3$ to replace what was removed. In industrial applications, this principle is often used by continuously removing products from the reaction mixture to drive the equilibrium toward higher yields.

The following YouTube video shows the effect of concentration change on a chemical equilibrium, utilizing Le Chatelier's principle:

https://www.youtube.com/watch?v=o9jPlyF6E1U.

#### 3.2.5.2 Pressure Changes

For gaseous reactions, changing the pressure can influence equilibrium if there is a difference in the number of moles of gases on each side of the reaction. Le Chatelier's principle states that increasing the pressure will favor the side with fewer moles of gas, as this reduces the pressure change. In the above example, there are four moles of gas on the reactant side ($N_2 + 3H_2$) and only two moles on the product side ($2NH_3$). Therefore, increasing the pressure would shift the equilibrium toward the production of ammonia, while decreasing the pressure would favor the dissociation of $NH_3$ back into $N_2$ and $H_2$.

For example, in the Haber process, the industrial process that produces ammonia from nitrogen and hydrogen, high pressures are used to favor the production of ammonia because the reaction to form ammonia has fewer moles of gas on the product side.

The following YouTube videos show the effect of pressure/volume change on a gas-phase chemical equilibrium, utilizing Le Chatelier's principle:

https://www.youtube.com/watch?v=fHlFTtn2gxU.
https://www.youtube.com/watch?v=RC8_Hz9h9g8.

#### 3.2.5.3 Temperature Changes

The effect of temperature on equilibrium depends on whether the reaction is: exothermic (releases heat; $\Delta H < 0$) or endothermic (absorbs heat; $\Delta H > 0$). According to Le Chatelier's principle, increasing the temperature of an exothermic reaction will shift the equilibrium toward the reactants, as the system tries to counteract the added heat by favoring the reverse, endothermic direction. Conversely, decreasing the temperature of an exothermic reaction favors product formation, as the system shifts toward releasing heat.

For endothermic reactions, the opposite is true: increasing temperature favors the forward reaction (products), while decreasing temperature favors the reverse reaction (reactants). For example, the endothermic decomposition of calcium carbonate:

$$CaCO_3(s) \rightleftharpoons CaO(s) + CO_2(g)$$

shifts toward more CaO and $CO_2$ when the temperature is raised, promoting decomposition. In contrast, cooling the system favors the formation of $CaCO_3$.

The following YouTube video shows the effect of temperature change on a chemical equilibrium, utilizing Le Chatelier's principle:

https://www.youtube.com/watch?v=esJ5jUWiAJA.

#### 3.2.5.4 Catalysts and Equilibrium

Catalysts speed up both the forward and reverse reactions by lowering the activation energy, allowing the system to reach equilibrium faster. However, they do not affect the position of equilibrium because they do not change the thermodynamic properties, such as Gibbs free energy, of the reaction. Therefore, while catalysts are crucial for increasing reaction rates, they do not shift the equilibrium. Table 3.1 shows the cause of disturbance and the response of the system under dynamic equilibrium.

**Table 3.1** Cause of disturbance to the dynamic equilibrium state and the system response.

| Imposed Disturbance | System Response | Direction of Shift |
|---|---|---|
| Addition of reactants or removal of products | Consumption of reactants and formation of products | Toward products side |
| Decrease of reactants or accumulation of products | Production of reactants by backward reaction | Toward reactants side |
| Increase in system $P$ or decrease in system $V$ | Decrease system $P$ or increase system $V$ | Toward the side with fewer number of moles |
| Decrease system $P$ or increase system $V$ | Increase system $P$ or decrease system $V$ | Toward the side with higher number of moles |
| Increase system $T$ by heating | Reducing system $T$ by consuming the added heat | Toward reactant side in an exothermic reaction and toward product side in an endothermic reaction |
| Decrease system $T$ by cooling | Increasing system $T$ by heat generation | Toward product side in an exothermic reaction and toward reactant side in an endothermic reaction |
| Catalyst addition | Reaching equilibrium faster | No change or no shift in equilibrium condition |

#### 3.2.5.5 Practical Applications of Le Chatelier's Principle in Industry

In chemical engineering, Le Chatelier's principle is leveraged to optimize yields and energy efficiency in processes involving reversible reactions. For instance, in the synthesis of sulfuric acid through the contact process:

$$2SO_2(g) + O_2(g) \rightleftharpoons 2SO_3(g)$$

High pressures and specific temperatures are selected to maximize $SO_3$ production. Similarly, in the esterification reaction between acetic acid and ethanol to form ethyl acetate and water, removing water as it forms shifts the equilibrium toward ester production, significantly increasing yield.

## 3.3 MATLAB Code for Calculation of $\Delta H^0$, $\Delta G^0$, and $K_{eq}$

Appendix A | A.1 shows MATLAB (**DelG_Keq.m**) code for the calculation of $\Delta H^0$, $\Delta G^0$, and $K_{eq}$ for a given chemical reaction. A human-friendly algorithm of the MATLAB Code of Appendix | A.1 section is shown there, at the start.

Here is a demonstration showing how to run **DelG_Keq.m** using ammonia reaction synthesis: $N_2(g) + 3H_2(g) \rightleftharpoons 2NH_3(g)$

Input parameters and results are copied from the Command Window.

```
>> DelG_Keq
Enter the number of reactants: 2
Input "1" to search by Compound Name, "2" to search by Formula: 1
Input the name of the Compound: Nitrogen
206: Nitrogen atom (g) (N(g))
```

```
207: Nitrogen molecules (g) (N2(g))
211: Dinitrogen tetroxide (g) (N2O4(g))
212: Dinitrogen tetroxide (l) (N2O4(l))
236: Nitrogen trifluoride (g) (NF3(g))
247: Nitrogen dioxide (g) (NO2(g))
Select the index for reactant 1: 207
Enter the number of moles for reactant Nitrogen molecules (g): 1
Input "1" to search by Compound Name, "2" to search by Formula: 1
Input the name of the Compound: Hydrogen
128: Hydrogen atom (g) (H(g))
129: Hydrogen ion (aq) (H+(aq))
130: Hydrogen molecule (g) (H2(g))
134: Hydrogen peroxide (g) (H2O2(g))
135: Hydrogen peroxide (l) (H2O2(l))
136: Hydrogen sulfide (g) (H2S(g))
137: Hydrogen selenide (g) (H2Se(g))
139: Hydrogen phosphate ion (aq) (H3PO4(l))
143: Hydrogen bromide (g) (HBr(g))
145: Hydrogen chloride (g) (HCl(g))
146: Hydrogen cyanide (g) (HCN(g))
147: Hydrogen cyanide (l) (HCN(l))
151: Hydrogen fluoride (g) (HF(g))
160: Hydrogen iodide (g) (HI(g))
165: Hydrogen sulfite ion (aq) (HSO4- (aq))
Select the index for reactant 2: 130
Enter the number of moles for reactant Hydrogen molecule (g): 3
Enter the number of products: 1
Input "1" to search by Compound Name, "2" to search by Formula: 1
Input the name of the Compound: Ammonia
237: Ammonia (g) (NH3(g))
Select the index for product 1: 237
Enter the number of moles for product Ammonia (g): 2
Enter the temperature in Kelvin (K): 300
Selected Reactants:
1 mol of Nitrogen molecules (g) (N2(g))
3 mol of Hydrogen molecule (g) (H2(g))
Selected Products:
2 mol of Ammonia (g) (NH3(g))
Temperature (K): 300
Molar Gibbs free energy of reaction (kJ/mol): -32.8
Molar heat of reaction @ 298 K, (kJ/mol): -91.8
Equilibrium constant (dimensionless): 514272.8197
```

Notice that $\Delta H^0$ is evaluated at $T = 298$ K (25 °C). Heat capacity data for both reactants and products must be provided should we calculate it at any $T$ other than $T^0 = 25$ °C.

## 3.4 Expressions of $K_{eq}$

The following YouTube video is about determining the equilibrium expression:

https://youtu.be/ZK9cMIWFerY.

### 3.4.1 Liquid-Phase Reactions

Let us have a general *liquid-phase* reaction:

$$aA_{(l)} + bB_{(l)} \rightleftharpoons cC_{(l)} + dD_{(l)}$$

$$-r_A = -\frac{dC_A}{dt} = k_f[A]^a[B]^b - k_b[C]^c[D]^d \tag{3.10}$$

At equilibrium, the forward rate of consumption of $A$ is equal to the backward rate of generation of $A$:

$$-r_A = -\frac{dC_A}{dt} = 0 = k_f[A]^a[B]^b - k_b[C]^c[D]^d \rightarrow k_f[A]^a[B]^b = k_b[C]^c[D]^d \tag{3.11}$$

The true (i.e., thermodynamic) equilibrium constant, $K_{eq}$, which is an activity-based constant, is given by:

$$K_{eq} = \frac{k_f}{k_b} = \frac{[a_C]^c[a_D]^d}{[a_A]^a[a_B]^b} = \frac{[\gamma_C]^c[\gamma_D]^d}{[\gamma_A]^a[\gamma_B]^b} \times \frac{[C]^c[D]^d}{[A]^a[B]^b} = K_\gamma \times K_C \tag{3.12}$$

For an ideal liquid mixture or a slightly polar mixture (i. e. , $K_\gamma \approx 1$), we have:

$$K_{eq} \cong K_C \equiv \frac{[C]^c[D]^d}{[A]^a[B]^b} \tag{3.13}$$

Notice that nonideal behavior arises due to molecular or ionic interactions, particularly in solutions where electrostatic forces, hydrogen bonding, or van der Waals forces are present. Some common causes include:

a) **High Concentrations:** In concentrated solutions, molecules or ions are close enough to experience significant intermolecular forces, causing deviations from ideality.
b) **Electrolyte Solutions:** Charged ions in electrolytes interact through electrostatic forces, significantly impacting activity.
c) **Polar Interactions:** Polar molecules experience strong permanent-permanent dipole van der Waals forces that change their effective concentration.

Handling nonunity activity coefficient mixtures is beyond the scope of this book. However, activity coefficients are critical in accurately predicting equilibrium for nonideal systems commonly encountered in:

a) **Electrolyte solutions** (e.g., salts in water).
b) **Gas mixtures at high pressures**.
c) **Complex biochemical systems** where proteins, enzymes, and other molecules interact in nonideal ways.

**Example 3.7** Consider the following elementary (see Chapter 1 | Elementary Reactions) reaction:

$$A \underset{k_{bA}}{\overset{k_{fA}}{\rightleftarrows}} 2B$$

Here, $k_{fA}$ and $k_{bA}$ are the forward and backward reaction rate constants, both defined with respect to $A$. Express $K_C$ in terms of the equilibrium concentration of species $A$ and $B$.

$$A \xrightarrow{k_{fA}} 2B \qquad -r_{Af} = k_{fA}\, C_A$$

$$2B \xrightarrow{k_{bA}} A \qquad r_{Ab} = k_{bA}\, C_B^2$$

$$Rate_{net} = Rate_{forward} + Rate_{backward} = -k_{fA}\, C_A + k_{bA}\, C_B^2$$

$$-r_A = k_{fA}\, C_A - k_{bA}\, C_B^2$$

$$-r_A = k_{fA}\left[C_A - \frac{C_B^2}{k_{fA}/k_{bA}}\right]$$

$$-r_A = k_{fA}\left[C_A - \frac{C_B^2}{K_c}\right]$$

$$\text{At equilibrium: } -r_A = 0, \quad \left[C_A - \frac{C_B^2}{K_c}\right] = 0, \quad C_A = \frac{C_B^2}{K_c}$$

$$\text{Alternatively, } K_c = \left[\frac{C_B^2}{C_A}\right]_{Equilibrium} = \left[\frac{C_{Beq}^2}{C_{Aeq}}\right]$$

### 3.4.2 Gas-Phase Reactions

On the other hand, for a general gas-phase reaction:

$$aA_{(g)} + bB_{(g)} \rightleftharpoons cC_{(g)} + dD_{(g)}$$

If we have an ideal gas (low-pressure and high-temperature) mixture governed by the ideal gas law:

$$PV = nRT \tag{3.14}$$

Then, for a multicomponent gas mixture, we can write the component version as:

$$p_i V = n_i RT \tag{3.15}$$

We can define the concentration in the gas phase as:

$$C_i = \frac{\textit{Moles of species i}}{\textit{Total volume of the gas}} = \frac{n_i}{V} = \frac{p_i}{RT} \tag{3.16}$$

Substitute the value of $C_i$ in each bracket appearing in Eq. (3.13) and simplify:

$$K_C \equiv \frac{\left[\frac{p_C}{RT}\right]^c \left[\frac{p_D}{RT}\right]^d}{\left[\frac{p_A}{RT}\right]^a \left[\frac{p_B}{RT}\right]^b} \equiv \frac{[p_C]^c[p_D]^d}{[p_A]^a[p_B]^b} \times \left(\frac{1}{RT}\right)^{c+d-a-b} \equiv \frac{[p_C]^c[p_D]^d}{[p_A]^a[p_B]^b} \times \left(\frac{1}{RT}\right)^{\Delta n} \tag{3.17}$$

where $\Delta n = (c + d - a - b)$ is the net change in total number of moles as a result of a chemical reaction. Moreover, the partial pressure-based term is basically called $K_P$. Equation (3.17) becomes:

$$K_C \equiv K_P \times \left(\frac{1}{RT}\right)^{\Delta n} \tag{3.18}$$

Let us rearrange to get:

$$K_P \equiv (RT)^{\Delta n} K_C \tag{3.19}$$

We can further express $K_P$ in terms of the mole fraction-based equilibrium constant, $K_Y$:

$$K_P \equiv \frac{[p_C]^c[p_D]^d}{[p_A]^a[p_B]^b} = \frac{[P \times Y_C]^c[P \times Y_D]^d}{[P \times Y_A]^a[P \times Y_B]^b} = (P)^{c+d-a-b} \times \frac{[Y_C]^c[Y_D]^d}{[Y_A]^a[Y_B]^b} = (P)^{\Delta n} \times K_Y \tag{3.20}$$

Equating Eqs. (3.19) and (3.20), we reach at:

$$(RT)^{\Delta n} K_C = (P)^{\Delta n} \times K_Y \tag{3.21}$$

Let us rearrange Eq. (3.21) to reach at:

$$K_Y \equiv \left(\frac{RT}{P}\right)^{\Delta n} \times K_C \tag{3.22}$$

Table 3.2 shows the equilibrium constant, $K_{eq}$, for different gas-phase reactions, evaluated at a given temperature, T.

Notice that $K_{eq}$ can span a vast range of values. For instance, the values shown in Table 3.2 vary by as much as 66 orders of magnitude. Since products appear in the numerator of the equilibrium constant expression and reactants in the denominator, values of $K_{eq}$ greater than 1 indicate a strong tendency for reactants to convert into products. In such cases, chemists and their cousins, chemical engineers, describe the equilibrium as lying to the right, favoring product formation. An example is the reaction between $H_2$ and $O_2$ to produce $H_2O$, which has a very large equilibrium constant at 500 K. Here, $H_2$ acts as a strong reductant and $O_2$ as a strong oxidant, causing the reaction to proceed nearly to completion. Conversely, values of $K_{eq}$ less than 1 indicate that the ratio of products to reactants at equilibrium is very low, meaning reactants have little tendency to form products, and the equilibrium lies to the left, favoring reactant formation.

**Table 3.2** The equilibrium constant, $K_{eq}$, for different gas-phase reactions.

| **Reaction** | **Temperature (*K*)** | **Equilibrium Constant ($K_{eq}$)** |
|---|---|---|
| $S(s) + O_2(g) \rightleftharpoons SO_2(g)$ | 300 | $4.4 \times 10^{53}$ |
| $2H_2(g) + O_2(g) \rightleftharpoons 2H_2O(g)$ | 500 | $2.4 \times 10^{47}$ |
| $H_2(g) + Cl_2(g) \rightleftharpoons 2HCl(g)$ | 300 | $1.6 \times 10^{33}$ |
| $H_2(g) + Br_2(g) \rightleftharpoons 2HBr(g)$ | 300 | $4.1 \times 10^{18}$ |
| $2NO(g) + O_2(g) \rightleftharpoons 2NO_2(g)$ | 300 | $4.2 \times 10^{13}$ |
| $3H_2(g) + N_2(g) \rightleftharpoons 2NH_3(g)$ | 300 | $2.7 \times 10^{8}$ |
| $H_2(g) + D_2(g) \rightleftharpoons 2HD(g)$ | 100 | 1.92 |
| $H_2(g) + I_2(g) \rightleftharpoons 2HI(g)$ | 300 | $2.9 \times 10^{-1}$ |
| $I_2(g) \rightleftharpoons 2I(g)$ | 800 | $4.6 \times 10^{-7}$ |
| $Br_2(g) \rightleftharpoons 2Br(g)$ | 1,000 | $4.0 \times 10^{-7}$ |
| $Cl_2(g) \rightleftharpoons 2Cl(g)$ | 1,000 | $1.8 \times 10^{-9}$ |
| $F_2(g) \rightleftharpoons 2F(g)$ | 500 | $7.4 \times 10^{-13}$ |

*Source:* Adapted from https://chem.libretexts.org/Bookshelves/General_Chemistry/Map%3A_General_Chemistry_(Petrucci_et_al.)/15%3A_Princi ples_of_Chemical_Equilibrium/15.2%3A_The_Equilibrium_Constant_Expression.

Many reactions have equilibrium constants that fall within the range ($0.01 \le K_{eq} \le 100$), meaning they are neither very large nor very small. In these systems at equilibrium, there are substantial amounts of both products and reactants, suggesting no strong preference for converting reactants into products or vice versa. An example of such a system is the reaction between gaseous hydrogen and deuterium to form HD, which is used in high-stability fiber-optic light sources for ocean studies.

Moreover, observe in Table 3.2 that $K_{eq}$'s are unitless, even though $K_{eq}$ equations suggest that concentration units might not always cancel due to varying exponents. This is because equilibrium constants are determined using "effective concentrations," or activities, of reactants and products; these are ratios of the measured concentrations to a standard state of 1 M. The dimensionless $K_C$ is the true thermodynamic equilibrium constant, or I call it the chemist's approach. On the other hand, using dimensional $K_C$ is considered an apparent as opposed to true thermodynamic $K_C$, or I call it a chemical engineer's approach. The scalar value of $K_C$ will be the same; nevertheless, the dimensions/units are different.

The equilibrium composition for each species will be the same for both approaches. This is very important.

**Example 3.8** ***$N_2O_4$ Dissociation into $NO_2$***
Dinitrogen tetroxide ($N_2O_4$) reversibly dissociates into nitrogen dioxide ($NO_2$) as given by the following gas-phase reaction:

$$N_2O_4(g) \leftrightarrow 2NO_2(g) \tag{3.23}$$

0.50 mol of dinitrogen tetroxide is initially heated in a 1 L flask and when equilibrium is reached, the concentrations of dinitrogen tetroxide and nitrogen dioxide are measured to be 0.035 and 0.285 mol/L, respectively [2]. Calculate the equilibrium constant, $K_C$.

Based on Eq. (3.23), we have:

$$K_c = \left[\frac{C^2_{NO_2}}{C_{N_2O_4}}\right]_{Equilibrium} = \left[\frac{C^2_{NO_{2,eq}}}{C_{N_2O_{4,eq}}}\right] = \frac{[0.285]^2}{[0.035]} = 2.3027 \tag{3.24}$$

**NOTE**

The equilibrium constant for a reaction written in reverse is the reciprocal of the equilibrium constant for the reaction in its original direction. Moreover, in general, if all coefficients in a balanced chemical equation are multiplied by *n*, the new equilibrium constant becomes the original equilibrium constant raised to the power of *n*.

**Example 3.9** ***The Inverse of $K_{eq}$***
Consider the formation of water:

$$2H_2(g) + O_2(g) \rightleftharpoons 2H_2O(g) \tag{3.25}$$

Since $H_2$ is a good reductant and $O_2$ is a good oxidant, this reaction has a very large equilibrium constant ($K_C = 1.22 \times 10^{44}$ at 298 K) [https://www.vaia.com/en-us/textbooks/chemistry/general-chemistry-the-essential-concepts-5-edition/chapter-15/problem-58-the-equilibrium-constant-kp-for-the-reaction-2-ma/]. Consequently, the equilibrium constant for the reverse reaction, that is, the decomposition of water into $H_2$ and $O_2$, is very small: $K'_C = \frac{1}{K_C} = \frac{1}{1.22\times10^{44}} = 8.18 \times 10^{-45}$.

As indicated by this extremely small equilibrium constant, and luckily for sustaining life of "Animalia" and "Plantae" kingdom as we know it, a substantial amount of energy is also required to dissociate water into its constituents $H_2$ and $O_2$. Thus, water exists in nature as a highly stable compound. The hydrogen and oxygen gas mixture does not readily react at room temperature, even though water formation is highly favorable, because the reaction requires an activation energy higher than what room temperature can supply. Thus, an ignition or starting point is needed to trigger the reaction to spontaneously proceed.

### 3.4.3 Heterogeneous Reactions

When the products and reactants of an equilibrium reaction exist in a single phase, either gas or liquid, the system is termed a homogeneous equilibrium. In such systems, the concentrations of reactants and products can vary widely. Conversely, if the reactants, products, or both are in different phases, the system is known as a heterogeneous equilibrium; for example, a reaction involving a gas and a solid or liquid.

For pure solids and liquids, molar concentrations remain relatively constant with temperature, so they are treated as constants in equilibrium expressions, simplifying the calculations. The reference states for pure solids and liquids are their stable forms at 1 bar (roughly 1 atm), where they are assigned an activity of 1.

Consider the following reaction:

$$CO_2(g) + C(s) \rightleftharpoons 2CO(g) \tag{3.26}$$

$$K'_C = \frac{[CO(g)]^2}{[CO_2(g)][C(s)]} \tag{3.27}$$

However, since graphite is a solid, its molar concentration being calculated from its density and molar mass remains essentially constant, with a value of:

$$[C(s)] = \frac{mass\ density}{molar\ mass} = \frac{2.3\ \frac{g}{cm^3}}{12\ \frac{g}{mol}} = 0.19167\ \frac{mol}{cm^3} \times \frac{1{,}000\ cm^3}{1\ L} = 191.67\frac{mol}{L} \tag{3.28}$$

We can define a new $K_C$ such that:

$$K_C = K'_C[C(s)] = K'_C * 191.67\ M = \frac{[CO(g)]^2}{[CO_2(g)]} \tag{3.29}$$

The equilibrium constant for this reaction can also be expressed in terms of the gases' partial pressures:

$$K_P = \frac{P^2_{CO}}{P_{CO_2}} \tag{3.30}$$

Although the concentrations of pure liquids or solids are not explicitly included in the equilibrium constant expression, these substances must still be present in the reaction mixture for chemical equilibrium to be established. Regardless of the concentrations of the other components, the reaction described in Eq. (3.26) will only reach equilibrium if a stoichiometric or excess amount of solid carbon is present, ensuring that some remains when equilibrium is achieved. The quantity of solid carbon, whether 1 g or 1,000 g, does not really affect the equilibrium composition of the gaseous components in the system.

**Example 3.10** ***Write Expressions for $K_C$ and/or $K_P$ if Applicable, for the Following Heterogeneous Reversible Reactions at Equilibrium:***

a)

$$HF_{(aq)} + H_2O_{(l)} \rightleftharpoons H_3O^+_{(aq)} + F^-_{(aq)} \tag{3.31}$$

$$K'_C = \frac{\left[H_3O^+_{(aq)}\right]\left[F^-_{(aq)}\right]}{[HF_{(aq)}][H_2O_{(l)}]} \tag{3.32}$$

Notice that $[H_2O_{(l)}]$, as a medium or solvent, is sufficiently large and at the same time constant, then we can redefine

$$[H_2O(l)] = \frac{\text{mass density}}{\text{molar mass}} = \frac{1.0\,\frac{g}{cm^3}}{18.02\,\frac{g}{mol}} = 0.05549\,\frac{mol}{cm^3} \times \frac{1{,}000\,cm^3}{1\,L} = 55.5\frac{mol}{L} \tag{3.33}$$

$$K_C = K'_C[H_2O_{(l)}] = K'_C * 55.5\,M = \frac{\left[H_3O^+_{(aq)}\right]\left[F^-_{(aq)}\right]}{[HF_{(aq)}]} \tag{3.34}$$

b)

$$CH_3COOH(aq) + C_2H_5OH(aq) \rightleftarrows CH_3COOC_2H_5(aq) + H_2O(l) \tag{3.35}$$

$$K'_C = \frac{[H_2O(l)][CH_3COOC_2H_5(aq)]}{[CH_3COOH(aq)][C_2H_5OH(aq)]} \tag{3.36}$$

Since $[H_2O_{(l)}]$, as a medium or solvent, is sufficiently large and at the same time constant; then we can redefine

$$K_C = \frac{K'_C}{[H_2O_{(l)}]} = \frac{K'_C}{55.5\,M} = \frac{[CH_3COOC_2H_5(aq)]}{[CH_3COOH(aq)][C_2H_5OH(aq)]} \tag{3.37}$$

c)

$$CaCO_3(s) \rightleftarrows CaO(s) + CO_2(g) \tag{3.38}$$

$$K'_C = \frac{[CO_2(g)][CaO(s)]}{[CaCO_3(s)]} \tag{3.39}$$

Since both $[CaCO_3(s)]$ and $[CaO(s)]$ are available in appreciable quantities and their concentrations are constant, we can redefine:

$$K_C = K'_C \times \frac{[CaCO_3(s)]}{[CaO(s)]} = [CO_2(g)] = \frac{P_{CO2}}{RT} \tag{3.40}$$

**Example 3.11** ***$CaCO_3$ Decomposition***

$$CaCO_3(s) \rightleftarrows CaO(s) + CO_2(g) \tag{3.41}$$

At 800 °C, the pressure of $CO_2$ is 0.24 atm. Calculate $K_P$ and $K_C$ for this reaction at the given temperature.

$$K_C = [CO_2(g)] = \frac{P_{CO_2}}{RT} = \frac{0.24\,atm}{0.08206\,\frac{atm\cdot L}{mol\cdot K} \times (1{,}073\,K)} = 0.0027\frac{mol}{L} \tag{3.42}$$

$$K_P = P_{CO_2} = 0.24\,atm \tag{3.43}$$

### 3.4.4 Equilibrium Constants for Multiple Reactions

In general, consider the following set of reactions:

$$A + B \rightleftarrows C + D \qquad K'_C = \frac{[C][D]}{[A][B]} \tag{3.44}$$

$$C + D \rightleftarrows E + F \qquad K''_C = \frac{[E][F]}{[C][D]} \tag{3.45}$$

$$A + B \rightleftarrows E + F \qquad K_C = \frac{[C][D]}{[A][B]} \times \frac{[E][F]}{[C][D]} = K'_C \times K''_C = \frac{[E][F]}{[A][B]} \tag{3.46}$$

**NOTE**

If a reaction can be represented as the sum of two or more reactions, the equilibrium constant for the overall reaction is the product of the equilibrium constants for each individual reaction.

**Example 3.12** ***$CaCO_3$ Dissociation in Acidic Medium***
Calculate the value of $K_{eq}$ for the reaction:

$$CaCO_3(s) + H^+_{aq} \rightleftarrows Ca^{++}_{aq} + HCO^-_{3(aq)} \tag{A}$$

Given the following equilibrium constants:

$$CaCO_3(s) \rightleftarrows Ca^{++}_{aq} + CO^{--}_{3(aq)} \qquad K_1 = 10^{-6.3} \tag{B}$$

$$HCO^-_{3(aq)} \rightleftarrows H^+_{aq} + CO^{--}_{3(aq)} \qquad K_2 = 10^{-10.3} \tag{C}$$

**Solution**
Let us use reaction B and switch reaction C to regenerate reaction A. We have:

$$CaCO_3(s) \rightleftarrows Ca^{++}_{aq} + CO^{--}_{3(aq)} \qquad K_1 = 10^{-6.3}$$

$$H^+_{aq} + CO^{--}_{3(aq)} \rightleftarrows HCO^-_{3(aq)} \qquad K'_2 = \frac{1}{K_2} = 10^{+10.3}$$

$$CaCO_3(s) + H^+_{aq} \rightleftarrows Ca^{++}_{aq} + HCO^-_{3(aq)} \qquad \mathrm{K} = \mathrm{K}_1 * \mathrm{K}'_2 = 10^{-6.3} * 10^{10.3} = 10^{4.0}$$

Notice that:

$$K = K_1 \times K'_2 = \left\{ \left[Ca^{++}_{aq}\right] \left[CO^{--}_{3(aq)}\right] \right\} \times \left\{ \frac{\left[HCO^-_{3(aq)}\right]}{\left[H^+_{aq}\right] \left[CO^{--}_{3(aq)}\right]} \right\} = \frac{\left[Ca^{++}_{aq}\right] \left[HCO^-_{3(aq)}\right]}{\left[H^+_{aq}\right]}$$

$[CaCO_3(s)]$ exists in an abundant amount as solid, thus it can be combined with the equilibrium constant, as we did earlier in previous heterogeneous reversible equilibrium cases.

## 3.5 Limiting and Excess Reactants

Two reactants, $A$ and $B$, are in stoichiometric proportion when the ratio of moles of $A$ to moles of $B$ matches the stoichiometric ratio specified by the balanced reaction equation.

For reactants in the following reaction:

$$2H_2 + O_2 \rightarrow 2H_2O \tag{3.47}$$

To be present in stoichiometric proportion, there must be 2 moles of $H_2$ for every mole of $O_2$ (so that $\frac{n_{H_2}}{n_{O_2}} = \frac{2}{1}$) present in the feed to the reactor.

If reactants are fed to a chemical reactor in stoichiometric proportion and the reaction proceeds to completion, it literally means all the reactants are completely consumed.

It follows that if we start with 1 mol of $O_2$ and less than 2 mol of $H_2$ (i.e., $[H_2] < 2$ mol, initially present in the reactor), $H_2$ will vanish first, while if more than 2 mol of $H_2$ is initially present, $O_2$ will disappear first.

When a reaction proceeds to completion, the reactant that is fully consumed first is known as the limiting reactant, while the remaining reactants are called excess reactants. A reactant is **limiting** if it is present in less than its stoichiometric proportion relative to each other reactant. However, if all reactants are present in stoichiometric proportions, any reactant can be considered limiting.

Let $(n_A)_{feed}$ represent the number of moles of an excess reactant, $A$, supplied to a reactor, and $(n_A)_{stoich}$ represent the stoichiometric amount of $A$ needed to fully react with the limiting reactant. Then, $(n_A)_{feed} - (n_A)_{stoich}$ gives the amount by which $A$ exceeds the quantity required for complete reaction (100% completion). The fractional excess of reactant A is the ratio of this excess amount to its stoichiometric requirement:

$$Fractional\ excess\ of\ A = \frac{(n_A)_{feed} - (n_A)_{stoich}}{(n_A)_{stoich}} \tag{3.48}$$

**The percentage excess of *A*** is 100 times the fractional excess:

$$Percent\ excess\ of\ A = \frac{(n_A)_{feed} - (n_A)_{stoich}}{(n_A)_{stoich}} \times 100\% \tag{3.49}$$

**Example 3.13** ***Fractional Excess of Reactant***

Consider the combustion of *n*-heptane:

$$C7H16 + 11O_2 \rightarrow 7CO_2 + 8H_2O \tag{3.50}$$

Suppose that 100.0 kmol/h of *n*-heptane and 800.0 kmol/h of oxygen are fed to a reactor. The *stoichiometric ratio of oxygen to n-hexane is 11:1* (the ratio of coefficients in the stoichiometric equation), and since the ratio of $O_2$ to $C_7H_{16}$ in the feed is 8:1 (800:100), Oxygen is fed in a smaller-than-stoichiometric proportion to *n*-heptane. **Oxygen is therefore the limiting reactant and *n*-heptane the excess reactant**. If the reaction goes to completion, then the molar flow rate of *n*-heptane that will be consumed will be given by:

$$(\dot{n}_{heptane})_{stoich} = 800 \frac{\text{kmol } O_2\ consumed}{\text{h}} \times \frac{1\ \text{kmol}\ n - heptane\ consumed}{11\ \text{kmol } O_2\ consumed}$$

$$= 72.73 \frac{\text{kmol}\ n - heptane}{\text{h}}$$

$$Fractional\ excess\ of\ n - heptane = \frac{(\dot{n}_A)_{feed} - (\dot{n}_A)_{stoich}}{(\dot{n}_A)_{stoich}} = \frac{100 - 72.73}{72.73} = 0.375$$

The *percent excess of n-heptane* = (*Fractional excess of n – heptane*) × 100 % = 37.5%

Notice that the *fractional excess of $O_2$ will be zero* as it is the *limiting reactant*. We say that there is 37.5% *excess n-heptane* in the feed.

## 3.6 Fractional Conversion

Chemical reactions generally do not occur instantaneously and often proceed at a slow rate. Therefore, designing a reactor for complete conversion of the limiting reactant may be impractical. Instead, the reactor effluent often contains some unreacted limiting reactant, which is typically separated from the product in a subsequent process. The separated reactant is then recycled to the reactor inlet. The fractional conversion of a reactant is defined as the ratio:

$$f = \frac{\textit{Moles reacted}}{\textit{Moles fed}} \tag{3.51a}$$

$$f = \frac{\textit{Molar flow rate reacted}}{\textit{Molar flow rate fed}} \tag{3.51b}$$

The unreacted fraction is accordingly $1-f$. If 100 moles of a reactant are fed and 95 moles react, the fractional conversion is 0.95 (the percentage conversion is 95%), and the fraction unreacted is 0.05. If 30 mol/min of a reactant is fed and the percentage conversion is 70%, then $(30)\times(0.70) = 21$ mol/min has reacted, and $(30)\times(1-0.70) = 9$ mol/min remains unreacted.

| NOTE |
|---|
| Definition of $v_i$: $v_i$ is a number that represents the stoichiometric coefficient of the $i$th species in a chemical reaction, making it negative for reactants and positive for products. |

**Example 3.14** ***The Stoichiometric Coefficient of Species i***
Consider the hydrogenation of propene into *n*-propane ($C_3H_8$):

$$CH_3 - CH = CH_2 + H_2 \rightarrow CH_3CH_2CH_3 \tag{3.52}$$

$$\nu_{CH_3CH_2CH_3} = +1 \qquad \nu_{CH_3-CH=CH_2} = -1 \qquad \nu_{H_2} = -1$$

## 3.7 The Extent of a Reaction

Consider the same reaction as: given by Eq. (3.52)

Suppose that we initially have 2 moles of $H_2$ ($n^o_{H_2} = 2$ mol), 1 mole of $CH_3$—$CH$=$CH_2$ ($n^o_{CH_3—CH=CH_2} = 1$ mol) and zero moles of $CH_3CH_2CH_3$ ($n^o_{CH_3CH_2CH_3} = 0$ mol).

If half a mole (1/2 mole) of $CH_3$—$CH$=$CH_2$ reacted, then the remaining $CH_3$—$CH$=$CH_2$ would be $1 - \frac{1}{2} = \frac{1}{2}$ mol. So, half a mole of $CH_3$—$CH$=$CH_2$ reacted and another half remained unreacted.

The number of moles of $H_2$ that reacted is given by:

$$\tfrac{1}{2}\text{ mol } CH_3—CH{=}CH_2 \text{ reacted} \times \left(\frac{1 \text{ mol } H_2 \text{ reacted}}{1 \text{ mol } CH_3 - CH = CH_2 \text{ reacted}}\right) = \frac{1}{2}\text{mol } H_2 \textit{ reacted} \tag{3.53}$$

So, half a mole of $H_2$ reacted, and 1½ moles of $H_2$ remained unreacted.

The number of moles of $CH_3CH_2CH_3$ (or $C_3H_8$) produced will be given by:

$$\tfrac{1}{2}\text{ mol } CH_3—CH{=}CH_2 \text{ reacted} \times \left(\frac{1 \text{ mol } CH_3CH_2CH_3 \text{ produced}}{1 \text{ mol } (CH_3—CH{=}CH_2) \text{ reacted}}\right) = \frac{1}{2}\text{ mol } C_3H_8 \textit{ produced} \tag{3.54}$$

**Table 3.3** Number of moles analysis for species involved in a chemical reaction given by Eq. (3.52).

| Component | $v_i$ | Initially, $n_i^o$ | Reacted | Remained, $n_i$ | $\Delta n_i$ | $\Delta n_i/v_i$ |
|---|---|---|---|---|---|---|
| $CH_3{-}CH{=}CH_2$ | −1 | 1 | ½ mol | ½ mol | − ½ mol | ½ mol |
| $H_2$ | −1 | 2 | ½ mol | 1½ mol | − ½ mol | ½ mol |
| $CH_3CH_2CH_3$ | +1 | 0 | ½ mol | ½ mol | ½ mol | ½ mol |

Let us define the change in number of moles of species $i$ as:

$$\Delta n_i = \textit{Finally Present (or Remained)} - \textit{Initially Present} = n_i - n_i^o \tag{3.55}$$

Table 3.3 summarizes the results of the reaction.

Notice that the value of the last column is the same for all species involved in the reaction and is equal to ½ mol. This term is named the **extent of reaction** and has the units as those of $n_i$. Thus, the **extent of reaction** is: defined as:

$$\xi = \frac{\Delta n_i}{v_i} = \frac{\textit{Remained} - \textit{Initially Present}}{v_i} = \frac{n_i - n_i^o}{v_i} \tag{3.56a}$$

$$\dot{\xi} = \frac{\Delta F_i}{v_i} = \frac{\textit{Remained} - \textit{Initially Present}}{v_i} = \frac{F_i - F_i^o}{v_i} \tag{3.56b}$$

Rearranging Eq. (3.56) gives:

$$n_i = n_i^o + v_i \xi \tag{3.57a}$$

Alternatively, for a flow-type reactor, we can write:

$$F_i = F_i^o + v_i \dot{\xi} \tag{3.57b}$$

For a multi-reaction system, we can generalize Eq. (3.57a) to be:

$$n_i = n_i^o + \sum_{j=1}^{\#of\,rxnx} v_{ij} \times \xi_j \tag{3.58a}$$

Alternatively, for a flow-type reactor, we can write:

$$F_i = F_i^o + \sum_{j=1}^{\#of\,rxnx} v_{ij} \times \dot{\xi}_j \tag{3.58b}$$

where

- $\Sigma_{j=1}^{\#of\,rxns}$ represents the summation over all $j$ reactions in the system.
- $v_{ij}$ is the stoichiometric coefficient of species $i$ in reaction $j$ (positive for products, negative for reactants).
- $\xi_j$ is the extent of reaction $j$, expressed in mole.
- $\dot{\xi}_j$ is the extent of reaction $j$, expressed in mole/time.

**Example 3.15** ***Calculation of the Extent of a Reaction***

Acetonitrile (also called cyano-methane or methyl cyanide) is produced via the following reaction:

$$C_2H_4 + NH_3 + O_2 \rightarrow C_2H_3N + 2\,H_2O \tag{3.59}$$

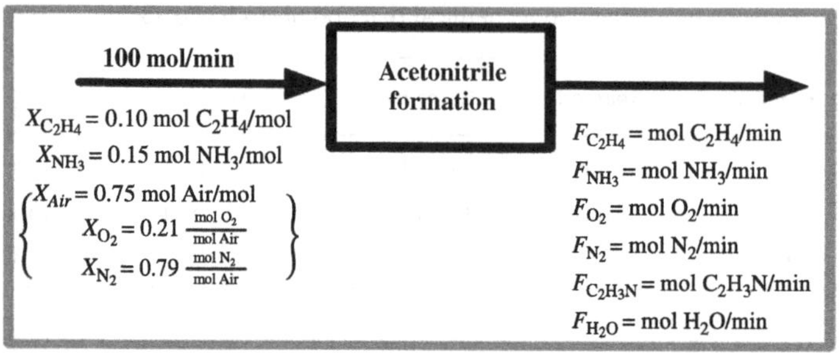


**Figure 3.3** Flow chart for a fluidized bed reactor that produces acetonitrile from ammonia, ethylene, and air.

The feed contains 10.0 mole% ethylene, 15.0% ammonia, and 75.0% air, as shown in Figure 3.3. A fractional conversion of 50.0% of the limiting reactant is achieved. Taking 100 mol/min of feed as a basis, determine which reactant is limiting, the percentage by which each of the other reactants is in excess, and the molar amounts of all product gas constituents for a 50% conversion of the limiting reactant.

**Solution**

**Basis: 100 $\frac{\text{mol}}{\text{min}}$ Feed $\left(F_0 = \dot{n}_0 = 100\ \frac{\text{mol}}{\text{min}}\right)$**

So, **initially** we have:

Feed molar flow rate of $C_2H_4$: $\left(F^o_{C_2H_4} = 10\ \frac{\text{mol}}{\text{min}}\right)$

Feed molar flow rate of $NH_3$: $\left(F^o_{NH_3} = 15\ \frac{\text{mol}}{\text{min}}\right)$

Feed molar flow rate of $O_2$: $\left(F^o_{O_2} = 0.21 * \left(F^o_{air} = 75\ \frac{\text{mol}}{\text{min}}\right)\right) = 15.75\ \frac{\text{mol}}{\text{min}}$

Feed molar flow rate of $N_2$: $\left(F^o_{O_2} = 0.79 * \left(F^o_{air} = 75\ \frac{\text{mol}}{\text{min}}\right)\right) = 59.25\ \frac{\text{mol}}{\text{min}}$

Feed molar flow rate of $H_2O$: $\left(F^o_{H_2O} = 0\ \frac{\text{mol}}{\text{min}}\right)$

Feed molar flow rate of $C_3H_3N$: $\left(F^o_{C_2H_3N} = 0\ \frac{\text{mol}}{\text{min}}\right)$

**NOTE**

To determine the *limiting reactant*, test for the *smallest ratio* of $\frac{F^o_i}{\|\nu_i\|}$

$$\frac{F^o_{C_2H_4}}{\|\nu_{C_2H_4}\|} = \frac{10\ \text{mol}}{1\ \text{min}} < \frac{F^o_{O_2}}{\|\nu_{O_2}\|} = \frac{15.75}{1.5} = 10.5\ \frac{\text{mol}}{\text{min}} < \frac{F^o_{NH_3}}{\|\nu_{NH_3}\|} = \frac{15}{1} = 15\frac{\text{mol}}{\text{min}}$$

Therefore, $C_2H_4$ is the limiting reactant.

$NH_3$ mol/min needed for 10 mol/min of reacting $C_2H_4$ are given by:

$$10\ \frac{\text{mol}}{\text{min}} C_2H_4\ \textit{react} \times \frac{1\ \text{mol } NH_3\ \textit{reacts}}{1\ \text{mol } C_2H_4\ \textit{reacts}} = 10\ \frac{\text{mol}}{\text{min}}\ NH_3\ \textit{react}$$

Thus, $NH_3$ is available in excess amount.

$$\textit{Fractional excess of}\, NH_3 = \frac{(F_{NH_3})_{feed} - (F_{NH_3})_{stoich}}{(F_{NH_3})_{stoich}} = \frac{15-10}{10} = \frac{5}{10} = 0.5$$

$$(\%\ excess)_{NH_3} = 100\% \times 0.5 = 50\%$$

$O_2$ mole/min needed for 10 mol/min of reacting $C_2H_4$ are given by:

$$10\,\frac{mol}{min} C_2H_4\ react \times \frac{1\ mol\ O_2\ reacts}{1\ mol\ C_2H_4\ reacts} = 10\,\frac{mol}{min} O_2\ react$$

Thus, $O_2$ is available in *excess* amount.

$$\textit{Fractional excess of}\, O_2 = \frac{(F_{O_2})_{feed} - (F_{O_2})_{stoich}}{(F_{O_2})_{stoich}} = \frac{15.75-10}{10} = \frac{5.75}{10} = 0.575$$

$$(\%\ excess)_{O_2} = 100\% \times 0.575 = 57.5\%$$

**Notice that** $(\%\ excess)_{C_2H_4} = 0.0\%$
The **fractional conversion** of the limiting reactant ($C_2H_4$) is given as 0.4. Using Eq. (3.51b):

$$f = \frac{\textit{Molar flow rate reacted}}{\textit{Molar flow rate fed}} = 0.5 \qquad \textit{Molar flow rate reacted}$$

$$= 0.5 \times 10\,\frac{mol}{min}\,fed = 5\,\frac{mol}{min}$$

$$\textit{molar flow rate remained} = \textit{Molar flow rate fed} - \textit{Molar flow rate reacted}$$

$$= 10\,\frac{mol}{min} - 5\,\frac{mol}{min} = 5\,\frac{mol}{min}$$

Based on the limiting reactant, $C_2H_4$, use Eq. (3.56b) to calculate the *extent of reaction*, $\dot{\xi}$:

$$\dot{\xi} = \frac{\Delta F_i}{\nu_i} = \frac{\textit{Remained} - \textit{Initially Present}}{\nu_i} = \frac{F_i - F_i^o}{\nu_i} = \frac{5-10}{-1} = 5\,\frac{mol}{min}$$

Use Eq. (3.57b) to calculate the number of moles for each species:

$$F_i = F_i^o + \nu_i\dot{\xi}$$

$$F_{C_2H_4} = F^o_{C_2H_4} + \nu_i\dot{\xi} = 10 + (-1) \times 5 = 5\,\frac{mol}{min}$$

$$F_{O_2} = F^o_{O_2} + \nu_i\dot{\xi} = 15.75 + (-1) \times 5 = 10.75\,\frac{mol}{min}$$

$$F_{NH_3} = F^o_{NH_3} + \nu_i\dot{\xi} = 15 + (-1) \times 5 = 10\,\frac{mol}{min}$$

$$F_{C_2H_3N} = C_2H_3N + \nu_i\dot{\xi} = 0 + (+1) \times 5 = 5\,\frac{mol}{min}$$

$$F_{H_2O} = F^o_{H_2O} + \nu_i\dot{\xi} = 0 + (+2) \times 5 = 10\,\frac{mol}{min}$$

$$F_{N_2} = F^o_{N_2} + \nu_i\dot{\xi} = 59.25 + (0) \times 5 = 59.25\,\frac{mol}{min}$$

$$F_{total} = 5 + 10.75 + 10 + 5 + 10 + 59.25 = 100\,\frac{mol}{min} = F_{feed}$$

Notice that the change in total number of moles between the feed and the exit stream is zero as $\Delta n = 2+1\text{-}1\text{-}1\text{-}1 = 0$.

## 3.8 The Extent of Reaction Calculation Using MATLAB Roots Finding Technique

Consider the reaction in a batch reactor between formic acid and methanol to produce methyl formate and water:

$$CHCOOH_{(l)} + CH_3OH_{(l)} \rightleftharpoons CH_3OOCH_{(l)} + H_2O_{(l)}$$

In symbolic form: FA + MOH $\rightleftharpoons$ MF + H2O

We have initially one mole of FA and 2 moles of MOH placed in $V$ (L) solution.
From Example 3.4, it was found that $K_C = 4.3\times 10^3$.
Calculate the equilibrium number of moles of all reacting species.
Make use of Eq. (3.57a) for the definition of $n_i$ as a function of the extent of reaction:

$$n_{FA} = n^o_{FA} + \nu_i\xi = 1 - \xi \qquad n_{MOH} = n^o_{MOH} + \nu_i\xi = 2 - \xi$$

$$n_{MF} = n^o_{MA} + \nu_i\xi = 0 + \xi \qquad n_{H_2O} = n^o_{H_2O} + \nu_i\xi = 0 + \xi$$

$$K_C = \frac{[MF][H_2O]}{[FA][MOH]} = \frac{\left[\frac{\xi}{V(L)}\right]\left[\frac{\xi}{V(L)}\right]}{\left[\frac{1-\xi}{V(L)}\right]\left[\frac{2-\xi}{V(L)}\right]} = \left[\frac{1}{V}\right]^{(1+1-1-1)=0} \frac{\xi^2}{(1-\xi)(2-\xi)} = 4.3\times 10^3$$

Simplify to end up with a quadrature (i.e., second-order polynomial) with two possible roots.

$$4.3\times 10^3 \times (2 - 3\xi + \xi^2) = \xi^2$$

Move the right side to the left and simplify.

$$8{,}600 - 12{,}900\xi + 4{,}299\xi^2 = 0$$

$$c + b\xi + a\xi^2 = 0$$

Figure 3.4 shows MATLAB code for finding the roots of a quadrature (second-order polynomial) with known coefficients.

```
% Define the coefficients of the quadratic equation: 2550 - 5100ξ + 2549ξ²
clear variables;
a = 4299;
b = -12900;
c = 8600;
% Solve for the roots
roots_quadratic = roots([a, b, c]);
% Display the roots
disp(['The first root of the equation is:', num2str(roots_quadratic(1))]);
disp(['The second root of the equation is:', num2str(roots_quadratic(2))]);
```

**Figure 3.4** MATLAB code for finding the roots of a quadrature (i.e., 2nd-order polynomial).

Below are the MATLAB results, copied from the Command Window:

```
The first root of the equation is: 2.0009
The second root of the equation is: 0.99977
```

Obviously, the second root is the feasible solution as $\xi$ can not be greater than 1; otherwise, we will have negative moles or concentration for FA.

The equilibrium number of moles of each species can be expressed in terms of $\xi$.

$$n_{FA} = 1 - \xi = 1 - 0.99977 = 0.00023 \text{ mol} \qquad n_{MOH} = 2 - \xi = 2 - 0.99977 = 1.00023 \text{ mol}$$

$$n_{MF} = \xi = 0.99977 \text{ mol} \qquad n_{H_2O} = \xi = 0.99977 \text{ mol}$$

$$n_{total} = 0.00023 + 0.00023 + 0.99977 + 0.99977 = 2 \text{ moles}$$

Of course, the equilibrium concentration of each species can be calculated by dividing the equilibrium number of moles of each species by the volume of the solution, $V$, expressed in liter. On the other hand, the equilibrium composition (i.e., mole fraction) can be calculated by dividing the number of moles of each species by the total number of moles, which is 2 moles as shown above.

**Example 3.16** ***Calculation of $K_C$ at $T \neq T_0$ and Equilibrium Compositions for $NH_3$ Synthesis***

Consider the gas-phase ammonia synthesis reaction, taking place in a batch reactor:

$$N_{2(g)} + 3H_{2(g)} \rightleftharpoons 2NH_{3(g)}$$

> **NOTE**
>
> For the Haber-Bosch process, the most widely used industrial method for ammonia synthesis, responsible for over 90% of global production, it operates at high temperature and pressure:
>
> **Temperature:** 350–550 °C (662–1022 °F) and pressure: 150–300 bar (2,175–4,350 psi). For the sake of simplicity and using the ideal gas law, we reduce the pressure down to 10 atm. By operating the Haber-Bosch process at higher temperatures, even though it may favor the reverse reaction at equilibrium, the rate of the forward reaction is: increased. This means that ammonia is produced more quickly in a reasonable timeframe, making the industrial process more efficient. The higher temperature helps overcome the activation energy barrier for the reaction, leading to a faster rate.

Let us assume that initially we have 5 moles of $H_2$ and 2 moles of $N_2$ at $T = 427\,°C$ and $P = 10$ atm. Find the equilibrium composition.

**Solution**

Refer to Eq. (3.4): $\Delta G^0 = \sum_{i=1}^{N} \nu_i \left(\Delta G_f^0\right)_{Products} + \sum_{i=1}^{N} \nu_i \left(\Delta G_f^0\right)_{Reactants}$

$$\Delta G^0 = 2\left(\Delta G_f^0\right)_{NH_3} - 1\left(\Delta G_f^0\right)_{N_2} - 3\left(\Delta G_f^0\right)_{H_2} =$$

$$\Delta G^0 = 2(-16.3) - 0 - 3(0) = -32.6 \frac{\text{kJ}}{\text{mol}}$$

From Eq. (3.3)

$$K_C = e^{-\left(\frac{\Delta G^0}{RT}\right)} = e^{-\left(\frac{-32{,}600\frac{\text{J}}{\text{mol}}}{8.314\frac{\text{J}}{\text{mol}\cdot\text{K}}(700.15)\text{K}}\right)} = 270.53$$

Make use of Eq. (3.57a) for the definition of $n_i$ as a function of the extent of reaction:

$$n_{\text{N}_2} = n^{o}_{\text{N}_2} + v_i\xi = 2 - \xi \qquad n_{H_2} = n^{o}_{H_2} + v_i\xi = 5 - 3\xi$$

$$n_{\text{NH}_3} = n^{o}_{\text{NH}_3} + v_i\xi = 0 + 2\xi \qquad n_{total} = 2 - \xi + 5 - 3\xi + 2\xi = 7 - 2\xi$$

From Eq. (3.22), we have:

$$K_Y \equiv \left(\frac{RT}{P}\right)^{\Delta n} \times K_C = \left(\frac{0.08206\,\frac{\text{atm}\cdot\text{L}}{\text{mol}\cdot\text{K}} \times 700.15\text{ K}}{10\text{ atm}}\right)^{2-3-1} \times K_C = 0.0303\left[\frac{\text{L}}{\text{mol}}\right]^{-2} \times K_C$$

Notice that $K_C = \frac{[\text{NH}_3]^2}{[\text{N}_2][\text{H}_2]^3} = 270.53\left[\frac{\text{mol}}{\text{L}}\right]^{-2}$

Thus, $K_Y = 0.0303\left[\frac{\text{L}}{\text{mol}}\right]^{-2} \times 270.53\left[\frac{\text{mol}}{\text{L}}\right]^{-2} = 8.1971[\textit{Dimensionless}]$

$$K_Y = \frac{[Y_{\text{NH}_3}]^2}{[Y_{\text{N}_2}][Y_{\text{H}_2}]^3} = \frac{\left[\frac{n_{\text{NH}_3}}{n_{total}}\right]^2}{\left[\frac{n_{\text{N}_2}}{n_{total}}\right]\left[\frac{n_{\text{H}_2}}{n_{total}}\right]^3} = \left(\frac{1}{n_{total}}\right)^{-2} \times \frac{(2\xi)^2}{(2-\xi)(5-3\xi)^3}$$

$$K_Y = (7-2\xi)^2\frac{(2\xi)^2}{(2-\xi)(5-3\xi)^3} = \frac{(7-2\xi)^2(2\xi)^2}{(2-\xi)(5-3\xi)^3} = 8.1971$$

Find $\xi$ that will satisfy the solution. Let us put it in the form of $f(x) = 0$:

$$f(x) = \frac{(7-2x)^2(2x)^2}{(2-x)(5-3x)^3} - 8.1971 = 0$$

Figure 3.5 shows MATLAB code for finding the roots of a defined $f(x) = 0$. MATLAB result is shown below, copied from the Command Window:

```
The value of x that satisfies the equation is: 0.9355
```

$$\xi = 0.9355$$

$$n_{\text{N}_2} = n^{o}_{\text{N}_2} + v_i\xi = 2 - 0.9355 = 1.0645\text{ mol}$$

```
% Define the equation: f(x) = (7-2x)^2(2x)^2/((2-x)(5-3x)^3) - 8.1971 = 0
clear variables;
f = @(x) (((7-2*x)^2)*(2*x)^2) / ((2-x)*(5-3*x)^3) - 8.1971;%x=ξ
% Find the root of the equation using 'fzero'
x = fzero(f, 0.5); % Initial guess for x
% Display the solution
disp(['The value of x that satisfies the equation is: ', num2str(x)]);
```

**Figure 3.5** Finding the root of $f(x) = 0$, using MATLAB.

$$n_{H_2} = n^o_{H_2} + v_i\xi = 5 - 3\xi = 5 - 3 \times 0.9355 = 2.1935 \text{ mol}$$

$$n_{NH_3} = n^o_{NH_3} + v_i\xi = 0 + 2\xi = 2 \times 0.9355 = 1.871 \text{ mol}$$

$$n_{total} = 7 - 2\xi = 7 - 2 \times 0.9355 = 5.129 \text{ mol}$$

Of course, the mole fraction can be calculated by dividing the number of moles of species $i$ by the total number of moles.

$$Y_{N_2} = \frac{n_{N_2}}{n_{total}} = \frac{n_{N_2} \text{ mol}}{5.129 \text{ mol}} = \frac{1.0645 \text{ mol}}{5.129 \text{ mol}} = 0.20754$$

$$Y_{H_2} = \frac{n_{H_2}}{n_{total}} = \frac{n_{H_2} \text{ mol}}{3.0563 \text{ mol}} = \frac{2.1935 \text{ mol}}{5.129 \text{ mol}} = 0.42767$$

$$Y_{NH_3} = \frac{n_{NH_3}}{n_{total}} = \frac{\text{n}_{NH_3} \text{ mol}}{5.129 \text{ mol}} = \frac{1.871 \text{ mol}}{5.129 \text{ mol}} = 0.36479$$

$$Y_{N_2} + Y_{H_2} + Y_{NH_3} = 0.20754 + 0.42767 + 0.36479 = 1.0000$$

Moreover, the equilibrium concentration can be calculated by:

$$C_i\left[\frac{\text{mol}}{\text{L}}\right] = \frac{P_i}{RT} = \frac{P_{total} \times Y_i}{RT} = \frac{10 \text{ atm} \times Y_i}{0.08206\frac{\text{atm}\cdot\text{L}}{\text{mol}\cdot\text{K}} \times 700.15 \text{ K}} = 0.17405 \times Y_i$$

$$C_{N_2} = 0.17405 \times 0.20754 = 0.0361 \text{ M}$$

$$C_{H_2} = 0.17405 \times 0.42767 = 0.0744 \text{ M}$$

$$C_{NH_3} = 0.17405 \times 0.36479 = 0.0635 \text{ M}$$

**NOTE**

If we start with the same conditions in terms of $P$, $T$, and the initial number of moles, but we use

$\frac{1}{2}N_{2(g)} + \frac{3}{2}H_{2(g)} \rightleftharpoons NH_{3(g)}$ (1), instead of $1N_{2(g)} + 3H_{2(g)} \rightleftharpoons 2NH_{3(g)}$ (2), we will end up with a root value that is double the value obtained here in Example 3.16; however, the equilibrium conditions will remain invariant. $K_C = 270.53$ for the second reaction is: double $K_C = 16.448$ for the first reaction. It is left as an exercise for the student to repeat Example 3.16 while using Eq. (1), instead of Eq. (2).

## 3.9 The Extent of Water-Gas Shift (WGS) Reaction Calculation Using MATLAB Roots Finding Technique

**Example 3.17** ***The Equilibrium Composition for Water-Gas Shift Reaction***

Consider the water-gas shift reaction (WGSR) taking place in a batch reactor:

$$CO(g) + H2O(g) \rightleftharpoons CO2(g) + H_2(g)$$

The equilibrium constant, $K_{eq}$, as $f(T)$ is given by [1]:

$$K_{eq} = e^{\left(\frac{4,577.8}{T} - 4.33\right)}$$

At $T = 550\,°\text{C} + 273 = 823\,\text{K}$, the equilibrium constant, $K_C = e^{\left(\frac{4,577.8}{823} - 4.33\right)} = 3.4292$.
Suppose the feed to a reactor contains 1.00 mol of CO, 2.00 mol of $H_2O$, and no $CO_2$ *or* $H_2$, and the reaction mixture comes to equilibrium at 823 K. Calculate the extent of reaction, the fractional conversion of the limiting reactant, an equilibrium composition of the gas mixture.

**Solution**

To determine the *limiting reactant*, test for the *smallest ratio* of $\frac{n_i^o}{\|v_i\|}$

$$\frac{n_{CO}^o}{\|v_{CO}\|} = \frac{1}{1} < \frac{n_{H_2O}^o}{\|v_{H_2O}\|} = \frac{2}{1} = 2$$

Therefore, **CO** is the limiting reactant.
Let us assume that the extent of reaction is: given by: $\xi$
Using Eq. (3.57a), we can express the number of moles for each species as:

$$n_{CO} = n_{CO}^o + v_{CO}\xi = 1 + (-1.0) \times \xi = (1 - \xi)\ \text{mol}$$

$$n_{H_2O} = n_{H_2O}^o + v_{H_2O}\xi = 2 + (-1.0) \times \xi = (2 - \xi)\ \text{mol}$$

$$n_{CO_2} = n_{CO_2}^o + v_{CO_2}\xi = 0 + (1.0) \times \xi = (\xi)\ \text{mol}$$

$$n_{H_2} = n_{H_2}^o + v_{H_2}\xi = 0 + (1.0) \times \xi = (\xi)\ \text{mol}$$

$$n_{tot} = \sum_{i=1}^{N} n_i = n_{CO} + n_{H_2O} + n_{CO_2} + n_{H_2} = (1 - \xi) + (2 - \xi) + (\xi) + (\xi) = 3$$

$$Y_{H_2} = \frac{n_{H_2}}{n_{tot}} = \frac{\xi}{3}$$

$$Y_{CO_2} = \frac{n_{CO_2}}{n_{tot}} = \frac{\xi}{3}$$

$$Y_{H_2O} = \frac{n_{H_2O}}{n_{tot}} = \frac{(2 - \xi)}{3}$$

$$Y_{CO} = \frac{n_{CO}}{n_{tot}} = \frac{(1 - \xi)}{3}$$

Since $\Delta n = (1 + 1 - 1 - 1) = 0$, then $K_C = K_P = K_Y$ (see Eqs. (3.18–3.21))

$$K_Y = \frac{Y_{H_2} \times Y_{CO_2}}{Y_{H_2O} \times Y_{CO}}$$

Substitute the value for each mole fraction in the previous equation, we have:

$$K_Y = \frac{Y_{H_2} \times Y_{CO_2}}{Y_{H_2O} \times Y_{CO}} = \frac{\frac{\xi}{3} \times \frac{\xi}{3}}{\frac{(2-\xi)}{3} \times \frac{(1-\xi)}{3}} = \frac{\xi \times \xi}{(2 - \xi) \times (1 - \xi)} = 3.4292$$

```
% Define the equation: f(x) = 6.8584 - 10.2876ξ+2.4292ξ² = 0
clear variables;
f = @(x) 6.8584-10.2876*x+2.4292*x^2;%x=ξ
% Find the root of the equation using 'fzero'
x = fzero(f, 0.5); % Initial guess for x
% Display the solution
disp(['The value of x that satisfies the equation is: ', num2str(x)]);
```

**Figure 3.6** Finding the root of $f(x) = 0$, using MATLAB.

Simplifying both the numerator and denominator, we get:

$$K_Y = \frac{\xi^2}{2 - 3\xi + \xi^2} = 3.4292$$

It can be further simplified into: $6.8584 - 10.2876\xi + 3.4292\xi^2 = \xi^2 \rightarrow$

$$6.8584 - 10.2876\xi + 2.4292\xi^2 = 0$$

Figure 3.6 shows MATLAB code for finding the roots of a defined $f(x) = 0$.
MATLAB result is shown below, copied from the Command Window:

```
The value of x that satisfies the equation is: 0.82891
```

$\xi = 0.829$

Substituting back into the number of moles equations, we get:

$$n_{CO} = n^o_{CO} + \nu_{CO}\xi = 1 + (-1.0) \times \xi = (1 - 0.829) = 0.171 \text{ mol}$$

$$n_{H_2O} = n^o_{H_2O} + \nu_{H_2O}\xi = 2 + (-1.0) \times \xi = (2 - 0.829) = 1.171 \text{ mol}$$

$$n_{CO_2} = n^o_{CO_2} + \nu_{CO_2}\xi = 0 + (1.0) \times \xi = 0.829 \text{ mol}$$

$$n_{H_2} = n^o_{H_2} + \nu_{H_2}\xi = 0 + (1.0) \times \xi = 0.829 \text{ mol}$$

The **fractional conversion** of the limiting reactant **(CO)** is given by Eq. (3.51a):

$$f = \frac{\textit{moles reacted}}{\textit{moles fed}} = \frac{0.829}{1} = 0.829$$

$$y_{CO} = \frac{n_{CO}}{n_{tot}} = \frac{(1 - \xi)}{3} = \frac{0.171}{3} = 0.0570$$

$$y_{H_2O} = \frac{n_{H_2O}}{n_{tot}} = \frac{(2 - \xi)}{3} = \frac{1.171}{3} = 0.3903$$

$$y_{CO_2} = \frac{n_{CO_2}}{n_{tot}} = \frac{\xi}{3} = \frac{0.829}{3} = 0.2763$$

$$y_{H_2} = \frac{n_{H_2}}{n_{tot}} = \frac{\xi}{3} = \frac{0.829}{3} = 0.2763$$

$$Y_{Total} = \sum_{i=1}^{N} Y_i = 0.057 + 0.3903 + 0.2763 + 0.2763 = 0.9999 \cong 1.0$$

## 3.10 MATLAB Code for $NH_3$ $K_{eq}$ Under High-Pressure, Assuming Ideal Gas Mixture

**Example 3.18** ***Equilibrium Composition of Ammonia at T = 500 °C and P = 200 atm***
The synthesis of ammonia from nitrogen and hydrogen is a reversible reaction:

$$N_{2(g)} + 3H_{2(g)} \rightleftharpoons 2NH_{3(g)}$$

This reaction is: exothermic and influenced by pressure, temperature, and reactant concentrations. The equilibrium constant $K_p$ is given in terms of partial pressures and depends on temperature. The reaction takes place at a temperature, where $K_p = 1.45 \times 10^{-5}$ (@$T = 500\,°C$). An ideal gas mixture assumption is assumed here. Notice that $K_\phi$ was found to be close to unity even at $P = 300$ bar (see a subsequent section: MATLAB Code for $NH_3$ $K_{eq}$ Under Nonideal High-Pressure Conditions).

**References:**

1) https://brainly.com/question/34836428
2) https://byjus.com/question-answer/for-the-reversible-reaction-n-2-g-3-h-2-g-rightleftharpoons-2-nh-3/

The total initial pressure of the system is $P = 200$ atm. The initial molar ratio of reactants is 1:3 for $N_2$, with no $NH_3$ present initially. Initially, we have 100 moles of $N_2$ and $H_2$ feed mixture. Determine the equilibrium partial pressures and number of moles of $N_2$, $H_2$, and $NH_3$, considering a batch reactor.

**Solution**
To determine the *limiting reactant*, test for the *smallest ratio* of $\frac{n_i^o}{\|\nu_i\|}$

$$\frac{n^o_{N_2}}{\|\nu_{N_2}\|} = \frac{25}{1} = \frac{n^o_{H_2}}{\|\nu_{H_2}\|} = \frac{75}{3} = 25$$

Therefore, both $H_2$ and $N_2$ are limiting reactants.
Let $\xi$ represent the change in moles of $N_2$ that react to form $NH_3$. At equilibrium:

Moles of $N_2$: $25 - \xi$
Moles of $H_2$: $75 - 3\,\xi$
Moles of $NH_3$: $2\,\xi$

---

Total Moles: $100 - 2\,\xi$

Since the total pressure is fixed at 200 atm, the partial pressures at equilibrium can be expressed in terms of $\xi$.

$$P_{N_2} = Y_{N_2}P = \left(\frac{25 - \xi}{100 - 2\xi}\right) 200\text{ atm}$$

$$P_{H_2} = Y_{H_2}P = \left(\frac{75 - 3\xi}{100 - 2\xi}\right) 200\text{ atm}$$

$$P_{NH_3} = Y_{NH_3}P = \left(\frac{2\xi}{100 - 2\xi}\right) 200\text{ atm}$$

```
% Given data
clear variables;
Kp = 1.45e-5;        % Equilibrium constant for ammonia synthesis
P_total = 200;       % Total pressure in atm
% Define the function for fzero
Numerator = @(x) (((2 * x) / (100 - 2 * x)) * P_total)^2;
Denom = @(x) (((25 - x) / (100 - 2 * x)) * P_total) * (((75 - 3 * x) / (100 - 2 * x)) * P_total)^3;
equilibrium_eq = @(x) Kp - Numerator(x) / Denom(x); % Equation to find the root
% Adjusted interval to avoid division by zero issues
x_lower = 0 + 1e-6;  % Small increment to avoid zero in the denominator
x_upper = 25 - 1e-6; % Slightly below 25 to keep denominators valid
% Solve for x using fzero with an interval
x = fzero(equilibrium_eq, [x_lower, x_upper]);
% Compute equilibrium partial pressures
P_N2 = (25 - x) * P_total / (100 - 2 * x);
P_H2 = (75 - 3 * x) * P_total / (100 - 2 * x);
P_NH3 = (2 * x) * P_total / (100 - 2 * x);
% Display results
fprintf('The extent of reaction zeta : %.2f moles\n', x);
fprintf('Equilibrium partial pressure of N2: %.2f atm\n', P_N2);
fprintf('Equilibrium partial pressure of H2: %.2f atm\n', P_H2);
fprintf('Equilibrium partial pressure of NH3: %.2f atm\n', P_NH3);
fprintf('Equilibrium moles of N2: %.2f moles\n', 25-x);
fprintf('Equilibrium moles of H2: %.2f moles\n', 75-3*x);
fprintf('Equilibrium moles of NH3: %.2f moles\n', 2*x);
fprintf('Equilibrium YN2: %.2f \n', P_N2/P_total);
fprintf('Equilibrium YH2: %.2f \n', P_H2/P_total);
fprintf('Equilibrium YNH3: %.2f \n', P_NH3/P_total);
```

**Figure 3.7** Solving the extent of reaction for ammonia synthesis and finding equilibrium compositions.

At equilibrium, the relationship between the partial pressures is:

$$K_P = \frac{[p_{NH_3}]^2}{[p_{N_2}][p_{H_2}]^3} = 1.45 \times 10^{-5} = \frac{\left[\left(\frac{2\xi}{100-2\xi}\right)200\right]^2}{\left[\left(\frac{25-\xi}{100-2\xi}\right)200\right]\left[\left(\frac{75-3\xi}{100-2\xi}\right)200\right]^3}$$

$$f(\xi) = 1.45 \times 10^{-5} - \frac{\left[\left(\frac{2\xi}{100-2\xi}\right)200\right]^2}{\left[\left(\frac{25-\xi}{100-2\xi}\right)200\right]\left[\left(\frac{75-3\xi}{100-2\xi}\right)200\right]^3} = 0$$

Figure 3.7 shows **MATLAB code** for solving the above equation. This expression can be challenging to simplify by hand, so using MATLAB's numerical solvers like **fsolve** can help solve for $\xi$.

Results are copied from the Command Window:

```
>> Fig3_7
The extent of reaction zeta: 7.27 moles
Equilibrium partial pressure of N2: 41.49 atm
Equilibrium partial pressure of H2: 124.46 atm
Equilibrium partial pressure of NH3: 34.05 atm
Equilibrium moles of N2: 17.73 moles
```

*(Continued)*

| (Continued) |
|---|
| **Equilibrium moles of H2: 53.18 moles** |
| **Equilibrium moles of NH3: 14.55 moles** |
| **Equilibrium YN2: 0.21** |
| **Equilibrium YH2: 0.62** |
| **Equilibrium YNH3: 0.17** |

## 3.11 $N_2O_4$ Decomposition into $NO_2$ Using MATLAB for Finding the Equilibrium Conversion

**Example 3.19** ***Decomposition of $N_2O_4$ into $NO_2$***

$$N_2O_4(g) \leftrightarrow 2NO_2(g) \qquad \text{Or: } A(g) \leftrightarrow 2B(g)$$

$N_2O_4$ is a potent oxidizer that reacts on contact with various hydrazine compounds, making this combination widely used as a rocket propellant.

Figure 3.8 shows the transition between the two substances at various temperatures from left to right (−196 °C, 0 °C, 23 °C, 35 °C, and 50 °C, respectively), which is described by the above reversible reaction. The forward reaction (dissociation of $N_2O_4(g)$ to $2NO_2(g)$ is endothermic ($\Delta H^0 > 0$). This means that heat is consumed when $N_2O_4(g)$ breaks apart [https://chem.libretexts.org/@api/deki/files/126175/imageedit_5_3015884689.jpg?revision=].

The first-order gaseous decomposition of $N_2O_4$ (colorless) into $NO_2$ (brown) has a rate constant value $k_f = 1.15 \times 10^{16} exp^{-\left(\frac{6{,}460}{T\,(K)}\right)}$ $s^{-1}$ and $k_b = 6.02\times10^8$ L/(mol.S) [3].

$$K_C = \frac{[NO_2]^2}{[N_2O_4]} = \frac{k_f}{k_b} = \frac{1.15 \times 10^{16} exp^{-\left(\frac{6{,}460}{T\,(K)}\right)} s^{-1}}{6.02 \times 10^8 \frac{L}{mol \cdot s}} = 1.91 \times 10^7 \times exp^{-\left(\frac{6{,}460}{T\,(K)}\right)} \frac{mol}{L}$$

The feed consists of pure $N_2O_4$ at $T_0 = 350$ K and $P_0 = 151.987$ kPa (1.5 atm).

1) Calculate the equilibrium conversion of $N_2O_4$ in a constant-volume batch reactor.
2) Assuming the reaction is: elementary, determine the CSTR volume necessary to achieve 99.9% of the equilibrium conversion. Assume constant $T$ and $P$. Assume $F_{A0} = 100$ kmol/hr.

**Solution**

1) At $T = 350$ K, we have:
   In symbolic form:

$$K_C = \frac{[NO_2]^2}{[N_2O_4]} = \frac{[B]^2}{[A]} = 1.91 \times 10^7 \times exp^{-\left(\frac{6{,}460}{350}\right)} \frac{mol}{L} = 0.1842$$

   For a batch reactor, we have:

$$C_{Aeq} = \left(\frac{N_A}{V}\right)_{eq} = \frac{N_{Ao}(1 - X_{eq})}{V} = \frac{N_{Ao}(1 - X_{eq})}{V_0} = C_{A0}(1 - X_{eq})$$

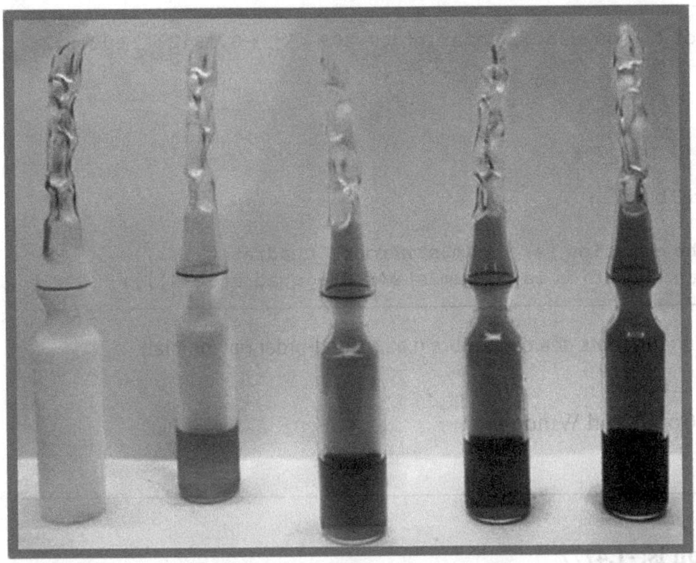

**Figure 3.8** The brown nitrogen dioxide ($NO_2$) undergoes a reversible conversion to colorless dinitrogen tetroxide ($N_2O_4$) at low temperatures and shifts back to $NO_2$ as the temperature increases. From left to right, the temperature is −196 °C, 0 °C, 23 °C, 35 °C, and 50 °C. *Source:* Eframgoldberg / Nitrogen dioxide at different temperatures / CC BY-SA 3.0 / https://commons.wikimedia.org/wiki/File:Nitrogen_dioxide_at_different_temperatures.jpg/last accessed on 7 January 2026.

From Eq. (1.28b) and replacing $C$ by $B$ as $B$ is the product, we have:

$$C_{Beq} = C_{A0}\left[\left(\theta_{B0} + \frac{b}{a}Xeq\right)\right] = C_{A0}\left[\left(0 + \frac{2}{1}Xeq\right)\right] = 2\,C_{A0}Xeq$$

Substitute for both $C_{Aeq}$ and $C_{Beq}$ in the $K_C$ equation, we have:

$$K_C = \frac{[B]^2}{[A]} = \frac{[2C_{A0}(X_{eq})]^2}{[C_{A0}(1 - X_{eq})]} = \frac{4C_{A0}[(X_{eq})]^2}{[(1 - X_{eq})]} = 0.1842\,\frac{\text{mol}}{\text{L}}$$

Using the ideal gas law, we have:

$$C_{A0} = \frac{P_{A0}}{RT_0} = \frac{Y_{A0}P_0}{RT_0} = \frac{1 * 1.5\,\text{atm}}{0.08206\,\frac{\text{atm}\cdot\text{L}}{\text{mol}\cdot\text{K}} * 350\,\text{K}} = 0.0522\,\frac{\text{mol}}{\text{L}}$$

Substitute in the $K_C$ equation, we have:

$$\frac{4 * 0.0522\frac{\text{mol}}{\text{L}}\,[(X_{eq})]^2}{[(1 - X_{eq})]} = 0.1842\,\frac{\text{mol}}{\text{L}}$$

Solve for $X_{eq}$.

$$0.1842 * (1 - X_{eq}) = 0.209 * X_{eq}^2 \rightarrow 0.209 * X_{eq}^2 + 0.1842X_{eq} - 0.1842 = 0$$

Figure 3.9 shows MATLAB code for finding the roots of the above quadrature.

```
% Define the coefficients of the quadratic equation: 0.209 * X_eq^2 + 0.1842X_eq - 0.1842
clear variables;
a = 0.209;
b = 0.1842;
c = -0.1842;
% Solve for the roots
roots_quadratic = roots([a, b, c]);
% Display the roots
disp(['The first root of the equation is:', num2str(roots_quadratic(1))]);
disp(['The second root of the equation is:', num2str(roots_quadratic(2))]);
```

**Figure 3.9** MATLAB code for finding the roots of a quadrature (i.e., second-order polynomial).

Results are copied from the Command Window:

```
Figure3_9
The first root of the equation is: -1.4777
The second root of the equation is: 0.59641
```

2) $V_{CSTR}$ for 99.9% of $X_{eq}$.

$$-r_A = k_f C_A - k_b C_B^2 = k_f \left[C_A - \frac{k_b}{k_f} C_B^2\right] = k_f \left[C_A - \frac{1}{K_C} C_B^2\right]$$

At equilibrium, $-r_A = 0 \rightarrow K_C = \frac{C_{Beq}^2}{C_{Aeq}} = 0.1842 \, \frac{\text{mol}}{\text{L}}$.
For a flow reactor, refer to equations (2.74) up to (2.77).

$$\varepsilon = \delta \times \frac{F_{A0}}{F_0} = \delta \times y_{A0} = \left(\frac{b}{a} - 1\right) y_{A0} = \left(\frac{2}{1} - 1\right) 1 = 1$$

For constant temperature and pressure, the volumetric flow rate is $\upsilon = \upsilon_0 \, (1 + \varepsilon \, X)$, and the resulting concentrations of species $A$ (reactant) and $B$ (as a product) are:

$$C_{Aeq} = \frac{F_A}{\upsilon} = \frac{F_{A0}(1 - X)}{\upsilon_0(1 + \varepsilon X)} \frac{P}{P_0} \frac{T_0}{T} = C_{A0} \frac{(1 - X)}{(1 + X)}$$

$$C_{Beq} = C_{A0} \frac{\left(\theta_{B0} + \frac{b}{a} X\right)}{(1 + \varepsilon X)} \frac{P}{P_0} \frac{T_0}{T} = C_{A0} \frac{\left(0 + \frac{2}{1} X\right)}{(1 + X)} = C_{A0} \frac{(2X)}{(1 + X)}$$

Substitute in the $K_C$ equation, we have:

$$\frac{C_{Beq}^2}{C_{Aeq}} = \frac{\left[C_{A0} \frac{(2X)}{(1+X)}\right]^2}{\left[C_{A0} \frac{(1-X)}{(1+X)}\right]} = \frac{\left[0.0522 * \frac{(2X_{eq})}{(1+X_{eq})}\right]^2}{\left[0.0522 * \frac{(1-X_{eq})}{(1+X_{eq})}\right]} = 0.1842 \, \frac{\text{mol}}{\text{L}}$$

Solve for $X_{eq}$ $f(X_{eq}) = \frac{\left[0.0522 * \frac{(2X_{eq})}{(1+X_{eq})}\right]^2}{\left[0.0522 * \frac{(1-X_{eq})}{(1+X_{eq})}\right]} - 0.1842 = 0$

```
% Define the function f(X_eq) as given
f = @(X_eq) (0.0522 * (2 * X_eq) / (1 + X_eq))^2 / (0.0522 * (1 - X_eq) / (1 + X_eq)) - 0.1842;
% Set the initial guess within the interval [0, 1]
initial_guess = 0.5;
% Use fzero to find the root of the function within the specified range
X_eq_solution = fzero(f, initial_guess);
% Display the result
fprintf('The value of X_eq is: %.4f\n', X_eq_solution);
```

**Figure 3.10** Finding the root of an implicit equation in terms of $X_{eq}$.

Figure 3.10 shows MATLAB code for finding the root of an implicit equation in terms of Xeq. Results are copied from the Command Window:

```
The value of X_eq is: 0.6846
```

$$X = 0.999*0.6846 = 0.6839$$

$$V_{CSTR} = \frac{F_{A0}X}{k_f\left[C_A - \frac{1}{K_C}C_B^2\right]} = \frac{27.78\ \frac{\text{mol}}{\text{s}} * 0.6839}{\left(1.15\times10^{16} exp^{-\left(\frac{6,460}{350}\right)}\right)\text{s}^{-1}\left[C_A - \frac{1}{K_C}C_B^2\right]\frac{\text{mol}}{\text{L}}}$$

Simplify to have:

$$V_{CSTR} = \frac{1.7134\times10^{-7}}{\left[C_A - \frac{1}{K_C}C_B^2\right]} = \frac{1.7134\times10^{-7}\ \text{mol}}{\left[C_{A0}\frac{(1-X)}{(1+X)} - \frac{1}{K_C}\left[C_{A0}\frac{(2X)}{(1+X)}\right]^2\right]\frac{\text{mol}}{\text{L}}} =$$

$$V_{CSTR} = \frac{1.7134\times10^{-7}}{\left[C_A - \frac{1}{K_C}C_B^2\right]} = \frac{1.7134\times10^{-7}}{\left[0.0522 * \frac{(1-0.6839)}{(1+0.6839)} - \frac{1}{0.1842} * \left[0.0522 * \frac{(2*0.6839)}{(1+0.6839)}\right]^2\right]}$$

$$= \frac{1.7134\times10^{-7}\ \text{mol}}{3.8613\times10^{-5}\ \frac{\text{mol}}{\text{L}}} = 0.0044\ \text{L}$$

**NOTE**

The reported $k_f$ value is exceptionally large. In fact, it is reported as $1,15\times10^{16} exp^{-\left(\frac{6,460}{T}\right)}$, not $1.15\times10^{16} exp^{-\left(\frac{6,460}{T}\right)}$. I treated it as a mistake compared with other reported values of $k_0$ up to $k_4$ [3].

## 3.12 MATLAB Code for $NH_3$ $K_{eq}$ Under Nonideal High-Pressure Conditions

Below is a step-by-step algorithm using Peng–Robinson EOS [4] for predicting the behavior of equilibrium under nonideal conditions:

1) **Step 1:** Define constants and inputs
**Gas constant:** $R = 0.08314$ (L·bar/mol·K)
**Input parameters:** Temperature $T$ [K], Pressure $P$ [bar], Mole fractions of species $y = [y_{N_2}, y_{H_2}, y_{NH_3}]$, Critical temperature $T_c$ [K], Critical pressure $P_c$ [bar], Acentric factor $\omega$
**Step 2: Calculate component parameters $a_i$, $b_i$ for each species based on Peng-Robinson EOS [1].**

$$P = \frac{RT}{V_m - b} - \frac{a\alpha(T)}{V_m(V_m + b) + b(V_m - b)} = \frac{RT}{V_m - b} - \frac{a\alpha(T)}{V_m^2 + 2bV_m - b^2}$$

$$a_i = 0.45724 * \frac{R^2 T_{c,i}^2}{P_{c,i}}$$

$$b_i = 0.07780 * \frac{RT_{c,i}}{P_{c,i}}$$

Also, calculate the temperature correction $\alpha(T)$ using the acentric factor:

$$k_i = 0.37464 + 1.54226 * w_i - 0.26992 * w_i^2$$

$$\alpha_i = \left[1 + k_i\left(1 - \sqrt{\frac{T}{T_{c,i}}}\right)\right]^2$$

$$a_{i,corr} = a_i * \alpha_i(T)$$

**Step 3:** Compute mixture parameters using the mixing rules:

$$a_{mix} = \sum_i y_i \sum_j y_j \sqrt{a_i a_j}$$

$$b_{mix} = \sum_j y_i b_i$$

**Step 4:** Calculate $A$ and $B$ for the cubic EOS:

$$A = \frac{a_{mix} * P}{(RT)^2} \qquad B = \frac{b_{mix} * P}{RT}$$

**Step 5:** Solve the Peng–Robinson cubic equation of state using the cubic form:

$$Z^3 + (B - 1)Z^2 + (A - 3B^2 - 2B)Z + (B^3 + B^2 - AB) = 0$$

Solve for real roots only. Choose the largest real root $Z$ to represent the vapor-phase compressibility factor.

**Step 6:** Compute fugacity coefficients $\phi_i$:

$$\ln \phi_i = \frac{b_i}{b_{mix}}(Z - 1) - \ln(Z - B) + \frac{A}{2\sqrt{2}B}\left[\sum_j y_j\left(\frac{2\sqrt{a_i a_j}}{a_{mix}}\right) - \frac{b_i}{b_{mix}}\right]\ln\left(\frac{Z + (1 + \sqrt{2})B}{Z + (1 - \sqrt{2})B}\right)$$

Then: $\phi_i = exp(\ln\phi_i)$
**Step 7:** Compute nonideal equilibrium constant:
From an ideal equilibrium constant $K_{ideal}$, calculate the fugacity-corrected (nonideal) constant:

$$K_\phi = \frac{\phi_{NH_3}^2}{\phi_{N_2} \cdot \phi_{H_2}^3} \quad K_{non-ideal} = K_\phi * K_{ideal}$$

**Table 3.4** Moles of reacting species as $f(X)$. $X$ is the extent of reaction.

| Species | Initial | Change | Final |
|---|---|---|---|
| $N_2$ | 1 | $-X$ | $1-X$ |
| $H_2$ | 3 | $-3X$ | $3-3X$ |
| $NH_3$ | 0 | $2X$ | $0+2X$ |
| Total | 4 | $-2X$ | $4-2X$ |

**Step 8:** Apply to the Haber process

For the reaction: $N_2 + 3H_2 \rightleftharpoons 2NH_3$

Assuming mixture): $y = [y_{N_2}, y_{H_2}, y_{NH_3}] = [0.25, 0.75, 0]$, the code evaluates how conversion of $N_2$ to $NH_3$ depends on pressure using a range: $P = 100$ to 300 bar. For each pressure:

Recalculate fugacity coefficients $\boldsymbol{\phi_i}$

Recalculate $K_\phi$ and $K_{nonideal}$

Table 3.4 shows the moles of reacting species in a batch reactor as a function of the extent of reaction.

The mole fraction of each reacting species is:

$$Y_{N_2} = \frac{1-X}{4-2X} \qquad Y_{H_2} = \frac{3-3X}{4-2X} \qquad Y_{NH_3} = \frac{2-X}{4-2X}$$

$$K_{non-ideal} = K_\phi * K_{ideal} = \frac{\phi^2_{NH_3}}{\phi_{N_2} * \phi^3_{H_2}} * K_p$$

$$K_P = \frac{p^2_{NH_3}}{p_{N_2} * p^3_{H_2}} = \frac{(PY_{NH_3})^2}{(PY_{N_2})(PY_{H_2})^3} = P^{\Delta n} * \frac{(Y_{NH_3})^2}{(Y_{N_2})(Y_{H_2})^3} = P^{-2} * \frac{(Y_{NH_3})^2}{(Y_{N_2})(Y_{H_2})^3}$$

$$K_{non-ideal} = \frac{\phi^2_{NH_3}}{\phi_{N_2} * \phi^3_{H_2}} * \left(P^{-2} * \frac{(Y_{NH_3})^2}{(Y_{N_2})(Y_{H_2})^3}\right) ***$$

Solve this equation numerically (using **fzero**) to find the extent of reaction or conversion, X.

Notice that if $n_{A0}$ is 1.0, which is the case here, then the extent of reaction and conversion will be numerically equal but physically different.

**Step 9: Plot results:** Plot % conversion of $N_2$ versus pressure: Conversion (%) = $X \times 100\%$

Figure 3.11 shows MATLAB code (**KeqNH3_RG.m**) to evaluate the equilibrium constant under nonideal conditions, described by parameters based on the Peng-Robinson EOS.

Results are copied from the Command Window.

```
>> KeqNH3_RG
Ideal KP at 300 bar: 0.0850
K_ϕ at 300 bar: 0.97792
Non-ideal KP at 300 bar: 0.0831
```

Figure 3.12 shows the generated plot of X as f(P) at T = 700 K carried out under nonideal gas conditions. Based on Le Chatelier's principle, increasing the pressure will shift the reaction to the right, in favor of $NH_3$ formation.

```
% Application: NH3 at 700K/300 bar
% The main calling script.
clear variables;
y0 = [0.25; 0.75; 0.00]; % [N2; H2; NH3]
Tc = [126.2; 33.2; 405.6]; % K
Pc = [33.9; 13.0; 112.8]; % bar
w = [0.037; -0.216; 0.250]; % Acentric factors
T=700; %T in K
[phi, Z] = peng_robinson(700, 300, y0, Tc, Pc, w);
% Non-ideal equilibrium constant
DelG0=-32800; % DelG0=-32.8 kJ/mol
Patm=300/1.01325;
Deln=-2; % Dn= 2-3-1=-2
Rgas=0.082057; % Rgas=0.08206 (atm.L)/(mol.K)
KC=exp(-DelG0/(8.314*T));
K_ideal =(((Rgas*T))^Deln)*KC;
K_phi = (phi(3)^2) / (phi(1) * phi(2)^3);
K_nonideal = K_ideal* K_phi;
fprintf('Ideal KP at 300 bar: %.4f\n', K_ideal)
% fprintf('K_{phi} at 300 bar: %.4f\n', K_phi)
phi_char = char(981);
disp(['K_', phi_char, ' at 300 bar: ', num2str(K_phi)])
fprintf('Non-ideal KP at 300 bar: %.4f\n', K_nonideal);
% Haber Process
P_range = 100:50:300;
conversions = zeros(size(P_range));
for i = 1:length(P_range)
   [phi, ~] = peng_robinson(700, P_range(i), y0, Tc, Pc, w);
   K_phi = (phi(3)^2) / (phi(1) * phi(2)^3);
   YN2 = @(X) (1-X)/(4-2*X);
   YH2 = @(X) (3-3*X)/(4-2*X);
   YNH3 = @(X) (2*X)/(4-2*X);
   fun = @(X) K_nonideal - ...
       ((YNH3(X))^2 / (YN2(X) * (YH2(X))^3)) * (P_range(i)^-2) * K_phi;
   conversions(i) = fzero(fun, [0.001, 0.9999]); % reasonable physical domain
end
figure;
plot(P_range, conversions*100, 'LineWidth', 2);
xlabel('Pressure (bar)'); ylabel('Conversion (%)');
title('X = f(P): High-P NH_3 Synthesis @ T=700 K');
ax = gca;
ax.FontSize = 16;
ax.FontWeight="bold";
function [phi, Z] = peng_robinson(T, P, y, Tc, Pc, w)
   % Corrected Peng-Robinson EOS for fugacity coefficients
   R = 0.08314; % L·bar/(mol·K) [CRITICAL UNIT FIX]
   a = 0.45724 * (R^2 .* Tc.^2) ./ Pc;
   b = 0.07780 * R .* Tc ./ Pc;
   k = 0.37464 + 1.54226.*w - 0.26992.*w.^2;
   alpha = (1 + k.*(1 - sqrt(T./Tc))).^2;
   ac = a .* alpha; % Temperature-corrected a
```

**Figure 3.11** MATLAB code (**KeqNH3_RG.m**) to evaluate the equilibrium constant under nonideal condition, described by parameters based on the Peng-Robinson EOS.

```
    % Mixture parameters with correct combining rules
    [i_idx, j_idx] = meshgrid(1:length(y), 1:length(y));
    a_mix = sum(y(i_idx) .* y(j_idx) .* sqrt(ac(i_idx) .* ac(j_idx)), 'all');
    b_mix = sum(y .* b);
    A = a_mix * P / (R^2 * T^2);
    B = b_mix * P / (R * T);
    % Solve cubic EOS (Z^3 + pZ^2 + qZ + r = 0)
    coeffs = [1, (B-1), (A - 3*B^2 - 2*B), (B^3 + B^2 - A*B)];
    Z_roots = roots(coeffs);
    Z_roots = Z_roots(imag(Z_roots) == 0); % Real roots only
    Z = max(Z_roots); % Vapor root
    % Fugacity coefficient calculation
    sqrt2 = sqrt(2);
    log_term = log((Z + (1+sqrt2)*B) / (Z + (1-sqrt2)*B));
    bi_b = b / b_mix;
    phi = zeros(size(y));
    for i = 1:length(y)
        sum_term = 2/ a_mix * sum(y .* sqrt(ac(i) .* ac));
        ln_phi = bi_b(i)*(Z-1) - log(Z - B) + ...
             (A/(2*sqrt2*B)) * (sum_term - bi_b(i)) * log_term;
        phi(i) = exp(ln_phi);
    end
end
```

**Figure 3.11** (*Continued*)

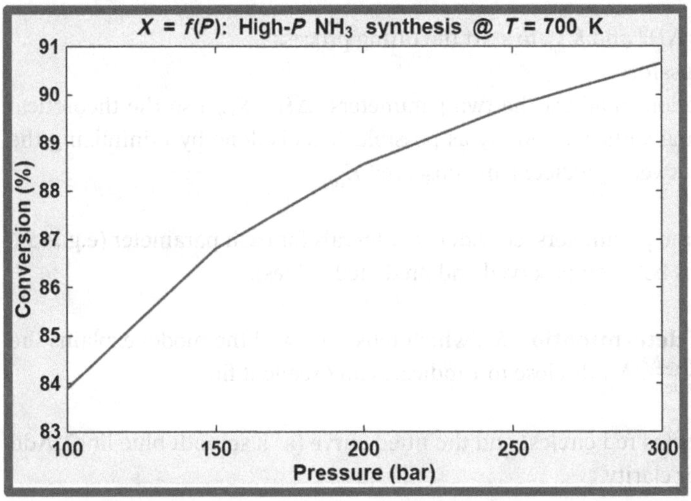


**Figure 3.12** The generated plot of $X$ as $f(P)$ at $T = 700$ K, carried out under nonideal gas conditions as: $P$ goes up, the reaction will shift toward $NH_3$ formation, in harmony with Le Chatelier's principle.

## 3.13 MATLAB Code for Estimating $\Delta H^0$ from Experimental Data

Often, $\Delta H^0$ is not directly measurable and must be inferred from experimental data using parameter estimation techniques. Suppose we have experimentally determined $K_{eq}$ values at several temperatures, as shown in Table 3.5.

**Table 3.5** Experimental data for $K_{eq}$ as $f(T)$.

| $T$ (K) | $K_{eq}$ | $T$ (K) | $K_{eq}$ |
|---|---|---|---|
| 300 | 1.20 | 350 | 0.84 |
| 310 | 1.10 | 360 | 0.79 |
| 320 | 1.02 | 370 | 0.74 |
| 330 | 0.95 | 380 | 0.70 |
| 340 | 0.89 | 390 | 0.66 |

**Below is the human-language algorithm using Van't Hoff fit, augmented by a statistical analysis:**

1) **Collect Experimental Data**
Gather pairs of temperature values (in Kelvin) and their corresponding experimentally measured equilibrium constants, $K_{eq}$, as shown in Table 3.5.
2) **Define the Theoretical Model**
Use the Van't Hoff equation (see Eqs. (3.7–3.9)), which relates temperature $T$ to the equilibrium constant $K_{eq}$, assuming constant enthalpy change $\Delta H^0$. The model has two unknowns: $\Delta H^0$ = standard enthalpy change [J/mol] and $K_{eq0}$, the reference equilibrium constant at $T_0 = 298$ K.
3) **Provide Initial Guesses**
Make reasonable guesses for $\Delta H^0$ and $K_{eq0}$ to start the fitting process.
4) **MATLAB Nonlinear Regression**
Use MATLAB's **nlinfit** function to adjust the two parameters ($\Delta H^0$, $K_{eq0}$) so the theoretical values match the experimental values as closely as possible. This is done by minimizing the sum of squared differences between predicted and observed $K_{eq}$.
5) **Extract Results**
Obtain the best-fit values of the parameters, confidence intervals for each parameter (e.g., 95% CI), and residuals (differences between observed and predicted values).
6) **Compute Goodness of Fit**
Calculate the **coefficient of determination** $R^2$, which tells how well the model explains the data: $R^2 = 1 - \frac{\textit{Sum of squared residuals}}{\textit{Total variance of the data}}$. An $R^2$ close to 1 indicates an excellent fit.
7) **Visualize Results**
Plot experimental data points (as red circles) and the fitted curve (as a smooth blue line). Add labels, a title, and a legend for clarity.
8) **Interpret Results**
Use the fitted parameters to understand the thermodynamics of the reaction. If the reaction is: exothermic, $\Delta H^0$ should be negative.

Figure 3.13 shows MATLAB code (**Keq_T_NLR.m**) which simultaneously estimates both parameters: $\Delta H^0$ and $K_{eq0}$, at reference temperature $T_0 = 298$ K. In addition, it calculates statistical metrics, such as Estimated parameters with standard errors and confidence intervals and the coefficient of determination $R^2$.

```
clear variables;
% Experimental data: [Temperature (K), Keq]
exp_data = [ ...
  300, 1.20;
  310, 1.10;
  320, 1.02;
  330, 0.95;
  340, 0.89;
  350, 0.84;
  360, 0.79;
  370, 0.74;
  380, 0.70;
  390, 0.66 ];
T_data = exp_data(:,1);      % Temperatures [K]
Keq_exp = exp_data(:,2);     % Experimental Keq values
% Constants
R = 8.314;                  % Universal gas constant [J/mol/K]
T0 = 298;                   % Reference temperature [K]
% Define model function: x(1) = ΔH0, x(2) = Keq0
model_fun = @(x, T) x(2) .* exp((x(1)/R) .* (1/T0 - 1./T));
% Initial guess: [ΔH0, Keq0]
x0 = [-40000, 1.2];
% Perform nonlinear regression using nlinfit
[x_fit, R, J, covB, mse] = nlinfit(T_data, Keq_exp, model_fun, x0);
% Extract fitted parameters
dH0_opt = x_fit(1);
Keq0_opt = x_fit(2);
% Confidence intervals (95%)
ci = nlparci(x_fit, R, 'jacobian', J);
% Compute predicted Keq and R-squared
Keq_pred = model_fun(x_fit, T_data);
SS_res = sum((Keq_exp - Keq_pred).^2);
SS_tot = sum((Keq_exp - mean(Keq_exp)).^2);
R_squared = 1 - SS_res / SS_tot;
% Display results
fprintf('— Parameter Estimation Results —\n');
fprintf('Optimized ΔHo = %.2f J/mol\n', dH0_opt);
fprintf('95%% CI for ΔHo: [%.2f, %.2f]\n', ci(1,1), ci(1,2));
fprintf('Optimized Keqo = %.4f\n', Keq0_opt);
fprintf('95%% CI for Keqo: [%.4f, %.4f]\n', ci(2,1), ci(2,2));
fprintf('R^2 = %.5f\n', R_squared);
% Plot results
figure;
plot(T_data, Keq_exp, 'ro', 'MarkerSize', 8, 'DisplayName', 'Experimental Data');
hold on;
plot(T_data, Keq_pred, 'b-', 'LineWidth', 2, 'DisplayName', 'Fitted Model');
xlabel('Temperature (K)');
ylabel('K_{eq}');
title('Nonlinear Fit: Van''t Hoff Equation with Statistics');
legend('Location', 'northeast');
grid on;
ax = gca;
ax.FontSize = 16;
ax.FontWeight="bold";
```

**Figure 3.13** MATLAB code (**Keq_T_NLR.m**) for estimating both parameters: ΔH∘ and $K_{eq0}$, at reference temperature $T_0 = 298$ K, equipped with statistical metrics.

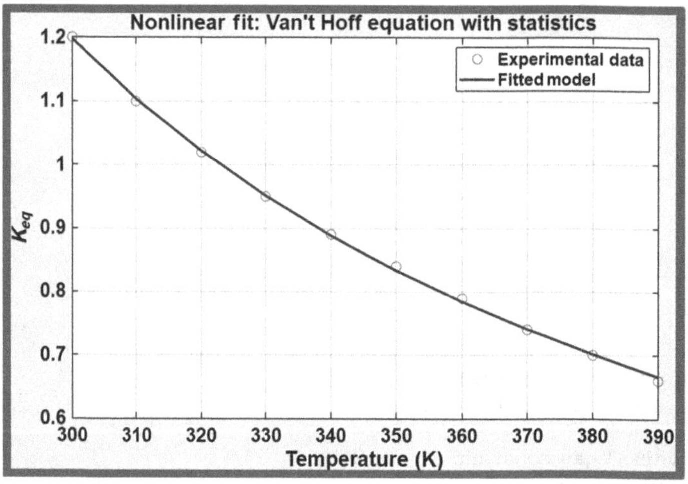


**Figure 3.14** The generated plot for both the experimental data and the fitting curve.

Results are copied from the Command Window:

```
>> Keq_T_NLR
--- Parameter Estimation Results ---
Optimized ΔHo = -6341.23 J/mol
95% CI for ΔHo: [-6459.23, -6223.22]
Optimized Keq0 = 1.2189
95% CI for Keq0: [1.2116, 1.2262]
R2 = 0.99948
```

Figure 3.14 shows the generated plot for both the experimental and the fitting curve.

## 3.14 MATLAB Code for Coupled Equilibrium Reactions in Complex Systems

Many chemical systems involve multiple reactions occurring simultaneously and sharing common species. These are known as *coupled equilibrium reactions*. Examples include:

- Ammonia synthesis (involves $N_2$, $H_2$, $NH_3$, and dissociation equilibria)
- Fischer-Tropsch synthesis (involves chain-growth reactions producing hydrocarbons)
- Methanol synthesis (involving CO, $CO_2$, $H_2$, and $CH_3OH$)

Such systems require solving nonlinear algebraic equations derived from molar balances and equilibrium expressions, often involving the extent of reaction ($\xi$) as a convenient variable.

**Example 3.20** ***Multi-equilibrium Reversible Reactions***
We consider the following two reversible gas-phase reactions:

1) $A + B \rightleftharpoons C K_{eq1} = \frac{[C]}{[A][B]}$
2) $C + B \rightleftharpoons D K_{eq2} = \frac{[D]}{[C][B]}$

Let the initial moles be: $[A]_0 = 1\frac{\text{mol}}{\text{L}}$ $[B]_0 = 2\frac{\text{mol}}{\text{L}}$
We define the extent of reaction for each step:

$\xi_1$: extent of reaction 1 $\xi_2$: extent of reaction 2

Using stoichiometry:

- $[A] = [A]_0 - \xi_1$
- $[B] = [B]_0 - \xi_1 - \xi_2$
- $[C] = [C]_0 + \xi_1 - \xi_2 = \xi_1 - \xi_2$
- $[D] = [D]_0 + \xi_2 = \xi_2$

Figure 3.15 shows MATLAB code (**Multi_Keq.m**) for finding the equilibrium number of moles of reacting species for the abovementioned reactions. The code to analyze the impact of initial $[B]_0$ on $D$ yield or extent of reaction #2 is also included. We assume here that $D$ is the desired product.

Results are copied from the Command Window:

```
>> Multi_Keq
Equilibrium extents: ξ1 = 0.8000 ξ2 = 0.5333
Equilibrium concentrations:
[A] = 0.2000
[B] = 0.6667
[C] = 0.2667
[D] = 0.5333
```

Figure 3.16 shows the generated plot of $C$ yield (i.e., $\xi_2$) as a function of $[B]_0$.

## 3.15 MATLAB Code for Dynamic Equilibrium in CSTRs

In chemical engineering, the concept of equilibrium typically arises in the context of batch systems, where a reaction proceeds over time until no further net change in species concentrations occurs. However, in continuous-flow reactors such as CSTRs and PFRs, the idea of equilibrium must be adapted: it depends not just on thermodynamics, but also on residence time, flow rate, and reactor volume or length.

### 3.15.1 A Dynamic Equilibrium in CSTR

A CSTR operates under steady-state conditions: fresh reactants enter, products leave, and the reactor contents are ideally well mixed. While a batch reactor needs infinite time to reach equilibrium, a CSTR achieves dynamic equilibrium when the rate of formation equals the rate of exit, and the residence time is sufficiently long. Consider the following reaction:

$$A \rightleftharpoons B \qquad K_{eq} = \frac{k_f}{k_b} = \frac{[B]}{[A]}$$

Steady-State Molar Balance of $A$:

$$\upsilon_0 C_{A0} - \upsilon_0 C_A - (-r_A) V_{CSTR} = 0$$

```
clc; clear;
% Adjusted Input Parameters
A0 = 1.0;
B0 = 2.0;
Keq1 = 2.0;
Keq2 = 3.0;
% Nonlinear system
function F = coupled_eqns(xi, A0, B0, Keq1, Keq2)
   xi1 = xi(1);
   xi2 = xi(2);
   A = A0 - xi1;
   B = B0 - xi1 - xi2;
   C = xi1 - xi2;
   D = xi2;
   if A <= 0 || B <= 0 || C <= 0
      F = [1e6; 1e6];
      return;
   end
   F(1) = Keq1 - C / (A * B);
   F(2) = Keq2 - D / (C * B);
end
% Initial guess with C = ξ1 - ξ2 > 0
xi0 = [0.3, 0.1];
% Solve
options = optimoptions('fsolve','Display','none');
xi_sol = fsolve(@(xi) coupled_eqns(xi, A0, B0, Keq1, Keq2), xi0, options);
% Results
xi1 = xi_sol(1);
xi2 = xi_sol(2);
A_eq = A0 - xi1;
B_eq = B0 - xi1 - xi2;
C_eq = xi1 - xi2;
D_eq = xi2;
fprintf('Equilibrium extents:\n ξ1 = %.4f\n ξ2 = %.4f\n', xi1, xi2);

fprintf('Equilibrium concentrations:\n[A] = %.4f\n[B] = %.4f\n[C] = %.4f\n[D] = %.4f\n', ...
   A_eq, B_eq, C_eq, D_eq);
% Product D sensitivity to initial number of moles of B.
B0_range = linspace(1, 5, 50);
D_eq_values = zeros(size(B0_range));

for i = 1:length(B0_range)
   B0 = B0_range(i);
   xi_sol = fsolve(@(xi) coupled_eqns(xi, A0, B0, Keq1, Keq2), xi0,options);
   D_eq_values(i) = xi_sol(2); % D = ξ2
end
plot(B0_range, D_eq_values, 'LineWidth', 2);
xlabel('[B]_0 (mol/L)');
ylabel('[D]_{eq} (mol/L)');
title('Effect of [B]_0 on D Product Yield');
grid on;
ax = gca;
ax.FontSize = 16;
ax.FontWeight="bold";
```

**Figure 3.15** MATLAB code (**Multi_Keq.m**) for finding the equilibrium number of moles of reacting species for the set of two reactions, while appending the code to analyze the impact of initial $[B]_0$ on $D$ yield or extent of reaction #2. $D$ is the desired product.

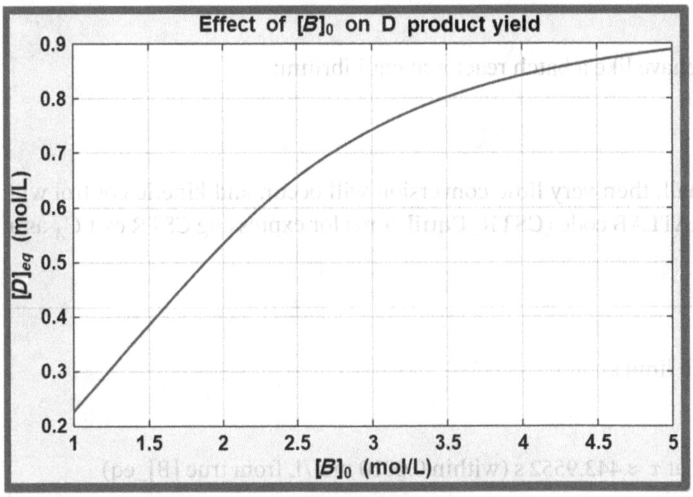


**Figure 3.16** Plot of $C$ yield (i.e., $\xi_2$) as a function of $[B]_0$.

The last equation is: valid for a constant-volume flow rate, which is valid for either a liquid-phase or a gas-phase with a constant $P$, $T$, and no change in the number of total moles as a result of the chemical reaction. A steady-state condition is: a must for an equilibrium condition, and the other condition is: uniformity to have an equilibrium that can be dynamically maintained in a CSTR.

$$-r_A = k_f C_A - k_b C_B$$

$$v_0 C_{A0} - v_0 C_A - (k_f C_A - k_b C_B) V_{CSTR} = 0$$

$$C_{A0} - C_A - (k_f C_A - k_b C_B) \frac{V_{CSTR}}{v_0} = 0$$

Given that $\tau = \frac{V_{CSTR}}{v_0}$, we have:

$$C_{A0} - C_A - (k_f C_A - k_b C_B)\tau = 0$$

Replace $C_B$ by $(C_{A0} - C_A)$, we have:

$$C_{A0} - C_A - (k_f C_A - k_b (C_{A0} - C_A))\tau = 0$$

Split combined terms in the bracket, we have:

$$C_{A0} - C_A - k_f \tau C_A - k_b \tau C_A + k_b \tau C_{A0} = 0$$

Move all $C_A$-containing terms to one side, we have:

$$C_{A0} + k_b \tau C_{A0} = k_f \tau C_A + k_b \tau C_A + C_A$$

Rearrange, we have:

$$C_A = \frac{C_{A0}(1 + k_b \tau)}{(1 + k_f \tau + k_b \tau)}$$

$$C_B = (C_{A0} - C_A)$$

### 3.15.2 Limiting Cases

If $\tau \to \infty$, then the CSTR will behave like a batch reactor at equilibrium:

$$K_{eq} = \frac{k_f}{k_b} = \frac{[B]}{[A]}$$

On the other hand, if $\tau$ is small, then very little conversion will occur, and kinetic control will dominate. Figure 3.17 shows MATLAB code (**CSTR_Equilib.m**) for expressing CSTR exit $C_A$ as a function of its residence time.

```
>> CSTR_Equilib
True equilibrium values (batch limit):
[A]_eq = 0.3333 mol/L
[B]_eq = 0.6667 mol/L
Effective equilibrium reached at τ ≈ 443.9552 s (within 0.0010 mol/L from true [B]_eq)
```

Figure 3.18 shows the generated plot of the molar concentration of $B$ as a function of CSTR residence time. It also reports both $C_A$ and $C_B$ of the exit stream concentration. The larger $\tau$, the larger $C_B$, or the lower $C_A$, until a certain value of $\tau$ is reached; at such a point, $C_B$ will hit the ceiling limit, that is, $C_{Beq}$. In light of the above results, at $\tau \approx 444$ s, $[B]_{out} = 0.6667 - 0.001 = 0.6657$ mol/L. In practice, however, a value of $\tau$ between 100 and 150 s (see Figure 3.18) will be sufficiently large to hit the ceiling limit of $B$, which is $[B]_{eq}$.

## 3.16 MATLAB Code for Liquid-Phase Activity-Based $K_{eq}$

In real-world chemical systems, especially liquid-phase reactions and high-pressure gas-phase reactions, the assumption of ideality often fails. Nonideal behavior arises due to intermolecular interactions, non-negligible molecular volumes, or strong polarity differences between components. In such cases, concentrations or mole fractions are not sufficient to describe the thermodynamic driving forces. Instead, we use activities.

The activity ($a_i$) of a species represents its "effective concentration" and is defined as:

$$a_i = \gamma_i x_i$$

where

- $x_i$ is the mole fraction of species $i$.
- $\gamma_i$ is the activity coefficient, accounting for non-ideality.

For an ideal system, $\gamma_i = 1$, so activity reduces to mole fraction. However, in nonideal systems, $\gamma_i \neq 1$, and deviations become significant.

The equilibrium constant based on activities is:

$$K_{eq} = \prod_{i=1}^{n} a_i^{\nu_i} = \prod_{i=1}^{n} (\gamma_i x_i)^{\nu_i}$$

```
clear;
% === Parameters ===
CA0 = 1.0;         % Inlet concentration of A (mol/L)
kf = 1.0;        % Forward rate constant (1/s)
kr = 0.5;        % Reverse rate constant (1/s)
Keq = kf / kr;     % Equilibrium constant
% === True equilibrium values (batch reactor or τ →∞) ===
CA_eq_true = CA0 / (1 + Keq);
CB_eq_true = CA0 * Keq / (1 + Keq);
fprintf('True equilibrium values (batch limit):\n');
fprintf(' [A]_eq = %.4f mol/L\n', CA_eq_true);
fprintf(' [B]_eq = %.4f mol/L\n\n', CB_eq_true);
% === Parameters for scanning τ  ===
tol = 1e-3;                     % Tolerance to consider as "equilibrium"
tau_range = linspace(0.1, 500, 1000); % Range of residence times
CB_vals = zeros(size(tau_range)); % To store [B] values
CA_vals = zeros(size(tau_range)); % To store [A] values
tau_eq = NaN;                   % To store effective τ at equilibrium
% === Loop over τ values ===
for i = 1:length(tau_range)
   tau = tau_range(i);
   % Steady-state expression for [A] in CSTR
   CA_tau = (CA0 * (1 + tau * kr)) / (1 + tau * (kf + kr));
   CB_tau = CA0 - CA_tau;
   CA_vals(i) = CA_tau;
   CB_vals(i) = CB_tau;
   % Check for first τ where [B] is within tolerance of batch equilibrium
   if isnan(tau_eq) && abs(CB_tau - CB_eq_true) < tol
      tau_eq = tau;
   end
end
% === Report τ at which equilibrium is effectively reached ===
if ~isnan(tau_eq)
   fprintf('Effective equilibrium reached at τ ≈ %.4f s (within %.4f mol/L from true
     [B]_eq)\n', ...
        tau_eq, tol);
else
   fprintf('Equilibrium not reached within τ range [0.1, 500] under given
     tolerance.\n');
end
% === Plotting [B] vs τ  ===
figure;
plot(tau_range, CB_vals, 'r-', 'LineWidth', 2); hold on;
yline(CB_eq_true, 'k-', 'LineWidth', 1.5);
xlabel('\tau (Residence Time, s)');
ylabel('[B]_{out} (mol/L)');
title('CSTR: Product B Concentration vs. Residence Time');
legend('[B]_{out}', '[B]_{eq, true}', 'Location', 'best');
grid on;
set(gca, 'FontSize', 14, 'FontWeight', 'bold');
```

**Figure 3.17** MATLAB code (**CSTR_Equilib.m**) for expressing CSTR exit $C_A$ as a function of its residence time.

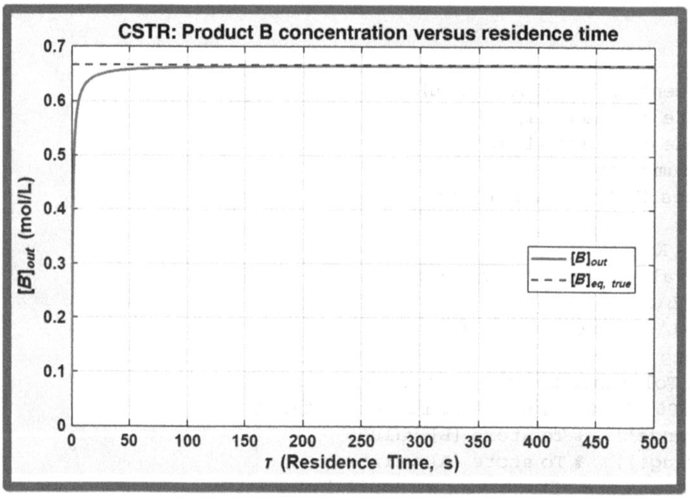


**Figure 3.18** The generated plot of the exit molar concentration of $B$ as a function of CSTR residence time. The plateau value represents the ceiling limit for $B$.

where

- $\nu_i$ is the stoichiometric coefficient (positive for products, negative for reactants).
- The product is done over all species involved in the equilibrium.

To calculate $\gamma_i$ for liquid systems, we may use:

- **NRTL**, **Wilson**, or **UNIQUAC** models for binary/ternary systems
- **UNIFAC** for predictive modeling (group contribution method)

These models can be implemented in MATLAB either directly or using external thermodynamic packages. For more information on activity coefficient-based thermodynamic models, see my textbook [5].

**Example 3.21** ***Gamma-based $K_{eq}$***
Consider the following reaction:

$$A + B \rightleftharpoons 2C$$

**Given:**

- **Mole Fractions:**

$$x = [0.25, 0.25, 0.5] \rightarrow [A, B, C]$$

- **Activity Coefficients:**

$$\gamma = [1.2, 1.0, 0.9]$$

- **Stoichiometric Coefficients:**

$$\nu = [-1, -1, 2]$$

Figure 3.19 shows MATLAB code (**Keq_Gamma.m**) to evaluate the equilibrium constant, $K_{eq}$, as well as the reaction quotient ($Q$), which is calculated using the same form as the equilibrium constant expression ($K_{eq}$), but with the current (nonequilibrium) concentrations, mole fractions, or

```
clear variables;
% Set 1: Used to calculate Keq (e.g., from literature or initial equilibrium conditions)
x1 = [0.25, 0.25, 0.5];        % Mole fractions [A, B, C]
gamma1 = [1.2, 1.0, 0.9];      % Activity coefficients
nu = [-1, -1, 2];              % Stoichiometric coefficients A + B ⇌ 2C
% Calculate Keq from Set 1
Keq = calc_Keq_activity(x1, gamma1, nu);
fprintf('Calculated activity-based Keq = %.4f\n\n', Keq);
% Set 2: Used to calculate reaction quotient Q
x2 = [0.30, 0.20, 0.45];       % New mole fractions [A, B, C]
gamma2 = [1.1, 1.05, 0.95];    % New activity coefficients
% Calculate Q from Set 2
Q = calc_Keq_activity(x2, gamma2, nu);
fprintf('Calculated activity-based Q = %.4f\n', Q);
% Compare Q to Keq
tolerance = 0.05; % acceptable tolerance
if abs(Q - Keq) < tolerance
   disp('=> The system is very close to equilibrium.');
elseif Q < Keq
   disp('=> The reaction will proceed forward (more products will form).');
else
   disp('=> The reaction will proceed in reverse (more reactants will form).');
end
% Function to compute activity-based equilibrium constant or Q
function K = calc_Keq_activity(x, gamma, nu)
   a = gamma .* x;           % Compute activities
   K = prod(a .^ nu);        % Apply: K = Π a_i^v_i
end
```

**Figure 3.19** MATLAB code (**Keq_Gamma.m**) to evaluate the equilibrium constant, $K_{eq}$, as well as the reaction quotient ($Q$) for a liquid-phase reaction.

activity coefficients. The relationship between $K_{eq}$ and $Q$ will demonstrate Le Chatelier's principle. The comparison of $Q$ with $K_{eq}$ tells the direction the system will move to reach equilibrium:

- $Q < K_{eq} \rightarrow$ forward reaction favored
- $Q > K_{eq} \rightarrow$ reverse reaction favored
- $Q = K_{eq} \rightarrow$ system is already at equilibrium

Results are copied from the Command Window:

```
≫ Keq_Gamma
Calculated activity-based Keq = 2.7000
Calculated activity-based Q = 2.6372
=> The reaction will proceed forward (more products will form).
```

## 3.17 MATLAB Code for Gas-Phase Fugacity-Based $K_{eq}$

For a gas-phase reversible reaction:

$$2A \rightleftharpoons B$$

**Given:**

- **Mole Fractions:** $y = [0.6, 0.4]$ for $A$ and $B$

- **Fugacity Coefficients:** $\phi = [0.95, 1.05]$
- **Stoichiometric Coefficients:** $\nu = [-2, 1]$
- **Pressure:** $P = 5$ atm

Express and calculate $K_{eq}$ in terms of fugacity coefficients.

$$K_{eq} = \frac{(a_B)}{(a_A)^2} = \frac{(\phi_B Y_B P)}{(\phi_A Y_A P)^2}$$

Figure 3.20 shows MATLAB code (**Keq_FuG.m**) to evaluate the equilibrium constant, $K_{eq}$, as well as the reaction quotient ($Q$), which is calculated using the same form as the equilibrium constant expression ($K_{eq}$), but with the current (nonequilibrium) concentrations, mole fractions, or fugacities.

Results are copied from the Command Window:

```
>> Keq_Fug
Calculated Keq (from assumed equilibrium): 0.8726
Calculated Q (from current state): 0.2093
=> Q < Keq: The reaction will proceed forward (toward products).
```

```
clear variables;
% Equilibrium conditions (to calculate Keq)
y_eq = [0.4, 0.6];          % Equilibrium mole fractions [A, B]
phi_eq = [0.95, 1.05];        % Equilibrium fugacity coefficients
nu = [-2, 1];               % Stoichiometric coefficients: 2A ↔ B
P = 5;                  % Pressure in atm
% Calculate equilibrium constant Keq (based on assumed equilibrium state)
Keq = calc_Keq_fugacity(y_eq, phi_eq, nu, P);
% Current (non-equilibrium) conditions (to calculate Q)
y_current = [0.6, 0.4];       % Current mole fractions [A, B]
phi_current = [1.02, 0.98];   % Current fugacity coefficients
% Calculate reaction quotient Q
Q = calc_Keq_fugacity(y_current, phi_current, nu, P);
% Display results
fprintf('Calculated Keq (from assumed equilibrium): %.4f\n', Keq);
fprintf('Calculated Q (from current state): %.4f\n', Q);
% Determine direction of reaction
if abs(Q - Keq) < 1e-3
   disp('=> The system is at equilibrium.');
elseif Q < Keq
   disp('=> Q < Keq: The reaction will proceed forward (toward products).');
else
   disp('=> Q > Keq: The reaction will proceed in reverse (toward reactants).');
end
% === Function Definition ===
function Keq = calc_Keq_fugacity(y, phi, nu, P)
   f = phi .* y * P;          % Fugacities: f_i = phi_i * y_i * P
   Keq = prod(f .^ nu);        % Keq or Q = Π(f_i)^ν_i
end
```

**Figure 3.20** MATLAB code (**Keq_FuG.m**) to evaluate the equilibrium constant, $K_{eq}$, as well as the reaction quotient ($Q$) for a gas-phase reaction.

## References

1 Nandagopal, N.S. (2023). *Chemical Engineering Principles and Applications*, 139. Springer Nature Switzerland AG. ISBN 978-3-031-27878-5. ISBN 978-3-031-27879-2 (eBook). https://doi.org/10.1007/978-3-031-27879-2.
2 Byron Smith, R.J., Loganathan, M., and Shantha, M.S. (2010). A review of the water gas shift reaction kinetics. *International Journal of Chemical Reactor Engineering* 8: https://doi.org/10.2202/1542-6580.2238. Review R4. https://www.researchgate.net/publication/228506095.
3 Głowiński, J., Hoffmann, J., and Wilk, M. Mechanism of the oxidation of nitric oxide with oxygen. *Polish Journal of Chemical Technology* 23 (1): 31–36. https://doi.org/10.2478/pjct-2021-0005. https://bibliotekanauki.pl/articles/1849326.pdf.
4 Al-Malah, K. (2025). MATLAB-based computations of chemical engineering principles, chapter 5.3. In: *Real Gases*. Wiley. https://www.wiley.com/en-us/MATLAB-based+Computations+of+Chemical+Engineering+Principles-p-9781394308828.
5 Al-Malah, K. (2022). *Aspen Plus: Chemical Engineering Applications*, 2e. Wiley. https://www.wiley.com/en-ae/Aspen+Plus:+Chemical+Engineering+Applications,+2nd+Edition-p-9781119868699.

## End of Chapter Problems

**CHW 3.1 Express $K_{eq}$ for the Following Reversible Reactions**

a) $2H_2\,(g) + O_2(g) \rightleftharpoons 2H_2O(g)$

b) $H_2\,(g) + \frac{1}{2}O_2(g) \rightleftharpoons H_2O(g)$

**CHW 3.2 The Reaction Quotient**

At the start of a reaction, there are 0.24 mol $N_2$, 0.036 mol $H_2$, and 0.00069 mol $NH_3$ in a 3.0-L vessel at 375 °C. If the equilibrium constant $K_C$ for the reaction: $N_2(g)+3H_2(g) \rightleftharpoons 2NH_3(g)$

**CHW 3.3 Calculation of $K_C$ for Ethyl Acetate Formation**

**Consider the reaction:**

$$CH_3COOH + C_2H_5OH \rightleftharpoons CH_3COOC_2H_5 + H_2O$$

**In symbolic form:**

$$AA + EOH \rightleftharpoons EA + H_2O$$

We have initially one mole of AA and one mole of EOH placed in $V$ (L) solution. Calculate $K_C$ for this reaction.

$$\left(\Delta G_f^0\right)_{EA} = -327.6\frac{kJ}{mol}$$

(https://www.chemeo.com/cid/57 – 231 – 2/Ethyl – Acetate)

$$\left(\Delta G_f^0\right)_{H_2O(l)} = -237.14\frac{kJ}{mol}$$

(https://en.wikipedia.org/wiki/Standard_Gibbs_free_energy_of_formation)

$$\left(\Delta G_f^0\right)_{EOH} = -168.4\frac{kJ}{mol}$$

(https://www.chemeo.com/cid/35 – 653 – 8/Ethanol)

$$\left(\Delta G_f^0\right)_{AA} = -376.9\frac{\text{kJ}}{\text{mol}}$$

(https://www.chemeo.com/cid/51 – 574 – 8/Acetic – acid)

**CHW 3.4 The Van't Hoff Equation**

For the following reversible gas-phase reaction:

$$NO\ (g) + \frac{1}{2}O_2(g) \rightleftharpoons NO_2\ (g)$$

$K_P$ is $2.2\times10^{12}$ at $T = 25\,°C$. If $K_P$ is $5.6\times10^{-4}$ at $T = 1{,}000\,°C$, what will be $\Delta H^0_{rxn}$? Are the observed values of $K_P$ in harmony with Le Chatelier's principle?

**CHW 3.5 Le Chatelier's Principle (I)**

Consider the following equilibrium systems:

(a) $2PbS(s) + 3O_2(g) \rightleftharpoons 2PbO(s) + 2SO_2(g)$

(b) $PCl_5(g) \rightleftharpoons PCl_3(g) + Cl_2(g)$

(c) $H_2(g) + CO_2(g) \rightleftharpoons H_2O(g) + CO(g)$

Determine the direction of the net reaction in each case when the pressure is increased (volume is decreased) at constant temperature.

**CHW 3.6 Le Chatelier's Principle (II)**

Consider the following equilibrium process between dinitrogen tetrafluoride, $N_2F_4$, and nitrogen difluoride, $NF_2$:

$$N_2F_4(g) \rightleftarrows 2NF_2(g) \qquad \Delta H^0_{rxn} = 38{,}500\frac{\text{kJ}}{\text{mol}}$$

Predict the changes in equilibrium in each of the following scenarios:

(a) The reaction equilibrium mixture is heated at constant volume.

(b) Some $N_2F_4$ gas is removed from the reaction equilibrium mixture at constant temperature and volume.

(c) The pressure of the reaction equilibrium mixture is decreased at constant temperature.

(d) A catalyst is added to the reaction equilibrium mixture.

**CHW 3.7 Using DelG_Keq.m**

Calculate $\Delta H^0$, $\Delta G^0$, and $K_{eq}$ for the following chemical reactions:

1) $2NH_3(g) \rightleftharpoons N_2(g) + 3H_2(g)$@ $T = 300\,K$

2) $H_2(g) + \frac{1}{2}O_2(g) \rightleftharpoons H_2O(g)$@ $T = 300\,K$

3) $H_2(g) + \frac{1}{2}O_2(g) \rightleftharpoons H_2O(l)$@ $T = 300\,K$

**CHW 3.8 Using DelG_Keq.m**

Calculate $\Delta H^0$, $\Delta G^0$, and $K_{eq}$ for the following reversible water-gas shift reaction (WGSR):

1) $CO\ (g) + H_2O(g) \rightleftharpoons H_2(g) + CO_2\ (g)$ @ $T = 1{,}900\,K$

**CHW 3.9 Calculation of $K_C$ for $NH_3$ Formation**

a) Calculate the equilibrium constant, $K_{eq}$, for ammonia synthesis at 700 K?

$$N_2(g) + 3H_2(g) \rightleftharpoons 2NH_3(g)$$

b. What is $K'_{eq}$ for: $2NH_3(g) \rightleftharpoons N_2(g) + 3H_2(g)$?
c. What is $K''_{eq}$ for: $\frac{1}{2}N_2(g) + \frac{3}{2}H_2(g) \rightleftharpoons NH_3(g)$?
d. What is $K'''_{eq}$ for: $NH_3(g) \rightleftharpoons \frac{1}{2}N_2(g) + \frac{3}{2}H_2(g)$?

**CHW 3.10 $SO_2$ Oxidation**
At $T = 527\,°C$, $K_{eq}$ for the reaction:

$$2SO_2(g) + O_2(g) \rightleftharpoons 2SO_3(g)$$

is equal to $7.9{\times}10^4$. $K'_{eq}$ for the following reaction: $SO_3(g) \rightleftharpoons SO_2(g) + \frac{1}{2}O_2(g)$, is:

**A** $\times 10^{-5}$
**B** 281.07
**C** 0.0036
**D** 16.765

**CHW 3.11 Phosphorus Pentachloride ($PCl_5$) Partial Pressure**
The equilibrium constant $K_P$ for the decomposition of phosphorus pentachloride to phosphorus trichloride and molecular chlorine:

$$PCl_5(g) \rightleftharpoons PCl_3(g) + Cl_2(g)$$

is found to be 1.05 atm at 250 °C. If the equilibrium partial pressures of $PCl_5$ and $PCl_3$ are 0.875 atm and 0.463 atm, respectively, what will be the equilibrium partial pressure of $Cl_2$ at 250 °C?

**CHW 3.12 Heterogeneous Reactions**
Write the equilibrium constant expression $K_C$ and/or $K_P$ if applicable, for each of the following heterogeneous systems:
(a) $(NH_4)_2Se(s) \rightleftharpoons 2NH_3(g) + H_2Se(g)$
(b) $AgCl(s) \rightleftharpoons Ag^+_{(aq)} + Cl^-_{(aq)}$
(c) $P_4(s) + 6\,Cl_2(g) \rightleftharpoons 4PCl_3(l)$

**CHW 3.13 HBr Formation**
The synthesis of HBr from hydrogen and liquid bromine

$$H_{2(g)} + Br_{2(l)} \rightleftharpoons 2HBr(g) \qquad K_P = 4.5 \times 10^{15}\ \text{atm} \tag{A}$$

has an equilibrium constant, at 25 °C, $K_P = 4.5 \times 10^{15}$ atm. Given that the saturation vapor pressure of liquid bromine at $T = 25\,°C$ is 0.28 atm

$$Br_{2(l)} \rightleftharpoons Br_{2(g)} \qquad K'_P = 0.28\ \text{atm} \tag{B}$$

Find $K_P$ for the homogeneous gas-phase reaction at the same temperature

$$H_{2(g)} + Br_{2(g)} \rightleftharpoons 2HBr(g) K''_P = ? \tag{C}$$

**CHW 3.14 $NO_2$ Formation**
Given the equilibrium constant for the following two reversible gas-phase reactions:

$$N_2(g) + O_2(g) \rightleftharpoons 2NO(g) \qquad K_1 = 2.5 \times 10^{-25} \tag{A}$$

$$2NO(g) + O_2(g) \rightleftharpoons 2NO_2(g) \qquad K_2 = 6.0 \times 10^{9} \tag{B}$$

What will be the equilibrium constant for the following gas-phase reaction?

$$N_2(g) + 2O_2(g) \rightleftharpoons 2NO_2(g) K_3 = ? \quad (C)$$

**CHW 3.15 $CS_2$ Formation**

Given the equilibrium constant for the following two reversible gas-phase reactions:

$$CO(g) + 3H_2(g) \rightleftharpoons CH_4(g) + H_2O(g) \quad K_1 = 9.2 \times 10^{-2} \quad (A)$$

$$CS_2(g) + 4H_2(g) \rightleftharpoons 2H_2S(g) + CH_4(g) \quad K_2 = 3.03 \times 10^{-5} \quad (B)$$

What will be the equilibrium constant for the following gas-phase reaction?

$$CO(g) + 2H_2S(g) \rightleftharpoons CS_2(g) + H_2O(g) + H_2(g) \quad K_3 = ? \quad (C)$$

**CHW 3.16 The Extent of Reaction**

Acrylonitrile is produced via the following reaction:

$$C_3H_6 + NH_3 + \frac{3}{2}O_2 \rightarrow C_3H_3N + 3\,H_20$$

The feed contains 10.0 mole% propylene, 15.0% ammonia, and 75.0% air. *A fractional conversion of 60.0% of the limiting reactant is achieved.* Taking 100 mol/min of feed as a basis, determine which reactant is limiting, the percentage by which each of the other reactants is in excess, and the molar amounts of all product gas constituents for a 40% conversion of the limiting reactant (Figure CHW 3.16).

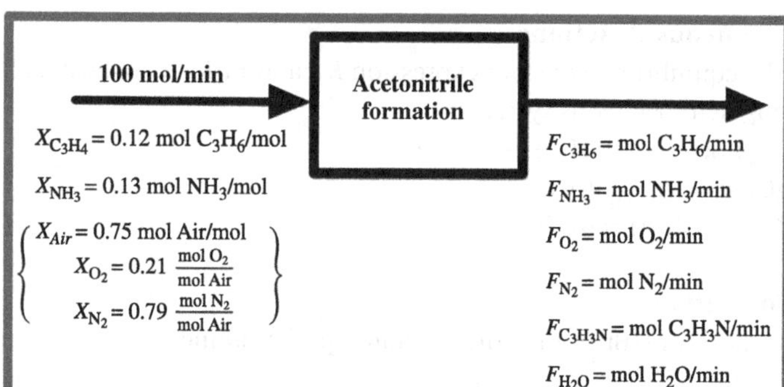


**CHW 3.16** Flow chart for a fluidized bed reactor that produces acrylonitrile from ammonia, propylene, and air.

**CHW 3.17 Calculation of the Equilibrium Extent of Reaction**

Consider the reaction:

$$CH_3COOH + C_2H_5OH \rightleftharpoons CH_3COOC_2H_5 + H_2O$$

In symbolic form:

$$AA + EOH \rightleftharpoons EA + H_2O$$

We have initially one mole of AA and one mole of EOH placed in $V$ (L) solution. Given $K_C = 2.55 \times 10^3$ for this reaction, calculate the equilibrium number of moles of reacting species.

**CHW 3.18 Calculation of Equilibrium Compositions for Water-Gas Shift Reaction (WGSR)**

If the WGSR,

$$CO(g) + H_2O(g) \rightleftharpoons CO_2(g) + H_2(g) \qquad (A)$$

proceeds to **equilibrium** at $T = 1{,}000$ (K). Initially, there exist two moles of $H_2O$ and one mole of CO. Using **DelG_Keq.m**, find Keq at $T = 1{,}000$ K and then find the equilibrium composition.

**CHW 3.19 Hydrogen Iodide Formation**

Consider the reversible gas-phase formation reaction of hydrogen iodide gas:

$$H_2(g) + I_2(g) \Leftrightarrow 2HI(g) \text{ at } T = 440\,°C = 713\text{ K}.$$

The reaction starts with 2 moles of $H_2$, 1 mole of $I_2$, and 3 moles of HI in a 1-L flask. Use **DelG_Keq.m** to find Keq at $T = 713$ K and then find the equilibrium concentration and mole fraction of $H_2$, $I_2$, and HI?

What are the equilibrium mole fractions of $H_2$, $I_2$, and HI?

**CHW 3.20 $SO_2$ Oxidation into $SO_3$**

The reversible gas-phase reaction and the equilibrium composition of the mixture at 323 °C are shown below:

$$2SO_2(g) + O_2(g) \leftrightarrow 2SO_3(g)$$

$$[SO_3]_{eq} = 1.50 \times 10^{-2}\text{ M}, [SO_2]_{eq} = 1.25 \times 10^{-2}\text{ M}, [O_2]_{eq} = 1.75 \times 10^{-2}\text{ M}$$

Determine $K_C$ and $K_P$ for this reaction.

**CHW 3.21 Phosphorus Pentachloride ($PCl_5$) Formation**

Phosphorus trichloride ($PCl_3$) gas reversibly reacts with chlorine gas ($Cl_2$) to produce phosphorus pentachloride ($PCl_5$) gas as in the following reaction:

$$PCl_3(g) + Cl_2(g) \leftrightarrow PCl_5(g)$$

The initial and equilibrium concentrations of $PCl_3$ are 0.40 mol/L and 0.2 mol/L, respectively, while the initial concentration of chlorine is 0.35 mol/L. Calculate the equilibrium constant, $K_C$.

**CHW 3.22 Formation of Ethanol from Ethene Using Acid Catalyst**

$$C_2H_4(g) + H_2O(g) \leftrightarrow C_2H_5OH(g) \qquad \text{Or: } A(g) + B(g) \leftrightarrow C(g)$$

The temperature between 300 and 350 °C is typically used in industry, specifically in the acid-catalyzed hydration process (using phosphoric acid on a silica catalyst). The reaction is: exothermic, but higher temperatures are needed to increase the reaction rate. On the other hand, the process is conducted at high pressure, typically around 60–70 atm. This high pressure helps increase the solubility of ethene in water and drives the reaction toward the production of ethanol. The feed consists of a mixture of 50 mol% $C_2H_4$ and 50 mol% $H_2O$ at $T_0 = 600$ K and $P_0 = 60$ atm.

1) Calculate the equilibrium conversion of ethanol in a constant-volume batch reactor. Use (**DelG_Keq.m**) MATLAB code to calculate $K_{eq}$.

2) Assuming the reaction is: elementary, determine the CSTR volume necessary to achieve 99% of the equilibrium conversion. Assume constant $T$ and $P$. Assume $F_{A0}$ = 100 mol/s. Given that a typical $kf$ value, at the above conditions, is 0.50 mol/L·s.

**CHW 3.23 Oxidation of Carbon Monoxide**

Consider the reaction:

$$CO(g) + \frac{1}{2}O_2 \leftrightarrow CO_2(g)$$

**Given:**

- **Standard Reaction Enthalpy:** $\Delta H^0 = -283$ kJ/mol
- **Standard Reaction Entropy:** $\Delta S^0 = -86$ J/mol·K
- Ideal gas behavior is assumed.
- **Initial Molar Feed:** 1 mol of CO and 0.5 mol of $O_2$
- **Temperature Range:** 500–2000 K

Derive and compute the equilibrium constant $K_{eq}(T)$. Write the equilibrium expression in terms of the extent of reaction or conversion of CO. Using MATLAB, for each temperature in the specified range, solve for the equilibrium conversion of CO. Plot the equilibrium conversion of CO versus temperature.

**CHW 3.24 Formation of Nitrogen Dioxide**

Consider the reaction:

$$NO(g) + \frac{1}{2}O_2 \leftrightarrow NO_2(g)$$

**Given:**

- **Standard Reaction Enthalpy:** $\Delta H^0 = -57$ kJ/mol
- **Standard Reaction Entropy:** $\Delta S^0 = -146$ J/mol·K
- Ideal gas behavior is assumed.
- **Initial Molar Feed:** 1 mol of NO and 0.5 mol of $O_2$
- **Temperature Range:** 300–1500 K

Compute the temperature-dependent equilibrium constant $K_{eq}(T)$ using $\Delta H^0$ and $\Delta S^0$. Express the equilibrium relationship in terms of the conversion of NO. Using MATLAB, for each temperature, determine the equilibrium conversion of NO. Plot the equilibrium conversion of NO as a function of temperature.

**CHW 3.25 Fugacity-based Keq for Water-Gas Shift Reaction (WGSR)**

Consider the WGSR:

$$CO(g) + H_2O(g) \rightleftharpoons CO_2(g) + H_2(g)$$

Compute the nonideal equilibrium constant using fugacity coefficients and calculate conversion as a function of pressure. Use MATLAB code to implement the Peng–Robinson fugacity coefficient method for the WGSR. Test the effect of pressure on the conversion or extent of reaction is: the pressure effect on the equilibrium conversion in line with Le Chatelier's principle?

See Section 3.11 MATLAB Code for $NH_3$ $K_{eq}$ Under Nonideal High-pressure Conditions.

Use the following data for setting up the code:

$T = 1{,}000$; %K

$P_0 = 50$; %bar

$y_0 = [0.25; 0.25; 0.25; 0.25]$; %Initial equal mole fractions

%Critical data: $[CO, H_2O, CO_2, H_2]$

$T_c = [132.9; 647.1; 304.2; 33.2]$; %K

$P_c = [34.99; 220.6; 73.8; 13.0]$; %bar

$w = [0.045; 0.344; 0.225; -0.216]$;

**CHW 3.26 Estimation of Heat of Reaction Using Nonlinear Regression**

We have experimentally determined $K_{eq}$ values at several temperatures, as shown in Table 3.6.

**Table 3.6** Experimental data for $K_{eq}$ as $f(T)$.

| *T* (K) | $K_{eq}$ | *T* (K) | $K_{eq}$ |
|---|---|---|---|
| 300 | 1.25 | 350 | 0.80 |
| 310 | 1.15 | 360 | 0.75 |
| 320 | 1.01 | 370 | 0.70 |
| 330 | 0.90 | 380 | 0.65 |
| 340 | 0.85 | 390 | 0.60 |

Find the standard heat of reaction within the given temperature range, assuming a constant heat of reaction. Report statistical parameters showing the goodness of the fitting model.

**CHW 3.27 Coupled Parallel Equilibrium Reactions**

Consider the following two reversible parallel gas-phase reactions:

1) $A + B \rightleftharpoons C K_{eq1} = \frac{[C]}{[A][B]} = 2$

2) $A + B \rightleftharpoons D K_{eq2} = \frac{[D]}{[A][B]} = 3$

Let the initial moles be: $[A]_0 = 2\frac{\text{mol}}{\text{L}}$ $[B]_0 = 1\frac{\text{mol}}{\text{L}}$

Find the extent of reaction for both reactions and the equilibrium composition:

$\xi_1$: extent of reaction 1 $\quad$ $\xi_2$: extent of reaction 2

**CHW 3.28 Dynamic Equilibrium in CSTR**

Consider the following liquid-phase second-order reversible reaction to be carried out in an isothermal CSTR:

$$2A \rightleftharpoons 2B \qquad K_{eq} = \frac{k_f}{k_b} = \frac{[B]^2}{[A]^2} = \frac{C_B^2}{C_A^2}$$

$C_{A0} = 1.0$; $\quad$ Inlet concentration of $A$ (mol/L)

$k_f = 2.0;$ Forward rate constant (L/mol · s)

$k_b = 1.0;$ Backward rate constant (L/mol · s)

Derive the equilibrium concentration for both $A$ and $B$. Carry out steady-state molar balance for $A$ to express the exit $C_A$ and $C_B$ as $f(t)$ of the CSTR. Write MATLAB code to plot exit $C_B$ as $f(t)$ and find the minimum $t$ such that exit $C_B$ practically approaches the equilibrium $C_B$.

**CHW 3.29 Liquid-phase g-based $K_{eq}$ Calculation**

Consider the following liquid-phase reaction:

$A + B \rightleftharpoons C + D$

**At Equilibrium:**

**Mole Fractions:** $x = [0.30, 0.30, 0.20, 0.20] \rightarrow [A, B, C, D]$

**Activity Coefficients:** $\gamma = [1.3, 1.2, 1.1, 0.9]$

**At Nonequilibrium:**

**Mole Fractions:** $x = [0.40, 0.30, 0.15, 0.15] \rightarrow [A, B, C, D]$

**Activity Coefficients:** $\gamma = [1.2, 1.1, 1.05, 0.95]$

Determine whether the reaction will move to the right or left, based on Le Chatelier's principle.

**CHW 3.30 Gas-phase Fugacity-based $K_{eq}$ Calculation**

Consider the gas-phase reversible reaction:

$$2A \rightleftharpoons 2B$$

**At Equilibrium:**

- **Mole Fractions:** $y = [0.55, 0.45]$ for $A$ and $B$
- **Fugacity Coefficients:** $\phi = [0.90, 1.1]$

**At Nonequilibrium:**

- **Mole Fractions:** $y = [0.45, 0.55]$ for $A$ and $B$
- **Fugacity Coefficients:** $\phi = [0.95, 1.2]$

Express and calculate $K_{eq}$ in terms of fugacity coefficients for $P = 10$ atm.

# 4

# Selectivity and Yield of Multiple Reactions

When multiple chemical reactions occur together, they often compete for the same reactants, making it a challenge to maximize the formation of the desired product. In such cases, simply tracking overall conversion is insufficient because high conversion of reactants does not necessarily translate into high production of the target (or desired) compound; side reactions may consume significant portions of the reactants, leading to the formation of undesired by-products. This is where the concepts of selectivity and yield become essential analytical tools. Selectivity helps quantify the preference of a process toward producing the desired product over competing products, while yield measures the actual amount of desired product obtained relative to either the feed or the reactant consumed. These metrics guide both the design and optimization of chemical processes, enabling engineers to make informed decisions about operating conditions, catalyst choice, and reactor configuration.

Furthermore, in industrial practice, selectivity and yield are directly tied to process economics, environmental impact, and sustainability. High selectivity minimizes waste streams, reduces energy and separation costs, and improves raw material efficiency, which is especially critical when dealing with expensive or hazardous feedstocks. Yield, on the other hand, reflects the overall effectiveness of the entire process, from reaction kinetics to downstream purification, and is often a key performance indicator in production targets. Understanding the interplay between selectivity and yield is therefore vital not only for achieving technical optimization but also for meeting profitability and regulatory requirements. This chapter will explore the different types of multiple reactions, such as parallel, series, and complex networks, and present methods to quantify, analyze, and enhance selectivity and yield through both theoretical and practical approaches.

## 4.1 A Selectivity of the Desired Product

Selectivity is a crucial performance parameter in chemical reaction engineering because most industrial processes do not yield a single product but rather a mixture of desired and undesired products. The choice of how selectivity is expressed depends on the reaction pathway and the available experimental or process data. For instance, in parallel reactions where reactant $A$ can form either the desired product $D$ or the undesired product $U$, selectivity may be expressed as the molar ratio $S_{D/U} = n_D/n_U$ or in terms of reaction rates $S_{D/U} = r_D/r_U$. In series reactions, where the desired product may further react to form by-products, selectivity can instead be defined relative to the initial reactant, such as $S_D = \frac{n_D}{n_A^{Converted}}$.

The form of selectivity chosen influences how reactor design, operating conditions, and catalyst formulation are optimized, making it essential to understand the context in which it is applied.

Moreover, selectivity often serves as a decision-making metric for economic and environmental optimization. High selectivity toward the desired product reduces raw material waste, minimizes downstream separation costs, and decreases the generation of harmful by-products. In kinetic studies, selectivity expressed in terms of rates provides real-time insight into competing reaction pathways and helps in identifying conditions such as temperature, pressure, and catalyst properties that favor the desired pathway. On the other hand, selectivity based on moles or yields is more relevant for process-scale assessments, where cumulative production over a time period is considered. Since real processes are often affected by side reactions, mass transfer limitations, and catalyst deactivation, improving selectivity is rarely a matter of reaction chemistry alone; it also involves reactor configuration, residence time distribution control, and heat/mass transfer optimization. Here are some common definitions of selectivity where either moles or rates can be used.

## 4.2 Relative to a Specific Undesired Product

$$Selectivity = \frac{Moles\ (or\ Rate)\ of\ Desired\ Product}{Moles\ (or\ Rates)\ of\ a\ Specfic\ Undesired\ Product} \tag{4.1}$$

Equation (4.1) compares the desired product to a particular undesired by-product. Equation (4.1) definition is useful when a particular undesired product is of primary concern, either because it poses operational challenges (e.g., fouling, catalyst deactivation) or has regulatory/compliance implications (e.g., hazardous by-products). Moreover, if the undesired product results from a key side reaction that competes with the main reaction, this definition helps evaluate how effectively the side reaction is being suppressed. For example, in a reaction system where methane ($CH_4$) is the primary undesired by-product, this definition allows focus on suppressing its formation relative to the desired product.

## 4.3 Relative to All Undesired Products

$$Selectivity = \frac{Moles\ (or\ Rate)\ of\ Desired\ Product}{Moles\ (or\ Rates)\ of\ All\ Undesired\ Products} \tag{4.2}$$

Equation (4.2) considers the total formation of all undesired products. Equation (4.2) definition provides a comprehensive assessment of the system by comparing the desired product to all undesired by-products collectively. This definition can be used when the goal is to minimize the total formation of by-products across the system, especially in complex reactions with multiple side reactions. For example, in industrial-scale chemical production, where undesired by-products include multiple gases or liquids (e.g., methane, ethane, and other hydrocarbons), this definition helps assess overall by-product minimization.

## 4.4 Relative to All Products (Excluding Desired Product)

$$Selectivity = \frac{Moles\ (or\ Rate)\ of\ Desired\ Product}{Moles\ (or\ Rates)\ of\ All\ Products,\ Except\ Moles\ (or\ Rate)\ of\ Desired\ Product} \tag{4.3}$$

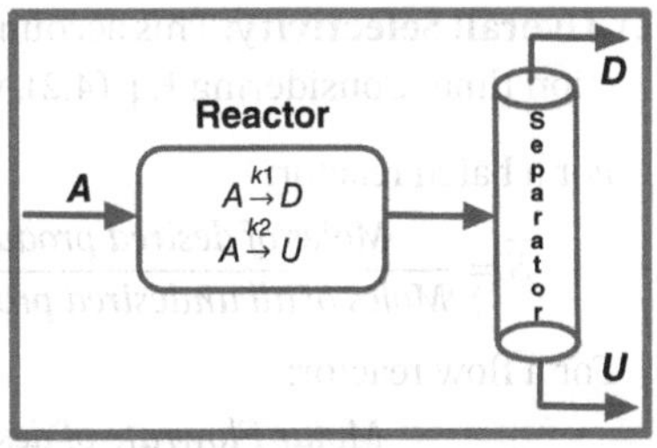


**Figure 4.1** A reaction-separation system that generates both desired and undesired products.

Equation (4.3) compares the desired product to all other products combined. Equation (4.3) definition is particularly useful for benchmarking or comparing different processes or reactor designs since it considers the desired product in relation to everything else produced. It is useful in processes where the total product distribution is critical for downstream separation or environmental compliance. This definition offers insight into the overall process composition. For example, in a process that produces a mixture of gases, including the desired product, this definition highlights the proportion of the desired product relative to the total output, excluding itself.

Regardless of selecting any of the above three definitions, a higher selectivity value always indicates better process performance or more favorable operating conditions, as it reflects greater efficiency in forming the desired product while minimizing unwanted by-products. We should always be mindful of which definition is being used to ensure proper interpretation of results. The aim is to minimize the formation of $U$ and maximize the formation of $D$ because a higher amount of the undesired product $U$ increases the cost of separating it from the desired product $D$, as shown in Figure 4.1.

### 4.4.1 Types of Selectivity

1) **Instantaneous Selectivity:** It represents selectivity at a given point in time and is often used in analyzing reaction kinetics. Referring to Eq. (4.2), it is defined as:

$$S_{D/U} = \frac{r_D}{r_U} = \frac{\textit{Rate of formation of desired product}}{\textit{Rate of formation of undesired products}} \tag{4.4}$$

**Example 4.1** ***Selectivity of Two Parallel Reactions***

Consider the following two parallel reactions:

$$A + B \xrightarrow{k1} D \tag{A}$$

$$A + B \xrightarrow{k2} U \tag{B}$$

The instantaneous selectivity of the desired over the undesired product is given by their molar production rate per unit volume. Based on Eq. (4.1), we have:

$$S_{D/U} = \frac{r_D}{r_U} = \frac{k_1 C_A C_B}{k_2 C_A C_B} = \frac{k_1}{k_2} C_A \tag{C}$$

To maximize $D$ selectivity, one must keep a high concentration of $A$; hence, PFR is recommended. Reactant $A$ can be supplied at the feed with a high $A/B$ molar ratio. Alternatively, $A$ can be supplied as the mainstream, and $B$ can be supplied on the spot as a multi-port side stream along the length of the tubular reactor.

2) **Overall Selectivity:** This accounts for the total amount of products formed over the entire reaction time. Considering Eq. (4.2), we have:

For a batch reactor:

$$\overline{S} = \frac{\textit{Moles of desired product}}{\textit{Moles of all undesired products}} \tag{4.5a}$$

For a flow reactor:

$$\overline{S} = \frac{\textit{Molar Flowrate of desired product}}{\textit{Molar Flowrates of all undesired products}} \tag{4.5b}$$

The moles for both the numerator and the denominator are evaluated at the end of the batch time or the end of the residence time for flow reactors.

### 4.4.2 Factors Influencing Selectivity

- Reaction pathways (parallel vs. consecutive reactions)
- Reaction conditions (temperature, pressure, and catalysts)
- Concentration of reactants and intermediates
- Reactor type (batch, PFR, and CSTR)

Maximizing selectivity often involves manipulating the above factors to favor the desired reaction pathway while suppressing others.

Table 4.1 shows the definition of both instantaneous and overall selectivity and yield terms for the desired product. The yield is defined as the molar reaction rate or ratio between the desired and the total rates or moles of products. The total rate of the limiting reactant $A$ is the sum of all rates that result in the consumption of reactant $A$. Alternatively, the total number of moles of $A$ consumed includes moles of $A$ consumed toward the production of desired and undesired products. More on the reaction yield of the desired product will be explained in a subsequent section.

Throughout this chapter, selectivity $S$ refers to instantaneous, unless otherwise specified.

Let us examine the selectivity considering the following scenarios of multiple reactions.

1) **Series Reactions:**

$$A \xrightarrow{k1} B \xrightarrow{k2} C \tag{4.6}$$

An example of a series reaction is the hydrolysis of an ester into the production of carboxylic acid. This is seen in the conversion of esters to carboxylic acids via intermediate alcohols.

**Step 1 (Desired Reaction):** Ester is hydrolyzed to produce carboxylic acid and an alcohol:

$$\mathrm{RCOOR'} + \mathrm{H_2O} \leftrightarrow \mathrm{RCOOH} + \mathrm{R'OH} \qquad \text{(Desired)}$$

**Table 4.1** Definition of both instantaneous and overall selectivity and yield term.

| Item | Instantaneous | Overall |
|---|---|---|
| Selectivity | $S_{D/U} = \frac{r_D}{r_U}$ | $\overline{S}_{D/U} = \frac{F_D}{F_U}$ |
| Yield | $Y_D = \frac{r_D}{-r_A}$ | $\overline{Y}_D = \frac{F_D}{F_{Ao} - F_A}$ |

**Step 2:** If not controlled, the produced alcohol can further oxidize into aldehyde:

$$R'OH + O_2 \leftrightarrow R'CHO + H_2O \qquad \text{(Undesired)}$$

2) **Parallel Reactions:**

$$A \xrightarrow{k1} B \tag{4.7a}$$

$$A \xrightarrow{k2} C \tag{4.7b}$$

An industrially significant example of parallel reactions is the oxidation of ethylene into ethylene oxide, with the simultaneous goal of minimizing its complete combustion to carbon dioxide and water.

$$CH_2 = CH_2 + O_2 \rightarrow CH_2 - (O) - CH_2$$

$$CH_2 = CH_2 + O_2 \rightarrow 2CO_2 + 2H_2O$$

Another example of a parallel reaction can be found in the nitration of toluene, where toluene reacts with a nitrating mixture to produce multiple isomers simultaneously:

a) The formation of para-nitrotoluene:

$$C_6H_5CH_3 + HNO_3 \rightarrow p - C_6H_4NO_2CH_3 + H_2O \qquad \text{(Desired)}$$

b) The formation of ortho-nitrotoluene:

$$C_6H_5CH_3 + HNO_3 \rightarrow o - C_6H_4NO_2CH_3 + H_2O \qquad \text{(Undesired)}$$

c) The formation of meta-nitrotoluene:

$$C_6H_5CH_3 + HNO_3 \rightarrow m - C_6H_4NO_2CH_3 + H_2O \qquad \text{(Undesired)}$$

3) **Complex/Mixed Reactions: Series and Parallel:**

$$A \xrightarrow{k1} B + C \tag{4.8a}$$

$$A + C \xrightarrow{k2} D \tag{4.8b}$$

An example of a set of complex/mixed reactions is phenol ($A$) hydrogenation to cyclohexanone ($B$):

$$C_6H_5OH + H_2 \rightarrow C_6H_{10}O \qquad \text{(Desired)}$$

followed serially by cyclohexanone ($B$) hydrogenation into cyclohexanol ($C$):

$$C_6H_{10}O + H_2 \rightarrow C_6H_{11}OH \qquad \text{(Undesired)}$$

At the same time, the above two reactions are parallel with respect to $H_2$ consumption.

4) **Independent Reactions:**

Independent reactions are those that happen at the same time and are not influenced by each other. None of the substances involved in one reaction affects the substances in the other.

$$A \xrightarrow{k1} C \tag{4.9a}$$

$$B \xrightarrow{k2} D \tag{4.9b}$$

An example of independent reactions is:
Sulfur dioxide is oxidized to sulfur trioxide

$$SO_2 + \frac{1}{2}O_2 \xrightarrow{k_1} SO_3$$

And simultaneously, sulfur monoxide reacts with hydrogen to form hydrogen sulfide and water.

$$SO + 2H_2 \xrightarrow{k_2} H_2S + H_2O$$

### 4.4.3 MATLAB Code for Selectivity of Parallel Reactions

**Example 4.2** ***Selectivity of Three Parallel Reactions***
Consider the following three parallel reactions:

$$A + B \xrightarrow{k1} D \qquad r_D = k_1 C_A^2 C_B \tag{A}$$

$$A + B \xrightarrow{k2} U_1 \qquad r_{U_1} = k_2 C_A C_B \tag{B}$$

$$A + B \xrightarrow{k3} U_2 \qquad r_{U_2} = k_3 C_A^3 C_B \tag{C}$$

The instantaneous selectivity (Eq. 4.2) is defined as:

$$S_{D/(U_1+U_2)} = \frac{r_D}{r_{U_1} + r_{U_2}} = \frac{k_1 C_A^2 C_B}{k_2 C_A C_B + k_3 C_A^3 C_B} = \frac{k_1 C_A}{k_2 + k_3 C_A^2} \tag{D}$$

Finding the optimum concentration:

$$\frac{dS}{dC_A} = 0 = \frac{k_1\{k_2 + k_3 C_A^2\} - k_1 C_A\{2k_3 C_A\}}{[k_2 + k_3 C_A^2]^2} \rightarrow [k_1\{k_2 + k_3 C_A^2\} - k_1 C_A\{2k_3 C_A\}] = 0 \tag{E}$$

Further simplification yields to:

$$[\{k_2 + k_3 C_A^2\} - C_A\{2k_3 C_A\}] = 0 \rightarrow k_2 + k_3 C_A^2 - 2k_3 C_A^2 = 0 \tag{F}$$

$$k_2 - k_3 C_A^2 = 0 \rightarrow C_A^2 = \frac{k_2}{k_3} \rightarrow C_A^{Optim} = \sqrt{\frac{k_2}{k_3}} \tag{G}$$

We use a CSTR with exit concentration $C_{A,exit} = C_A^{Optim} = \sqrt{\frac{k_2}{k_3}}$.

Figure 4.2 shows MATLAB code (**SD_U.m**) for making a plot of the desired product selectivity as a function of $C_A$, using Eq. (D).

```
clear variables;
% Define constants
k1 = 0.1; % L^2.mol^-2.s^-1
k2 = 0.2; % L.mol^-1.s^-1
k3 = 0.1; % L^3.mol^-3.s^-1
% Define the range of CA
CA = linspace(0.0, 10, 1001); % mol/L
```

**Figure 4.2** MATLAB code (**SD_U.m**) for generating a plot of the desired product selectivity, given by Eq. (D), as a function of $C_A$.

```
% Selectivity equation
Selectivity = (k1 * CA) ./ (k2 + k3 * CA.^2);
[max_selectivity, index] = max(Selectivity);
% Display the maximum selectivity and CA_opt
disp(['Maximum Selectivity: ', num2str(max_selectivity)]);
disp(['CA_opt occurs at CA = ', num2str(CA(index))]);
% Plot
figure;
plot(CA, Selectivity, 'b-', 'LineWidth', 1.5);
% Add labels, title, and grid
xlabel('C_A (mol/L)', 'FontSize', 12);
ylabel('Selectivity (S_{D/(U_1+U_2)})', 'FontSize', 12);
title('Selectivity as a Function of C_A', 'FontSize', 14);
grid on;
ax = gca; % Get handle to the current axes
% Set new font size for all axes text (including labels)
ax.FontSize = 16;
ax.FontWeight="bold";
```

**Figure 4.2** *(Continued)*

Results are copied from the Command Window:

```
>> SD_U
Maximum Selectivity: 0.35355
CA_opt occurs at CA = 1.41
```

Figure 4.3 shows the generated plot of the desired product selectivity, given by Eq (D), as a function of $C_A$. $k_1 = 0.1$ L$^2$/mol$^2$·s, $k_2 = 0.2$ L/mol·s; and $k_3 = 0.1$ L$^3$/mol$^3$·s were used. The peak of $S_{D/(U_1+U_2)} = 0.35355$ occurs at $C_A = C_A^{Optim} = \sqrt{\frac{k_2}{k_3}} = \sqrt{\frac{0.2}{0.1}} = \sqrt{2} = 1.414\,\text{mol/L}$.

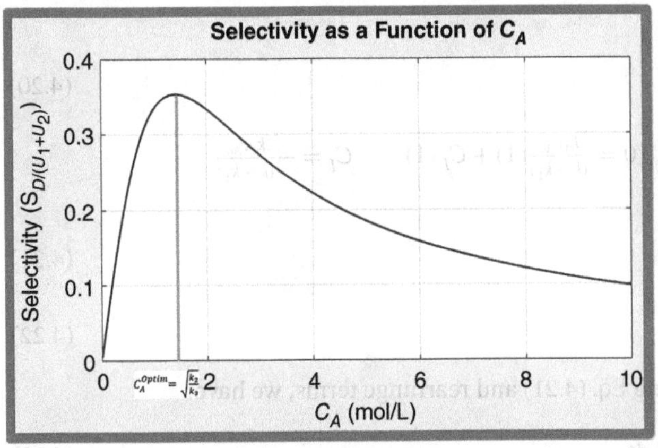


**Figure 4.3** A plot of desired product selectivity as a function of concentration of *A*. The peak of $S_{D/(U_1+U_2)}$ occurs at $C_A = C_A^{Optim} = \sqrt{\frac{k_2}{k_3}} = \sqrt{\frac{0.2}{0.1}} = \sqrt{2} = 1.414\,\text{mol/L}$.

### 4.4.4 MATLAB Code for Selectivity of Series Reactions

$$A \xrightarrow{k1} B \xrightarrow{k2} C \tag{4.10}$$

$$\text{Species } A: \; -r_A = -\frac{dC_A}{dt} = k_1 C_A \tag{4.11}$$

$$C_A = c_{A0} e^{-k_1 t} \tag{4.12}$$

$$\text{Species B: } r_B = \frac{dC_B}{dt} = k_1 C_A - k_2 C_B \tag{4.13}$$

Rearrange and make use of Eq. (4.12) to replace $C_A$, we have:

$$\frac{dC_B}{dt} + k_2 C_B = k_1 C_A = k_1 c_{A0} e^{-k_1 t} \tag{4.14}$$

Notice that Eq. (4.14) is first-order and nonhomogeneous. The integrating factor is given by:

$$\text{Integrating factor: } \mu = e^{k_2 t} \tag{4.15}$$

Multiply both sides of Eq. (4.14) by $\mu$, we have:

$$\frac{dC_B}{dt} e^{k_2 t} + k_2 C_B e^{k_2 t} = k_1 C_A e^{k_2 t} = k_1 c_{A0} e^{-k_1 t} e^{k_2 t} \tag{4.16}$$

It is not difficult to realize that the two terms on the left side of Eq. (4.16) represent the differentiation of the term: $\frac{d[C_B e^{k_2 t}]}{dt}$. Hence, Eq. (4.16) becomes:

$$\frac{d\left[C_B e^{k_2 t}\right]}{dt} = k_1 c_{A0} e^{(k_2 - k_1)t} \tag{4.17}$$

Carrying out the indefinite integral while adding the constant of integration, we have:

$$C_B e^{k_2 t} = \int k_1 c_{A0} e^{(k_2 - k_1)t} dt + C_I \tag{4.18}$$

$$C_B e^{k_2 t} = \frac{k_1 c_{A0} e^{(k_2 - k_1)t}}{(k_2 - k_1)} + C_I \tag{4.19}$$

Dividing both sides by $e^{k_2 t}$, we have:

$$C_B = \frac{k_1 c_{A0}}{(k_2 - k_1)} e^{-k_1 t} + C_I e^{-k_2 t} \tag{4.20}$$

Initial condition: $@t = 0, C_B = 0$: $\quad 0 = \frac{k_1 c_{Ao}}{(k_2 - k_1)}(1) + C_I(1) \quad C_I = -\frac{k_1 c_{Ao}}{(k_2 - k_1)}$

Equation (4.20) becomes:

$$C_B = \frac{k_1 c_{A0}}{(k_2 - k_1)} \left[e^{-k_1 t} - e^{-k_2 t}\right] \tag{4.21}$$

$$\text{Species } C: \; r_C = \frac{dC_C}{dt} = k_2 C_B \tag{4.22}$$

Replace $C_B$ by its equivalence utilizing Eq. (4.21) and rearrange terms, we have:

$$dC_C = k_2 \frac{k_1 c_{A0}}{(k_2 - k_1)} \left[e^{-k_1 t} - e^{-k_2 t}\right] dt \tag{4.23}$$

Integrate both sides with $C_C(0) = 0$, we have:

$$\int_0^{C_C} dC_C = \int_0^t k_2 \frac{k_1 c_{Ao}}{(k_2 - k_1)} \left[e^{-k_1 t} - e^{-k_2 t}\right] dt \tag{4.24}$$

$$C_C = k_2 \frac{k_1 c_{A0}}{(k_2 - k_1)} \left[\frac{e^{-k_1 t}}{-k_1} - \frac{e^{-k_2 t}}{-k_2} - \left(\frac{1}{-k_1} - \frac{1}{-k_2}\right)\right] = \frac{k_2 k_1 c_{A0}}{(k_2 - k_1)} \left[\frac{e^{-k_2 t}}{k_2} - \frac{e^{-k_1 t}}{k_1} + \frac{1}{k_1} - \frac{1}{k_2}\right] \tag{4.25}$$

Thus, Eqs. (4.12), (4.21), and (4.25) each give the analytic solution for $C_A$, $C_B$, and $C_C$, respectively, as a function of time.

Figure 4.4 shows MATLAB code (**Ci_Series_Plot.m**) for creating a combined plot for $C_A$, $C_B$, and $C_C$ as a function of time with $C_{A0} = 100$ mol/L, $k_1 = 0.2\ s^{-1}$, and $k_2 = 0.1\ s^{-1}$.

```
clear variables;
% Define constants
k1 = 0.2; % rate constant k1 in s^-1
k2 = 0.1; % rate constant k2 in s^-1
C_Ao = 100.0; % initial concentration of A
% Time vector
t = linspace(0, 50, 1000); % time from 0 to 50 seconds with 1000 points
% Analytic solutions
C_A = C_Ao * exp(-k1 * t);
C_B = (k1 * C_Ao) / (k2 - k1) .* (exp(-k1 * t) - exp(-k2 * t));
C_C = (k2 * k1 * C_Ao) / (k2 - k1) .* (exp(-k2 * t) / k2 - exp(-k1 * t) / k1 +1/k1 -1/k2);
% Plot
figure;
hold on;
plot(t, C_A, 'r-', 'LineWidth', 1.5, 'DisplayName', 'C_A');
plot(t, C_B, 'b-', 'LineWidth', 1.5, 'DisplayName', 'C_B');
plot(t, C_C, 'g-', 'LineWidth', 1.5, 'DisplayName', 'C_C');
hold off;
% Add labels, title, and legend
xlabel('Time (s)', 'FontSize', 12);
ylabel('Concentration (mol/L)', 'FontSize', 12);
title('Concentration Profiles: C_A, C_B, and C_C vs Time', 'FontSize', 14);
legend('Location', 'best', 'FontSize', 10);
grid on;
ax = gca; % Get handle to the current axes
% Set new font size for all axes text (including labels)
ax.FontSize = 16;
ax.FontWeight = 'bold';
```

**Figure 4.4** MATLAB code (**Ci_Series_Plot.m**) for a plot creation of $C_A$, $C_B$, and $C_C$ as a function of time with $C_{A0} = 100$ mol/L, $k_1 = 0.2\ s^{-1}$, and $k_2 = 0.1\ s^{-1}$.

Figure 4.5 shows the generated plot where $C_B$ achieves a maximum concentration of $C_{B,max} =$ 50 mol/L, at $t_{opt} = 6.93$ s. Obviously, $C_A$ will exponentially decay with time until it hits zero concentration, and $C_C$ will exponentially increase until it hits 100 mol/L if time is sufficiently large (i.e., $t \geq 50$ s).

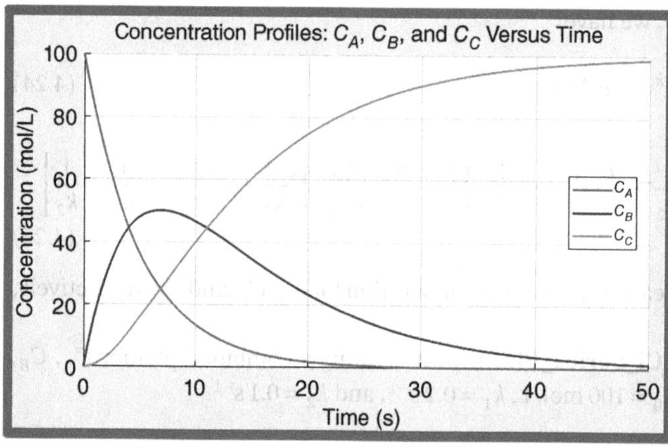


**Figure 4.5** A plot of $C_A$, $C_B$, and $C_C$ as a function of time with $C_{A0} = 100$ mol/L, $k_1 = 0.2$ s$^{-1}$, and $k_2 = 0.1$ s$^{-1}$.

When shall we stop the reaction to obtain the maximum amount of $B$?

$$\frac{dC_B}{dt} = 0 = \frac{k_1 c_{A0}}{(k_2 - k_1)}\left[-k_1 e^{-k_1 t} + k_2 e^{-k_2 t}\right] \rightarrow k_1 e^{-k_1 t_{opt}} = k_2 e^{-k_2 t_{opt}} \tag{4.26}$$

$$\left(\frac{k_2}{k_1}\right) = \frac{e^{k_2 t_{opt}}}{e^{k_1 t_{opt}}} = e^{(k_2 - k_1)t_{opt}} \tag{4.27}$$

Take the logarithmic value for both sides, we have:

$$\ln\left(\frac{k_2}{k_1}\right) = (k_2 - k_1)t_{opt} \tag{4.28}$$

Rearrange to have:

$$t_{opt} = \left(\frac{1}{k_2 - k_1}\right)\ln\left(\frac{k_2}{k_1}\right) \tag{4.29}$$

For example, as shown in Figures 4.4 and 4.5:

$$t_{opt} = \left(\frac{1}{k_2 - k_1}\right)\ln\left(\frac{k_2}{k_1}\right) = \left(\frac{1}{0.1 - 0.2}\right)\ln\left(\frac{0.1}{0.2}\right) = 6.9315 \text{ s}$$

One can then substitute the value of $t_{opt}$ in Eq. (4.21) to find $C_{B,max}$.

$$C_{B,max} = \frac{k_1 c_{A0}}{(k_2 - k_1)}\left[e^{-k_1 t} - e^{-k_2 t}\right] = \frac{0.2 * 100}{(0.1 - 0.2)}\left[e^{-0.2*6.9315} - e^{-0.1*6.9315}\right] = 50.0\ \frac{\text{mol}}{\text{L}}$$

On the other hand, if $k_1$ is equal to $k_2$, then we replace both $k$'s by $k$:

$$A \xrightarrow{k} B \xrightarrow{k} C \tag{4.30}$$

$$\text{Species } A:\ -r_A = -\frac{dC_A}{dt} = kC_A \qquad C_A = c_{A0}e^{-kt} \tag{4.31}$$

$$\text{Species } B:\ r_B = \frac{dC_B}{dt} = kC_A - kC_B \qquad \frac{dC_B}{dt} + kC_B = kC_A = kc_{A0}e^{-kt} \tag{4.32}$$

Multiply the last equation by $e^{kt}$(*the integrating factor*), we get:

$$\frac{dC_B}{dt}e^{kt} + kC_B e^{kt} = kc_{A0}e^{-kt}e^{kt} = kc_{A0} \tag{4.33}$$

The left hand-side can be put in the form of: $\frac{d[C_B e^{kt}]}{dt} = kc_{A0}$

Integrate both sides, we get: $\int d[C_B e^{kt}] = \int kc_{A0} dt + C$

$$\text{Simplify to reach at: } C_B e^{kt} = kc_{A0} t + C \qquad C_B = kc_{A0} t e^{-kt} + Ce^{-kt} \tag{4.34}$$

Apply I.C. @$t = 0$, $C_B = 0$, we have: $0 = 0 + C \times (1)$. This implies that $C = 0$.

$$C_B = kc_{A0} t e^{-kt} \tag{4.35}$$

To find the maximum $C_B$, differentiate and set the first derivative to zero to find the critical point(s).

$\frac{dC_B}{dt} = 0 = kc_{A0}[1 \times e^{-kt} + t(-ke^{-kt})] = 0$. This implies that the bracketed term is zero.

$\left[e^{-kt_{opt}} = t_{opt}\left(ke^{-kt_{opt}}\right)\right]$. That simply means:

$$t_{opt} = 1/k \tag{4.36}$$

Substitute $t = t_{opt} = \frac{1}{k}$ in Eq. (4.35), we have:

$$(C_B)_{max} = kc_{A0} \times \frac{1}{k} \times e^{-k \times \frac{1}{k}} = c_{A0} \times e^{-1} = \frac{c_{A0}}{e} = \frac{c_{A0}}{2.7182818} = 0.36788 \times c_{A0} \tag{4.37}$$

The maximum $C_B$ will be about 36.8% of $C_{A0}$ for the case $k_1 = k_2 = k$.

$$\text{Species } C\text{: } r_C = \frac{dC_C}{dt} = kC_B \tag{4.38}$$

Replace $C_B$ using Eq. (4.35), we have:

$$\frac{dC_C}{dt} = kC_B = kkc_{A0} t e^{-kt} \tag{4.39}$$

Separate variables and integrate both sides, we get:

$$\int dC_C = C_C = \int kkc_{A0} t e^{-kt} dt + C_I$$

$$\int kkc_{A0} t e^{-kt} dt = kkc_{A0}\left[t\frac{e^{-kt}}{-k} - \int \frac{e^{-kt}}{-k} dt\right] + C_{II}$$

$$\int kkc_{A0} t e^{-kt} dt = kkc_{A0}\left[t\frac{e^{-kt}}{-k} - \frac{1}{kk} e^{-kt}\right] + C_{II}$$

$$C_C = kkc_{A0}\left[t\frac{e^{-kt}}{-k} - \frac{1}{kk} e^{-kt}\right] + C_{III} \tag{4.40}$$

Apply I.C. @$t = 0$, $C_C = 0$, we have: $0 = kkc_{A0}\left[0 - \frac{1}{kk}\right] + C_{III}$. This implies that $C_{III} = c_{A0}$.

$$C_C = kkc_{A0}\left[t\frac{e^{-kt}}{-k} - \frac{1}{kk} e^{-kt}\right] + c_{A0}$$

Simplify to reach at:

$$C_C = c_{A0}[1 - e^{-kt}(1 + kt)] \tag{4.41}$$

Figure 4.6 shows MATLAB code (**Ci_Series_Plot2.m**) for creating a combined plot for $C_A$, $C_B$, and $C_C$ as a function of time with $C_{A0} = 100$ mol/L and $k_1 = k_2 = k = 0.1\ \text{s}^{-1}$.

Figure 4.7 shows the generated plot where $C_B$ achieves a maximum concentration of $C_{B,max} =$ 36.8 mol/L at $t_{opt} = 1/k = (1/0.1\ \text{s}^{-1}) = 10$ s.

```
clear variables;
% Define constants
CAo = 100; % Initial concentration of A in mol/L
k = 0.1; % Rate constant in s^-1
% Time vector
t = linspace(0, 50, 1000); % Time from 0 to 50 seconds with 1000 points
% Concentration equations
C_A = CAo * exp(-k * t);
C_B = k * CAo * t .* exp(-k * t);
C_C = CAo * (1 - exp(-k * t) .* (1 + k * t));
% Plot
figure;
hold on;
plot(t, C_A, 'r-', 'LineWidth', 1.5, 'DisplayName', 'C_A');
plot(t, C_B, 'b-', 'LineWidth', 1.5, 'DisplayName', 'C_B');
plot(t, C_C, 'g-', 'LineWidth', 1.5, 'DisplayName', 'C_C');
hold off;
% Add labels, title, and legend
xlabel('Time (s)', 'FontSize', 12);
ylabel('Concentration (mol/L)', 'FontSize', 12);
title('Concentration Profiles: C_A, C_B, and C_C vs Time', 'FontSize', 14);
legend('Location', 'best', 'FontSize', 10);
grid on;
grid on;
ax = gca; % Get handle to the current axes
% Set new font size for all axes text (including labels)
ax.FontSize = 16;
ax.FontWeight = 'bold';
```

**Figure 4.6** MATLAB code (**Ci_Series_Plot2.m**) for a plot creation of $C_A$, $C_B$, and $C_C$ as a function of time with $C_{A0} = 100$ mol/L and $k_1 = k_2 = k = 0.1\ s^{-1}$.

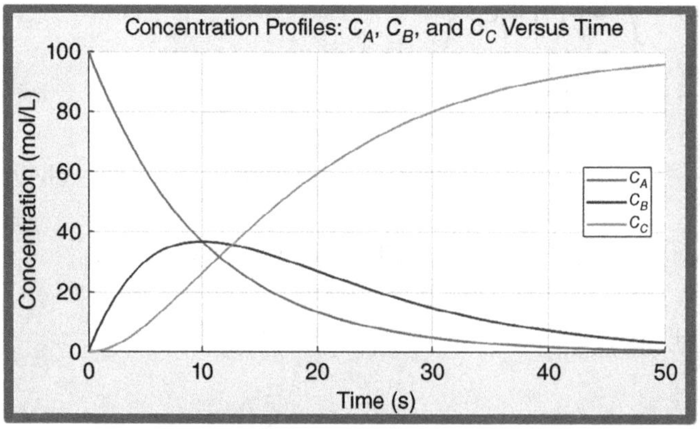


**Figure 4.7** A plot of $C_A$, $C_B$, and $C_C$ as a function of time with $C_{A0} = 100$ mol/L and $k_1 = k_2 = k = 0.1\ s^{-1}$. $C_{B,max} = 36.8$ M occurs at $t_{opt} = 1/k = (1/0.1\ s^{-1}) = 10$ s.

### 4.4.5 Optimum Reactor Feed Concentration for Reactants *A* and *B*

Consider the following two parallel reactions:

$$A + B \xrightarrow{k1} D \tag{4.42}$$

$$A + B \xrightarrow{k2} U \tag{4.43}$$

$$S_{D/U} = \frac{r_D}{r_U} = \frac{k_1 C_A^{\alpha 1} C_B^{\beta 1}}{k_2 C_A^{\alpha 2} C_B^{\beta 2}} = \frac{k_1}{k_2} C_A^{\alpha 1 - \alpha 2} C_B^{\beta 1 - \beta 2} \tag{4.44}$$

For the effect of concentration on $S_{D/U}$, we will have the following cases:

**Case #1:** $(\alpha 1 - \alpha 2) > 0$ and $(\beta 1 - \beta 2) > 0$. *Run the reactor at both high* $C_A$ *and high* $C_B$.
**Case #2:** $(\alpha 1 - \alpha 2) < 0$ and $(\beta 1 - \beta 2) < 0$. *Run the reactor at both low* $C_A$ *and low* $C_B$.
**Case #3:** $(\alpha 1 - \alpha 2) > 0$ and $(\beta 1 - \beta 2) < 0$. *Run the reactor at high* $C_A$ *and low* $C_B$.
**Case #4:** $(\alpha 1 - \alpha 2) < 0$ and $(\beta 1 - \beta 2) > 0$. *Run the reactor at low* $C_A$ *and high* $C_B$.

Such different permutations are demonstrated in Figure 4.8, where it shows interaction patterns for different combinations of high and low reactant concentrations in batch operations. Case #1: Both $C_A$ and $C_B$ are high; *A* and *B* are simultaneously added together. Case #2: Both $C_A$ and $C_B$ are low; *A* and *B* are slowly (dropwise) added to one another. Case #3: $C_A$ is high, and $C_B$ is low; *A* is abundantly placed inside the batch reactor and *B* is slowly added to *A*. Notice that for Case #4, that is, $C_B$ is high and $C_A$ is low; *B* is abundantly placed inside the batch reactor, and *A* is slowly added to *B*, which can be achieved in Figure 4.8 (Case #3) by switching the blue and red fluid.

For a PFR, it cannot be used for Case #2, where the feed case is for both low $C_A$ and low $C_B$. Else, it can be used for the other three feed scenarios. Figure 4.9 (left) shows Case #1, where the feed case is for both high $C_A$ and high $C_B$. Figure 4.9 (right) shows Case #3, where the feed case is for high $C_A$ and low $C_B$. Of course, for Case #4, the feed case of low $C_A$ and high $C_B$, the fluids are interchanged in terms of feed location in Figure 4.9 (right).

For CSTR, Figure 4.10 shows the three different configurations for the first three scenarios mentioned above.

**Figure 4.8** Interaction patterns for different combinations of high and low reactant concentrations in batch operations. (Case #1) Both $C_A$ and $C_B$ are high. (Case #2) Both $C_A$ and $C_B$ are low. (Case #3) $C_A$ is high, and $C_B$ is low.

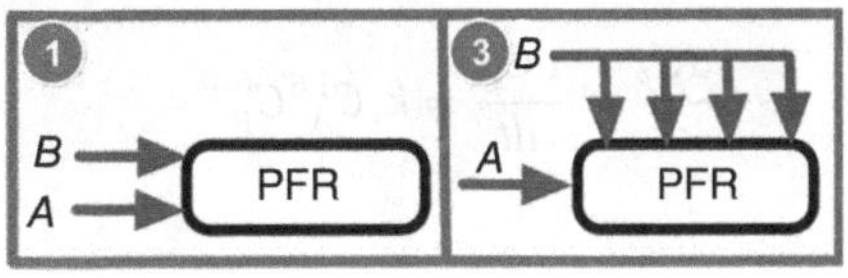


**Figure 4.9** The optimum feed concentration of *A* and *B*. (Case #1) The feed case for both high $C_A$ and high $C_B$. (Case #3) The feed case for high $C_A$ and low $C_B$.

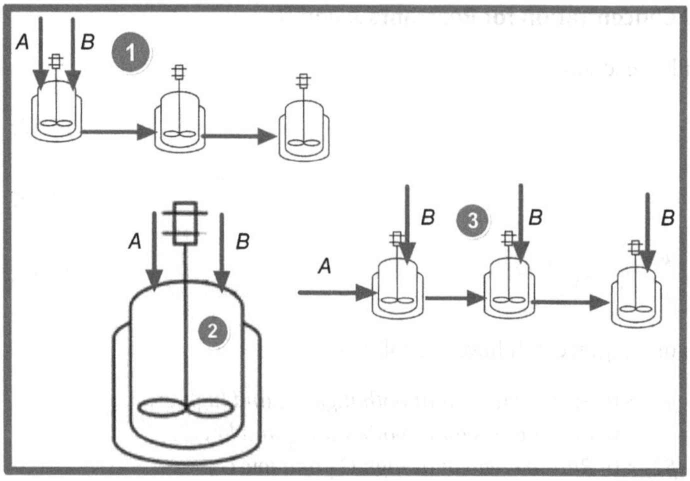


**Figure 4.10** Interaction patterns for different combinations of high and low reactant concentrations in CSTR operations. (Case #1) Both $C_A$ and $C_B$ are high. (Case #2) Both $C_A$ and $C_B$ are low. (Case #3) $C_A$ is high, and $C_B$ is low.

**Case #1** shows three small-sized consecutive CSTRs to maintain both $C_A$ and $C_B$ at high levels, because a single large CSTR would likely result in a significant drop in concentrations due to mixing and reaction progression. Thus, multiple small-sized reactors allow partial conversion in each stage, preserving higher concentrations of $C_A$ and $C_B$ in the earlier reactors. By adjusting the size or residence time in each small reactor, the system can better control reaction conditions, including concentration profiles, temperature, and mixing. Moreover, for exothermic or endothermic reactions, dividing into smaller reactors can help manage heat removal or supply more effectively, preventing hot spots or inefficient heat transfer.

**Case #2** shows a large-sized CSTR where the concentration of both *A* and *B* will significantly drop to low values.

**Case #3** shows a high-feed concentration of *A*, while *B* is supplied on the spot as a multi-port side stream for small-size consecutive CSTRs.

**Case #4** is basically the opposite of **Case #3** as far as the feed location of *A* and *B* is concerned.

> **NOTE**
>
> If any or both terms, $(\alpha 1 - \alpha 2)$ and $(\beta 1 - \beta 2)$, *are equal to zero*, then the selectivity of the desired product will become independent of $C_A$, $C_B$, or both. In any of the zero cases, we run the reactor at the maximum concentration of *A*, *B*, or both.

**Example 4.3** ***Feed Location and Concentration for Two Parallel Reactions***

The desired liquid-phase reaction:

$$A + B \xrightarrow{k_1} R + D \quad \text{(A)}$$

$$\frac{dC_R}{dt} = \frac{dC_D}{dt} = k_1 C_A^{1.6} C_B^{0.4} \quad \text{(B)}$$

is accompanied by the unwanted side reaction:

$$A + B \xrightarrow{k_2} S + U \quad \text{(C)}$$

$$\frac{dC_S}{dt} = \frac{dC_U}{dt} = k_2 C_A^{0.6} C_B^{1.9} \quad \text{(D)}$$

What is the optimum configuration for a batch reactor, PFR, and CSTR in terms of feed concentration of $A$ and $B$ and feed location?

**Solution**

Dividing Eq. (B) by Eq. (D) gives the ratio:

$$\frac{r_D}{r_U} = \frac{k_1}{k_2} C_A^{1.6-0.6} C_B^{0.4-1.9} = \frac{k_1}{k_2} C_A^{1} C_B^{-1.5} = \frac{k_1}{k_2} \frac{C_A}{C_B^{1.5}} \quad \text{(E)}$$

Comparing Eq. (E) with the four cases mentioned earlier, we see that it fits Case #3: $(\alpha 1 - \alpha 2) > 0$ and $(\beta 1 - \beta 2) < 0$. *Hence, run the reactor at high $C_A$ and low $C_B$.*

**For a Batch Reactor:** Figure 4.8, Case #3, represents the optimum configuration in terms of feed concentration and location.

**For a PFR:** Figure 4.9, Case #3, represents the optimum configuration in terms of feed concentration and location.

**For a CSTR:** Figure 4.10, Case #3, represents the optimum configuration in terms of feed concentration and location.

### 4.4.6 Effect of Temperature

Consider the following two parallel reactions:

$$A + B \xrightarrow{k1} D$$

$$A + B \xrightarrow{k2} U$$

$$S_{DU} = \frac{r_D}{r_U} = \frac{k_1 C_A^{\alpha 1} C_B^{\beta 1}}{k_2 C_A^{\alpha 2} C_B^{\beta 2}} = \frac{k_1}{k_2} C_A^{\alpha 1-\alpha 2} C_B^{\beta 1-\beta 2} = \frac{k_1^* \times e^{-\frac{E1}{RT}}}{k_2^* \times e^{-\frac{E2}{RT}}} C_A^{\alpha 1-\alpha 2} C_B^{\beta 1-\beta 2}$$

$$S_{DU} = \frac{r_D}{r_U} = \frac{k_1^*}{k_2^*} e^{-\frac{(E1-E2)}{RT}} C_A^{\alpha 1-\alpha 2} C_B^{\beta 1-\beta 2} \quad (4.45)$$

There are three possible cases:

1) If $(E_1 - E_2) > 0$, then run the reactor at the *maximum possible temperature*.
2) If $(E_1 - E_2) = 0$, then the selectivity is independent of temperature, and we can run the reactor at the *maximum possible temperature*.
3) If $(E_1 - E_2) < 0$, then run the reactor at a *moderate temperature*. To run the reactor at low temperature, it will quench both the desired and undesired reactions, as well; hence, we must elevate the temperature, but not so high.

**Example 4.4** ***Selectivity of the Liquid-phase System Involving Three Parallel Reactions***
Reactant $A$ decomposes by three simultaneous liquid-phase reactions to form three products, one that is desired, $D$, and two that are undesired, $U_1$ and $U_2$.

$$A \xrightarrow{k_1} D \qquad -r_{1A} = r_D = k_1 C_A = 0.002\frac{1}{s}\, C_A(\text{First} - \text{order}) \tag{A}$$

$$A \xrightarrow{k_2} U1 \qquad -r_{12} = r_{U1} = k_2 = 0.0002\frac{mol}{L \cdot s}\ (\text{Zero} - \text{order}) \tag{B}$$

$$A \xrightarrow{k_3} U2 \qquad -r_{13} = r_{U2} = k_3 C_A^2 = 0.01\frac{\text{L}}{\text{mol} \cdot \text{s}} C_A^2\ (\text{Second} - \text{order}) \tag{C}$$

The specific reaction rates are given at 400 K, and the activation energies for reactions (A)–(C) are $E_1 = 65{,}000$ J/mol, $E_2 = 45{,}000$ J/mol, and $E_3 = 85{,}000$ J/mol, respectively.

1) Under what conditions (e.g., reactor type and concentrations) should the reaction be conducted to maximize the selectivity of species $D$, given an initial concentration of species $A$ of 1.0 M and a volumetric flow rate of 3.0 L/s? Use both the analytic and MATLAB codes to find the optimum $C_A$ at which $S_{D/(U_1+U_2)}$ is maximum.
2) Find the residence time and volume requirement for the selected reactor. Calculate the overall conversion of the reactions, based on the consumption of $A$.
3) What is the effect of temperature on both the reaction conversion and the selectivity of the desired product?
4) Considering the effect of $T$ on selectivity, carry out two optimization cases for the desired product selectivity: One for CSTR and another for PFR, but both with different operating conditions in terms of $T$ and $C_A$ range. Find both VCSTR and VPFR to achieve 98% conversion.

**Solution**

$$S_{D/(U_1+U_2)} = \frac{r_D}{r_{U_1} + r_{U_2}} = \frac{k_1 C_A}{k_2 + k_3 C_A^2} \tag{D}$$

Finding the optimum concentration:

$$\frac{dS}{dC_A} = 0 = \frac{k_1\left\{k_2 + k_3 C_A^2\right\} - k_1 C_A\{2k_3 C_A\}}{\left[k_2 + k_3 C_A^2\right]^2} \rightarrow \left[k_1\left\{k_2 + k_3 C_A^2\right\} - k_1 C_A\{2k_3 C_A\}\right] = 0 \tag{E}$$

Further simplification yields to:

$$\left[\left\{k_2 + k_3 C_A^2\right\} - C_A\{2k_3 C_A\}\right] = 0 \qquad \rightarrow k_2 + k_3 C_A^2 - 2k_3 C_A^2 = 0 \tag{F}$$

$$k_2 - k_3 C_A^2 = 0 \qquad \rightarrow C_A^2 = \frac{k_2}{k_3}$$

$$C_A^{Optim} = \sqrt{\frac{k_2}{k_3}} = \sqrt{\frac{0.0002\frac{\text{mol}}{\text{L}\cdot\text{s}}}{0.01\frac{\text{L}}{\text{mol}\cdot\text{s}}}} = 0.1414\frac{\text{mol}}{\text{L}} \tag{G}$$

### 4.4.7 MATLAB Code for $S$ as $f(C_A)$ @ Fixed $T$

Figure 4.11 shows MATLAB code (**SD_U_CA.m**) for making a plot of the desired product selectivity as a function of $C_A$, using Eq. (D). It also displays the maximum selectivity at $C_A^{Optim}$.

```
clear variables;
% Define constants
k1 = 0.002; % s^-1
k2 = 0.0002; % mol.L^-1.s^-1
k3 = 0.01; % L.mol^-1.s^-1
% Define the range of CA
CA = linspace(0.0, 10, 1000); % mol/L
% Selectivity equation
Selectivity = (k1 * CA) ./ (k2 + k3 * CA.^2);
CA_opt = sqrt(k2/k3);
% Find the maximum selectivity and corresponding CA
[max_selectivity, index] = max(Selectivity);
% Display the maximum selectivity and CA_opt
disp(['Maximum Selectivity: ', num2str(max_selectivity)]);
disp(['CA_opt occurs at CA = ', num2str(CA(index))]);
disp(['Analytic CA_opt occurs at CA = ', num2str(CA_opt)]);
% Plot
figure;
plot(CA, Selectivity, 'b-', 'LineWidth', 1.5);
% Add labels, title, and grid
xlabel('C_A (mol/L)', 'FontSize', 12);
ylabel('Selectivity (S_{D/(U_1+U_2)})', 'FontSize', 12);
title('Selectivity as a Function of C_A', 'FontSize', 14);
grid on;
ax = gca; % Get handle to the current axes
% Set new font size for all axes text (including labels)
ax.FontSize = 16;
ax.FontWeight = 'bold';
```

**Figure 4.11** MATLAB code (**SD_U_CA.m**) for generating a plot of the desired product selectivity, given by Eq. (D), as a function of $C_A$.

The maximum selectivity and $C_A^{Optim}$ are copied from the Command Window:

```
>> SD_U_CA
Maximum Selectivity: 0.70708
CA_opt occurs at CA = 0.14014
Analytic CA_opt occurs at CA = 0.14142
```

Figure 4.12 shows the generated plot of the desired product selectivity, given by Eq (D), as a function of $C_A$. $k_1 = 0.002\,\text{s}^{-1}$, $k_2 = 0.0002$ mol/L·s; and $k_3 = 0.01$ L/mol·s were used. The peak of $S_{D/(U_1+U_2)} = 0.7071$, at $T = 400$ K, occurs at $C_A^{Optim} = 0.1414$ mol/L (Eq. G).

For a CSTR, $\overline{S}_{D/(U_1+U_2)} = S_{D/(U_1+U_2)} = 0.7071$

To maximize selectivity $S_{D/(U_1+U_2)}$, the reaction should be conducted such that the concentration of $A$ in the CSTR is maintained at $C_A^{Optim}$. The selectivity corresponding to $C_A^{Optim}$ is equal to 0.707. We use a CSTR with exit concentration $C_{A,exit} = C_A^{Optim} = \sqrt{\frac{k_2}{k_3}}$.

2) Let us calculate the residence and volume requirement to drop $C_A$ down to $C_{A,exit} = C_A^{Optim} = \sqrt{\frac{k_2}{k_3}} = 0.1414$ mol/L

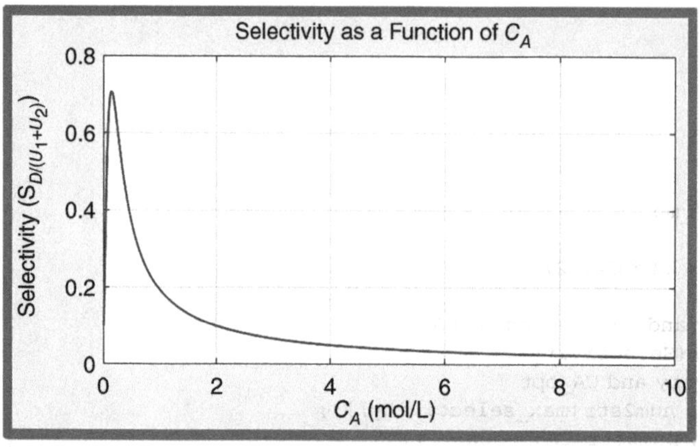


**Figure 4.12** The generated plot of selectivity as $f(C_A)$ showing the peak of 0.707 at $C_A^{Optim} = 0.1414$ mol/L, at $T = 400$ K.

For a constant-volume flow rate (i.e., liquid-phase), we can start with Eq. (2.18c), we have:

$$\tau = \frac{V_{CSTR}}{v_0} = \left(\frac{C_{A0} - C_A}{k_1 C_A + k_2 + k_3 C_A^2}\right) = \frac{1\ \text{M} - 0.1414\ \text{M}}{[0.002 * 0.1414 + 0.0002 + 0.01 * (0.1414)^2]\frac{\text{mol}}{\text{L·s}}}$$

$$= 1{,}257.6\ s = 20.96\ \text{min}.$$

$$V_{CSTR} = \tau * v_0 = 1{,}257.6\ \text{s} * 3\ \frac{\text{L}}{\text{s}} = 3.7727 \times 10^3\ \text{L} = 3.773\ \text{m}^3$$

The conversion is given as: $X = 1 - \frac{C_A}{C_{A0}} = 1 - \frac{0.1414}{1} = 0.8586$

### 4.4.8 MATLAB Code for $S$ as $f(C_A, T)$: 3-D Plot

3) The effect of T on both the conversion and selectivity can be visualized by plotting a 3-D plot for the selectivity as a function of both $C_A$ and $T$. Figure 4.13 shows the MATLAB code (**SD_U_T_CA_effects.m**) for generating a 3-D plot for $S_{D/(U_1+U_2)} = f(C_A, T)$.

```
clear variables;
% Define constants
T0 = 400; % Reference temperature in Kelvin
R = 8.314; % Universal gas constant in J/(mol*K)
E1 = 65000; % Activation energy for k1 in J/mol
E2 = 45000; % Activation energy for k2 in J/mol
E3 = 85000; % Activation energy for k3 in J/mol
% Define ranges for CA and T
CA_range = linspace(0, 1, 100); % CA varies between 0 and 1 mol/L
T_range = linspace(T0 - 100, T0 + 300, 100); % T varies between T0-100 and T0+300 K
% Create meshgrid for CA and T
[CA, T] = meshgrid(CA_range, T_range);
% Calculate rate constants
k1 = 0.002 * exp(-(E1 / R) * (1 ./ T - 1 / T0)); % s^-1
```

**Figure 4.13** MATLAB code (**SD_U_T_CA_effects.m**) for generating a 3-D plot for $S_{D/(U_1+U_2)} = f(C_A, T)$.

```
k2 = 0.0002 * exp(-(E2 / R) * (1 ./ T - 1 / T0)); % mol.L^-1.s^-1
k3 = 0.01 * exp(-(E3 / R) * (1 ./ T - 1 / T0)); % L.mol^-1.s^-1
% Selectivity equation
Selectivity = (k1 .* CA) ./ (k2 + k3 .* CA.^2);
% Plot 3-D surface
figure;
surf(CA, T, Selectivity);
% Add labels, title, and grid
xlabel('C_A (mol/L)', 'FontSize', 12);
ylabel('T [K]', 'FontSize', 12);
zlabel('Selectivity (S_{D/(U_1+U_2)})', 'FontSize', 12);
title('Selectivity as a Function of C_A and T', 'FontSize', 14);
grid on;
% Set font size for the axes
ax = gca;
ax.FontSize = 16;
ax.FontWeight = 'bold';
```

**Figure 4.13** *(Continued)*

Figure 4.14 shows the generated 3-D plot of $S_{D/(U_1+U_2)} = f(C_A, T)$. One can notice the following trends:

1) At a given exit in $C_A$, there will be a bell-shaped curve where there exists a peak for the selectivity as $f(T)$. This is represented as a slice out of the 3-D plot shown in Figure 4.14.
2) At a given $T$, there will be a bell-shaped curve where there exists a peak for the selectivity as $f(C_A)$. This behavior is demonstrated earlier in Figure 4.12, where $k_1$, $k_2$, and $k_3$ are reported at $T = 400$ K. See also Figure 4.16.
   Notice that in Figure 4.14, $X$ represents $C_A$; $Y$ represents $T$; and $Z$ represents selectivity.

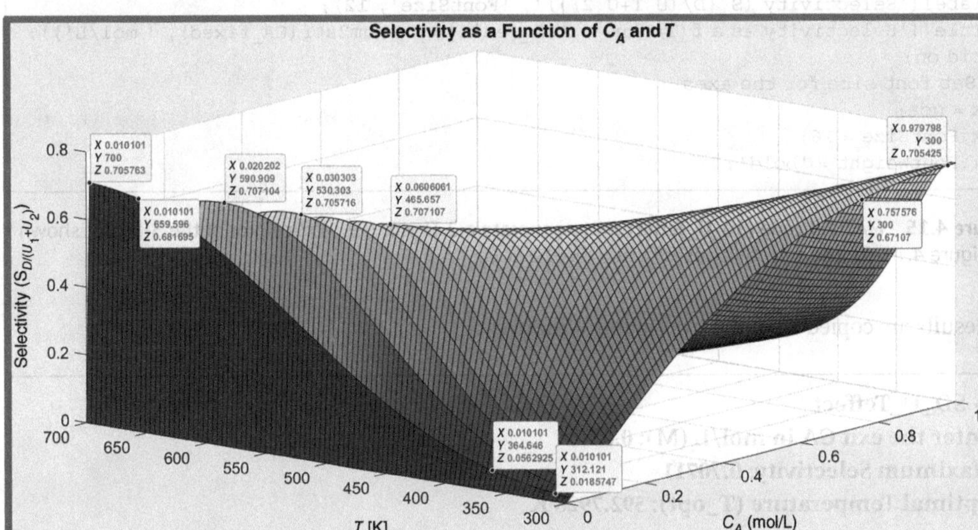


**Figure 4.14** The generated 3-D plot of $S_{D/(U_1+U_2)} = f(C_A, T)$. $X$ represents $C_A$, $Y$ represents $T$, and $Z$ represents selectivity.

### 4.4.9 MATLAB Code for $S$ as $f(T)$ @ Fixed $C_A$: 2-D Plot

Figure 4.15 shows MATLAB code (**SD_U_Teffect.m**) to generate a 2-D plot (i.e., a slice plot of the 3-D plot shown in Figure 4.14) for a given exit $C_A$, where there will be a bell-shaped curve with a peak for the selectivity as $f(T)$.

```
clear variables;
CA_fixed = str2double(input('Enter the exit CA in mol/L (M): ', 's'));
% Define constants
T0 = 400; % Reference temperature in Kelvin
R = 8.314; % Universal gas constant in J/(mol*K)
E1 = 65000; % Activation energy for k1 in J/mol
E2 = 45000; % Activation energy for k2 in J/mol
E3 = 85000; % Activation energy for k3 in J/mol
% Define temperature range
T_range = linspace(T0 - 100, T0 + 350, 1000); % Temperature range from T0-100 to T0+100
% Calculate rate constants and selectivity as a function of T
k1 = 0.002 * exp(-(E1 / R) * (1 ./ T_range - 1 / T0)); % s^-1
k2 = 0.0002 * exp(-(E2 / R) * (1 ./ T_range - 1 / T0)); % mol.L^-1.s^-1
k3 = 0.01 * exp(-(E3 / R) * (1 ./ T_range - 1 / T0)); % L.mol^-1.s^-1
% Selectivity equation
Selectivity = (k1 .* CA_fixed) ./ (k2 + k3 .* CA_fixed.^2);
% Find the maximum selectivity and corresponding temperature
[max_selectivity, index] = max(Selectivity);
T_opt = T_range(index); % Optimal temperature
% Display the results
disp(['Maximum Selectivity: ', num2str(max_selectivity)]);
disp(['Optimal Temperature (T_opt): ', num2str(T_opt), ' K']);
% Plot selectivity as a function of temperature
figure;
plot(T_range, Selectivity, 'r-', 'LineWidth', 1.5);
% Add labels, title, and grid
xlabel('Temperature (T) [K]', 'FontSize', 12);
ylabel('Selectivity (S_{D/(U_1+U_2)})', 'FontSize', 12);
title(['Selectivity as a f(T) for ', 'CA_{exit} = ', num2str(CA_fixed), ' mol/L']);
grid on;
% Set font size for the axes
ax = gca;
ax.FontSize = 16;
ax.FontWeight = 'bold';
```

**Figure 4.15** MATLAB code (**SD_U_Teffect.m**) to generate a 2-D plot (i.e., a slice plot of the 3-D plot shown in Figure 4.14) for the selectivity as $f(T)$, at a fixed exit $C_A$.

Results are copied from the Command Window:

```
>> SD_U_Teffect
Enter the exit CA in mol/L (M): 0.02
Maximum Selectivity: 0.70711
Optimal Temperature (T_opt): 592.7928 K
```

Figure 4.16 shows the generated 2-D plot of $S_{D/(U_1+U_2)} = f(T)@\textit{fixed}\ C_{A,exit} = 0.02\frac{\text{mol}}{\text{L}}$. The user may select, however, the exit $C_A$ and selectivity will be plotted as $f(T)$, accordingly.

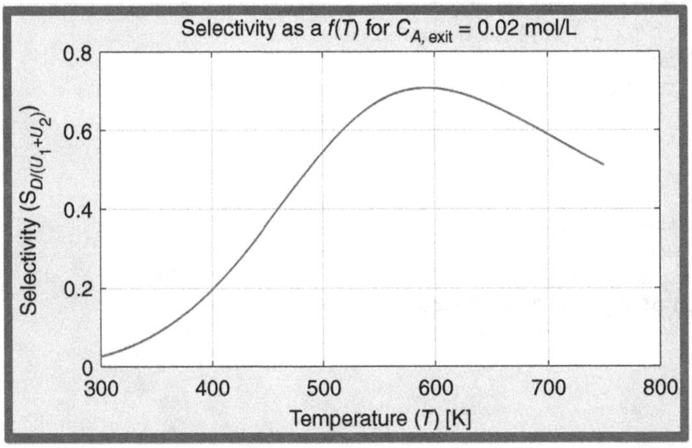


**Figure 4.16** The generated 2-D plot of $S_{D/(U_1+U_2)} = f(T)@fixed\ C_{A,\ exit} = 0.02\ \text{mol/L}$.

One can see that for a given $C_{A,exit}$ of CSTR, we can maximize the selectivity via manipulating the temperature. For example, at $C_{A,exit} = 0.02\frac{\text{mol}}{\text{L}}$, which amounts to a very high conversion value of $X = 0.98$, the optimum $T$ is found to be $\approx$593 K with a selectivity of 0.7071. Apparently, we can run the CSTR for any conversion we wish simply by manipulating T without sacrificing the selectivity of the desired product.

Although the previous argument is valid and justifiable for any conversion at the CSTR exit concerning desired product selectivity, the residence time/volume requirement is a different matter. See the next section.

### 4.4.10 MATLAB Code for VCSTR as $f(C_A,T)$: 3-D Plot

Figure 4.17 shows the MATLAB code (**VCSTR_T_CA_effects**) for generating a 3-D plot for $V_{CSTR} = f(C_A, T)$.

```
clear variables;
% Define constants
TO = 400; % Reference temperature in Kelvin
R = 8.314; % Universal gas constant in J/(mol*K)
E1 = 65000; % Activation energy for k1 in J/mol
E2 = 45000; % Activation energy for k2 in J/mol
E3 = 85000; % Activation energy for k3 in J/mol
CA0 = 1; % Initial concentration of A in mol/L
v0 = 3; % Volumetric flow rate in L/s
% Define ranges for CA and T
CA_range = linspace(0, 1, 100); % CA varies between 0 and 1 mol/L
T_range = linspace(TO - 100, TO + 300, 400); % T varies between TO-100 and TO+300 K
% Create meshgrid for CA and T
[CA, T] = meshgrid(CA_range, T_range);
% Calculate rate constants
k1 = 0.002 * exp(-(E1 / R) * (1 ./ T - 1 / TO)); % s^-1
k2 = 0.0002 * exp(-(E2 / R) * (1 ./ T - 1 / TO)); % mol.L^-1.s^-1
```

**Figure 4.17** MATLAB code (**VCSTR_T_CA_effects**) for generating a 3-D plot for $V_{CSTR} = f(C_A, T)$.

```
k3 = 0.01 * exp(-(E3 / R) * (1 ./ T - 1 / T0)); % L.mol^-1.s^-1
% Calculate VCSTR (volume of CSTR)
VCSTR = ((CA0 - CA) ./ (k1 .* CA + k2 + k3 .* CA.^2)) * v0; % Volume in L
% Plot 3-D surface
figure;
surf(CA, T, VCSTR);
% Add labels, title, and grid
xlabel('C_A (mol/L)', 'FontSize', 12);
ylabel('T [K]', 'FontSize', 12);
zlabel('V_{CSTR} (L)', 'FontSize', 12);
title('V_{CSTR} as a Function of C_A and T', 'FontSize', 14);
grid on;
% Set font size for the axes
ax = gca;
ax.FontSize = 16;
ax.FontWeight = 'bold';
```

**Figure 4.17** (*Continued*)

Figure 4.18 shows the generated 3-D plot of $V_{CSTR}=f(C_A, T)$. One can notice the following trends:

1) $V_{CSTR}$ will extremely grow as $C_A$ approaches low values closer to zero.
2) $V_{CSTR}$ will be relatively acceptable at high $C_A$ values closer to $C_{A0}$.
3) $V_{CSTR}$ may look as if it approaches zero mainly at high $T$ and high $C_A$ (i.e., the black sheet in Figure 4.16); this is an illusion which has to deal with the scaling issue due to the extremely large volume requirement at low $C_A$ and relatively low $T$.

Considering the above trends, although one can operate at very low concentrations of $A$ (i.e., very high conversion $X$), the penalty will be a significantly large volume requirement, thereby increasing the fixed cost of equipment, such as a CSTR. As shown in Figure 4.16, $V_{CSTR}$ reaches an order of magnitude of $1.0\times10^6$ L (1,000 m$^3$). Consequently, a tradeoff

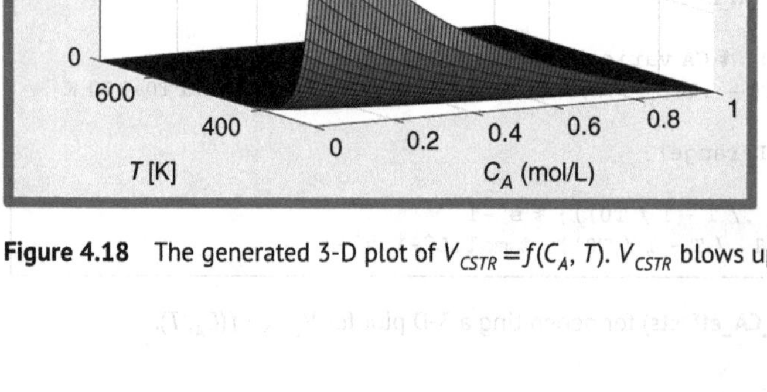


**Figure 4.18** The generated 3-D plot of $V_{CSTR}=f(C_A, T)$. $V_{CSTR}$ blows up as $C_A \approx 0$.

must be made between achieving high conversion and managing large volume requirements. In practice, this is a typical optimization case, merely dictated by process economics, which will tradeoff between two extreme cases: high operating cost and high fixed cost. In this regard, see Chapter 17: Aspen Process Economic Analyzer (APEA) for more details on process economics [1].

To avoid an excessively large volume requirement when aiming for high conversion near unity, one approach is to operate a CSTR at the optimal concentration for maximizing the desired product selectivity, followed by a PFR to achieve the desired exit concentration, optimized for a different concentration range. To do that for the given example, we operate the CSTR at one temperature and the PFR at a different temperature.

### 4.4.11 MATLAB Code for $S$ as $f(T)$ for CSTR and PFR: 2-D Plot

Figure 4.19 shows MATLAB code (**VCSTR_VPFR_CA.m**) for carrying out two optimization cases for the desired product selectivity: One for CSTR and another for the PFR but both with different operating conditions in terms of $T$ and $C_A$ range.

Notice that since this is a liquid-phase reaction, PFR has to be run under a pressure higher than that of CSTR in order to keep the reacting species in the liquid phase to offset the effect of the elevated temperature. The exit concentration of the CSTR will be at $C_{A,exit} = C_A^{Optim1}$ and the exit concentration of the PFR will be at $C_{A,exit} = C_A^{Optim2}$.

Results are copied from the Command Window.

```
≫ VCSTR_VPFR_CA
Enter CSTR temperature (T) in K: 570
Enter PFR temperature (T) in K: 600
Enter the initial concentration of A (CA0) in mol/L: 1
Enter the volumetric flow rate (v0) in L/s: 3
Maximum CSTR Selectivity: 0.70694
Optimal CA (CA_opt): 0.024048 mol/L
VCSTR: 74.1248 L
Maximum PFR Selectivity2: 0.70711
Optimal CA (CA_opt2): 0.019036 mol/L
VPFR: 0.21392 L
Total Volume (VTot): 74.3387 L
Final CSTR CAexit, mol/L: 0.024048
Final CSTR XCSTR: 0.97595
Final PFR CAexit, mol/L: 0.019036
Final PFR XPFR: 0.98096
```

Figure 4.20 (Left) shows $S_{D/(U_1+U_2)} = f(T)$ for CSTR over the entire range $[C_{A0}, 0]$ of $C_A$ to search for the first optimum $C_A$, and (right) shows $S_{D/(U_1+U_2)} = f(T)$ for PFR over the subset range $\left[C_{A,CSTR\ exit} = C_A^{Optim1}, 0\right]$. Notice that the selectivity is maintained at a high level at $S_{D/(U_1+U_2)} = 0.707$ for both reactors.

```
clear variables;
% Prompt user for inputs
T = str2double(input('Enter CSTR temperature (T) in K: ', 's'));
T2 = str2double(input('Enter PFR temperature (T) in K: ', 's'));
CA0 = str2double(input('Enter the initial concentration of A (CA0) in mol/L: ', 's'));
v0 = str2double(input('Enter the volumetric flow rate (v0) in L/s: ', 's'));
% Define constants
T0 = 400; % Reference temperature in Kelvin
R = 8.314; % Universal gas constant in J/(mol*K)
E1 = 65000; % Activation energy for k1 in J/mol
E2 = 45000; % Activation energy for k2 in J/mol
E3 = 85000; % Activation energy for k3 in J/mol
% Define range for CA
CA_range = linspace(0, CA0, 500); % CA from CAf to CA0
% Calculate rate constants
k1 = 0.002 * exp(-(E1 / R) * (1 / T - 1 / T0)); % s^-1
k2 = 0.0002 * exp(-(E2 / R) * (1 / T - 1 / T0)); % mol.L^-1.s^-1
k3 = 0.01 * exp(-(E3 / R) * (1 / T - 1 / T0)); % L.mol^-1.s^-1
% Selectivity equation for CSTR
Selectivity = (k1 .* CA_range) ./ (k2 + k3 .* CA_range.^2);
% Find the maximum selectivity and corresponding CA under CSTR condition
[max_selectivity, index] = max(Selectivity);
CA_opt = CA_range(index); % Optimal concentration of A
% Calculate VCSTR
VCSTR = ((CA0 - CA_opt) / (k1 * CA_opt + k2 + k3 * CA_opt^2)) * v0; % Volume in L
figure (1);
plot(CA_range, Selectivity, 'b-', 'LineWidth', 1.5);
hold on;
plot(CA_opt, max_selectivity, 'ro', 'MarkerSize', 8, 'MarkerFaceColor', 'r'); % Highlight maximum selectivity
xlabel('C_A (mol/L)', 'FontSize', 12);
ylabel('CSTR Selectivity (S_{D/(U_1+U_2)})', 'FontSize', 12);
title(['CSTR Selectivity f(C_A) at T_{CSTR} = ', num2str(T), ' K'], 'FontSize', 14);
grid on;
ax = gca;
```

**Figure 4.19** MATLAB code (**VCSTR_VPFR_CA.m**) for carrying out two optimization cases for the desired product selectivity: One for the CSTR, followed by another for the PFR, but both with different operating conditions in terms of $T$ and $C_A$ range.

```
ax.FontSize = 16;
ax.FontWeight="bold";
% Calculate rate constants under PFR condition
k12 = 0.002 * exp(-(E1 / R) * (1 / T2 - 1 / T0)); % s^-1
k22 = 0.0002 * exp(-(E2 / R) * (1 / T2 - 1 / T0)); % mol.L^-1.s^-1
k32 = 0.01 * exp(-(E3 / R) * (1 / T2 - 1 / T0)); % L.mol^-1.s^-1
% Selectivity equation for PFR
CA_range2 = linspace(0, CA_opt, 500); % CA from CAf to CA0
Selectivity2 = (k12 .* CA_range2) ./ (k22 + k32 .* CA_range2.^2);
% Find the maximum selectivity and corresponding CA under PFR condition
[max_selectivity2, index] = max(Selectivity2);
CA_opt2 = CA_range2(index); % Optimal concentration of A
% Calculate VPFR using numerical integration
rA = @(CA) k12 .* CA + k22 + k32 .* CA.^2; % Rate of reaction
VPFR = v0 * integral(@(CA) 1 ./ rA(CA), CA_opt2, CA_opt); % Volume in L
% Total volume
VTot = VCSTR + VPFR;
% Calculate XCSTR and XPFR.
XCSTR=1.0-CA_opt/CA0;
XPFR=1.0-CA_opt2/CA0;
% Display results
disp(['Maximum CSTR Selectivity: ', num2str(max_selectivity)]);
disp(['Optimal CA (CA_opt): ', num2str(CA_opt), ' mol/L']);
disp(['VCSTR: ', num2str(VCSTR), ' L']);
disp(['Maximum PFR Selectivity2: ', num2str(max_selectivity2)]);
disp(['Optimal CA (CA_opt2): ', num2str(CA_opt2), ' mol/L']);
disp(['VPFR: ', num2str(VPFR), ' L']);
disp(['Total Volume (VTot): ', num2str(VTot), ' L']);
disp(['Final CSTR CAexit, mol/L: ', num2str(CA_opt)]);
disp(['Final CSTR XCSTR: ', num2str(XCSTR)]);
disp(['Final PFR CAexit, mol/L: ', num2str(CA_opt2)]);
disp(['Final PFR XPFR: ', num2str(XPFR)]);
% Plot VPFR as a function of CA
figure (2);
```

**Figure 4.19** (*Continued*)

```
plot(CA_range2, Selectivity2, 'b-', 'LineWidth', 1.5);
hold on;
plot(CA_opt2, max_selectivity2, 'ro', 'MarkerSize', 8, 'MarkerFaceColor', 'r'); % Highlight maximum selectivity
xlabel('C_A (mol/L)', 'FontSize', 12);
ylabel('PFR Selectivity (S_{D/(U_1+U_2)})', 'FontSize', 12);
title(['PFR Selectivity f(C_A) at T_{PFR} = ', num2str(T2), ' K'], 'FontSize', 14);
grid on;
ax = gca;
ax.FontSize = 16;
ax.FontWeight="bold";
hold off;
```

**Figure 4.19** *(Continued)*

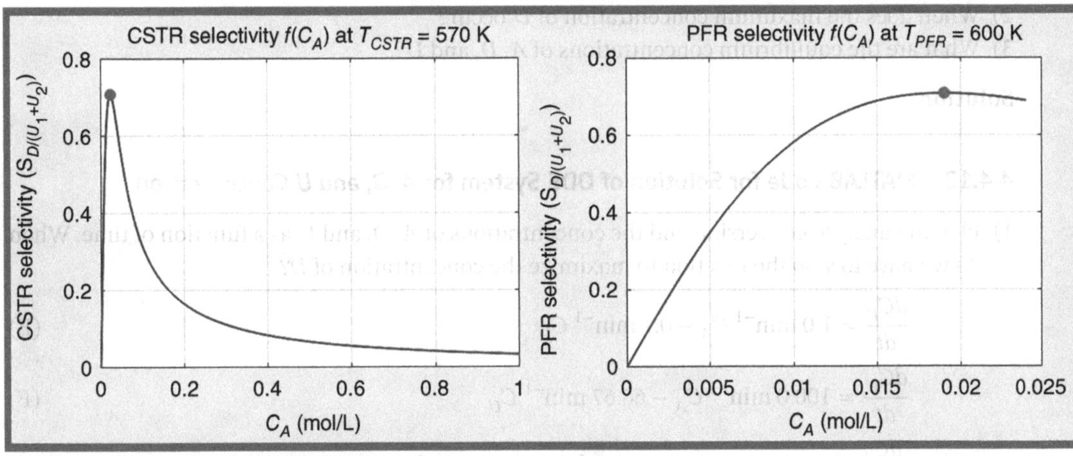


**Figure 4.20** (Left) $S_{D/(U_1+U_2)} = f(T)$ for CSTR over the entire range of $C_A$ to search for the first optimum $C_A$, and (right) $S_{D/(U_1+U_2)} = f(T)$ for PFR over the subrange $\left[C_{A,CSTR\ exit} = C_A^{Optim1} = 0.024\ M, 0\right]$.

**NOTE**

1) Any combination of CSTR and PFR temperature will give different values of $V_{CSTR}$ and $V_{PFR}$; hence, different conversions at the exit of both reactors.
2) Increasing the temperature generally increases the rate constants, leading to faster reaction rates and potentially reducing reactor volume requirements, provided the reaction kinetics and equilibrium favor higher temperatures without significant side effects.
3) The upper limit of temperature is limited by constraints, like: mechanical strength of the reactor material and of its accessories and measuring devices, the safety regulation (i.e., explosive limit), and reactant and product thermal stability (i.e., some materials are heat sensitive).

**Example 4.5** ***Selectivity of Parallel Equilibrium Reactions in BR***

Consider the following reactions taking place in a constant-volume batch reactor:

$$A \underset{k_{bD}}{\overset{k_{fD}}{\rightleftarrows}} D \tag{A}$$

$$-r_{1A} = k_{fD}\,C_A - k_{bD}\,C_D = 1.0\ \text{min}^{-1}C_A - 0.1\ \text{min}^{-1}\ C_D \tag{B}$$

$$A \underset{k_{bU}}{\overset{k_{fU}}{\rightleftarrows}} U \tag{C}$$

$$-r_{2A} = k_{fU}\,C_A - k_{bU}\,C_U = 100.0\ \text{min}^{-1}\ C_A - 66.67\ \text{min}^{-1}\ C_U \tag{D}$$

$C_{A0} = 1\ \text{mol/L} = 1\ \text{M}$

1) Plot and analyze conversion and the concentrations of $A$, $D$, and $U$ as a function of time. When do we have to stop the reaction to maximize the concentration of $D$?

2) When does the maximum concentration of $U$ occur?
3) What are the equilibrium concentrations of $A$, $D$, and $U$?

**Solution**

### 4.4.12 MATLAB Code for Solution of ODE System for *A*, *D*, and *U* Concentration

1) Plot and analyze conversion and the concentrations of $A$, $D$, and $U$ as a function of time. When do we have to stop the reaction to maximize the concentration of $D$?

$$\frac{dC_D}{dt} = 1.0\ \text{min}^{-1}\ C_A - 0.1\ \text{min}^{-1}\ C_D \tag{E}$$

$$\frac{dC_U}{dt} = 100.0\ \text{min}^{-1}\ C_A - 66.67\ \text{min}^{-1}\ C_U \tag{F}$$

$$-\frac{dC_A}{dt} = -r_{1A} + -r_{2A} = 1.0\ \text{min}^{-1}\ C_A - 0.1\ \text{min}^{-1}\ C_D + 100.0\ \text{min}^{-1}\ C_A - 66.67\ \text{min}^{-1}\ C_U$$

Combine similar terms, we have:

$$-\frac{dC_A}{dt} = +101.0\ \text{min}^{-1}\ C_A - 0.1\ \text{min}^{-1}\ C_D - 66.67\ \text{min}^{-1}\ C_U$$

Multiply both sides by −1, we have:

$$\frac{dC_A}{dt} = -101.0\ \text{min}^{-1}\ C_A + 0.1\ \text{min}^{-1}\ C_D + 66.67\ \text{min}^{-1}\ C_U \tag{G}$$

We have three differential equations: (E), (F), and (G). The initial conditions are:

$$C_{A0} = 1\ \text{mol/L} = 1\ \text{M}; \qquad C_{D0} = 0\ \text{mol/L} = 0\ \text{M}; \qquad \text{and } C_{U0} = 0 \qquad \text{mol/L} = 0\ \text{M}$$

Figure 4.21 shows MATLAB code **(solveODEs.m)** for numerically solving three simultaneous ordinary differential equations with known initial conditions, as well as making a single plot of $C_A$, $C_D$, and $C_U$ as $f(t)$.

```
% Define the system of ODEs
function solveODEs()
   % Initial conditions
   CA0 = 1; % mol/L
   CD0 = 0; % mol/L
   CU0 = 0; % mol/L
   % Time span
   tspan = [0 20]; % minutes
   % Solve the system using ode45
   [t, y] = ode45(@odeSystem, tspan, [CA0, CD0, CU0]);
   % Extract solutions
   CA = y(:, 1);
   CD = y(:, 2);
   CU = y(:, 3);
   [max_Undesired, index] = max(CU);
   t_Undesire=t(index);
   disp(['CU_max, CU,max = ', num2str(max_Undesired), ' mol/L']);
   disp(['CA, @ max Undesired = ', num2str(CA(index)), ' mol/L']);
```

**Figure 4.21** MATLAB code (**solveODEs.m**) for numerically solving three simultaneous ordinary differential equations with known initial conditions and making a single plot of $C_A$, $C_D$, and $C_U$ as $f(t)$.

```
    disp(['CD, @ max Undesired = ', num2str(CD(index)), ' mol/L']);
    disp(['t, @ max Undesired = ', num2str(t_Undesire), ' minutes']);
    % Plot the results
    figure;
    plot(t, CA, 'r', 'LineWidth', 1.5); hold on;
    plot(t, CD, 'b', 'LineWidth', 1.5);
    plot(t, CU, 'g', 'LineWidth', 1.5);
    xlabel('Time (minutes)');
    ylabel('Concentration (mol/L)');
    legend('C_A', 'C_D', 'C_U');
    title('Concentration Profiles');
    grid on;
    ax = gca; % Get handle to the current axes
    % Set new font size for all axes text (including labels)
    ax.FontSize = 16;
    ax.FontWeight = 'bold';
end
% Define the ODE system
function dydt = odeSystem(~, y)
    % Extract variables
    CA = y(1);
    CD = y(2);
    CU = y(3);
    % Parameters
    k1 = 1.0; % min^-1
    k2 = 0.1; % min^-1
    k3 = 100.0; % min^-1
    k4 = 66.67; % min^-1
    k5 = 101.0; % min^-1
    % Define the ODEs
    dCDdt = k1 * CA - k2 * CD;
    dCUdt = k3 * CA - k4 * CU;
    dCAdt = -k5 * CA + k2 * CD + k4 * CU;
    % Return derivatives
    dydt = [dCAdt; dCDdt; dCUdt];
end
```

**Figure 4.21** (*Continued*)

Figure 4.22 shows the generated plot of $C_A$, $C_D$, and $C_U$ as $f(t)$. One can notice at about 10 min, the reaction system practically achieves what is called equilibrium, and the concentration of the desired products also reaches the maximum (i.e., the plateau value). So, we can stop the batch reactor by quenching the reaction at $t = 10$ min.

2) When does the maximum concentration of $U$ occur?
Change the following code in line #8 of Figure 4.21 from:

```
tspan = [0 20]; % minutes
```

to:

```
tspan = [0 1]; % minutes
```

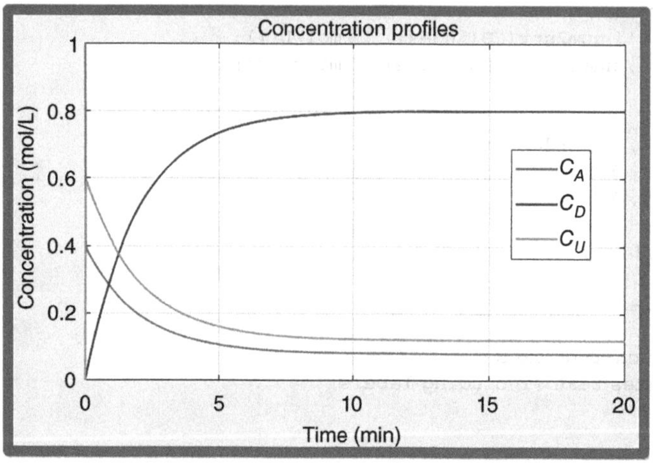


**Figure 4.22** The generated plot of $C_A$, $C_D$, and $C_U$ as $f(t)$, showing the ultimate values of $C_i$.

Then rerun the code of Figure 4.21. Figure 4.23 plot is generated. In addition, the following numerical results are copied from the Command Window:

```
≫ solveODEs
CU_max, CU,max = 0.58925 mol/L
CA, @ max Undesired = 0.39335 mol/L
CD, @ max Undesired = 0.017401 mol/L
t, @ max Undesired = 0.035097 minutes
```

The maximum $C_{U,max} = 0.589$ M, occurs at $t = 0.035$ min. However, $C_A$ at this point in time is 0.393 M (i.e., it amounts to the conversion $X = 1 - 0.393 = 0.607$, which is relatively low to medium). On the other hand, $C_D$ at this point in time is 0.017 M. Of course, if we stop the batch at this point in time, it will be in the worst condition in terms of maximizing the selectivity of the desired product, $S_{\{D/U\}}$.

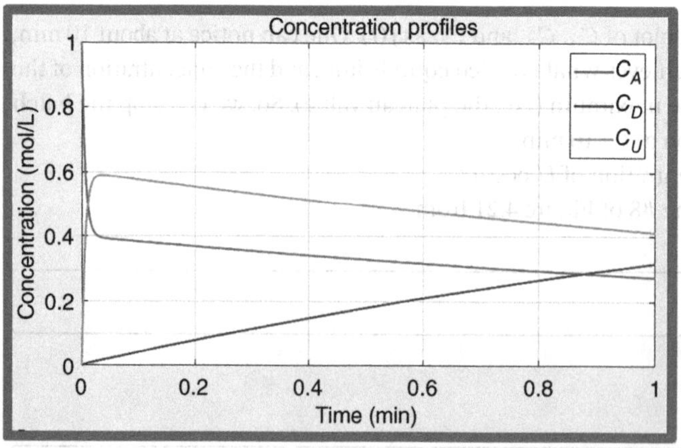


**Figure 4.23** The plot showing the maximum $C_U = 0.589$ M occurs at $t = 0.035$ min.

### 4.4.13 MATLAB Code for Solution of Algebraic Equations

3) What are the equilibrium concentrations of $A$, $D$, and $U$?
   Figure 4.24 shows MATLAB code (**solveAlgebraicEquations.m**) for simultaneously solving the three algebraic equations by setting the first derivative of each of the three ordinary differential equations (E, F, and G) to zero. In addition, a fourth equation is added as a constraint to avoid the trivial solution, that is, all concentrations are zero.

```
% Solve the system of algebraic equations
function solveAlgebraicEquations()
    % Define symbolic variables
    syms CA CD CU
    % Define the system of equations
    eq1 = 1.0 * CA - 0.1 * CD == 0;
    eq2 = 100.0 * CA - 66.67 * CU == 0;
    eq3 = -101.0 * CA + 0.1 * CD + 66.67 * CU == 0;
    % Add a normalization constraint to avoid trivial solutions
    eq4 = CA + CD + CU == 1; % Total concentration equals 1 mol/L
    % Solve the system
    solutions = solve([eq1, eq2, eq3, eq4], [CA, CD, CU]);
    % Display the solutions
    fprintf('Solution:\n');
    fprintf('CAeq = %.4f mol/L\n', double(solutions.CA));
    fprintf('CDeq = %.4f mol/L\n', double(solutions.CD));
    fprintf('CUeq = %.4f mol/L\n', double(solutions.CU));
end
```

**Figure 4.24** MATLAB code (**solveAlgebraicEquations.m**) for simultaneously solving the three algebraic equations by setting the first derivative of each of the three ordinary differential equations (E, F, and G) to zero.

The results are copied from the Command Window:

```
>> solveAlgebraicEquations
Solution:
CAeq = 0.0800 mol/L
CDeq = 0.8000 mol/L
CUeq = 0.1200 mol/L
```

Such equilibrium values match with the plateau values shown in Figure 4.22.

**Example 4.6** ***Parallel Irreversible Gas-phase Reactions with Desired and Undesired Products***

Reactant $A$ decomposes by three simultaneous gas-phase reactions to form three products, one that is desired, $D$, and two that are undesired, $U_1$ and $U_2$.

$$A \xrightarrow{k_1} U_1 \qquad -r_{1A} = r_{U1} = k_1 C_A^{0.5} = 0.004 \frac{\text{mol}^{1/2}}{\text{L}^{1/2} \cdot \text{min}} C_A^{0.5} (\text{Half} - \text{order}) \tag{A}$$

$$A \xrightarrow{k_2} D \qquad -r_{12} = r_D = k_2 C_A = 0.3 \frac{1}{\text{min}} C_A\ (\text{First} - \text{order}) \tag{B}$$

$$A \xrightarrow{k_3} U_2 \qquad -r_{13} = r_{U2} = k_3 C_A^2 = 0.25 \frac{\text{L}}{\text{mol} \cdot \text{min}} C_A^2\ (\text{Second} - \text{order}) \tag{C}$$

$D$ is the desired product, while $U_1$ and $U_2$ are undesired pollutants that are costly to eliminate. The specific reaction rates are determined at 27 °C. The reaction system will be operated under conditions of 27 °C and 4 atm. Pure $A$ is introduced into the system at a volumetric flow rate of 10 L/min.

Let us calculate $C_{A0}$. The overall change in number of moles can be expressed in terms of the change in number of moles for each reaction.

$$\Delta n = \Delta n_1 + \Delta n_2 + \Delta n_3 = (1-1) + (1-1) + (1-1) = 0$$

$$\varepsilon = \Delta n y_{A0} = 0 * 1 = 0$$

$$C_{A0} = \frac{P y_{A0}}{RT} = \frac{4\,\text{atm} * 1}{0.08206\,\frac{\text{atm.L}}{\text{mol.K}} * 300\,K} = 0.1625\frac{\text{mol}}{\text{L}}$$

1) Analytically, find $C_A^{optim}$ at which $S_{D/(U_1+U_2)}$ is maximum.

$$S_{D/(U_1+U_2)} = \frac{r_D}{r_{U_1} + r_{U_2}} = \frac{k_2 C_A}{k_1 C_A^{0.5} + k_3 C_A^2} \tag{D}$$

Finding the optimum concentration:

$$\frac{dS}{dC_A} = 0 = \frac{\{k_1 C_A^{0.5} + k_3 C_A^2\}\, k_2 - k_2 C_A \{0.5 * k_1 C_A^{-0.5} + 2k_3 C_A\}}{[k_1 C_A^{0.5} + k_3 C_A^2]^2} = 0 \tag{E}$$

This implies that

$$k_2 \{k_1 C_A^{0.5} + k_3 C_A^2\} - k_2 C_A \{0.5 * k_1 C_A^{-0.5} + 2k_3 C_A\} = 0 \tag{F}$$

Further simplification yields to

$$k_1 k_2 C_A^{0.5} + k_2 k_3 C_A^2 - 0.5 k_1 k_2 C_A^{0.5} - 2k_2 k_3 C_A^2 = 0 \tag{G}$$

$$0.5 k_1 k_2 C_A^{0.5} - k_2 k_3 C_A^2 = 0 \quad \rightarrow 0.5 k_1 C_A^{0.5} - k_3 C_A^2 = 0 \tag{H}$$

$$C_A^{0.5}\left(0.5k_1 - k_3 C_A^{1.5}\right) = 0 \quad \rightarrow C_A^{0.5} = 0\,\text{(trivial solution), or } k_3 C_A^{1.5} = 0.5 k_1$$

$$C_A^{optim} = \left(\frac{0.5 k_1}{k_3}\right)^{\frac{2}{3}} = \left(\frac{0.5 * 0.004\frac{\text{mol}^{1/2}}{\text{L}^{1/2}\cdot\text{min}}}{0.25\frac{\text{L}}{\text{mol}\cdot\text{min}}}\right)^{\frac{2}{3}} = \left(0.008\frac{\text{mol}^{3/2}}{\text{L}^{3/2}}\right)^{\frac{2}{3}} = 0.04\frac{\text{mol}}{\text{L}} \tag{I}$$

2) **Sketch the Selectivity:** $S_{D/(U_1+U_2)}$ as a function of the concentration of $C_A$.

### 4.4.14 MATLAB Code for $S_{D/(U_1+U_2)} = f(C_A)$

MATLAB code (**SD_U_CA.m**) in Figure 4.11 is used here, except for changing kinetic parameters and $C_A$ range and defining $C_A^{optim}$. Figure 4.25 shows a portion, made of the first ten lines of code, which replaces the first ten lines of code in Figure 4.11 (**SD_U_CA.m**). The entire code, however, is saved under the **SD_U_Fig4_25.m** file.

```
clear variables;
% Define constants
k1 = 0.004; % mol^0.5*L^-0.5*min^-1
k2 = 0.3;% min^-1
k3 = 0.25; % L.mol^-1.min^-1
% Define the range of CA
CA = linspace(0.0, 0.1625, 100); % mol/L
% Selectivity equation
Selectivity = (k2 * CA) ./ (k1*CA.^0.5 + k3 * CA.^2);
CA_opt = (0.5*k1/k3)^(2/3);
```

**Figure 4.25** The first 10-line portion of the **SD_U_Fig4_25.m** file.

Results are copied from the Command Window:

```
>> SD_U_Fig4_25
Maximum Selectivity: 9.9994
CA_opt occurs at CA = 0.039394
Analytic CA_opt occurs at CA = 0.04
```

Figure 4.26 shows the generated plot of the desired product selectivity as a function of $C_A$, where $C_A^{optim} = 0.04\frac{\text{mol}}{\text{L}}$.

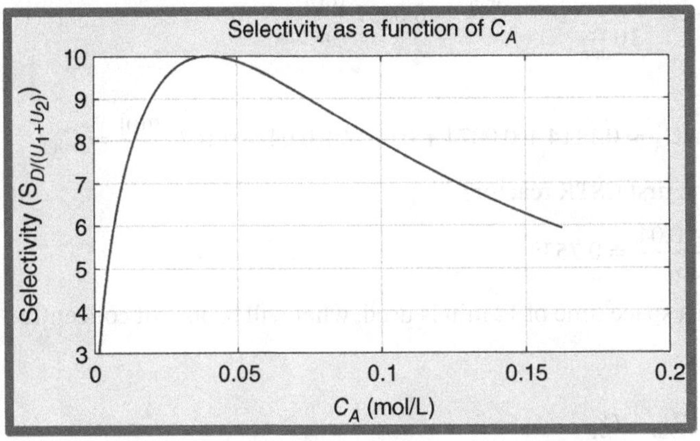


**Figure 4.26** The generated plot of $S_{D/(U_1+U_2)}$ as a function of $C_A$, where $C_A^{optim} = 0.04\,\text{mol/L}$.

3) Consider a series of reactors. What should be the type and volume of the first reactor?
   The first reactor must be a CSTR to run it at the exit concentration such that $C_{A,exit} = C_A^{optim} = 0.04\frac{\text{mol}}{\text{L}}$ to maximize the selectivity of the desired product.
   Assume the feed temperature and pressure are the same as those of the reactor, then $\upsilon = \upsilon_0$
   From Eq. (2.18b), we have:

$$V_{CSTR} = \frac{\upsilon_0 C_{Ao} - \upsilon C_A}{-r_A} = \frac{\upsilon_0 C_{Ao} - \upsilon C_A}{k_1 C_A^{0.5} + k_2 C_A + k_3 C_A^2}$$

$$= \frac{10\frac{\text{L}}{\text{min}}(0.1625 - 0.04)\frac{\text{mol}}{\text{L}}}{\{0.004 * (0.04)^{0.5} + 0.3 * 0.04 + 0.25 * (0.04)^2\}\frac{\text{mol}}{\text{L.min}}} = 92.8\,\text{L} \quad \text{(J)}$$

4) What are the effluent concentrations of $A$, $D$, $U1$, and $U2$ from the first reactor?
**D-component Balance:**

$$In - Out + Generation - Consumption = Accumulation$$

We have a steady-state condition for the CSTR, which means zero accumulation. In addition, there is no input of $D$ and no consumption of $D$ as well. Thus, we have:

$$Generation = Out$$

$$k_2 C_A V_{CSTR} = F_D = C_D * v = C_D * v_0 \quad \text{(K)}$$

$$C_D = \frac{k_2 C_A V_{CSTR}}{v_0} = \frac{(0.3 * 0.04 * 92.8)\frac{\text{mol}}{\text{min}}}{10\frac{\text{L}}{\text{min}}} = 0.1114\ \frac{\text{mol}}{\text{L}} \quad \text{(L)}$$

The same can be said about $U_1$:

$$k_1 C_A^{0.5} V_{CSTR} = F_{U1} = C_{U1} * v = C_{U1} * v_0 \quad \text{(M)}$$

$$C_{U1} = \frac{k_1 C_A^{0.5} V_{CSTR}}{v_0} = \frac{(0.004 * (0.04)^{0.5} * 92.8)\frac{\text{mol}}{\text{min}}}{10\frac{\text{L}}{\text{min}}} = 0.0074\ \frac{\text{mol}}{\text{L}} \quad \text{(N)}$$

So is the case for $U_2$:

$$k_3 C_A^2 V_{CSTR} = F_{U2} = C_{U2} * v = C_{U2} * v_0 \quad \text{(O)}$$

$$C_{U2} = \frac{k_3 C_A^2 V_{CSTR}}{v_0} = \frac{(0.25 * (0.04)^2 * 92.8)\frac{\text{mol}}{\text{min}}}{10\frac{\text{L}}{\text{min}}} = 0.0037\ \frac{\text{mol}}{\text{L}} \quad \text{(P)}$$

Notice that:

$$C_{Total} = C_D + C_{U1} + C_{U2} + C_A = 0.1114 + 0.0074 + 0.0037 + 0.04 = 0.1625\frac{\text{mol}}{\text{L}} = C_{A0}$$

5) What is the conversion of $A$ in the first CSTR reactor?

$$X = \frac{C_{A0} - C_A}{C_{A0}} = \frac{0.1625 - 0.04}{0.1625} = 0.7538$$

6) Suppose that a single CSTR with a space time of 12 min is used, what will be its exit concentration of $A$, at $T = 300$ K?
Refer to Eq. (J), we have:

$$\frac{V_{CSTR}}{v_0} = 12\ \text{min} = \frac{C_{Ao} - C_A}{k_1 C_A^{0.5} + k_2 C_A + k_3 C_A^2}$$

$$= \frac{(0.1625 - C_A)\frac{\text{mol}}{\text{L}}}{\{0.004 * (C_A)^{0.5} + 0.3 * C_A + 0.25 * (C_A)^2\}\ \frac{\text{mol}}{\text{L. min}}} \quad \text{(R)}$$

Solve for $C_A$.

$$f(C_A) = 12\ \text{min} - \frac{(0.1625 - C_A)}{\{0.004 * (C_A)^{0.5} + 0.3 * C_A + 0.25 * (C_A)^2\}} = 0 \quad \text{(S)}$$

### 4.4.15 MATLAB Code for Root Finding of $f(C_A) = 0$

Figure 4.27 shows MATLAB code (**C_A_Root.m**) for finding the root of Eq. (S). Notice that the rate constants $k_1$, $k_2$, and $k_3$ are evaluated at $T = 300$ K.

```
% Define the function f(C_A) as an anonymous function
f = @(C_A) 12 - (0.1625 - C_A) / (0.004 * sqrt(C_A) + 0.3 * C_A + 0.25 * C_A^2);
% Provide an initial guess for C_A (based on expected root range)
initial_guess = 0.1; % Modify as needed
% Use fzero to find the root
C_A_root = fzero(f, initial_guess);
% Display the result
fprintf('The root of the function f(C_A) is C_A = %.5f\n', C_A_root);
```

**Figure 4.27** MATLAB code (**C_A_Root.m**) for finding the exit CA of CSTR with a known residence time.

The results are copied from the Command Window:

**≫ C_A_Root**
**The root of the function f(C_A) is C_A = 0.03274**

### 4.4.16 MATLAB Code for $S_{D/(U_1+U_2)} = f(\mathrm{T})$

7) Given that $E_1 = 50{,}000\,\mathrm{J/mol}$, $E_2 = 75{,}000\,\mathrm{J/mol}$, and $E_3 = 85{,}000\,\mathrm{J/mol}$. What temperature is recommended for a single CSTR with a space time of 12 min, as in part 6?
   We can find the optimum $T$ for a CSTR. MATLAB code (**SD_U_Teffect.m**) in Figure 4.15 is reused here, except for changing the following items: Reference temperature (T0), the activation energy values: $E_1$, $E_2$, and $E_3$, rate constants $k_1$, $k_2$, and $k_3$, and the selectivity function statement. Figure 4.28 shows a portion, made of the first sixteen lines of code, which replaces the first sixteen lines of code in Figure 4.15 (**SD_U_Teffect.m**). The entire code, however, is saved under the **SD_U_T_Fig4_28.m** file.
   The exit $C_A$ of CSTR is set to $C_A^{optim} = 0.03274\frac{\mathrm{mol}}{\mathrm{L}}$, which is the root of Eq. (S): $f(C_A) = 0$.

```
clear variables;
CA_fixed = str2double(input('Enter the exit CA in mol/L (M): ', 's'));
% Define constants
T0 = 300; % Reference temperature in Kelvin
R = 8.314; % Universal gas constant in J/(mol*K)
E1 = 50000; % Activation energy for k1 in J/mol
E2 = 75000; % Activation energy for k2 in J/mol
E3 = 85000; % Activation energy for k3 in J/mol
% Define temperature range
T_range = linspace(T0 - 100, T0 + 350, 1000); % Temperature range from T0-100 to T0+100
% Calculate rate constants and selectivity as a function of T
k1 = 0.004 * exp(-(E1 / R) * (1 ./ T_range - 1 / T0)); % mol^(1/2)/(L^(1/2)*min)
k2 = 0.3 * exp(-(E2 / R) * (1 ./ T_range - 1 / T0)); % 1/min
k3 = 0.25 * exp(-(E3 / R) * (1 ./ T_range - 1 / T0)); % L/(mol*min)
% Selectivity equation
Selectivity = (k2.*CA_fixed)./ (k1.*CA_fixed.^0.5 + k3.* CA_fixed.^2);
```

**Figure 4.28** The first 16-line portion of the **SD_U_T_Fig4_28.m** file.

Results are copied from the Command Window. The optimum $T$ for the given exit $C_A$ is 347.3 K. This simply says that with such a given CSTR residence time of 12 min, which is equivalent to $C_{A,exit} = 0.03274\,\mathrm{M}$, the optimum $T$ ought to be 347.3, not 300 K.

```
≫ SD_U_T_Fig4_28
Enter the exit CA in mol/L (M): 0.03274
Maximum Selectivity: 15.1702
Optimal Temperature (T_opt): 347.2973 K
```

Figure 4.29 shows the 2-D plot of $S = f(T)$ for $C_{A,exit} = C_A^{optim} = 0.03274\frac{\text{mol}}{\text{L}}$, with the peak of $S$ at $T = 347.3$ K.

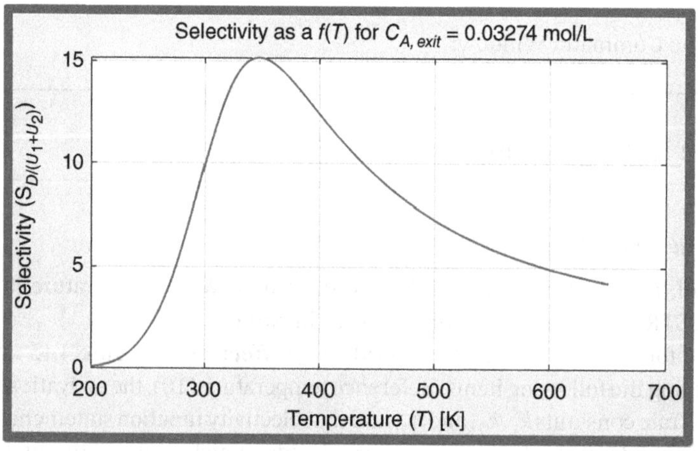


**Figure 4.29** CSTR $S = f(T)$ @ $C_{A,exit} = C_A^{optim} = 0.03274\frac{\text{mol}}{\text{L}}$, with $T_{opt} = 347.3$ K.

8) If 97% conversion of $A$ is desired, what reaction scheme and reactor sizes shall we use to maximize $S_{D/(U_1+U_2)}$?
   Notice that the given flow rate of 10 L/min is considered to be very low. Low-flow applications (e.g., instrumentation, sampling) have typical flow rates that range from 0.1 to 10 L/min. Medium flow applications (e.g., laboratory equipment, small-scale mixing) have flow rates that are often between 10 and 100 L/min. High-flow industrial applications (e.g., pneumatic tools, large-scale operations) have flow rates that can range from 100 to several thousand L/min. So, we will increase the flow rate from 10 to 500 L/min and elevate the initial concentration of $A$ to 1 M. MATLAB code (**VCSTR_VPFR_CA_Fig4_30.m**) file is shown in Figure 4.30, which is similar to Figure 4.19 MATLAB code (**VCSTR_VPFR_CA.m**).
   Results are copied from the Command Window:

```
≫ VCSTR_VPFR_CA_Fig4_30
Enter CSTR temperature (T) in K: 300
Enter PFR temperature (T) in K: 310
Enter the initial concentration of A (CA0) in mol/L: 1
Enter the volumetric flow rate (v0) in L/min: 500
```

```
Maximum CSTR Selectivity: 10
Optimal CA (CA_opt): 0.04008 mol/L
VCSTR: 36287.8755 L
Maximum PFR Selectivity2: 11.882
Optimal CA (CA_opt2): 0.029558 mol/L
VPFR: 177.3552 L
Total Volume (VTot): 36465.2307 L
Final CSTR CAexit, mol/L: 0.04008
Final CSTR XCSTR: 0.95992
Final PFR CAexit, mol/L: 0.029558
Final PFR XPFR: 0.97044
```

### 4.4.17 Selectivity in a Parallel + Consecutive Reaction Network ($A \rightarrow D$ Desired; $A \rightarrow U$ Undesired; $D \rightarrow W$ Waste)

A single reactant $A$ can form the desired product $D$ or the undesired by-product $U$ in parallel, and $D$ can further degrade to $W$ in series:

$$A \xrightarrow{k_1} D \qquad A \xrightarrow{k_2} U \qquad D \xrightarrow{k_3} W$$

All steps are first order in their own species. Feed is pure $A$ with $C_{A0} = 1.0$ mol/L.

$$k_1 = 2 \times 10^6 exp\left(-\frac{75{,}000}{8.314 * T}\right) \qquad k_2 = 1 \times 10^8 exp\left(-\frac{95{,}000}{8.314 * T}\right)$$

$$k_3 = 5 \times 10^6 exp\left(-\frac{80{,}000}{8.314 * T}\right)$$

A) Compare PFR and CSTR performance at two temperatures, $T_1 = 500$ K and $T_2 = 600$ K, as far as $k_1$, $k_2$, and $k_3$ are concerned.
B) Evaluate the exit selectivity $S_{D/U} = C_D/C_U$ for:
   a) A fixed residence $\tau = [2\ 50]$ s and fixed temperature $T = [600\ 500]$ K
   b) The PFR residence time ($\tau^*$) that maximizes $C_D$.

Comment on temperature effects and reactor choice.

**Step 1 Rate Constants (Arrhenius, §4.A.4)**

**Rate constants ($s^{-1}$):**

| *T* (K) | $k_1$ | $k_2$ | $k_3$ |
|---|---|---|---|
| 500 | 0.029211 | 0.011887 | 0.021934 |
| 600 | 0.59083 | 0.53606 | 0.54213 |

Because $E_2 > E_1$, the undesired path accelerates faster with $T$; we therefore anticipate lower selectivity at higher temperature.

```
clear variables;
% Prompt user for inputs
T = str2double(input('Enter CSTR temperature (T) in K: ', 's'));
T2 = str2double(input('Enter PFR temperature (T) in K: ', 's'));
CA0 = str2double(input('Enter the initial concentration of A (CA0) in mol/L: ', 's'));
v0 = str2double(input('Enter the volumetric flow rate (v0) in L/min: ', 's'));
% Define constants
T0 = 300; % Reference temperature in Kelvin
R = 8.314; % Universal gas constant in J/(mol*K)
E1 = 50000; % Activation energy for k1 in J/mol
E2 = 75000; % Activation energy for k2 in J/mol
E3 = 85000; % Activation energy for k3 in J/mol
% Define range for CA
CA_range = linspace(0, CA0, 500); % CA from CAf to CA0
% Calculate rate constants
k1 = 0.004 * exp(-(E1 / R) * (1 ./ T - 1 / T0)); % time^-1
k2 = 0.3 * exp(-(E2 / R) * (1 ./ T - 1 / T0)); % mol.L^-1.time^-1
k3 = 0.25 * exp(-(E3 / R) * (1 ./ T - 1 / T0)); % L.mol^-1.time^-1
% Selectivity equation
Selectivity = (k2.*CA_range)./ (k1.*CA_range.^0.5 + k3.* CA_range.^2);
% Find the maximum selectivity and corresponding CA under CSTR condition
[max_selectivity, index] = max(Selectivity);
CA_opt = CA_range(index); % Optimal concentration of A
% Calculate VCSTR
VCSTR = ((CA0 - CA_opt) / (k1 * CA_opt.^0.5 + k2*CA_opt+ k3 * CA_opt^2)) * v0; % Volume in L
figure (1);
plot(CA_range, Selectivity, 'b-', 'LineWidth', 1.5);
hold on;
plot(CA_opt, max_selectivity, 'ro', 'MarkerSize', 8, 'MarkerFaceColor', 'r'); % Highlight maximum selectivity
xlabel('C_A (mol/L)', 'FontSize', 12);
ylabel('CSTR Selectivity (S_{D/(U_1+U_2)})', 'FontSize', 12);
title(['CSTR Selectivity f(C_A) at T_{CSTR} = ', num2str(T), ' K'], 'FontSize', 14);
grid on;
ax = gca;
ax.FontSize = 16;
ax.FontWeight="bold";
% Calculate rate constants under PFR condition
k12 = 0.004 * exp(-(E1 / R) * (1 / T2 - 1 / T0)); % time^-1
k22 = 0.3 * exp(-(E2 / R) * (1 / T2 - 1 / T0)); % mol.L^-1.time^-1
```

**Figure 4.30** MATLAB code (**VCSTR_VPFR_CA_Fig4_30.m**) to calculate $V_{CSTR}$ and $V_{PFR}$ at given conditions of $T$ and exit $C_A$.

```
k32 = 0.25 * exp(-(E3 / R) * (1 / T2 - 1 / T0)); % L.mol^-1.time^-1
% Selectivity equation for PFR
CA_range2 = linspace(0, CA_opt, 500); % CA from CAf to CA0
Selectivity2 = (k22.*CA_range2)./ (k12.*CA_range2.^0.5 + k32.* CA_range2.^2);
% Find the maximum selectivity and corresponding CA under PFR condition
[max_selectivity2, index] = max(Selectivity2);
CA_opt2 = CA_range2(index); % Optimal concentration of A
% Calculate VPFR using numerical integration
rA = @(CA) k12 .* CA.^0.5 + k22.*CA + k32 .* CA.^2; % Rate of reaction
VPFR = v0 * integral(@(CA) 1 ./ rA(CA), CA_opt2, CA_opt); % Volume in L
% Total volume
VTot = VCSTR + VPFR;
% Calculate XCSTR and XPFR.
XCSTR=1.0-CA_opt/CA0;
XPFR=1.0-CA_opt2/CA0;
% Display results
disp(['Maximum CSTR Selectivity: ', num2str(max_selectivity)]);
disp(['Optimal CA (CA_opt): ', num2str(CA_opt), ' mol/L']);
disp(['VCSTR: ', num2str(VCSTR), ' L']);
disp(['Maximum PFR Selectivity2: ', num2str(max_selectivity2)]);
disp(['Optimal CA (CA_opt2): ', num2str(CA_opt2), ' mol/L']);
disp(['VPFR: ', num2str(VPFR), ' L']);
disp(['Total Volume (VTot): ', num2str(VTot), ' L']);
disp(['Final CSTR CAexit, mol/L: ', num2str(CA_opt)]);
disp(['Final CSTR XCSTR: ', num2str(XCSTR)]);
disp(['Final PFR CAexit, mol/L: ', num2str(CA_opt2)]);
disp(['Final PFR XPFR: ', num2str(XPFR)]);
% Plot VPFR as a function of CA
figure (2);
plot(CA_range2, Selectivity2, 'b-', 'LineWidth', 1.5);
hold on;
plot(CA_opt2, max_selectivity2, 'ro', 'MarkerSize', 8, 'MarkerFaceColor', 'r'); % Highlight maximum selectivity
xlabel('C_A (mol/L)', 'FontSize', 12);
ylabel('PFR Selectivity (S_{D/(U_1+U_2)})', 'FontSize', 12);
title(['PFR Selectivity f(C_A) at T_{PFR} = ', num2str(T2), ' K'], 'FontSize', 14);
grid on;
ax = gca;
ax.FontSize = 16;
ax.FontWeight="bold";
hold off;
```

**Figure 4.30** (*Continued*)

**Step 2 PFR (ODE System, §4.A.10)**

With space time $\tau$ (isothermal, constant $v$):

$$\frac{dC_A}{d\tau} = -(k_1 + k_2)C_A \qquad \rightarrow C_A = C_{A0}e^{-(k_1+k_2)\tau} \tag{A}$$

$$\frac{dC_D}{d\tau} = (k_1)C_A - (k_3)C_D \qquad \rightarrow \frac{dC_U}{d\tau} = (k_2)C_A \rightarrow C_D(0) = C_U(0) = 0$$

$$\rightarrow C_D(\tau) = \frac{k_1 C_{A0}}{k_1 + k_2 - k_3}\left(e^{-(k_3)\tau} - e^{-(k_1+k_2)\tau}\right) \tag{B}$$

$$\rightarrow C_U(\tau) = \frac{k_2 C_{A0}}{k_1 + k_2}\left(1 - e^{-(k_1+k_2)\tau}\right) \tag{C}$$

$$\text{Hence, } S_{D/U}(\tau) = \frac{C_D(\tau)}{C_U(\tau)} \qquad \text{(used in§§4.A.5 – 4.A.9 for 2 – D/3 – D plotting)} \tag{D}$$

**NOTE:** Calculations here are borrowed from MATLAB results, shown below.

a) **Fixed Residence Time for PFR:**

Choosing two sets of values of $\tau$ and $T$:

- **1st Set:** PFR @ 500 K and $\tau = 50$ s, using Eqs. (A)–(D) $\Rightarrow C_A = 0.128$
  $C_D = 0.314$, $C_U = 0.252$, $S_{D/U} = 0.314/0.252 = 1.244$
- **2nd Set:** PFR @ 600 K and $\tau = 2.0$ s, using Eqs. (A)–(D) $\Rightarrow C_A = 0.105$
  $C_D = 0.236$, $C_U = 0.426$, $S_{D/U} = 0.236/0.426 = 0.553$

***T* Effect:** Higher $T$ sharply reduces selectivity because $k_2$ and $k_3$ catch up to/overcome $k_1$.

b) **Residence time that maximizes $C_D$ (analytical optimum)**

For this first-order network of reactions, the PFR $C_D(\tau)$ is unimodal, with

$$\tau^* = \frac{1}{k_1 + k_2 - k_3}\ln\left(\frac{k_1 + k_2}{k_3}\right) \qquad \text{Valid for: } k_1 + k_2 > k_3 \tag{E}$$

Notice that $\tau^*$ is $f(k_1, k_2, k_3) = f(T)$. So, $\tau^*$ is evaluated for a given $T$.

$$\text{PFR}@T = 500\text{ K}, \tau\star = 32.7\text{ s (Eq.E)} \Rightarrow$$

$$C_D = 0.346, C_U = 0.214, S_{D/U} = 0.346/0.214 = 1.619 \text{ (Equations B – D)}$$

$$\text{PFR}@T = 600\text{ K}, \tau\star = 1.251\text{ s (Eq.E)} \Rightarrow$$

$$C_D = 0.266, C_U = 0.360, S_{D/U} = 0.266/0.360 = 0.740 \text{ (Equations B – D)}$$

**Design Insight:** Even at its own optimum $\tau\star$, the high-$T$ case remains less selective.

**Step 3 CSTR (Algebraic Model, §4.A.11)**

At steady state (space time $\tau$):

$$C_A = \frac{C_{A0}}{1 + (k_1 + k_2)\tau}, \tag{F}$$

$$C_D = \frac{k_1 C_A \tau}{1 + (k_3)\tau}, \tag{G}$$

$$C_U = (k_2)C_A\tau \tag{H}$$

$$S_{D/U}(\tau) = \frac{C_D(\tau)}{C_U(\tau)} \tag{I}$$

Using the same two sets of values of $\tau$ and $T$:

**1st Set:** CSTR @ 500 K, $\tau = 50.00$ s ⇒

$$C_A = 0.327, C_D = 0.228, C_U = 0.195, S_{D/U} = 0.228/0.195 = 1.172 \text{ (Equations F – I)}$$

**2nd Set:** CSTR @ 600 K, $\tau = 2.00$ s ⇒

$$C_A = 0.307, C_D = 0.174, C_U = 0.330, S_{D/U} = 0.174/0.339 = 0.529 \text{ (Equations F – I)}$$

**Reactor Effect on Selectivity:** With consecutive loss of $D$ ($k_3$), a PFR typically outperforms a CSTR in selectivity, because the PFR's high-$C_A$ inlet region forms $D$ quickly before much degradation occurs.

Figure 4.31 shows MATLAB code (**Parall_Series_SD_U.m**) for predicting concentration as well as selectivity of the desired product as a function of $T$ and $\tau$ for both types: PFR and CSTR.

```
%% Step 1: Arrhenius Rate Constants (Section 4.A.4)
clear;
% Given data
A = [2.0e6, 1.0e8, 5.0e6];      % Pre-exponential factors, s^-1
E = [75, 95, 80];               % Activation energies, kJ/mol
R = 8.314e-3;                   % kJ/mol-K
T = [500, 600];                 % Temperatures to compare, K
CA0 = 1.0;                      % mol/L
% Calculate rate constants at both temperatures
k = zeros(length(T),3);
for iT = 1:length(T)
    for i = 1:3
        k(iT,i) = A(i)*exp(-E(i)/(R*T(iT)));
    end
end
fprintf('Rate constants (s^-1):\n');
disp(array2table(k, 'VariableNames',{'k1','k2','k3'}, 'RowNames',{'500K','600K'}));
%% Step 2: PFR Selectivity (Sections 4.A.5-4.A.9)
% Analytical solutions for PFR exit concentrations
% Fixed residence times (chosen for moderate conversion)
tau_fixed = [50, 2.0]; % s
for iT = 1:length(T)
    k1 = k(iT,1); k2 = k(iT,2); k3 = k(iT,3);
    tau = tau_fixed(iT);
    CA = CA0 * exp(-(k1+k2)*tau);
    CD = (k1*CA0/(k1+k2-k3)) * (exp(-k3*tau) - exp(-(k1+k2)*tau));
    CU = (k2*CA0/(k1+k2)) * (1 - exp(-(k1+k2)*tau));
    S  = CD / CU;
    fprintf('\nPFR @ %d K, tau=%.2f s:\n', T(iT), tau);
    fprintf('CA=%.3f, CD=%.3f, CU=%.3f, S_D/U=%.3f\n', CA, CD, CU, S);
end
% Optimal tau for maximum CD
for iT = 1:length(T)
    k1 = k(iT,1); k2 = k(iT,2); k3 = k(iT,3);
    if k1+k2 > k3
        tau_opt = (1/(k1+k2-k3))*log((k1+k2)/k3);
```

**Figure 4.31** MATLAB code (**Parall_Series_SD_U.m**) for predicting concentration as well as selectivity of the desired product as a function of $T$ and $\tau$ for both types: PFR and CSTR.

```
        CD_opt = (k1*CA0/(k1+k2-k3))*(exp(-k3*tau_opt) - exp(-(k1+k2)*tau_opt));
        CU_opt = (k2*CA0/(k1+k2))*(1 - exp(-(k1+k2)*tau_opt));
        S_opt  = CD_opt / CU_opt;
        fprintf('\nPFR @ %d K, tau_opt=%.3f s:\n', T(iT), tau_opt);
        fprintf('CD_opt=%.3f, CU_opt=%.3f, S_D/U=%.3f\n', CD_opt, CU_opt, S_opt);
    else
        fprintf('\nNo optimum tau exists for T=%d K (k1+k2 <= k3)\n', T(iT));
    end
end
%% Step 3: CSTR Selectivity (Section 4.A.11)
for iT = 1:length(T)
    k1 = k(iT,1); k2 = k(iT,2); k3 = k(iT,3);
    tau = tau_fixed(iT);
    CA = CA0 / (1 + tau*(k1+k2));
    CD = (tau*k1*CA) / (1 + tau*k3);
    CU = tau*k2*CA;
    S  = CD / CU;
    fprintf('\nCSTR @ %d K, tau=%.2f s:\n', T(iT), tau);
    fprintf('CA=%.3f, CD=%.3f, CU=%.3f, S_D/U=%.3f\n', CA, CD, CU, S);
end
%% Step 4: Plot Selectivity vs. tau for PFR and CSTR (Sections 4.A.6-4.A.9)
tau_vec = linspace(0.1,60,200);
figure; hold on;
for iT = 1:length(T)
    k1 = k(iT,1); k2 = k(iT,2); k3 = k(iT,3);

    % PFR selectivity curve
    CD_pfr = (k1*CA0/(k1+k2-k3)) .* (exp(-k3*tau_vec) - exp(-(k1+k2)*tau_vec));
    CU_pfr = (k2*CA0/(k1+k2)) .* (1 - exp(-(k1+k2)*tau_vec));
    S_pfr = CD_pfr ./ CU_pfr;
    plot(tau_vec, S_pfr, 'LineWidth', 2, 'DisplayName', sprintf('PFR %dK',T(iT)));
    % CSTR selectivity curve
    CA_cstr = CA0 ./ (1 + tau_vec*(k1+k2));
    CD_cstr = (tau_vec*k1 .* CA_cstr) ./ (1 + tau_vec*k3);
    CU_cstr = tau_vec*k2 .* CA_cstr;
    S_cstr = CD_cstr ./ CU_cstr;
   plot(tau_vec, S_cstr, '-', 'LineWidth', 2, 'DisplayName', sprintf('CSTR %dK',T(iT)));
end
xlabel('\tau (s)'); ylabel('S_{D/U}');
legend('Location','best');
title('Selectivity vs Residence Time');
grid on;
ax = gca; ax.FontSize = 14;ax.FontWeight = 'bold';
```

**Figure 4.31** (*Continued*)

The MATLAB results are copied from the Command Window.

```
>> Parall_Series_SD_U
Rate constants (s^-1):
             k1          k2          k3
        ________    ________    ________
500K    0.029211    0.011887    0.021934
600K    0.59083     0.53606     0.54213
```

```
PFR @ 500 K, tau=50.00 s:
CA=0.128, CD=0.314, CU=0.252, S_D/U=1.244
PFR @ 600 K, tau=2.00 s:
CA=0.105, CD=0.236, CU=0.426, S_D/U=0.553
PFR @ 500 K, tau_opt=32.765 s:
CD_opt=0.346, CU_opt=0.214, S_D/U=1.619
PFR @ 600 K, tau_opt=1.251 s:
CD_opt=0.266, CU_opt=0.360, S_D/U=0.740
CSTR @ 500 K, tau=50.00 s:
CA=0.327, CD=0.228, CU=0.195, S_D/U=1.172
CSTR @ 600 K, tau=2.00 s:
CA=0.307, CD=0.174, CU=0.330, S_D/U=0.529
```

The generated plot for the desired product selectivity as a function of residence time and temperature for both PFR and CSTR is shown in Figure 4.32. PFR at low $T$ gives the best $S_{D/U}$ over the entire range of $\tau$.

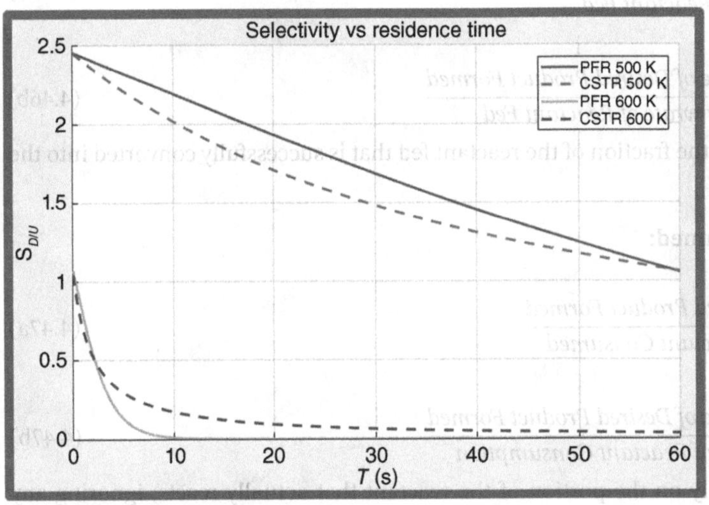


**Figure 4.32** The generated plot for the desired product selectivity as a function of residence time and temperature for both PFR and CSTR.

## 4.5 The Yield of the Desired Product

The yield of a desired product in a system with multiple chemical reactions is a fundamental performance metric in chemical reaction engineering, as it directly reflects how effectively raw materials are being transformed into the target compound. Yield can be expressed in several ways, such as the fraction of the theoretical maximum obtainable from stoichiometry, the ratio of desired product formed to the reactants consumed, or the proportion of desired product in the total mixture of products. In systems involving multiple reactions, yield becomes particularly significant because it captures the impact of competing reaction pathways, including primary reactions that directly form the desired product and secondary reactions that consume it or form unwanted by-products. Factors such as reaction temperature, pressure, catalyst selectivity, feed composition, and reactor type (batch, CSTR, PFR, or combinations) all play a role in shaping the final yield.

Importantly, yield is not a universally fixed concept; its definition depends on the reference basis chosen, which can lead to different numerical values even for the same physical process. For example, yield might be reported relative to the limiting reactant, the total feed, or the total products, and each perspective serves a different analytical or industrial purpose. This definitional flexibility is well captured in the classic text by Felder and Rousseau [2], which emphasizes that the interpretation of yield should always be accompanied by a clear statement of its basis. Yield essentially measures the efficiency of converting a reactant into a desired product, but without a consistent definition, comparisons between processes or studies can be misleading. In the following sections, we will present common yield definitions used in chemical reaction engineering, highlight their applications, and discuss how the choice of definition can influence both process evaluation and optimization strategies.

**Fractional Yield:**

For a batch reactor, we have:

$$Yield = \frac{Moles\ of\ Desired\ Product\ Formed}{Moles\ of\ Reactant\ Fed} \tag{4.46a}$$

For a flow reactor, we have:

$$Yield = \frac{Molar\ Flowrate\ of\ Desired\ Product\ Formed}{Molar\ Flowrate\ of\ Reactant\ Fed} \tag{4.46b}$$

Equations (4.46a–b) consider the fraction of the reactant fed that is successfully converted into the desired product.

**Relative to Reactant Consumed:**

For a batch reactor, we have:

$$Yield = \frac{Moles\ of\ Desired\ Product\ Formed}{Moles\ of\ Reactant\ Consumed} \tag{4.47a}$$

For a flow reactor, we have:

$$Yield = \frac{Molar\ Flowrate\ of\ Desired\ Product\ Formed}{Molar\ Rate\ of\ Reactant\ Consumption} \tag{4.47b}$$

Equations (4.47a–b) focus only on the portion of the reactant that actually reacts, ignoring any excess reactant that remains unreacted.

**Relative to Other Products:**

Yield can also be expressed relative to undesired products, linking closely to the concept of selectivity, which measures how efficiently the desired product forms compared to by-products. Equation (4.46) will be adopted in this chapter.

No matter how yield is defined, a higher value always indicates improved process performance or more favorable operating conditions.

### 4.5.1 Multiple Reactions: Yield and Selectivity

Consider propylene production by the dehydrogenation of propane:

$$C_3H_8 \rightarrow C_3H_6 + H_2 \tag{4.48}$$

Once some hydrogen is produced, it can react with propane to produce methane and ethane:

$$C_3H_8 + H_2 \rightarrow C_2H_6 + CH_4 \tag{4.49}$$

The second reaction is typical of a hydrocracking process, often occurring in the presence of a catalyst (e.g., zeolites or metallic catalysts like platinum or nickel) under high temperature and pressure.

Observe that the second reaction proceeds in parallel with the first reaction, competing directly with it. Initially, the second reaction is relatively weak due to the limited presence of hydrogen in the reaction medium. However, as the process continues and more hydrogen is generated, the second reaction gains momentum and becomes a strong contender against the first.

Moreover, propylene can react with ethane to mainly form 2-butylene (*cis*- or *trans*-), which is generally more thermodynamically stable than 1-butylene due to hyperconjugation and alkyl group stabilization of the double bond, and methane:

$$C_3H_6 + C_2H_6 \rightarrow C_4H_8 + CH_4 \tag{4.50}$$

Similarly, the same reasoning applies to the third reaction regarding its progression over time.

As the primary goal of the process is to produce propylene, only the first reaction is considered desirable. The second reaction depletes the reactant without generating the target product, while the third reaction consumes the desired product ($C_3H_6$). The reactor engineer must design the reactor and define its operating conditions to maximize the production of the desired product ($C_3H_6$) while minimizing the formation of undesired by-products ($CH_4$, $C_2H_6$, and $C_4H_8$).

The term selectivity (defined in the previous section) is coined to describe the degree to which a desired reaction predominates over competing side reactions. Equations (4.46a–b) can also be put in the following form:

$$Yield = \frac{\textit{net moles of desired product formed}}{\textit{moles formed without side reactions, assuming a complete main reaction}} \tag{4.51a}$$

$$Yield = \frac{\textit{net molar flowrate of desired product formed}}{\textit{molar flowrate formed without side reactions, assuming a complete main reaction}} \tag{4.51b}$$

The yield, defined by Eqs. (4.51a–b), is always a fraction; it may also be expressed as a percentage by multiplying by 100%. High yield and selectivity indicate effective suppression of undesired side reactions relative to the desired reaction under specified reactor conditions, such as temperature, pressure, feed composition, and the presence of inerts or impurities.

**Example 4.7** ***Calculation of Yield and Selectivity I***

Consider the set of reactions given by Eqs. (4.48–4.50). Assume that the extent of reaction for the first is $\xi_1$; the second $\xi_2$; and the third $\xi_3$. Consider a batch reactor.

Notice that propylene ($C_3H_6$) is mainly produced by the first reaction and consumed by the third reaction; propylene is not involved in the second reaction. For a multi-reaction system, refer to Eq. (3.58a) for expressing the yield in terms of the reaction extent:

$$Yield = \frac{\left(n_{C_3H_6}\right)_{Actual\ Condition}}{\left(n_{C_3H_6}\Big|_{f_1=1}\right)_{Ideal\ Condition}} = \frac{n_{C_3H_6} = n^o_{C_3H_6} + (1)\,\xi_1 + (-1)\,\xi_3}{n_{C_3H_6}\Big|_{f_1=1}} = \frac{0 + \xi_1 - \xi_3}{n_{C_3H_6}\Big|_{f_1=1}} = \frac{\xi_1 - \xi_3}{n^o_{C_3H_6}} \tag{A}$$

The denominator in Eq. (A) is equal to the produced number of moles of propylene ($C_3H_6$ under the assumption of a complete conversion for the first (main) reaction (i.e., $f_1 = 1.0$) while assuming zero conversion for all other competing reactions (i.e., $f_2 = f_3 = 0.0$). This means that all initial moles of propane ($C_3H_8$) will be completely converted into propylene ($C_3H_6$), the desired product. The overall selectivity ($S$), based on Eq. (4.3), is:

$$S = \frac{\left(n_{C_3H_6}\right)_{Actual\ Condition}}{All\ moles\ of\ undesired\ products}$$

$$= \frac{\xi_1 - \xi_3}{(moles\ of\ H_2) + (moles\ of\ CH_4) + (moles\ of\ C_2H_6) + (moles\ of\ C_4H_8)} \quad \text{(B)}$$

The denominator in Eq. (B) represents the waste (or a by-product like $H_2$) of competing materials formed at the expense of propylene. Equation (B) says that not all propane ($C_3H_8$) moles are converted into propylene ($C_3H_6$), but some of them are converted into $H_2$, $CH_4$, $C_2H_6$, and $C_4H_8$.

$$S = \frac{\xi_1 - \xi_3}{\{\xi_1 + (-1)\,\xi_2\} + \{\xi_2 + \xi_3\} + \{\xi_2 - \xi_3\} + \{\,\xi_3\}} = \frac{\xi_1 - \xi_3}{\{\xi_1 - \xi_2\} + \{\xi_2 + \xi_3\} + \{\xi_3\}}$$

$$= \frac{\xi_1 - \xi_3}{\xi_1 + \xi_2 + 2\,\xi_3} \quad \text{(C)}$$

For example, if the initial amount for propane is 100 mol; $\xi_1$ is 84 mol; $\xi_2$ is 12 mol; and $\xi_3$ is 8 mol. Then

$$Yield = \frac{84 - 8}{100} = \frac{76}{100} = 0.76 \quad \text{(D)}$$

$$Selectivity = \frac{\xi_1 - \xi_3}{\xi_1 + \xi_2 + 2\,\xi_3} = \frac{84 - 8}{84 + 12 + (2 \times 8)} = 0.68 \quad \text{(E)}$$

**Example 4.8** ***CSTR Yield and Selectivity Calculations for Gas-phase Reaction***

Reactant $A$ decomposes by three simultaneous gas-phase reactions to form three products, one that is desired, $D$, and two that are undesired, $U_1$ and $U_2$. The activation energy terms are: $E_1 = 50{,}000$ J/mol, $E_2 = 75{,}000$ J/mol, and $E_3 = 85{,}000$ J/mol.

$$A \xRightarrow{k_1} U_1 \qquad -r_{1A} = r_{U1} = k_1 C_A^{0.5} = 0.004 \frac{mol^{1/2}}{L^{1/2} \cdot min} C_A^{0.5} (\text{Half} - \text{order}) \quad \text{(A)}$$

$$A \xRightarrow{k_2} D \qquad -r_{12} = r_D = k_2 C_A = 0.3 \frac{1}{min} \ (\text{First} - \text{order}) \quad \text{(B)}$$

$$A \xRightarrow{k_3} U_2 \qquad -r_{13} = r_{U2} = k_3 C_A^2 = 0.25 \frac{L}{mol \cdot min} C_A^2 \ (\text{Second} - \text{order}) \quad \text{(C)}$$

The reaction is carried out in a CSTR with constant $T = 300$ K and $P = 10.774$ atm, equal to those of the feed. The exit concentration of $A$ is 0.1 mol/L, and the volume flow rate, $\upsilon_0 = 100$ L/min.

a) Find both $C_{A0}$ and $F_{A0}$.
b) Find species concentrations: $D$, $U_1$, $U_2$, and $A$
c) Find species molar flow rates: $F_D$, $F_{U1}$, $F_{U2}$, and $F_A$
d) The yield, as defined by Eq. (4.51b)
e) Find the overall selectivity, as defined by Eq. (4.5b)

f) Find instantaneous selectivity, defined by: $S_{D/(U_1+U_2)} = \frac{r_D}{r_{U_1}+r_{U_2}}$

g) Use MATLAB to carry out the previous steps from $a$ to $f$.

**Solution**

a) **Find Both $C_{A0}$ and $F_{A0}$**

$$C_{A0} = \frac{Py_{A0}}{RT} = \frac{10.774\ \text{atm} * 1}{0.08206\ \frac{\text{atm.L}}{\text{mol.K}} * 300\ \text{K}} = 0.4376\frac{\text{mol}}{\text{L}}$$

$$F_{A0} = C_{A0} * v_0 = 0.4376\frac{\text{mol}}{\text{L}} * 100\frac{\text{L}}{\text{min}} = 43.76\ \frac{\text{mol}}{\text{min}}$$

b) **Species Concentrations**
***A*-Component Molar Balance:**

*In – Out + Generation – Consumption = Accumulation*

We have a steady-state condition for a CSTR, which means zero accumulation. In addition, the input of $A$ is $F_{A0}$, and there is no generation of $A$ as well. Thus, we have:

*In – Out – Consumption = 0*

$$C_{A0}v_0 - C_A * v_0 - \left(k_1C_A^{0.5} + k_2C_A + k_3C_A^2\right)V_{CSTR} = 0$$

$$V_{CSTR} = \frac{v_0(C_{A0} - C_A)}{\left(k_1C_A^{0.5} + k_2C_A + k_3C_A^2\right)} = \frac{100\ \frac{\text{L}}{\text{min}}\ (0.4376 - 0.1)\frac{\text{mol}}{\text{L}}}{(0.004 * 0.1^{0.5} + 0.3 * 0.1 + 0.25 * 0.1^2)\frac{\text{mol}}{\text{L. min}}} = 1{,}000\ \text{L}$$

***D*-Component Molar Balance:**

*In – Out + Generation – Consumption = Accumulation*

We have a steady-state condition for the CSTR, which means zero accumulation. In addition, there is no input of $D$, and there is no consumption of $D$ as well. Thus, we have:

*Generation = Out*

$$k_2C_AV_{CSTR} = F_D = C_D * v = C_D * v_0$$

$$C_D = \frac{k_2C_AV_{CSTR}}{v_0} = \frac{(0.3 * 0.1 * 1{,}000)\frac{\text{mol}}{\text{min}}}{100\frac{\text{L}}{\text{min}}} = 0.3\ \frac{\text{mol}}{\text{L}}$$

The same can be said about $U_1$:

*Generation = Out*

$$k_1C_A^{0.5}V_{CSTR} = F_{U1} = C_{U1} * v = C_{U1} * v_0$$

$$C_{U1} = \frac{k_1C_A^{0.5}V_{CSTR}}{v_0} = \frac{(0.004 * (0.1)^{0.5} * 1{,}000)\frac{\text{mol}}{\text{min}}}{100\frac{\text{L}}{\text{min}}} = 0.0126\ \frac{\text{mol}}{\text{L}}$$

The same can also be said about $U_2$:

$$Generation = Out$$

$$k_3 C_A^2 V_{CSTR} = F_{U2} = C_{U2} * v = C_{U2} * v_0$$

$$C_{U2} = \frac{k_3 C_A^2 V_{CSTR}}{v_0} = \frac{(0.25 * (0.1)^2 * 1{,}000)\frac{\text{mol}}{\text{min}}}{100\frac{\text{L}}{\text{min}}} = 0.025\,\frac{\text{mol}}{\text{L}}$$

Notice that:

$$C_{Total} = C_D + C_{U1} + C_{U2} + C_A = 0.30 + 0.0126 + 0.025 + 0.1 = 0.4376\frac{\text{mol}}{\text{L}} = C_{A0}$$

c) **The Species Molar Flow rate**
Refer to Eq. (3.58b), we have:

$$F_{U1} = F_{U1}^o + \dot{\xi}_1 = 0 + \dot{\xi}_1 = C_{U1} v_0 = 0.0126\,\frac{\text{mol}}{\text{L}} * 100\frac{\text{L}}{\text{min}} = 1.26\,\frac{\text{mol}}{\text{min}} \rightarrow \dot{\xi}_1 = 1.26\,\frac{\text{mol}}{\text{min}}$$

$$F_D = F_D^o + \dot{\xi}_2 = 0 + \dot{\xi}_2 = C_D v_0 = 0.3\,\frac{\text{mol}}{\text{L}} * 100\frac{\text{L}}{\text{min}} = 30\,\frac{\text{mol}}{\text{min}} \rightarrow \dot{\xi}_2 = 30\,\frac{\text{mol}}{\text{min}}$$

$$F_{U2} = F_{U2}^o + \dot{\xi}_3 = 0 + \dot{\xi}_3 = C_{U2} v_0 = 0.025\,\frac{\text{mol}}{\text{L}} * 100\frac{\text{L}}{\text{min}} = 2.5\,\frac{\text{mol}}{\text{min}} \rightarrow \dot{\xi}_3 = 2.5\,\frac{\text{mol}}{\text{min}}$$

$$F_A = F_A^o - \dot{\xi}_1 - \dot{\xi}_2 - \dot{\xi}_3 = 43.76 - 1.26 - 30 - 2.5 = 43.76 - 33.51 = 10.00\frac{\text{mol}}{\text{min}}$$

d) **The Yield**

$$Yield = \frac{(F_D)_{Actual\ Condition}}{\left(F_D\big|_{f_2=1}\right)_{Ideal\ Condition}} = \frac{F_D = F_D^o + (1)\,\xi_2}{F_D\big|_{f_2=1}} = \frac{0 + \dot{\xi}_2}{F_D\big|_{f_2=1}} = \frac{\dot{\xi}_2}{F_A^o} = \frac{30}{43.76} = 0.686$$

e) **The Overall Selectivity ($\overline{S}$)**
The overall selectivity ($\overline{S}$), based on Eq. (4.5b), is:

$$\overline{S} = \frac{(F_D)_{Actual\ Condition}}{All\ moles\ of\ undesired\ products} = \frac{\dot{\xi}_2}{(Molar\ Flowrate\ of\ U_1) + (Molar\ Flowrate\ of\ U_2)}$$

$$\overline{S} = \frac{\dot{\xi}_2}{(F_{U1}) + (F_{U2})} = \frac{\dot{\xi}_2}{\dot{\xi}_1 + \dot{\xi}_3} = \frac{30}{1.26 + 2.5} = 7.968$$

f) **Instantaneous Selectivity**

$$S_{D/(U_1+U_2)} = \frac{r_D}{r_{U_1} + r_{U_2}} = \frac{k_2 C_A}{k_1 C_A^{0.5} + k_3 C_A^2} = \frac{0.3 * 0.1\,\frac{\text{mol}}{\text{L.min}}}{(0.004 * (0.1)^{0.5} + 0.25 * (0.1)\text{^}2)\frac{\text{mol}}{\text{L.min}}} = 7.968$$

Notice that for a CSTR, both the overall and instantaneous selectivity are the same.

g) **Use MATLAB to carry out the previous steps from *a* to *f*** (See section 4.5.2 below).

### 4.5.2 MATLAB Code for Multi-Reaction CSTR Yield and Selectivity

Figure 4.33 shows MATLAB code (**CSTR_Yield_S_Ex4_8.m**) for calculating the CSTR volume requirement and the yield and selectivity of the desired product.

```
%% Calculation of Yield and Selectivity for a Multi-reaction system carried out in CSTR
clear variables;
CAf = str2double(input('Enter Exit or Final CA in mol/L: ', 's'));
P0 = str2double(input('Enter Initial Pressure in atm: ', 's'));
T0 = str2double(input('Enter Initial Temperature in K: ', 's'));
v0 = str2double(input('Enter volume flowrate in L/min: ', 's'));
yA0 = str2double(input('Enter initial mole-fraction of A: ', 's'));
T = str2double(input('Enter Reactor Temperature in K: ', 's'));
% Define constants
E1 = 50000; % Activation energy for k1 in J/mol
E2 = 75000; % Activation energy for k2 in J/mol
E3 = 85000; % Activation energy for k3 in J/mol
R=8.314;% J/(mol.K)
% Calculate rate constants as a function of T
k1 = 0.004 * exp(-(E1 / R) * (1 ./ T - 1 / 300));% mol^(1/2)/(L^(1/2)*min)
k2 = 0.3 * exp(-(E2 / R) * (1 ./ T - 1 / 300)); % 1/min
k3 = 0.25 * exp(-(E3 / R) * (1 ./ T - 1 / 300)); % L/(mol*min)
CA0=P0*yA0/(0.08206*T0);
FA0 = CA0 * v0; % mol/min
% Define the rate of reaction function as an anonymous function
rA = -(k1 * CAf.^0.5 + k2 * CAf + k3 * CAf.^2); % mol/(L*min)
% Compute the volume of the CSTR using algebraic equation
VCSTR=v0*(CA0-CAf)/-rA;
% Exit concentration of D
CD=(k2*CAf* VCSTR)/v0;
% Exit concentration of U1
CU1=(k1*CAf^0.5*VCSTR)/v0;
% Exit concentration of U2
CU2=(k3*CAf^2*VCSTR)/v0;
CTotal=CU1+CD+CU2+CAf;
% Compute the extents of reactions using numerical integration
xi1 = v0*CU1; %Extent of reaction #1 in mol/min
xi2 = v0*CD; % Extent of reaction #2 in mole/min
xi3 = v0*CU2; % Extent of reaction #3 in mole/min
% The yield for the desired product, D
Yield=xi2/FA0;
% The overall selectivity
Soverall=xi2/(xi1+xi3);
% Instantaneous Selectivity S_D_{U1_U2}
SD_U1U2=(k2 * CAf)./(k1*CAf.^0.5 + k3 * CAf.^2);
% Display the results
disp(['Volume of CSTR (VCSTR): ', num2str(VCSTR), ' L']);
disp(['Extent of reaction #1 (xi1): ', num2str(xi1), ' mole/min']);
disp(['Extent of reaction #2 (xi2): ', num2str(xi2), ' mole/min']);
disp(['Extent of reaction #3 (xi3): ', num2str(xi3), ' mole/min']);
disp(['U1 concentration, CU1: ', num2str(CU1), ' mole/L']);
disp(['D concentration, CD: ', num2str(CD), ' mole/L']);
disp(['U2 concentration, CU2: ', num2str(CU2), ' mole/L']);
disp(['Total concentration, CTotal: ', num2str(CTotal), ' mole/L']);
disp(['Initial A concentration, CA0: ', num2str(CA0), ' mole/L']);
disp(['The Yield of Desired Product: ', num2str(Yield)])
disp(['The Overall Selectivity of Desired Product: ', num2str(Soverall)])
disp(['The Instantaneous Selectivity of Desired Product: ', num2str(SD_U1U2)])
```

**Figure 4.33** MATLAB code (**CSTR_Yield_S_Ex4_8.m**) for $V_{CSTR}$ calculation and yield and selectivity of the desired product.

Results are copied from the Command Window.

```
>> CSTR_Yield_S_Ex4_8
Enter Exit or Final CA in mol/L: 0.1
Enter Initial Pressure in atm: 10.774
Enter Initial Temperature in K: 300
Enter volume flowrate in L/min: 100
Enter initial mole-fraction of A: 1
Enter Reactor Temperature in K: 300
Volume of CSTR (VCSTR): 999.9945 L
Extent of reaction #1 (xi1): 1.2649 mole/min
Extent of reaction #2 (xi2): 29.9998 mole/min
Extent of reaction #3 (xi3): 2.5 mole/min
U1 concentration, CU1: 0.012649 mole/L
D concentration, CD: 0.3 mole/L
U2 concentration, CU2: 0.025 mole/L
Total concentration, CTotal: 0.43765 mole/L
Initial A concentration, CA0: 0.43765 mole/L
The Yield of Desired Product: 0.68548
The Overall Selectivity of Desired Product: 7.9683
The Instantaneous Selectivity of Desired Product: 7.9683
```

**Example 4.9** ***PFR Yield and Selectivity Calculations***

Repeat Example 4.8 for a PFR with an exit concentration of $A$ equal to: $C_{Af} = 0.1$ mol/L.

We will use MATLAB code to find the relevant yield and selectivity parameters.

**NOTE:** $C_A$ is a variable that changes its value from $C_{A0}$ down to $C_{Af}$. The following equations are used in the MATLAB code of Figure 4.34.

```
%% PFR Yield and Selectivity Calculations
clear variables;
CAf = str2double(input('Enter Exit or Final CA in mol/L: ', 's'));
P0 = str2double(input('Enter Initial Pressure in atm: ', 's'));
T0 = str2double(input('Enter Initial Temperature in K: ', 's'));
v0 = str2double(input('Enter volume flowrate in L/min: ', 's'));
yA0 = str2double(input('Enter initial mole-fraction of A: ', 's'));
T = str2double(input('Enter Reactor Temperature in K: ', 's'));
% Input parameters
CA0=P0*yA0/(0.08206*T0);% Initial concentration in mol/
E1 = 50000; E2 = 75000; E3 = 85000; % Activation energies in J/mol
R = 8.314; % Gas constant in J/(mol.K)
% Rate constants at given temperature
k1 = 0.004 * exp(-E1 / R * (1 / T - 1 / 300)); % mol^(1/2)/(L^(1/2)*min)
k2 = 0.3 * exp(-E2 / R * (1 / T - 1 / 300)); % 1/min
k3 = 0.25 * exp(-E3 / R * (1 / T - 1 / 300)); % L/(mol*min)
% Rate of reaction function
rA = @(CA) -(k1 * CA.^0.5 + k2 * CA + k3 * CA.^2); % mol/(L*min)
```

**Figure 4.34** MATLAB code (**PFR_Yield_S_Ex4_9.m**) for $V_{PFR}$ calculation and yield and selectivity of the desired product.

```
% Define ODE for dCA/dV
odefun = @(V, CA) rA(CA) / v0;
% Event function to stop if CA becomes unrealistic
options = odeset('Events', @(V, CA) eventStop(V, CA, CAf));
% Solve the ODE from CA0 to CAf
[V, CA] = ode45(odefun, [0, 5000], CA0, options);
% Check and interpolate for VPFR at CAf
if ~isempty(CA) && all(isreal(CA)) && all(isreal(V))
   VPFR = interp1(CA, V, CAf); % Interpolate for exact volume at CAf
else
   error('ODE solver produced invalid results. Check inputs or rA definition.');
end
% Extent of reactions
xi1 = trapz(V, k1 * CA.^0.5); % Extent of reaction #1 in mole/min
xi2 = trapz(V, k2 * CA);      % Extent of reaction #2 in mole/min
xi3 = trapz(V, k3 * CA.^2);   % Extent of reaction #3 in mole/min
CU1=xi1/v0; % CU1 concentration, mol/L
CD=xi2/v0; % CD concentration, mol/L
CU2=xi3/v0; % CU2 concentration, mol/L
CTotal=CU1+CD+CU2+CAf;
% Yield and selectivity
FA0 = CA0 * v0; % Inlet molar flow rate of A
Yield = xi2 / FA0;
Soverall = xi2 / (xi1 + xi3);
SD_U1U2 = (k2 * CAf) / (k1 * CAf.^0.5 + k3 * CAf.^2);
% Display results
disp(['Volume of PFR (VPFR): ', num2str(VPFR), ' L']);
disp(['Extent of reaction #1 (xi1): ', num2str(xi1), ' mol/min']);
disp(['Extent of reaction #2 (xi2): ', num2str(xi2), ' mol/min']);
disp(['Extent of reaction #3 (xi3): ', num2str(xi3), ' mol/min']);
disp(['U1 concentration, CU1: ', num2str(CU1), ' mole/L']);
disp(['D concentration, CD: ', num2str(CD), ' mole/L']);
disp(['U2 concentration, CU2: ', num2str(CU2), ' mole/L']);
disp(['Total concentration, CTotal: ', num2str(CTotal), ' mole/L']);
disp(['Initial A concentration, CA0: ', num2str(CA0), ' mole/L']);
disp(['The Yield of Desired Product: ', num2str(Yield)]);
disp(['The Overall Selectivity of Desired Product: ', num2str(Soverall)]);
disp(['The Instantaneous Selectivity of Desired Product: ', num2str(SD_U1U2)]);
%% Event function to stop ODE if CA < CAf
function [value, isterminal, direction] = eventStop(~, CA, CAf)
   value = CA - CAf; % Stop when CA reaches CAf
   isterminal = 1; % Stop integration
   direction = -1; % Only stop when decreasing
end
```

**Figure 4.34** *(Continued)*

The volume of PFR, VPFR, is expressed in L and is given by:

$$V_{PFR} = 100\,\frac{\text{L}}{\text{min}}\int_{C_{A0}}^{0.1}\frac{dC_A}{r_A}$$

The extent of reaction #1, $\xi_1$, expressed in moles, is given by:

$$\dot{\xi}_1 = \int_0^{VPFR} k_1 C_A^{0.5} dV$$

The extent of reaction #2, $\dot{\xi}_2$, expressed in moles, is given by:

$$\dot{\xi}_2 = \int_0^{VPFR} k_2 C_A dV$$

The extent of reaction #3, $\dot{\xi}_3$, expressed in moles, is given by:

$$\dot{\xi}_3 = \int_0^{VPFR} k_3 C_A^2 dV$$

The yield of the desired product is given by:

$$Yield = \frac{(F_D)_{Actual\ Condition}}{\left(F_D|_{f_2=1}\right)_{Ideal\ Condition}} = \frac{F_D = F_D^o + (1)\,\xi_2}{F_D|_{f_2=1}} = \frac{0 + \dot{\xi}_2}{F_D|_{f_2=1}} = \frac{\dot{\xi}_2}{F_A^o}$$

The overall selectivity ($\overline{S}$), based on Eq. (4.5b), is:

$$\overline{S} = \frac{(F_D)_{Actual\ Condition}}{All\ moles\ of\ undesired\ products} = \frac{\dot{\xi}_2}{(Molar\ Flowrate\ of\ U_1) + (Molar\ Flowrate\ of\ U_2)}$$

$$\overline{S} = \frac{\dot{\xi}_2}{(F_{U1}) + (F_{U2})} = \frac{\dot{\xi}_2}{\dot{\xi}_1 + \dot{\xi}_3}$$

The instantaneous selectivity is given by:

$$S_{D/(U_1+U_2)} = \frac{r_D}{r_{U_1} + r_{U_2}} = \frac{k_2 C_{Af}}{k_1 C_{Af}^{0.5} + k_3 C_{Af}^2}$$

### 4.5.3 MATLAB Code for Multi-Reaction PFR Yield and Selectivity

Figure 4.34 shows MATLAB code (**PFR_Yield_S_Ex4_9.m**) for calculating the PFR volume requirement and the yield and selectivity of the desired product. Numerical integration is used for both:

1) $V_{PFR}$ as $f(C_A)$
2) Extent rates of reaction ($\dot{\xi}_1, \dot{\xi}_2, \dot{\xi}_3$) as $f(C_A, V)$.

The temperature, *T*, is implicitly embedded within the rate constants: $k_1$, $k_2$, and $k_3$. Results are copied from the Command Window:

```
>> PFR_Yield_S_Ex4_9
Enter Exit or Final CA in mol/L: 0.1
Enter Initial Pressure in atm: 10.774
Enter Initial Temperature in K: 300
Enter volume flowrate in L/min: 100
Enter initial mole-fraction of A: 1
Enter Reactor Temperature in K: 300
Volume of PFR (VPFR): 404.9398 L
Extent of reaction #1 (xi1): 0.74873 mol/min
Extent of reaction #2 (xi2): 27.1516 mol/min
Extent of reaction #3 (xi3): 5.9763 mol/min
```

U1 concentration, CU1: 0.0074873 mole/L
D concentration, CD: 0.27152 mole/L
U2 concentration, CU2: 0.059763 mole/L
Total concentration, CTotal: 0.43877 mole/L
Initial A concentration, CA0: 0.43765 mole/L
The Yield of Desired Product: 0.6204
The Overall Selectivity of Desired Product: 4.0374
The Instantaneous Selectivity of Desired Product: 7.9683

### 4.5.4 Overall and Per Pass Reactor Conversion: Recycle Loop

Recycle reactors are commonly used in cases where the reaction demonstrates autocatalytic behavior or when maintaining nearly isothermal conditions and enhancing specific selectivity are essential. They are also extensively utilized in biochemical processes. In these systems, the recycled stream is withdrawn at Point **2** and combined with the fresh feed at Point **1**. The recycle ratio $R$ is defined as the moles recycled per mole of product removed at Point **2**, as illustrated in Figure 4.35.

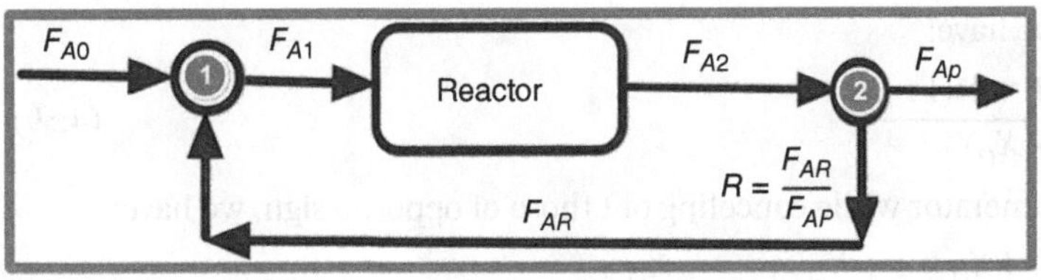


**Figure 4.35** A reactor equipped with a recycle stream.

In a chemical reactor, the overall conversion quantifies the extent to which reactants are transformed into products throughout the reaction process. It is defined as the ratio of the negative of the change in moles of the limiting reactant to its initial moles. For the limiting reactant, the overall conversion ($X_O$) is calculated using the following equation:

$$X_O = \frac{-\textit{Change in number of moles of the reactant across the process}}{\textit{Initial moles of reactant in the fresh feed}}$$

$$X_O = \frac{-(F_{AP} - F_{A0})}{F_{A0}} = \frac{F_{A0} - F_{AP}}{F_{A0}} = 1 - \frac{F_{AP}}{F_{A0}} \tag{4.52}$$

Introducing a stream split at Point **2** and recycling the unreacted portion back to the reactor inlet is a process known as recycle. This approach is employed to enhance the overall yield and conversion of the reactants.

On the other hand, considering the limiting reactant, the conversion per pass ($X_P$):

$$X_P = \frac{-\textit{Change in number of moles of the reactant across the reactor}}{\textit{Initial moles of reactant in the combined feed}}$$

$$X_P = \frac{-(F_{A2} - F_{A1})}{F_{A1}} = \frac{F_{A1} - F_{A2}}{F_{A1}} = 1 - \frac{F_{A2}}{F_{A1}} \tag{4.53}$$

$R$ is defined as:

$$R = \frac{F_{AR}}{F_{AP}} \tag{4.54}$$

$$\text{At Point 1, we have: } F_{A1} = F_{Ao} + F_{AR} = F_{Ao} + RF_{AP} \tag{4.55}$$

$$\text{At Point 2: we have: } F_{A2} = F_{AP} + F_{AR} = F_{AP} + RF_{AP} = (R+1)F_{AP} \tag{4.56}$$

Let us derive the relationship between the overall conversion ($X_O$), conversion per pass ($X_P$), and the recycle ratio, $R$.

Let us start with $X_P$ (Eq. 4.53):

Replace $F_{A1}$ by its equivalence using Eq. (4.55) and $F_{A2}$ by its equivalence using Eq. (4.56), we have:

$$X_P = 1 - \frac{F_{A2}}{F_{A1}} = 1 - \frac{(R+1)F_{AP}}{F_{Ao} + RF_{AP}} \tag{4.57}$$

Divide both numerator and denominator of the last quotient by $F_{A0}$ and given that $(1 - X_O) = \frac{F_{AP}}{F_{A0}}$ we have:

$$X_P = 1 - \frac{(R+1)F_{AP}}{F_{A0} + RF_{AP}} = 1 - \frac{(R+1)\frac{F_{AP}}{F_{A0}}}{\frac{F_{A0}}{F_{A0}} + R\frac{F_{AP}}{F_{A0}}} = 1 - \frac{(R+1)(1 - X_O)}{1 + R(1 - X_O)} \tag{4.58}$$

Combining the two terms, we have:

$$X_P = \frac{1 + R(1 - X_O) - (R+1)(1 - X_O)}{1 + R(1 - X_O)} \tag{4.59}$$

Splitting each term in the numerator while canceling out those of opposite sign, we have:

$$X_P = \frac{1 + R - RX_O - R + RX_O - 1 + X_O}{1 + R(1 - X_O)} = \frac{X_O}{1 + R(1 - X_O)} \tag{4.60}$$

Equation (4.60) gives $X_P$ as $f(X_O, R)$. Let us also express $X_O$ as $f(X_P, R)$.

Simplifying Eq. (4.60), we have:

$$X_O = X_P + RX_P(1 - X_O) = X_P + RX_P - RX_PX_O \tag{4.61}$$

Rearrangement of terms, we have:

$$X_O + RX_PX_O = X_P + RX_P \tag{4.62}$$

$$X_O = \frac{X_P + RX_P}{1 + RX_P} = \left\{\frac{(1+R)}{1 + RX_P}\right\} X_P \tag{4.63}$$

Since $X_P$ is, in general, less than or equal to 1.0, then the bracketed term $\left\{\frac{(1+R)}{1+RX_P}\right\}$ will always be greater than or equal to 1.0, rendering $X_O \geq X_P$.

If $R = 0$, then $X_O = X_P$.

If $R \to \infty$, then $X_O \to \frac{RX_P}{RX_P} \approx 1$.

**Example 4.10** ***Reactor with a Given* R**
Consider the reaction carried out in a CSTR with a recycle ratio, $R$, as shown in Figure 4.35:

$$aA \rightarrow Products$$

a) Given $R = 6$, $X_P = 0.7$, calculate the overall conversion, $X_O$.
b) Conversely, if $X_O = 0.9$ for the same $R$, what will be $X_P$?

**Solution**

a) Using Eq. (4.63), we have:

$$X_O = \left\{\frac{(1+R)}{1+RX_P}\right\} X_P = \left\{\frac{(1+6)}{1+6*0.7}\right\} 0.7 = \left\{\frac{7}{1+4.2}\right\} 0.7 = 0.94$$

b) Using Eq. (4.60), we have:

$$X_P = \frac{X_O}{1+R(1-X_O)} = \frac{0.9}{1+6(1-0.9)} = 0.56$$

**Example 4.11** ***Downstream Separation Followed by Recycle***
Consider a reaction with stoichiometry $A \rightarrow 2B$ that takes place in a process, as shown in Figure 4.36. With a basis of unit time, the molar flow rates are indicated on each stream.

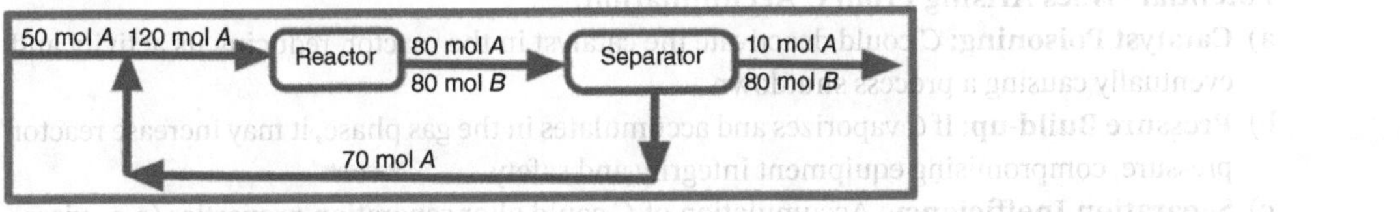


**Figure 4.36** A reactor followed by a separator to facilitate the recycle of $A$ to the inlet of the reactor.

In this setup, unlike the process shown in Figure 4.35, downstream separation is performed to isolate reactant $A$ from product $B$. The unreacted $A$ is primarily recycled to the reactor inlet. The concept behind this scheme is to recover the unreacted $A$, separate it from $B$, and reintroduce it into the reactor for further reaction. This approach aims to enhance the overall yield and conversion of the reactant.

Based on the indicated stream molar flow rates, calculate the overall conversion and the per-pass conversion considering the previous definitions given in Eqs. (4.63) and (4.60), respectively.

**Solution**

**Basis:** One unit of time for operation

Referring to Eq. (4.63), we have:

$$X_O = \frac{F_{A0} - F_{AP}}{F_{A0}} = 1 - \frac{F_{AP}}{F_{A0}} = 1 - \frac{10}{50} = 0.80$$

Referring to Eq. (4.60), we have:

$$X_P = \frac{F_{A1} - F_{A2}}{F_{A1}} = 1 - \frac{F_{A2}}{F_{A1}} = 1 - \frac{80}{120} = 0.33$$

Notice that $X_O = 0.80 > X_P = 0.33$.

**NOTE:** See the following questions and our reply:

1) The separation unit, along with the recycle pump and piping, contributes significantly to the overall costs. Could we justify omitting these components and selling the reactor effluent as it is? Alternatively, could we retain the separator but discontinue recycling the bottom stream?
   **Feasibility:** This approach could be viable if there is market demand for the unseparated mixture of $A$ and $B$. However, the product's value might be lower compared to selling pure $B$ due to the presence of the undesired component, $A$. Retaining the separator but discontinuing the recycling of $A$ could also be considered, particularly if an inert material, like $C$ accumulates in the bottom stream over time and creates challenges during recycling. However, discarding $A$ would result in a loss of reactant, reducing process efficiency and increasing costs.
2) If a small amount, such as 0.2% of inert material $C$, is present in the fresh feed and accumulates entirely in the recycled bottom effluent stream, what could ultimately lead to a process shutdown? What measures are necessary to maintain operational functionality?

   **Potential Issues Arising From $C$ Accumulation:**
   a) **Catalyst Poisoning:** $C$ could deactivate the catalyst in the reactor, reducing its activity and eventually causing a process shutdown.
   b) **Pressure Build-up:** If $C$ vaporizes and accumulates in the gas phase, it may increase reactor pressure, compromising equipment integrity and safety.
   c) **Separation Inefficiency:** Accumulation of $C$ could alter separation properties (e.g., viscosity or boiling point), reducing the efficiency of the separation process.

   **Remedies:** To mitigate these issues, the bottom stream must be periodically purged to remove a portion of $C$ and prevent it from reaching harmful levels. However, this also results in some loss of reactant $A$. Additional measures include selecting a catalyst that is less prone to poisoning by $C$ and exploring alternative separation techniques, such as adsorption, specifically designed to handle $C$ if purging proves insufficient.
3) Could the reactor be designed to directly produce ($A \rightarrow B$) 5 mol of $A$ and 25 mol of $B$ from an initial 30 mol of $A$ in a single pass, thereby eliminating the need for separation and recycling?

   **Limitations Preventing Use of a Single Pass**
   a) **Equilibrium Limitations:** Achieving a conversion (5 mol of $A$ to 25 mol of $B$) in a single pass may be constrained by thermodynamics. The reaction might reach an equilibrium state before the desired conversion is achieved, even with excess $A$ present.
   b) **Kinetic Limitations:** The reaction rate might be insufficient to reach the desired conversion within a single pass under practical reactor sizes and operating conditions. Extending the residence time would necessitate a larger reactor, while increasing temperature to enhance the reaction rate might not be viable due to cost and safety concerns.
   c) **Economic Limitations:** If the cost of $A$ is significantly higher than that of $B$, operating with multiple passes and recycling $A$ may be more economical, even with the added expense of separation and recycling equipment. This strategy allows for more efficient utilization of $A$ as a reactant.

### 4.5.5 The Yield of Desired Product in Multiple Reactions with Recycle

The gas-phase decomposition of species $A$ occurs via parallel and consecutive reactions:

$$A \xrightarrow{k_1} D \text{ (Desired)} \qquad A \xrightarrow{k_2} U \text{ (Undesired)} \qquad D \xrightarrow{k_3} W \text{ (Waste)}$$

**Feed:** Pure $A$, $C_{A0} = 2.0$ mol/L, isothermal $T = 550$ K, constant volumetric flow rate $\upsilon = 1.0$ L/s.
Kinetics are first order in the relevant species, with Arrhenius parameters:

$$k_1 = 3 \times 10^7 exp\left(-\frac{80{,}000}{8.314 * T}\right) \qquad k_2 = 1 \times 10^9 exp\left(-\frac{95{,}000}{8.314 * T}\right)$$

$$k_3 = 5 \times 10^6 exp\left(-\frac{75{,}000}{8.314 * T}\right)$$

Determine the overall yield when the reactor is operated as a function of the recycle loop ratio, $R_{loop}$ (ratio of the recycle stream to the fresh feed), assuming the same single-pass yield, with or without a purge stream. Assume a purge fraction equal to 5%. Assume that:

X_single = 0.35; % Per-pass conversion
S_single = 0.85; % Selectivity of desired product (Y_single / X_single)

**Solution**
We will use yield relative to reactant fed:

$$Y_D = \frac{Moles\ of\ D\ formed}{Moles\ of\ A\ fed} \tag{A}$$

With a recycle loop but with no purge, we have:

$$Y_D^{Overall} = \frac{Y_D^{Single}}{1 - \left(\frac{R_{Loop}}{1+R_{Loop}}\right)(1 - X_{Single})} \tag{B}$$

where

$\boldsymbol{X_{Single}}$: Single-pass conversion of $A$, based on reactor feed
$\boldsymbol{Y_D^{Single}}$: Single-pass yield of $D$, based on reactor feed
$\boldsymbol{\frac{R_{Loop}}{1+R_{Loop}}}$: Fraction of reactor feed coming from recycle

With a purge stream, the overall yield becomes:

$$Y_D^{Overall} = \frac{Y_D^{Single}}{1 - \left(\frac{R_{Loop}(1-Purge)}{1+R_{Loop}}\right)(1 - X_{Single})} \tag{C}$$

where

**Purge:** A fraction to be purged off the main loop. A purge fraction is a must, so inerts/by-products do not accumulate within the process. Keep in mind, however, that the purge stream will limit the maximum overall yield even with large recycle ratios.
Figure 4.37 shows MATLAB code (**Purge_Recyc_Yield.m**) written to estimate the overall yield with and without a purge stream.

Figure 4.38 shows the generated plot of the overall yield versus the recycle ratio, $R_{Loop}$. The ideal curve represents the case where purge = 0.0.

```
clear; close all;
% ====== INPUT DATA ======
X_single = 0.35;   % Per-pass conversion
S_single = 0.85;   % Selectivity of desired product (Y_single / X_single)
Y_single = S_single * X_single;
% Purge fraction (0 = no purge, 0.05 = 5% purge stream)
purge = 0.05;
% ====== RECYCLE RATIO RANGE ======
R_loop = linspace(0, 50, 500); % R_loop = m_recycle / m_fresh
% ====== IDEAL RECYCLE (Perfect Selectivity) ======
% For perfect selectivity: Y_single = X_single
Y_single_ideal = X_single;
r_ideal = R_loop ./ (1 + R_loop);
Y_overall_ideal = Y_single_ideal ./ (1 - r_ideal .* (1 - X_single));
% ====== REALISTIC RECYCLE (Finite Selectivity + Purge) ======
r_real = (R_loop .* (1 - purge)) ./ (1 + R_loop); % Adjusted for purge
Y_overall_real = Y_single ./ (1 - r_real .* (1 - X_single));
% ====== PLOT RESULTS ======
figure;
plot(R_loop, Y_overall_ideal, 'b-', 'LineWidth', 2); hold on;
plot(R_loop, Y_overall_real, 'r-', 'LineWidth', 2);
xlabel('Recycle Ratio, R_{loop}');
ylabel('Overall Yield of Desired Product');
legend('Ideal Recycle (Perfect Selectivity)', ...
    sprintf('Realistic Recycle (S=%.2f, Purge=%.0f%%)', S_single, purge*100), ...
    'Location', 'SouthEast');
grid on;
title('Effect of Recycle Ratio on Overall Yield');
ylim([0 1.05]);
ax = gca; ax.FontSize = 14;ax.FontWeight = 'bold';
% ====== TEXT ANNOTATION ======
text(30,0.98,'Ideal→Asymptote at 1.0','Color','b');
text(30,Y_overall_real(end),sprintf('Realistic→%.3f max', Y_overall_real(end)),'Color','r')
```

**Figure 4.37** MATLAB code (**Purge_Recyc_Yield.m**) to estimate the overall yield with and without a purge stream as a function of $R_{Loop}$.

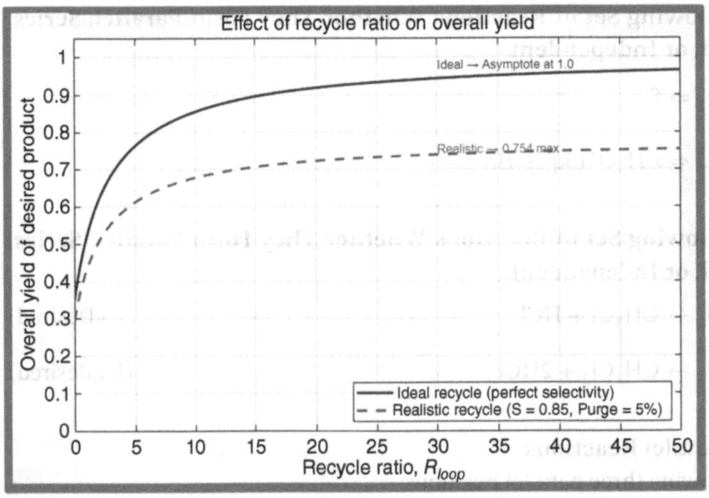


**Figure 4.38** The generated plot of the overall yield versus the recycle ratio, $R_{Loop}$.

## References


1 Al-Malah, K. (2022). *Aspen Plus: Chemical Engineering Applications*, 2e, 491–529. Wiley. https://www.wiley.com/en-ae/Aspen+Plus:+Chemical+Engineering+Applications,+2nd+Edition-p-9781119868699.

2 Felder, R.M. and Rousseau, R.W. (2005). Elementary Principles of Chemical Processes. In: 2e, 123. Wiley.


## End of Chapter Problems

**CHW 4.1 Classify the Following Set of Reactions Whether They Form Parallel, Series, Complex/Mixed, or Independent**

$$CaCO_3 \rightarrow CaO + CO_2 \qquad \text{(Desired)}$$

$$CaO + SiO_2 \rightarrow CaSiO_3 \qquad \text{(Undesired)}$$

**CHW 4.2 Classify the Following Set of Reactions Whether They Form Parallel, Series, Complex/Mixed, or Independent**

$$C_2H_2 + H_2 \rightarrow C_2H_4 \qquad \text{(Desired)}$$

$$C_2H_4 + H_2 \rightarrow C_2H_6 \qquad \text{(Undesired)}$$

**CHW 4.3 Classify the Following Set of Reactions Whether They Form Parallel, Series, Complex/Mixed, or Independent**

$$C_3H_4 + H_2 \rightarrow C_3H_6 \qquad \text{(Desired)}$$

$$C_3H_4 + 2H_2 \rightarrow C_3H_8 \qquad \text{(Undesired)}$$

**CHW 4.4 Classify the Following Set of Reactions Whether They Form Parallel, Series, Complex/Mixed, or Independent**

$$SO_2 + \frac{1}{2}O_2 \Rightarrow SO_3$$

$$H_2S + H_2O \Leftrightarrow H_3O^+(aq) + HS^-(aq)$$

**CHW 4.5 Classify the Following Set of Reactions Whether They Form Parallel, Series, Complex/Mixed, or Independent**

$$CH_4 + Cl_2 \rightarrow CH_3Cl + HCl \qquad \text{(Desired)}$$

$$CH_4 + 2Cl_2 \rightarrow CH_2Cl_2 + 2HCl \qquad \text{(Undesired)}$$

**CHW 4.6 Selectivity of Parallel Reactions**

Consider the following three parallel reactions:

$$A + B \xrightarrow{k1} D \qquad r_D = k_1 C_A C_B \qquad \text{(A)}$$

$$A + B \xrightarrow{k2} U_1 \qquad r_{U_1} = k_2 C_A^{1.5} C_B \qquad \text{(B)}$$

$$A + B \xrightarrow{k3} U_2 \qquad r_{U_2} = k_3 C_A^{0.5} C_B \qquad \text{(C)}$$

Express $S_{D/(U_1+U_2)}$ as $f(C_A)$ only. Carry out differentiation with respect to $C_A$ and find the optimum $C_A$ at which $S_{D/(U_1+U_2)}$ maximum. Make use of MATLAB code **(SD_U.m)** for generating a plot of $S_{D/(U_1+U_2)}$ versus $C_A$. Show both values of the optimum $C_A$ and maximum of $S_{D/(U_1+U_2)}$.

**CHW 4.7 The Optimum Time for a Series Reaction I**

Consider the reaction in series: $A \xrightarrow{k1} B \xrightarrow{k2} C$

with $-r_A = -\frac{dC_A}{dt} = k_1 C_A$ and $r_B = \frac{dC_B}{dt} = k_1 C_A - k_2 C_B$

If $k_1 = 0.30\,s^{-1}$ and $k_2 = 0.15\,s^{-1}$, what will be the optimum time at which $C_B$ is maximum?

**A** 13.86 s
**B** 6.93 s
**C** 4.62 s
**D** 1.93 s
**E** 0
**F** ∞

**CHW 4.8 The Optimum Time for a Series Reaction II**

Consider the reaction in series: $A \xrightarrow{k1} B \xrightarrow{k2} C$

with $-r_A = -\frac{dC_A}{dt} = k_1 C_A$ and $r_B = \frac{dC_B}{dt} = k_1 C_A - k_2 C_B$

If $k_1 \rightarrow \infty\ s^{-1}$ and $k_2 \rightarrow 0\,s^{-1}$, what will be the optimum time at which $C_B$ is maximum?

**A** $\rightarrow 0$ s
**B** 1 s
**C** Undefined
**D** $\rightarrow \infty$ s

**CHW 4.9 The Optimum Time for a Series Reaction III**

Consider the following two reactions in series:

$$A \xrightarrow{k} B \xrightarrow{k} C$$

Species $A$: $-r_A = -\frac{dC_A}{dt} = kC_A$ Species $B$: $r_B = \frac{dC_B}{dt} = kC_A - kC_B$

If $k = 0.40\ \text{s}^{-1}$, what will be the optimum time at which $C_B$ is maximum?

**A** 0.4 s
**B** 2.5 s
**C** 2.5 s$^{-1}$
**D** 0 s
**E** ∞ s

**CHW 4.10 CHW 4.10 The Optimum Time for a Series Reaction IV**

Consider the following two reactions in series:

$$A \xrightarrow{k} B \xrightarrow{k} C$$

Species $A$: $-r_A = -\frac{dC_A}{dt} = kC_A$ Species $B$: $r_B = \frac{dC_B}{dt} = kC_A - kC_B$

At what time does $C_A$ exactly equal $C_B$?

**A** 0
**B** $t_{opt}$
**C** $k$
**D** ∞

**CHW 4.11 Selectivity of Parallel Reactions**

Consider the following two parallel reactions:

$$A + B \xrightarrow{k1} D \qquad r_D = k_1 C_A^{1.6} C_B^{0.5}$$

$$A + B \xrightarrow{k2} U \qquad r_U = k_1 C_A^{0.6} C_B^{1.5}$$

Suggest a configuration of the feed concentration of both $C_A$ and $C_B$. for BR, PFR, and CSTR.

**CHW 4.12 Selectivity of Parallel Reactions**

Consider the following two parallel reactions:

$$A + B \xrightarrow{k1} D \qquad r_D = k_1 C_A^{1.5} C_B^{1.9}$$

$$A + B \xrightarrow{k2} U \qquad r_U = k_1 C_A^{0.5} C_B^{0.7}$$

Suggest a configuration of the feed concentration of both $C_A$ and $C_B$. For BR, PFR, and CSTR.

**CHW 4.13 Effect of Temperature on Selectivity**

Consider the following two parallel reactions:

$$A + B \xrightarrow{k1} D \qquad r_D = k_1 C_A^{1.6} C_B^{0.5} \qquad k_1 = k_1^* \times e^{-\frac{E1}{RT}} = k_1^* \times e^{-\frac{70{,}000}{8.314*T}}$$

$$A + B \xrightarrow{k2} U \qquad r_U = k_1 C_A^{0.6} C_B^{1.5} \qquad k_2 = k_2^* \times e^{-\frac{E2}{RT}} = k_1^* \times e^{-\frac{80{,}000}{8.314*T}}$$

The configuration of the feed concentration of both $C_A$ and $C_B$ for BR, PFR, and CSTR has been required in CHW 4.11 problem. Suggest what the best reactor temperature is.

**CHW 4.14 Optimum Reactant Concentration**

Reactant $A$ decomposes by three simultaneous liquid-phase reactions to form three products, one that is desired, $D$, and two that are undesired, $U_1$ and $U_2$.

$$A \xrightarrow{k_1} D \quad -r_{1A} = r_D = k_1 C_A^{1.1} = 0.05 \frac{\text{L}^{0.1}}{\text{mol}^{0.1} \cdot \text{s}} C_A^{1.1} \text{ (1.1th – order)} \tag{A}$$

$$A \xrightarrow{k_2} U1 \quad -r_{12} = r_{U1} = k_2 = 0.004 \frac{\text{mol}}{\text{L} \cdot \text{s}} \text{ (Zero – order)} \tag{B}$$

$$A \xrightarrow{k_3} U2 \quad -r_{13} = r_{U2} = k_3 C_A^{2.2} = 0.01 \frac{\text{L}^{1.2}}{\text{mol}^{1.2} \cdot \text{s}} C_A^{2.2} \text{ (2.2th – order)} \tag{C}$$

The specific reaction rates are given at 400 K, and the activation energies for reactions (A)–(C) are $E_1 = 70{,}000$ J/mol, $E_2 = 50{,}000$ J/mol, and $E_3 = 90{,}000$ J/mol, respectively.

1) Under what conditions (e.g., reactor type and concentrations) should the reaction be conducted to maximize the selectivity of species $D$, given an initial concentration of species $A$ of 10.0 M and a volumetric flow rate of 12.0 L/s? Use both the analytic and MATLAB codes to find the optimum $C_A$ at which $S_{D/(U_1+U_2)}$ is maximum. Modify the Figure 4.11 MATLAB code (**SD_U_CA.m**), considering the above given reaction kinetics of the three reactions.
2) Find the residence time and volume requirement for the selected reactor. Calculate the overall conversion of the reactions, based on consumption of $A$.
3) Modify Figure 4.19 code (**VCSTR_VPFR_CA.m**) for carrying out two optimization cases for the desired product selectivity: One for CSTR and another for the PFR, but both with different operating conditions in terms of $T$ and $C_A$ range. Find both VCSTR and VPFR to achieve 98% conversion.

**CHW 4.15 Selectivity of Parallel Equilibrium Reactions**

Consider the following reactions taking place in a constant-volume BR:

$$A \underset{k_{bD}}{\overset{k_{fD}}{\rightleftharpoons}} D \tag{A}$$

$$-r_{1A} = k_{fD}\, C_A - k_{bD}\, C_D = 2.0 \text{ min}^{-1}\; C_A - 0.15 \text{ min}^{-1}\; C_D \tag{B}$$

$$A \underset{k_{bU}}{\overset{k_{fU}}{\rightleftharpoons}} U \tag{C}$$

$$-r_{2A} = k_{fU}\, C_A - k_{bU}\, C_U = 10.0 \text{ min}^{-1}\; C_A - 7 \text{ min}^{-1}\; C_U \tag{D}$$

$C_{A0} = 10$ mol/L $= 10$ M

1) Plot and analyze conversion and the concentrations of $A$, $D$, and $U$ as a function of time. When do we have to stop the reaction to maximize the concentration of $D$?
2) When does the maximum concentration of $U$ occur?
3) What are the equilibrium concentrations of $A$, $D$, and $U$?

**CHW 4.16 Parallel Irreversible Reactions with Desired and Undesired Products**

Reactant $A$ decomposes by three simultaneous gas-phase reactions to form three products, one that is desired, $D$, and two that are undesired, $U_1$ and $U_2$.

$$A \xrightarrow{k_1} U_1 \qquad -r_{1A} = r_{U1} = k_1 C_A^{0.6} = 0.05 \frac{\text{mol}^{0.4}}{\text{L}^{0.4} \cdot \text{min}} C_A^{0.6} (0.6\text{th} - \text{order}) \quad \text{(A)}$$

$$A \xrightarrow{k_2} D \qquad -r_{12} = r_D = k_2 C_A = 0.5 \frac{1}{\text{min}} (\text{First} - \text{order}) \quad \text{(B)}$$

$$A \xrightarrow{k_3} U_2 \qquad -r_{13} = r_{U2} = k_3 C_A^{1.9} = 0.4 \frac{\text{L}^{0.9}}{\text{mol}^{0.9} \cdot \text{min}} C_A^{1.9} \ (1.9\text{th} - \text{order}) \quad \text{(C)}$$

$D$ is the desired product, while $U_1$ and $U_2$ are undesired pollutants that are costly to eliminate. The specific reaction rates are determined at 27 °C. The reaction system will be operated under conditions of 27 °C and 5 atm. Pure $A$ is introduced into the system at a volumetric flow rate of 100 L/min.

1) Analytically, find $C_A^{optim}$ at which $S_{D/(U_1+U_2)}$ is maximum.
2) Sketch the selectivity: $S_{D/(U_1+U_2)}$ as a function of the concentration of $C_A$.
3) Consider a series of reactors. What should be the volume of the first reactor?
4) What are the effluent concentrations of $A$, $D$, $U_1$, and $U_2$ from the first reactor?
5) What is the conversion of $A$ in the first reactor?
6) Suppose that a single CSTR with a space time of 20 min is used; what will be its exit concentration of $A$ at $T = 300$ K?
7) Given that $E_1 = 55{,}000$ J/mol, $E_2 = 80{,}000$ J/mol, and $E_3 = 90{,}000$ J/mol. What temperature is recommended for a single CSTR with a space time of 20 min, as in part 6?
8) If 76% conversion of $A$ is desired, what reaction scheme and reactor sizes shall we use to maximize $S_{D/(U_1+U_2)}$?

**CHW 4.17 Yield and Selectivity of Parallel Reactions**

Consider the following set of three reactions:
Ethylene (desired) production via catalytic dehydrogenation of ethane

$$C_2H_6 \rightarrow C_2H_4 + H_2 \quad \text{(A)}$$

Complete combustion of ethane (undesired due to loss of feedstock as $CO_2$):

$$C_2H_6 + 3.5O_2 \rightarrow 2CO_2 + 3H_2O \quad \text{(B)}$$

Cracking of ethane to form carbon and hydrogen (undesired due to solid carbon formation):

$$C_2H_6 \rightarrow 2C + 3H_2 \quad \text{(C)}$$

The desired product, ethylene ($C_2H_4$), is valuable as a precursor in polymer production (e.g., polyethylene). The undesired products include $CO_2$ and $H_2O$ in complete combustion, which waste the hydrocarbon feedstock, and solid carbon, which causes operational issues like catalyst deactivation.

Consider a batch reactor. Assume that the extent of reaction for the first is $\xi_1$; the second $\xi_2$; and the third $\xi_3$.

1) Express both the yield and selectivity of the desired product (i.e., ethylene) in terms of $\xi_1$, $\xi_2$, and $\xi_3$.

2) Calculate both the yield and selectivity of the desired product (i.e., ethylene) given that the initial amount for ethane is 100 mol; $\xi_1$ is 86 mol; $\xi_2$ is 10 mol; and $\xi_3$ is 6 mol.

**CHW 4.18 CSTR Yield and Selectivity Calculations**

Reactant $A$ decomposes by three simultaneous gas-phase reactions to form three products, one that is desired, $D$, and two that are undesired, $U_1$ and $U_2$. The specific reaction rates are determined at 27 °C. The activation energy terms are: $E_1 = 55{,}000$ J/mol, $E_2 = 80{,}000$ J/mol, and $E_3 = 90{,}000$ J/mol.

$$A \overset{k_1}{\Longrightarrow} U_1 \qquad -r_{1A} = r_{U1} = k_1 C_A^{0.6} = 0.05\frac{\text{mol}^{0.4}}{\text{L}^{0.4}\cdot\text{min}} C_A^{0.6} (0.6\text{th} - \text{order}) \quad \text{(A)}$$

$$A \overset{k_2}{\Longrightarrow} D \qquad -r_{12} = r_D = k_2 C_A = 0.5\frac{1}{\text{min}} \ (\text{First} - \text{order}) \quad \text{(B)}$$

$$A \overset{k_3}{\Longrightarrow} U_2 \qquad -r_{13} = r_{U2} = k_3 C_A^{1.9} = 0.4\frac{\text{L}^{0.9}}{\text{mol}^{0.9}\cdot\text{min}} C_A^{1.9} \ (1.9\text{th} - \text{order}) \quad \text{(C)}$$

The reaction is carried out in a CSTR with constant $T = 300$ K and $P = 11.0$ atm equal to those of the feed. The exit concentration of $A$ is 0.03 mol/L, and the volume flow rate, $\upsilon_0 = 300$ L/min. Pure $A$ is introduced into the system at a volumetric flow rate of 100 L/min.

**A** Find both $C_{A0}$ and $F_{A0}$.
**B** Find species concentrations: $D$, $U_1$, $U_2$, and $A$
**C** Find species molar flow rates: $F_D$, $F_{U1}$, $F_{U2}$, and $F_A$
**D** The yield, as defined by Eq. (4.51b)
**E** Find the overall selectivity, as defined by Eq. (4.5b)
**F** Find instantaneous selectivity, defined by: $S_{D/(U_1+U_2)} = \frac{r_D}{r_{U_1}+r_{U_2}}$
**G** Use MATLAB to carry out the previous step from $a$ to $f$.

**CHW 4.19 PFR Yield and Selectivity Calculations**

Repeat CHW 4.18 for a PFR with an exit concentration of $A$ equal to: $C_{Af} = 0.03$ mol/L. Use MATLAB code to find the relevant yield and selectivity parameters.

**CHW 4.20 Reactor with a Given $R$**

Consider the reaction carried out in a CSTR with a recycle ratio, $R$, as shown in Figure 4.34:

$$aA \rightarrow Products$$

**A** Given $R = 8$ and $X_P = 0.6$, calculate the overall conversion, $X_O$.
**B** Conversely, if $X_O = 0.95$ for the same $R$, what will be $X_P$?

**CHW 4.21 Reactor with Separation and Recycle**

Consider a reaction with stoichiometry $A \rightarrow 3B$ that takes place in a process, as shown in Figure CHW4.21. With a basis of a unit time, the molar flow rates are indicated on each stream.

Based on the indicated stream molar flow rates, calculate the overall conversion and the per-pass conversion considering the previous definitions given in Eqs. (4.63) and (4.60), respectively.

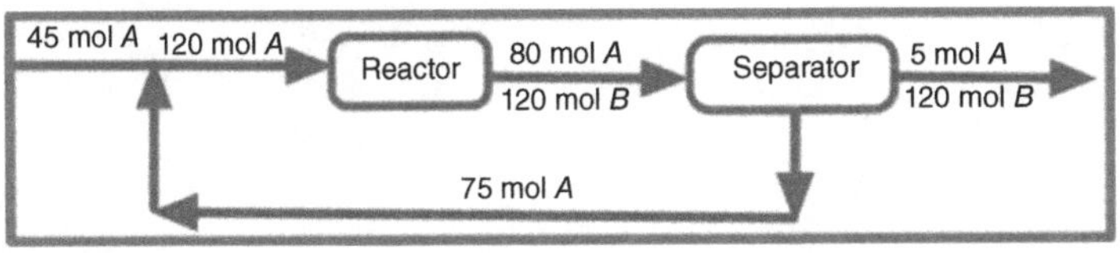


**CHW 4.21** A reactor followed by a separator to facilitate the recycling of *A* to the inlet of the reactor.

**CHW 4.22 Selectivity of Multiple Reactions**

A single reactant *A* can form the desired product *D* or the undesired by-product *U* in parallel, and *D* can further degrade to *W* in series:

$$A \xrightarrow{k_1} D \qquad A \xrightarrow{k_2} U \qquad D \xrightarrow{k_3} W$$

All steps are first order in their own species. Feed is pure *A* with $C_{A0} = 1.0\,\text{mol/L}$.

$$k_1 = 5\times 10^8 exp\left(-\frac{98{,}000}{8.314 * T}\right) k_2 = 6.8\times 10^5 exp\left(-\frac{75{,}000}{8.314 * T}\right)$$

$$k_3 = 5\times 10^2 exp\left(-\frac{40{,}000}{8.314 * T}\right)$$

**A** Compare PFR and CSTR performance at two temperatures, $T_1 = 500\,\text{K}$ and $T_2 = 600\,\text{K}$ as far as $k_1$, $k_2$, and $k_3$ are concerned.

**B** Evaluate the exit selectivity $S_{D/U} = C_D/C_U$ for:

a) A fixed residence $\tau = [2\ 30]$ s and fixed temperature $T = [600\ 500]$ K

b) The PFR residence time ($\tau^*$) that maximizes $C_D$.

Comment on temperature effects and reactor choice.

**CHW 4.23 The Yield of Desired Product in Multiple Reactions with Recycle**

The gas-phase decomposition of species *A* occurs via parallel and consecutive reactions:

$$A \xrightarrow{k_1} D \text{ (Desired)} \qquad A \xrightarrow{k_2} U \text{ (Undesired)} \qquad D \xrightarrow{k_3} W \text{ (Waste)}$$

**Feed:** Pure *A*, $C_{A0} = 2.0\,\text{mol/L}$, isothermal $T = 550\,\text{K}$, constant volumetric flow rate $\upsilon = 1.0\,\text{L/s}$.

Kinetics are first order in the relevant species, with Arrhenius parameters:

$$k_1 = 4\times 10^7 exp\left(-\frac{85{,}000}{8.314 * T}\right) \qquad k_2 = 2\times 10^9 exp\left(-\frac{90{,}000}{8.314 * T}\right)$$

$$k_3 = 6\times 10^6 exp\left(-\frac{70{,}000}{8.314 * T}\right)$$

Determine the overall yield when the reactor is operated as a function of the recycle loop ratio, $R_{loop}$ (ratio of the recycle stream to fresh feed), assuming the same single-pass yield, with or without a purge stream. Assume a purge fraction equal to 6%. Assume that:

X_single = 0.40; % Per-pass conversion

S_single = 0.90; % Selectivity to desired product (Y_single / X_single)

# 5

# Unsteady-State, Non-Isothermal Reactors

When dealing with non-isothermal or unsteady-state chemical reactors, the energy balance becomes crucial in understanding and predicting the reactor's behavior. Unlike isothermal reactors, where temperature remains constant, these reactors experience temperature variations that significantly influence reaction rates and overall performance.

The energy balance is based on the principle of conservation of energy, which states that energy cannot be created or destroyed, only transformed. In the context of a chemical reactor, this translates to accounting for all energy entering, leaving, and accumulating within the system.

**Notes on Applications of Energy Balance:**

a) **Reactor Design and Control:** The energy balance is essential for designing and operating non-isothermal reactors. It helps predict temperature profiles, prevent runaway reactions, and optimize reactor performance.
b) **Stability Analysis:** Analyzing the energy balance can reveal potential instability issues, such as multiple steady states or oscillations in temperature and conversion.
c) **Numerical Solutions:** Due to the complexity of the energy balance equation, especially in unsteady-state conditions, numerical methods are often employed to solve it.
d) **Coupling with Mole Balance:** The energy balance is often coupled with mole balance equations to fully describe the reactor behavior, as reaction rates depend on both temperature and concentration.

Let us start simple and grow up in the level of the system complexity being handled or analyzed in terms of writing a number of equations, both algebraic and differential, to describe an equal number of variables. In general, any system or material physical property can be a variable, parameter, or constant.

**NOTE #1**

What we do here in this chapter, through running examples, is implement the process modeling concept. This means writing enough number of linearly independent algebraic and differential equations to describe the variables of the process/system. In addition, we have to account for the physical constraints imposed on the process/system, including the process equipment and fluid and flow conditions. By doing so, we have what I call a fair and square (or just fit) description; that is, the number of equations is equal to the number of variables. Proceeding with the solution of such written equations is called simulation, which describes how such variables change with time, space, or both.

*(Continued)*

**(Continued)**

Finally, describing a physical or chemical process/system can involve varying levels of complexity. Greater depth in the description requires more equations, leading to a trade-off in simulation costs (\$1-, \$10-, \$100-, or \$1,000-solution), a concept known as the model's degree of rigor. However, increasing the number of equations also raises the likelihood of numerical stability or divergence issues during simulation or solution.

**NOTE #2**

**Dynamic Nature of a BR Chemical Process**

Batch reactors operate in a closed system where all reactants are added at the start, and no material is added or removed during the reaction. Conditions inside the reactor are inherently dynamic: concentrations, temperature, and reaction rates do continuously change over time as reactants convert to products, and the system never reaches a steady state (unlike flow reactors). For reversible reactions, the process may approach equilibrium conversion (the theoretical maximum under given conditions), at which point engineers typically halt the reaction or declare the batch complete. This dynamic behavior mirrors everyday processes; for example, imagine cooking a meal: ingredients are added at the beginning, and the food bowl is heated until it reaches the desired flavor, taste, texture, or color. During cooking, properties evolve dynamically until the process is stopped (e.g., turning off the heating source to prevent overcooking). Similarly, in a batch reactor, time-dependent changes persist until the reaction is terminated, either at equilibrium or a predetermined batch time.

A steady state implies time-invariant conditions, which is impossible in BR because the reaction progresses until completion or the end of the batch. Flow reactors, even if they have spatial variations, maintain temporal consistency once steady. Due to mixing, a CSTR rarely encounters spatial variation of such fluid properties. After a startup phase, a steady state can be achieved, where variables (e.g., concentration, temperature) stabilize at any given point in the reactor. Notice that a load (i.e., change in process feed properties or heat exchange properties) disturbance and a setpoint (change in controller's setpoint) disturbance are parameters that affect the steadiness of the CSTR and usually lead the CSTR to move from one to another steady-state condition. See the upcoming "Process Control and Dynamics" section.

On the other hand, for PFR, PBR, and FBR, concentration and temperature changes are driven by chemical reactions, mass transfer, and heat exchange. They can be steady or dynamic, based on reactor design and operational settings. Even under steady-state operation, we will often see spatial gradients in concentration and temperature due to the reaction, mass transfer, and heat effects within the reactor.

In this chapter, we will focus on the dynamics of a BR and on both the dynamics and the statics (i.e., steadiness) of the flow-type reactors.

## 5.1 Adiabatic Liquid-Phase Batch Reactor (BR) for Glucose Conversion

The isomerization of glucose to fructose is a reversible, slightly endothermic reaction widely utilized in the production of high-fructose corn syrup (HFCS). In aqueous solution, D-glucose is converted to D-fructose via an intramolecular rearrangement catalyzed by enzymes such as glucose isomerase or, less commonly, by alkaline catalysts in chemical processes. The stoichiometric reaction can be represented as: Glucose ⇌ Fructose.

Enzymatic isomerization is preferred industrially due to its high selectivity, mild operating conditions (typically 55–65 °C and near-neutral pH), and minimal by-product formation. The equilibrium conversion under standard conditions is typically around 42–50% fructose, after which separation or recycling strategies are used to achieve higher purity.

From a thermodynamic and kinetic perspective, the reaction exhibits a small positive enthalpy change (about +1.7 kJ/mol), indicating that higher temperatures slightly favor fructose formation. However, temperature increases are limited by enzyme deactivation in biocatalytic systems, necessitating an optimization between reaction rate and catalyst stability. Process designs often employ packed-bed reactors containing immobilized glucose isomerase to enable continuous operation, high throughput, and catalyst reuse. Careful control of substrate concentration, pH, and temperature ensures consistent product quality while minimizing the formation of degradation products such as colored compounds or organic acids. For simplicity, first-order irreversible kinetics will be adopted in this chapter.

**Example 5.1** ***Liquid-phase First-order Isomerization Reaction***

Consider the aqueous liquid-phase first-order reaction, which is the isomerization of glucose into fructose:

$$\text{Glugose } (C_6H_{12}O_6)_{(aq)} \rightarrow \text{Fructose } (C_6H_{12}O_6)_{(aq)} \tag{A}$$

Which can be put in the short form:

$$A \rightarrow B \tag{B}$$

With a reaction rate

$$-r_A = k_1 C_A = k^* e^{-\frac{E_A}{R}\left(\frac{1}{T}-\frac{1}{298}\right)} C_A = 1.2 \times 10^{-5} \frac{1}{s} e^{-\frac{60000}{8.314}\left(\frac{1}{T}-\frac{1}{298}\right)} C_A \frac{\text{mol}}{\text{L}} \tag{C}$$

The heat of reaction is given as:

$$\Delta H_r^0 = +1.7 \frac{\text{kJ}}{\text{mol of } A} = +1{,}700 \frac{\text{J}}{\text{mol of } A} \tag{D}$$

The heat capacity for 50 mol% glucose solution at 350 K is:

$$\text{Glucose: } C_P = 147.0 \frac{\text{J}}{\text{mol} \cdot \text{K}} \tag{E}$$

and for 50 mol% fructose solution at $T = 350$ K is:

$$\text{Fructose: } C_P = 145.0 \frac{\text{J}}{\text{mol} \cdot \text{K}} \tag{F}$$

**NOTE #2**

In general, for solids and incompressible fluids, the heat capacity at constant $P$ and that at constant $V$ are assumed equal. $C_P \cong C_V$. For a compressible gas, the ideal gas relationship can be assumed. $C_P = C_V + R$

Write enough number of linearly independent equations, including total material balance, $A$-component molar balance, and energy balance, to describe the adiabatic batch reactor system. Notice that both $C_A$ (or $X_A$) and temperature ($T$) do change with batch time. The initial concentration of $A$, $C_{A0} = 100$ mol/L, initial temperature, $T_0 = 350$ K, and $V_{BR} = 1{,}500$ L.

**Solution**

**Total Material Balance**

$$In - Out + Generation - Consumption = Accumulation = \frac{d(m_{sys})}{dt}$$

In = 0; Out = 0; Generation = 0.0; and Consumption = 0.

Hence, $Accumulation = \frac{d(m_{sys})}{dt} = 0 \quad \rightarrow \quad m_{sys} = m_{sys1} = m_{sys2} = constant$

**NOTE #3**

We will not carry out a total material balance for either a batch reactor or a steady-state flow reactor. It is done here once for demonstrating that for a batch reactor, the total mass is fixed.

**Total Molar Balance**

The change in total number of moles as a result of the chemical reaction is zero simply because $\Delta n = \sum_{i=1}^{N} \nu_i = \nu_A + \nu_B = -1 + 1 = 0$

This implies that:

$$dn = 0 \quad \rightarrow \quad n = constant = n_A + n_B = constant = n_{A0}$$

**NOTE #4**

In general, $\Delta n \neq 0$ for a system with a chemical reaction which results in a change in the total number of moles.

***A*-component Material Balance**

$$In - Out + Generation - Consumption = Accumulation = \frac{d(n_A)}{dt}$$

$$0 - 0 + 0 - k_1 C_A V_{BR} = \frac{d(n_A)}{dt} = \frac{d(V_{BR}C_A)}{dt} = \frac{V_{BR}d(C_A)}{dt} + \frac{C_A d(V_{BR})}{dt} \tag{G}$$

$V_{BR}$ is constant; hence, $\frac{C_A d(V_{BR})}{dt} = 0$. Equation (G) reduces to:

$$-k_1 C_A V_{BR} = \frac{V_{BR}d(C_A)}{dt} \quad \rightarrow \quad -k_1 C_A = r_A = \frac{dC_A}{dt} \tag{H}$$

**Energy Balance**

For simplicity, let us assume that the heat of reaction is constant (i.e., independent of $T$). Also, the heat exchange ($Q_{exch}$) with the surroundings is also zero (i.e., an adiabatic process), and no shaft work is imposed into or out of the system.

$$In - Out + Generation - Consumption = Accumulation$$

$$0 - 0 + 0 - k_1 C_A V_{BR} \left(\Delta H_r^0\right) = \frac{d\left(nC_V^{mix}T\right)}{dt} = nC_V^{mix}\frac{d(T)}{dt} + \frac{nTd\left(C_V^{mix}\right)}{dt} \tag{I}$$

For the given temperature range of sugar solutions, the heat capacity change with $T$ will be assumed to be negligible.

Using the chain rule differentiation, the time derivative of the heat capacity can be expressed as:

$$\frac{nTd\left(C_V^{mix}\right)}{dt} = \frac{nTd\left(C_V^{mix}\right)}{dT} \times \frac{dT}{dt} = \left(\frac{nTd\left(C_V^{mix}\right)}{dT} \approx 0\right) \times \frac{dT}{dt} = 0 \tag{J}$$

Later in this chapter, examples are shown where the heat capacity is allowed to change with $T$. Equation (I) reduces to:

$$-k_1 C_A V_{BR} \left(\Delta H_r^0\right) = \frac{d\left(nC_V^{mix}T\right)}{dt} = nC_V^{mix}\frac{d(T)}{dt} \tag{K}$$

Assuming an ideal liquid solution, the mixture's heat capacity, $C_V^{mix}$, is given by the following equation:

$$nC_V^{mix} = n_A C_V^A + n_B C_V^B \tag{L}$$

Dividing by $n$ and recalling the definition of mole fraction of $A$, $x_A$, and that of $B$, $x_B$, we have:

$$C_V^{mix} = \frac{n_A}{n}C_V^A + \frac{n_B}{n}C_V^B = x_A C_V^A + (1 - x_A)C_V^B \tag{M}$$

In general, $C_V^{mix}(T)$, $C_V^A(T)$, and $C_V^B(T)$ will have different functions of $T$.

We may also use the reaction conversion, $X_A$, with respect to $A$.

$$X_A = 1 - \frac{n_A}{n_{A0}} = 1 - \frac{n_A}{n_A + n_B} = 1 - x_A$$

$$\text{Or: } x_A = 1 - X_A \tag{N}$$

Equation (N) is valid for a chemical reaction described by Eq. (B).

In general, the following equation (see Appendix B | B6) can be used to relate conversion to mole fraction of $A$:

$$X_A = \frac{1}{x_{A0}}(x_{A0} - x_A) \tag{O}$$

$$C_V^{mix} = x_A C_V^A + (1 - x_A)C_V^B = x_A C_V^A + X_A C_V^B \tag{P}$$

If we have a constant heat capacity system in terms of T, but at the same time $C_V^A \neq C_V^B$, then the mixture's heat capacity, $C_V^{mix}$ can be explicitly expressed as a function of composition:

$$C_V^{mix} = x_A C_V^A + (1 - x_A)C_V^B = constant \neq f(T) \tag{Q}$$

Further simplification can be made if we have equal heat capacity values of $A$ and $B$. In other words, $C_V^A = C_V^B$. This means that $C_V^{mix}$ becomes constant but not a function of composition:

$$C_V^{mix} = x_A C_V^A + (1 - x_A)C_V^B = x_A C_V^A + (1 - x_A)C_V^A = C_V^A = C_V^B = constant \neq f(x_A) \tag{R}$$

Equation (R) can tolerate to have $C_V^{mix} = C_V^A = C_V^B = f(T)$

**NOTE #5**

In general, the pure component heat capacity is $f(T)$ only. On the other hand, $C_V^{mix} = f(T, x_A)$.

## 5.2 MATLAB Code for Adiabatic BR for Glucose Conversion: Constant $\Delta H_r$

MATLAB code is used to solve the above set of differential and algebraic equations.

The selected equations are: (E), (F), (H), (K), and (M).

We need to integrate both $C_A$ and $T$ from:

**Initial Condition:** $t_0 = 0$; $C_A = C_{A0} = 100$ mol/L, $T = T_0 = 350$ K
**To: Final Condition:** $t = t_f = 3{,}600$ s

The following algebraic equations are used in parallel with the above selected equations.

$$C_{Total} = C_{A0} = 100\frac{\text{mol}}{\text{L}}, \quad x_A = \frac{n_A}{n} = \frac{C_A}{C_{Total}}, \quad V_{BR} = 1{,}500\,L, \quad n = C_{Total} * V_{BR}$$

Figure 5.1 shows MATLAB code (**BR_T_CA.m**) for solving the set of differential/algebraic equations describing the adiabatic batch process for a liquid-phase first-order reaction.

```
%% MATLAB Code to Solve DAE for CA, T, and XA
clear variables;
% Constants
R = 8.314; % Universal gas constant, J/(mol*K)
C_V_A = 147.0; % Heat capacity of A, J/(mol*K)
C_V_B = 145.0; % Heat capacity of B, J/(mol*K)
C_Total = 100; % Total concentration, mol/L
DeltaH_r0 = 1700; % Reaction enthalpy change, J/(mol of A)
V_BR = input('Enter the BR Volume, (L): ');
%V_BR = 1500; % Volume of batch reactor, L
n = C_Total * V_BR; % Total moles in reactor, mol
% Initial conditions
t0 = 0; % Initial time, s
tf = input('Enter the final time (tf) in unit of time: '); % Final time
%tf = 3600; % Final time, s
CA0 = input('Enter the initial concentration CA0 (mol/L): ');
%CA0 = 100; % Initial concentration of A, mol/L
T0 = input('Enter initial temperature (T0) in K: '); % Initial temperature
%T0 = 350; % Initial temperature, K
% Define the system of ODEs
odefun = @(t, y) system_of_equations(t, y, R, C_V_A, C_V_B, C_Total, DeltaH_r0, V_BR, n);
% Initial state vector [CA, T]
y0 = [CA0, T0];
% Solve the ODE system
[t, y] = ode45(odefun, [t0 tf], y0);
% Extract results
CA = y(:, 1); % Concentration of A
T = y(:, 2); % Temperature
XA = 1 - CA / C_Total; % Conversion of A
% Plot results
figure;
subplot(3, 1, 1);
```

**Figure 5.1** MATLAB code (**BR_T_CA.m**) for solving the set of differential/algebraic equations describing the adiabatic batch process for a liquid-phase first-order reaction.

```
plot(t, CA, 'b-', 'LineWidth', 1.5);
grid on;
xlabel('Time (s)');
ylabel('C_A (mol/L)');
title('Concentration of A vs. Time');
ax = gca; % Get handle to the current axes
% Set new font size for all axes text (including labels)
ax.FontSize = 16;
ax.FontWeight = 'bold';
subplot(3, 1, 2);
plot(t, T, 'r-', 'LineWidth', 1.5);
grid on;
xlabel('Time (s)');
ylabel('T (K)');
title('Temperature vs. Time');
ax = gca; % Get handle to the current axes
% Set new font size for all axes text (including labels)
ax.FontSize = 16;
ax.FontWeight = 'bold';
subplot(3, 1, 3);
plot(t, XA, 'g-', 'LineWidth', 1.5);
grid on;
xlabel('Time (s)');
ylabel('X_A');
title('Conversion of A vs. Time');
ax = gca; % Get handle to the current axes
% Set new font size for all axes text (including labels)
ax.FontSize = 16;
ax.FontWeight = 'bold';
% Function defining the system of ODEs
function dydt = system_of_equations(~, y, R, C_V_A, C_V_B, C_Total, DeltaH_r0, V_BR, n)
   CA = y(1); % Concentration of A, mol/L
   T = y(2); % Temperature, K
   % Calculate k1
   k1 = 1.2e-5 * exp(-60000 / R * (1 / T - 1 / 298));
   % Mole fraction of A
   x_A = CA / C_Total;
   % Mixture heat capacity
   C_V_mix = x_A * C_V_A + (1 - x_A) * C_V_B;
   % System of equations
   dCA_dt = -k1 * CA;
   dT_dt = (-k1 * CA * V_BR * DeltaH_r0) / (n * C_V_mix);
   % Output derivatives
   dydt = [dCA_dt; dT_dt];
end
```

**Figure 5.1** (*Continued*)

The following input parameters are used, and the results are shown below:

```
≫ BR_T_CA
Enter the BR Volume, (L): 1500
Enter the final time (tf) in unit of time: 3600
Enter the initial concentration CA0 (mol/L): 100
Enter initial temperature (T0) in K: 350
```

Figure 5.2 shows the generated plot of $C_A$, $T$, and $X_A$ profiles with BR residence time. It is obvious that $C_A$ decreases and $X_A$ increases with $t_{BR}$. For $T$, since the reactor is adiabatically run and the heat of reaction is endothermic, then $T$ will drop with $t_{BR}$, as energy or heat is absorbed.

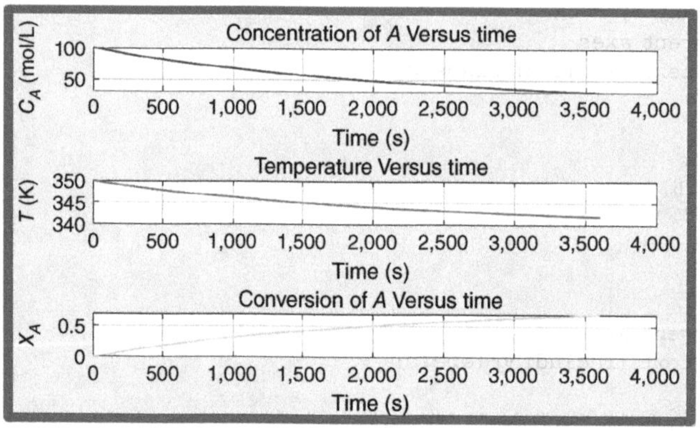


**Figure 5.2** The generated plot of $C_A$, $T$, and $X_A$ profiles with BR residence time.

## 5.3 The Heat of Reaction ($\Delta H_r(T)$) as $f(T)$

The heat of reaction at any $T$ can be related to that at a reference $T$, say, $T_0 = 25\,°\text{C}$, by:

$$\Delta H_r(T) = \Delta H_r(T_0) + \int_{T_0}^{T} \left( \sum_{i=1}^{N} \nu_i C_{P,i} \right) dT \tag{5.1}$$

In general, there are different functions that describe or fit the relationship between the heat capacity and temperature variation. For example, a polynomial can be used:

$$C_{P,i} = A_i + B_i T + C_i T^2 + D_i T^3 + E_i T^4 \tag{5.2}$$

Applying Eq. (5.1) to Example 5.1, we have:

$$\Delta H_r(T) = \Delta H_r(298) + \int_{T_0}^{T} \left( \sum_{i=1}^{N} \nu_i C_{P,i} \right) dT = 1{,}700 \frac{\text{J}}{\text{mol}} + \int_{298}^{T} \left( \sum_{i=1}^{N} \nu_i C_{P,i} \right) dT$$

$$\Delta H_r(T) = 1{,}700 \frac{\text{J}}{\text{mol}} + \int_{298}^{T} (-C_{P,A} + C_{P,B}) dT \tag{5.3}$$

Or, we can further simplify by substituting for numerical values of both $C_{V,A}$ and $C_{V,B}$ and carry out the addition of the two constant terms:

$$\Delta H_r(T) = 1{,}700 \frac{\text{J}}{\text{mol}} + \int_{298}^{T} (-147.0 + 145.0) dT = 1{,}700 \frac{\text{J}}{\text{mol}} + \{(-2.0)(T - 298)\} \frac{\text{J}}{\text{mol}}$$

$$\Delta H_r(T) = \{2{,}296 - 2.0 * T\} \frac{\text{J}}{\text{mol}} \tag{5.4}$$

## 5.4 MATLAB Code for Adiabatic BR for Glucose Conversion with $\Delta H_r(T) = f(T)$

Figure 5.3 shows MATLAB code (**BR_T_CA2.m**) for solving the set of differential/algebraic equations describing the adiabatic batch process for a liquid-phase first-order reaction while keeping the heat of reaction variable, not constant. Either Eq. (5.3) or (5.4) can be used to indicate that $\Delta H_r(T) = f(T)$.

```
% MATLAB Code to Solve DAE for CA, T, and XA
clear variables;
% Constants
R = 8.314; % Universal gas constant, J/(mol*K)
C_P_A = 147.0; % Heat capacity of A, J/(mol*K)
C_P_B = 145.0; % Heat capacity of B, J/(mol*K)
C_V_A = C_P_A; % Cv=Cp for liquids
C_V_B = C_P_B; % Cv=Cp for liquids
C_Total = 100; % Total concentration, mol/L
V_BR = input('Enter the BR Volume, (L) : ');
%V_BR = 1500; % Volume of batch reactor, L
n = C_Total * V_BR; % Total moles in reactor, mol
% Initial conditions
t0 = 0; % Initial time, s
tf = input('Enter the final time (tf) in unit of time: '); % Final time
%tf = 3600; % Final time, s
CA0 = input('Enter the initial concentration CA0 (mol/L) : ');
%CA0 = 100; % Initial concentration of A, mol/L
T0 = input('Enter initial temperature (T0) in K: '); % Initial temperature
%T0 = 350; % Initial temperature, K
% Define the system of ODEs
odefun = @(t, y) system_of_equations(t, y, R, C_P_A, C_P_B, C_Total, V_BR, n);
% Initial state vector [CA, T]
y0 = [CA0, T0];
% Solve the ODE system
[t, y] = ode45(odefun, [t0 tf], y0);
% Extract results
CA = y(:, 1); % Concentration of A
T = y(:, 2); % Temperature
XA = 1 - CA / C_Total; % Conversion of A
% Plot results
figure;
subplot(3, 1, 1);
plot(t, CA, 'b-', 'LineWidth', 1.5);
grid on;
xlabel('Time (s)');
ylabel('C_A (mol/L)');
title('Concentration of A vs. Time');
ax = gca; % Get handle to the current axes
% Set new font size for all axes text (including labels)
```

**Figure 5.3** MATLAB code (**BR_T_CA2.m**) for solving the set of differential/algebraic equations describing the adiabatic batch process for a liquid-phase first-order reaction.

```
ax.FontSize = 16;
ax.FontWeight = 'bold';
subplot(3, 1, 2);
plot(t, T, 'r-', 'LineWidth', 1.5);
grid on;
xlabel('Time (s)');
ylabel('T (K)');
title('Temperature vs. Time');
ax = gca; % Get handle to the current axes
% Set new font size for all axes text (including labels)
ax.FontSize = 16;
ax.FontWeight = 'bold';
subplot(3, 1, 3);
plot(t, XA, 'g-', 'LineWidth', 1.5);
grid on;
xlabel('Time (s)');
ylabel('X_A');
title('Conversion of A vs. Time');
ax = gca; % Get handle to the current axes
% Set new font size for all axes text (including labels)
ax.FontSize = 16;
ax.FontWeight = 'bold';
% Function defining the system of ODEs
function dydt = system_of_equations(~, y, R, C_P_A, C_P_B, C_Total, V_BR, n)
   CA = y(1); % Concentration of A, mol/L
   T = y(2); % Temperature, K
   % Calculate k1
   k1 = 1.2e-5 * exp(-60000 / R * (1 / T - 1 / 298));
   % Mole fraction of A
   x_A = CA / C_Total;
   % Mixture heat capacity
   C_V_A = C_P_A; % Cv=Cp for liquids
   C_V_B = C_P_B; % Cv=Cp for liquids
   C_V_mix = x_A * C_V_A + (1 - x_A) * C_V_B;
   % Temperature-dependent reaction enthalpy change. Eq. (5.3)
   DeltaH_r0 = 1700 + integral(@(T) (-C_P_A + C_P_B), 298, T, 'ArrayValued', true); % J/(mol of A)
   % Or, use Eq. (5.4)
   %DeltaH_r0 = 2296 - 2.0 * T; % J/(mol of A)
   % System of equations
   dCA_dt = -k1 * CA;
   dT_dt = (-k1 * CA * V_BR * DeltaH_r0) / (n * C_V_mix);
   % Output derivatives
   dydt = [dCA_dt; dT_dt];
   end
```

**Figure 5.3** *(Continued)*

Figure 5.4 shows the generated plot of $C_A$, $T$, and $X_A$ profiles with BR residence time. It is similar in behavior to that shown in Figure 5.2. The heat of reaction, however, becomes more endothermic because of expressing it as $f(T)$ via the heat capacity terms.

The following input parameters are used, and the results are shown below:

```
>> BR_T_CA2
Enter the BR Volume, (L): 1500
Enter the final time (tf) in unit of time: 3600
Enter the initial concentration CA0 (mol/L): 100
Enter initial temperature (T0) in K: 350
```

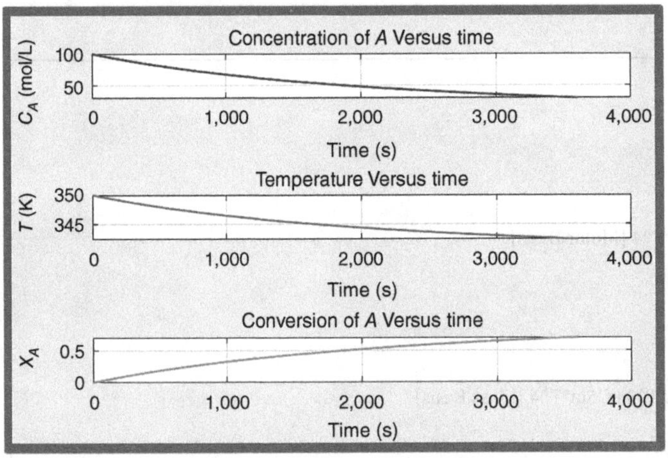


**Figure 5.4** The generated plot of $C_A$, $T$, and $X_A$ profiles with BR residence time. The heat of reaction is $f(T)$.

## 5.5 MATLAB Code for Heat Capacity of Liquids as $f(T)$

There are different functions that can be used to describe or fit the relationship between the heat capacity and temperature variation. Green and Perry [1] used a polynomial function to describe the heat capacity at constant pressure of inorganic and organic compounds in the liquid state:

$$C_P^{Liq}\left[\frac{\text{J}}{\text{kmol}\cdot\text{K}}\right] = C_1 + C_2T + C_3T^2 + C_4T^3 + C_5T^4 \tag{5.5}$$

Or:

$$C_P^{Liq}\left[\frac{\text{J}}{\text{kg}\cdot\text{K}}\right] = C_{1m} + C_{2m}T + C_{3m}T^2 + C_{4m}T^3 + C_{5m}T^4 \tag{5.6}$$

Appendix B | B.1 shows MATLAB code (**CpLiq_T.m**) for presenting the value of constants appearing in heat capacity polynomial Eqs. (5.5) and (5.6) for the selected compound, out of **335** compounds. In addition, it shows the calculated heat capacity both per unit mole and per unit mass of the selected compound at a given $T$, such that $T$ lies between $T_{min}$ and $T_{max}$, expressed in K. $T_{min}$ and $T_{max}$ are presented ahead of the user.

Below is a sample showing the results upon running MATLAB code (**CpLiq_T.m**). Water is selected at $T = 27\,°\text{C}$ as it is quite common with a known heat capacity equal to $1\,\text{cal/(g °C)} = 4.183\,\text{J/(g °C)}$.

```
≫ CpLiq_T
Search the database by:
1. Name
2. Formula
3. CAS_NO
Enter your choice (1, 2, or 3): 1
Enter a search keyword for Name: Water
Matching entries:
 - water (Row 332)
 - H2O (Row 332)
Enter the row number for the compound: 332
```

*(Continued)*

**(Continued)**

```
  {'Liquid Name: '}  {'water'}
  {'Liquid Formula: '}  {'H2O'}
Tmin: 273.16 K
Tmin: 0.01 oC
Tmax: 533.15 K
Tmax: 260 oC
Cp= C1+C2*T+C3*T^2+C4*T^3+C5*T^4 [Molar Basis]
C1: 276370 J/(kmol.K)
C2: -2090.1 J/(kmol.K^2)
C3: 8.125 J/(kmol.K^3)
C4: -0.014116 J/(kmol.K^4)
C5: 9.3701e-06 J/(kmol.K^5)
Cpm= C1m+C2m*T+C3m*T^2+C4m*T^3+C5m*T^4 [Mass Basis]
C1m: 15341.1046 J/(kg.K)
C2m: -116.02 J/(kg.K^2)
C3m: 0.45101 J/(kg.K^3)
C4m: -0.00078357 J/(kg.K^4)
C5m: 5.2013e-07 J/(kg.K^5)
Enter T (must be between Tmin and Tmax, expressed in K): 300
  {'Liquid Name: '}  {'water'}
  {'Liquid Formula: '}  {'H2O'}
At T = 300 K, Cp=75355.81 J/(kmol.K)
The mass-based heat capacity is the same by curve fitting and by conversion from mole to mass basis:
At T = 300 K, Cpm=4182.9481 J/(kg.K)
At T = 300 K, Cp2=4182.9481 J/(kg.K)
At T = 300 K, Cp3=75.3558 J/(mol.K)
At T = 300 K, Cp4=4.1829 J/(g.K)
```

## 5.6 MATLAB Code for Heat Capacity of Gases as $f(T)$

The same can be said about a gas heat capacity. There are different functions that can be used to describe or fit the relationship between the heat capacity and temperature variation. Green and Perry [2] used hyperbolic functions to describe the heat capacity at constant pressure of inorganic and organic compounds in the ideal gas state:

$$C_P^{IG}\left[\frac{\mathrm{J}}{\mathrm{kmol\cdot K}}\right] = C_1 + C_2\left[\frac{\frac{C_3}{T}}{\sin h\left(\frac{C_3}{T}\right)}\right]^2 + C_4\left[\frac{\frac{C_5}{T}}{\cos h\left(\frac{C_5}{T}\right)}\right]^2 \tag{5.7}$$

Or:

$$C_P^{IG}\left[\frac{\mathrm{J}}{\mathrm{kg\cdot K}}\right] = C_{1m} + C_{2m}\left[\frac{\frac{C_3}{T}}{\sin h\left(\frac{C_3}{T}\right)}\right]^2 + C_{4m}\left[\frac{\frac{C_5}{T}}{\cos h\left(\frac{C_5}{T}\right)}\right]^2 \tag{5.8}$$

Appendix B | B.2 shows MATLAB code (**CpGas_T.m**) for presenting the value of constants appearing in heat capacity hyperbolic Eqs. (5.7) and (5.8) for the selected compound, out of **341** compounds. In addition, it shows the calculated heat capacity both per unit mole and per unit mass of the selected compound at a given $T$, such that $T$ lies between $T_{min}$ and $T_{max}$, expressed in K. $T_{min}$ and $T_{max}$ are presented ahead of the user.

Below is a sample showing the results upon running MATLAB code (**CpGas_T.m**). Cyclopropane ($C_3H_6$) is selected on purpose to show the retrieved constants: C1A, C2A, C3A, C4A, and C5A, which can be plugged into Appendix B | B.4: MATLAB code (**BR_T_CA4_Gas.m**) file.

```
>> CpGas_T
Search the database by:
1. Name
2. Formula
3. CAS_NO
Enter your choice (1, 2, or 3): 2
Enter a search keyword for Formula: C3H6
Matching entries:
  - Acetone (Row 5)
  - C3H6O (Row 5)
  - Cyclopropane (Row 70)
  - C3H6 (Row 70)
  - 1,1-Dichloropropane (Row 90)
  - C3H6CL2 (Row 90)
  - 1,2-Dichloropropane (Row 91)
  - C3H6CL2 (Row 91)
  - Ethyl formate (Row 139)
  - C3H6O2 (Row 139)
  - Methyl acetate (Row 194)
  - C3H6O2 (Row 194)
  - Methyl vinyl ether (Row 243)
  - C3H6O (Row 243)
  - Propionaldehyde (Row 295)
  - C3H6O (Row 295)
  - Propionic acid (Row 296)
  - C3H6O2 (Row 296)
  - Propylene (Row 301)
  - C3H6 (Row 301)
Enter the row number for the compound: 70
   {'Gas Name:'} {'Cyclopropane'}
   {'Gas Formula:'} {'C3H6'}
 Tmin: 100 K
 Tmin: -173.15 °C
 Tmax: 1500 K
 Tmax: 1226.85 °C
 Cp= C1+C2*((C3/T)/sinh(C3/T))^2 + C4*((C5/T)/cosh(C5/T))^2[Molar Basis]
 C1= 33800 % J/(kmol.K)
 C2= 168940 % J/(kmol.K)
 C3= 1613.5 % K
 C4= 117680 % J/(kmol.K)
 C5= 722.8 % K
 Enter your choice (A, B, C, D) for compound to be used in BR_T_CA4_Gas.m file: A
 You selected case A.
 The following C1A,C2A,C3A,C4A, and C5A parameters can be pasted in BR_T_CA4_Gas.m file
 C1A= 33.8 ;% J/(mol.K)
 C2A= 168.94 ;% J/(mol.K)
 C3A= 1613.5 ;% K
 C4A= 117.68 ;% J/(mol.K)
 C5A= 722.8 ;% K
 CpAm= C1Am+C2Am*((C3Am/T)/sinh(C3Am/T))^2 + C4Am*((C5Am/T)/cosh(C5Am/T))^2[Mass Basis]
 C1Am= 803.2319 ;% J/(kg.K)
```

*(Continued)*

**(Continued)**

```
C2Am= 4014.7338 ;% J/(kg.K)
C3Am= 1613.5 ;% K
C4Am= 2796.5779 ;% J/(kg.K)
C5Am= 722.8 ;% K)
Enter T (must be between Tmin and Tmax, expressed in K): 300
  {'Gas Name: '} {'Cyclopropane'}
  {'Gas Formula: '} {'C3H6'}
At T = 300 K, Cp=55935.8521 J/(kmol.K)
The mass-based heat capacity is the same by curve fitting and by conversion from mole to mass basis:
At T = 300 K, Cpm=1329.2741 J/(kg.K)
At T = 300 K, Cp2=1329.2741 J/(kg.K)
At T = 300 K, Cp3=55.9359 J/(mol.K)
At T = 300 K, Cp4=1.3293 J/(g.K)
The mass-based heat capacity is the same by curve fitting and by conversion from mole to mass basis:
At T = 300 K, Cpm = 1,329.2741 J/(kg.K)
At T = 300 K, Cp2 = 1,329.2741 J/(kg.K)
At T = 300 K, Cp3 = 55.9359 J/(mol.K)
At T = 300 K, Cp4 = 1.3293 J/(g.K)
```

## 5.7 Adiabatic Gas-Phase BR for Cyclopropane Isomerization

The isomerization of cyclopropane to propene is a gas-phase, unimolecular rearrangement reaction in which the strained three-membered cyclopropane ring opens to form the more stable alkene, propene. The reaction is endothermic, driven by the relief of ring strain, and is typically studied in the temperature range of 400–600 °C. It proceeds through a concerted mechanism or via a diradical intermediate, depending on conditions, and exhibits first-order kinetics with respect to cyclopropane concentration. Because no catalyst is required, the process is often used as a model system for testing theories of unimolecular gas-phase reactions, particularly the Lindemann–Hinshelwood and RRKM (Rice–Ramsperger–Kassel–Marcus) theories.

From a kinetic and thermodynamic standpoint, the rate of cyclopropane isomerization increases significantly with temperature due to a relatively high activation energy (about 270 kJ/mol), reflecting the energy required to break the strained C—C bonds. The reaction's simple stoichiometry, absence of side products under controlled conditions, and well-characterized mechanism make it ideal for validating experimental techniques such as shock-tube studies, flow reactor measurements, and spectroscopic monitoring of reaction progress. Industrially, the isomerization itself has limited direct applications but serves as a benchmark for understanding reaction dynamics, thermal stability of hydrocarbons, and the influence of molecular structure on reactivity.

**Example 5.2** ***BR Irreversible Gas-phase First-order Cyclopropane Isomerization Reaction***
Consider an irreversible gas-phase first-order reaction, which is the isomerization of cyclopropane into propene (or propylene):

$$CH_2CH_2CH_2(g) \rightarrow CH_2 = CH - CH_3(g) \tag{A}$$

Which can be put in the short form:

$$A(g) \rightarrow B(g) \tag{B}$$

With a reaction rate (https://www.youtube.com/watch?v=amh-XSQpWb4&t=193s) and (https://www.chegg.com/homework-help/questions-and-answers/isomerization-cyclopropane-propene-activation-energy-273-kj-mol-reaction-run-750-k-rate-co-q22925871):

$$1.08\times10^{-2}\frac{1}{\text{min}} = k^* * e^{-\frac{273,000}{8.314}\left(\frac{1}{750}\right)} \rightarrow k^* = 1.116\times10^{17}\frac{1}{\text{min}}$$

$$-r_A = k_1 C_A = 1.116\times10^{17}\frac{1}{\text{min}}e^{-\frac{E_A}{RT}}C_A = 1.116\times10^{17}\frac{1}{\text{min}}e^{-\frac{273,000}{8.314*T}\left(\frac{1}{T}\right)}C_A\frac{\text{mol}}{\text{L}} \quad \text{(C)}$$

The standard heat of reaction is given as:

$$\Delta H_r^0 = -33\frac{\text{kJ}}{\text{mol of }A} \quad \text{(D)}$$

The ideal gas heat capacity data, namely, C1, C2, C3, C4, and C5, are given by running the **CpGas_T.m** file and searching for cyclopropane and, another time, for propylene. The constants are adjusted to J/(mol·K). Here are the results for both $A$ and $B$.

```
  {'Gas Name: '} {'Cyclopropane'}
  {'Gas Formula: '} {'C3H6'}
Tmin: 100 K
Tmin: -173.15 oC
Tmax: 1500 K
Tmax: 1226.85 oC
Cp= C1+C2[(C3/T)/sinh(C3/T)]^2 + C4[(C5/T)/cosh(C5/T)]^2[Molar Basis]
C1: 33800 J/(kmol.K)
C2: 168940 J/(kmol.K)
C3: 1613.5 K
C4: 117680 J/(kmol.K)
C5: 722.8 K
```

$$C_{P,A}^{IG}\left[\frac{\text{J}}{\text{mol}\cdot\text{K}}\right] = 33.8 + 168.94\left[\frac{\frac{1613.5}{T}}{\sin h\left(\frac{1613.5}{T}\right)}\right]^2 + 117.68\left[\frac{\frac{722.8}{T}}{\cos h\left(\frac{722.8}{T}\right)}\right]^2 \quad \text{(E)}$$

```
  {'Gas Name: '} {'Propylene'}
  {'Gas Formula: '} {'C3H6'}
Tmin: 130 K
Tmin: -143.15 oC
Tmax: 1500 K
Tmax: 1226.85 oC
Cp= C1+C2[(C3/T)/sinh(C3/T)]^2 + C4[(C5/T)/cosh(C5/T)]^2[Molar Basis]
C1: 43852 J/(kmol.K)
C2: 150600 J/(kmol.K)
C3: 1398.8 K
C4: 74754 J/(kmol.K)
C5: 616.46 K
```

$$C_{P,B}^{IG}\left[\frac{\text{J}}{\text{mol}\cdot\text{K}}\right] = 43.852 + 150.6\left[\frac{\frac{1398.8}{T}}{\sin h\left(\frac{1398.8}{T}\right)}\right]^2 + 74.754\left[\frac{\frac{616.46}{T}}{\cos h\left(\frac{616.46}{T}\right)}\right]^2 \tag{F}$$

Let us write both algebraic and differential equations to describe variables changing with BR time.

**Total Molar Balance**

The change in total number of moles as a result of the chemical reaction is zero simply because $\Delta n = \sum_{i=1}^{N} \nu_i = \nu_A + \nu_B = -1 + 1 = 0$

This implies that:

$$dn = 0 \quad \rightarrow \quad n = constant = n_A + n_B = constant = n_{A0}$$

***A*-Component Material Balance**

$$In - Out + Generation - Consumption = Accumulation = \frac{d(n_A)}{dt}$$

$$0 - 0 + 0 - k_1 C_A V_{BR} = \frac{d(n_A)}{dt} = \frac{d(V_{BR}C_A)}{dt} = \frac{C_A d(V_{BR})}{dt} + \frac{V_{BR} d(C_A)}{dt} \tag{G}$$

$V_{BR}$ is constant; hence, $\frac{C_A d(V_{BR})}{dt} = 0$. Equation (G) reduces to:

$$-k_1 C_A V_{BR} = \frac{V_{BR} d(C_A)}{dt} \quad \rightarrow -k_1 C_A = \frac{dC_A}{dt} \tag{H}$$

Assuming an ideal solution, the mixture heat capacity, $C_V^{mix}$, is given as:

$$nC_V^{mix} = n_A C_V^A(T) + n_B C_V^B(T) \tag{I}$$

Dividing by n and recalling the definition of mole fraction of $A$, $x_A$, and that of $B$, $x_B$, we have:

$$C_V^{mix} = x_A C_V^A(T) + (1 - x_A) C_V^B(T) \tag{J}$$

where;

$$C_V^A(T) = C_P^A(T) - R = C_P^A(T) - 8.314 \tag{K}$$

$$C_V^B(T) = C_P^B(T) - R = C_V^B(T) - 8.314 \tag{L}$$

**Energy Balance**

The heat exchange ($Q_{exch}$) with the surroundings is zero (i.e., adiabatic process), and no shaft work is imposed into or out of the system.

$$In - Out + Generation - Consumption = Accumulation$$

$$0 - 0 + 0 - k_1 C_A V_{BR}\left(\Delta H_r^0\right) = \frac{d\left(nC_V^{mix}T\right)}{dt} = n\overline{C_V^{mix}}\frac{d(T)}{dt} \tag{M}$$

One approach is to factor out the average mixture heat capacity, which is treated as a parameter, where its values will be updated by varying $T$. In the next modeling case, we allow variation of $C_V^{mix}$ with $T$. We basically promote $C_V^{mix}$ from a constant (as in Example 5.1), passing through a parameter (here), and eventually to a variable (next).

Let us define the average temperature, $T_{avg}$, as:

$$T_{avg} = \text{mean}\,(T) \tag{N}$$

$$\overline{C_V^{mix}} = x_A C_V^A(T_{avg}) + (1 - x_A) C_V^B(T_{avg}) \tag{O}$$

Notice that although $\overline{C_V^{mix}}$ is treated as constant; however, it is considered here as a parameter which keeps changing its value with changing $T$ from $T_0$ up to $T_f$.

$$\Delta H_r(T) = -33{,}000\frac{\text{J}}{\text{mol of}A} + \int_{298}^{T} \left(-C_{P,A}^{IG} + C_{P,B}^{IG}\right) dT \tag{P}$$

$$C_{A0} = \frac{Py_{A0}}{RT_0} \tag{Q}$$

MATLAB code (**BR_T_CA3_Gas.m**), shown below, is used to solve the above set of differential and algebraic equations.

The selected equations are: (C), (D), (E), (F), (H), (J), (K), (L), (M), (N), (O), (P), and (Q).

In addition, the following algebraic equations are also used in parallel with the above selected equations.

$$C_{Total} = C_{A0}, \quad x_A = \frac{n_A}{n} = \frac{C_A}{C_{Total}}, \quad V_{BR} = 1{,}500\,L, \quad n = C_{Total} * V_{BR}$$

We need to integrate both $C_A$ and $T$ from:

**Initial Condition:** $t_0 = 0$; $C_A = C_{A0}$, $T = T_0$
**To: Final Condition:** $t = t_f$

### 5.7.1 MATLAB Code for Adiabatic BR for Cyclopropane Gas-Phase Isomerization

Figure 5.5 shows MATLAB code (**BR_T_CA3_Gas.m**) for solving the set of differential/algebraic equations describing the adiabatic batch process for a gas-phase 1st-order reaction with variable heat capacity and heat of reaction as $f(T)$.

```
clear variables;
% Parameters
R = 8.314; % Universal gas constant, J/(mol·K)
V_BR = input('Enter the BR Volume, (L): ');
%V_BR = 1500; % Reactor volume, L
t0 = 0; % Initial time, min
% Prompt user for inputs
T0 = input('Enter initial temperature (T0) in K: '); % Initial temperature
P0 = input('Enter initial pressure (P0) in atm: '); % Initial pressure
tf=input('Enter the final time (tf) in unit of time: '); % Finla time
% Initial concentration of A
C_A0 = P0 / (0.08206 * T0); % mol/L
C_Total = C_A0; % Total concentration, mol/L
n = C_Total * V_BR; % Total moles
% Initial conditions
y0 = [C_A0; T0]; % Initial [C_A, T]
% Define the system of ODEs
odefun = @(t, y) reactor_odes(t, y, R, C_Total, V_BR, n);
% Solve ODEs
[t, y] = ode15s(odefun, [t0 tf], y0);
% Extract results
C_A = y(:, 1); % Concentration of A
```

**Figure 5.5** MATLAB code (**BR_T_CA3_Gas.m**) for solving the set of differential/algebraic equations describing the adiabatic batch process for a gas-phase, 1st-order reaction with variable heat capacity and heat of reaction as $f(T)$.

```
T = y(:, 2); % Temperature
X_A = 1 - C_A / C_A0; % Conversion of A
% Create subplots
figure;
% Subplot 1: Concentration of A
subplot(3, 1, 1);
plot(t, C_A, 'b', 'LineWidth', 2.5);
xlabel('Time (min)');
ylabel('C_A (mol/L)');
title('Concentration of A vs. Time');
grid on;
ax = gca;
ax.FontSize = 16;
ax.FontWeight = 'bold';
% Subplot 2: Temperature
subplot(3, 1, 2);
plot(t, T, 'r', 'LineWidth', 2.5);
xlabel('Time (min)');
ylabel('T (K)');
title('Temperature vs. Time');
grid on;
ax = gca;
ax.FontSize = 16;
ax.FontWeight = 'bold';
% Subplot 3: Conversion
subplot(3, 1, 3);
plot(t, X_A, 'g', 'LineWidth', 2.5);
xlabel('Time (min)');
ylabel('X_A');
title('Conversion of A vs. Time');
grid on;
ax = gca;
ax.FontSize = 16;
ax.FontWeight = 'bold';
% Adjust subplot spacing
sgtitle('Batch Reactor Simulation Results'); % Add a shared title
function dydt = reactor_odes(~, y, R, C_Total, V_BR, n)
   % Unpack variables
   C_A = y(1); % Concentration of A, mol/L
   T = y(2); % Temperature, K
   % Reaction rate constant
   k1= 1.116e+17*exp(-(273000 / R) * (1 / T));
   % Heat capacities (ideal gas) at T
   AsinhTerm=168.94 * ((1613.5 / T) ./ sinh(1613.5 / T)).^2;
   AcoshTerm=117.68 * ((722.8 / T) ./ cosh(722.8 / T)).^2;
   C_P_A = 33.8 +AsinhTerm+AcoshTerm;
   BsinhTerm=150.6 * ((1398.8 / T) ./ sinh(1398.8 / T)).^2;
   BcoshTerm=74.754 * ((616.46 / T) ./ cosh(616.46 / T)).^2;
   C_P_B = 43.852 + BsinhTerm + BcoshTerm;
   % Enthalpy change of reaction
   DeltaH_r = -33000 + integral(@(T) (-C_P_A + C_P_B), 298, T, 'ArrayValued', true);
   % Mole fraction of A
   x_A = C_A / C_Total;
```

**Figure 5.5** (*Continued*)

```
    % Average temperature for mixture heat capacity
    %Tavg = (T + y(2)) / 2; % Average temperature
    Tavg = mean(T); % Average temperature
    % Heat capacities at Tavg
    AsinhTerm=168.94 * ((1613.5 / Tavg) ./ sinh(1613.5 / Tavg)).^2;
    AcoshTerm=117.68 * ((722.8 / Tavg) ./ cosh(722.8 / Tavg)).^2;
    C_P_A_avg = 33.8 +AsinhTerm+AcoshTerm;
    BsinhTerm=150.6 * ((1398.8 / Tavg) ./ sinh(1398.8 / Tavg)).^2;
    BcoshTerm=74.754 * ((616.46 / Tavg) ./ cosh(616.46 / Tavg)).^2;
    C_P_B_avg = 43.852 + BsinhTerm + BcoshTerm;
    C_V_A_avg = C_P_A_avg - R; % Constant volume heat capacity of A at Tavg
    C_V_B_avg = C_P_B_avg - R; % Constant volume heat capacity of B at Tavg
    % Average mixture heat capacity
    C_V_mix_avg = x_A * C_V_A_avg + (1 - x_A) * C_V_B_avg;
    % Differential equations
    dC_A_dt = -k1 * C_A; % Component balance
    dT_dt = (-k1 * C_A * V_BR * DeltaH_r) / (n * C_V_mix_avg); % Energy balance
    % Output derivatives
    dydt = [dC_A_dt; dT_dt];
end
```

**Figure 5.5** (*Continued*)

The following input parameters were entered at the MATLAB Command prompt:

```
>> BR_T_CA3_Gas
Enter the BR Volume, (L): 1500
Enter initial temperature (T0) in K: 750
Enter initial pressure (P0) in atm: 12
Enter the final time (tf) in unit of time: 10
```

Figure 5.6 shows the generated plot showing $C_A$, $T$, and $X_A$ profiles with batch time. Notice that since the reactor is adiabatically operated and the heat of reaction is exothermic, then $T$ will increase with BR time.

Another, more fundamental approach is to keep the heat capacity variable as a function of $T$; hence, $C_V^{mix}$ is now a variable like $T$ itself. Appendix B | B.3 shows the detailed algorithm for the solution of Example 5.2, that is, the irreversible gas-phase first-order reaction of the isomerization of cyclopropane into propene (or propylene). Appendix B | B.4 shows MATLAB code (**BR_T_CA4_Gas.m**) for the solution of algebraic and differential equations describing the gas-phase reaction of the isomerization of cyclopropane into propene (or propylene) being carried out in an adiabatic BR.

The following input parameters are used for generating Figure 5.7 plot, which shows a similar behavior as in Figure 5.6, with BR time to reach a complete conversion (or $C_A$ reduces to zero) happening at about 6 min. Notice both approaches give almost identical results in terms of how $C_A$, $T$, and $X_A$ change with BR time. This has to deal with the fact that although $\overline{C_V^{mix}}$ is factored out of differentiation (see Eq. M); however, its value is stepwise updated in the algorithm as $f(T_{avg})$, where $T_{avg}$ keeps changing from one step to another (see Eqs. N and O).

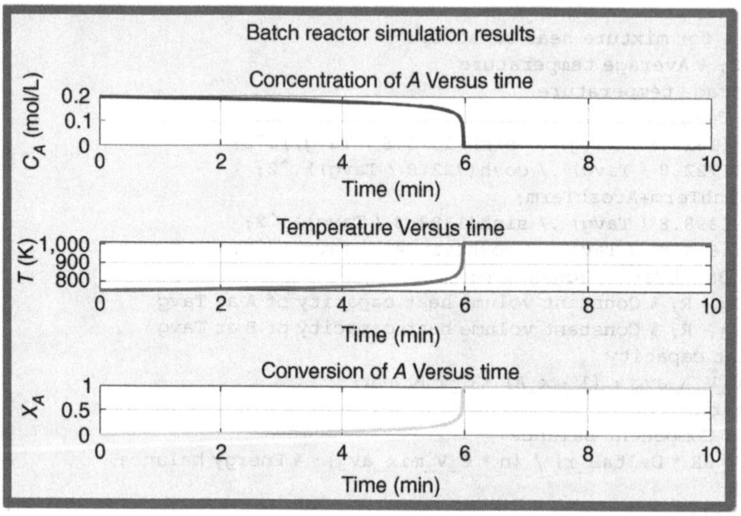


**Figure 5.6** The generated plot showing the $C_A$, $T$, and $X_A$ profiles with batch time for the reactor being adiabatically operated, and the heat of reaction is exothermic.

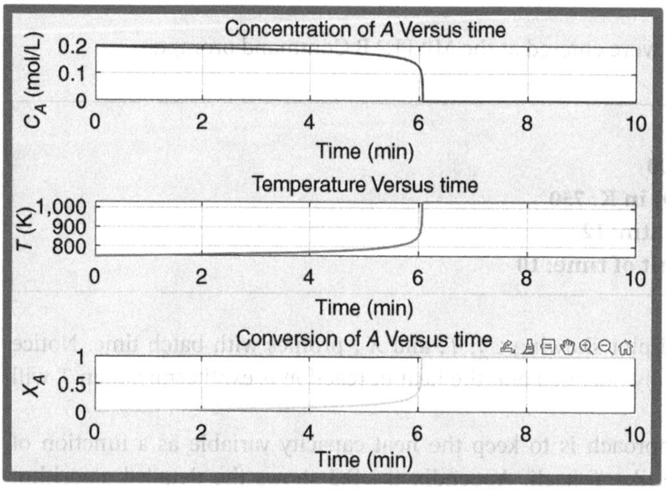


**Figure 5.7** The generated plot of $C_A$, $T$, and $X_A$ as $f$(*BR time*), based on the set of algebraic and differential equations listed in Appendix B | B.3 and using Appendix B | B.4 MATLAB code (**BR_T_CA4_Gas.m**).

```
>> BR_T_CA4_Gas
Enter initial temperature (T0) in K: 750
Enter initial pressure (P0) in atm: 12
Enter the final time (tf) in unit of time: 10
Enter the BR Volume, (L): 1500
```

**Note #6**

(**BR_T_CA3_Gas.m**) can be used for any gas-phase irreversible reaction of the form: $A \rightarrow B$ that takes place in an adiabatic BR. It will generate a plot showing how $C_A$, $T$, and $X_A$ change with BR time. However, the user has to modify/change the following data:

1) Kinetic data for his/her own chemical reaction.
2) Heat of reaction data
3) The heat capacity data, which can be retrieved using the **CpGas_T.m** file

On the other hand, (**BR_T_CA4_Gas.m**) can be used for any gas-phase irreversible reaction of the form: $A \rightarrow B$ that takes place in an adiabatic BR. The ideal heat capacity data for $A$ and $B$, in the form of constants: C1A, C2A, C3A, C4A, C5A, C1B, C2B, C3B, C4B, and C5B, can be copied from the results of running the **CpGas_T.m** file and pasted into the **BR_T_CA4_Gas.m** file.

## 5.8 Liquid-Phase Water Peroxide Decomposition

The liquid-phase decomposition of hydrogen peroxide ($H_2O_2$) is an exothermic reaction in which hydrogen peroxide breaks down into water and oxygen (see below). This reaction can proceed slowly in pure solutions but is significantly accelerated in the presence of catalysts such as soluble transition metal ions (e.g., $Fe^{2+}$, $Cu^{2+}$) or solid catalysts like platinum and manganese dioxide. In industrial and laboratory applications, careful control of temperature, catalyst concentration, and pH is critical, as excessive reaction rates can lead to rapid oxygen evolution, foaming, and even hazards due to localized overheating. The reaction is commonly employed in chemical processes, wastewater treatment, and disinfection, where oxygen generation or peroxide consumption is desired.

From a kinetic standpoint, the decomposition rate in the liquid phase often follows a first-order dependence on hydrogen peroxide concentration in the absence of added catalysts, although more complex rate laws can occur when catalyzed. The reaction mechanism involves the homolytic cleavage of the O—O bond, forming hydroxyl radicals that subsequently recombine to yield water and oxygen. Since the reaction is exothermic, heat release can cause temperature rise in poorly controlled systems, which in turn accelerates the decomposition rate, leading to runaway scenarios. Therefore, engineering design for hydrogen peroxide decomposition reactors, whether batch, continuous stirred tank (CSTR), or plug flow (PFR), must incorporate heat removal strategies, accurate kinetic modeling, and safety considerations for oxygen handling.

**Example 5.3** ***First-order $H_2O_2$ Decomposition***
Consider an irreversible liquid-phase first-order reaction which is the decomposition of hydrogen peroxide into water and oxygen:

$$H_2O_2(aq) \rightarrow H_2O(l) + \frac{1}{2}O_2(g) \tag{A}$$

Which can be put in the short form:

$$A \rightarrow B + \frac{1}{2}C \tag{B}$$

With a reaction rate

$$-r_A = k_1 C_A = k^* e^{-\frac{E_A}{R}\left(\frac{1}{T}-\frac{1}{298}\right)} C_A = 1.4\times 10^{-5}\frac{1}{s} e^{-\frac{75{,}000}{8.314}\left(\frac{1}{T}-\frac{1}{298}\right)} C_A \frac{\text{mol}}{\text{L}} \tag{C}$$

Notice that the maximum solubility of hydrogen peroxide ($H_2O_2$) in water at room temperature (25 °C) and atmospheric pressure is about 100% by weight, meaning $H_2O_2$ and water are completely miscible in all proportions. However, commercially available concentrated solutions typically max out at 90–98% $H_2O_2$ due to stabilization concerns, as pure $H_2O_2$ (>98%) becomes unstable and prone to decomposition. For practical applications:

- **Typical Lab-grade Solutions:** 30–50% $H_2O_2$ by weight.
- **Industrial-grade Solutions:** 70–90% $H_2O_2$ by weight.
- **Theoretical Upper Limit:** ~100% $H_2O_2$, but not stable under normal storage conditions.

We will initially have 800 L of 90 wt. % liquid hydrogen peroxide solution. 200 L will be dedicated as a headspace for $O_2$ gas.

Upon running Appendix A | A.1 MATLAB (**DelG_Keq.m**) Code for Calculation of $\Delta G^0$, $\Delta H^0$, and $K_{eq}$ for a Chemical Reaction, it is found that the heat of reaction is given as:

$$\Delta H_r^0(298\,K) = -98{,}000 \frac{\text{J}}{\text{mol of}\,A}$$

$$\Delta H_r(T) = -98{,}000 + \int_{298}^{T} \left(-C_{P,A} + C_{P,B} + \frac{1}{2} C_{P,C}\right) dT \tag{D}$$

The heat capacity data for liquid water and for gas $O_2$ can be obtained by running the **CpLiq_T.m** and **CpGas_T.m** files, respectively. For water peroxide aqueous solutions, the following approximations are used, based on experimental measurements [3, 4]:

Using MATLAB code (**H2O2CpFit.m**), we have:

```
>> H2O2CpFit
The slope (m) is: 14.8706
The intercept (b) is: 73.9198
```

$$Cp_A \left[\frac{\text{J}}{\text{mol}\cdot\text{K}}\right] = 73.92 + 14.87 * x_A \qquad 0 \le x_A \le 1$$

Combining both variables $x_{H2O2}$ and $T$, we have:

```
>> H2O2CpFit2
Linear Fit: Cp = 66.8568 + 14.9884*XH2O2 + 0.0229*T
```

$$Cp_A \left[\frac{\text{J}}{\text{mol}\cdot\text{K}}\right] = (66.86 + 14.988 * x_A) + 0.0229 * T \tag{E}$$

Notice that the heat capacity of aqueous peroxide solution varies with both $T$ and the concentration or mole fraction of the peroxide itself in water (solvent).

We will run the reactor under nonadiabatic conditions where the overall heat transfer coefficient for a typical metal reactor exposed to air is:

$$U_{overall} \approx h_{external} = 5 - 15\,\frac{\text{W}}{\text{m}^2\cdot\text{K}} \qquad \text{[free natural convection]} \tag{F}$$

If forced convection is used (e.g., with a fan), then

$$U_{overall} \approx h_{external} = 25 - 100\,\frac{\text{W}}{\text{m}^2\cdot\text{K}} \qquad \text{[forced convection]}$$

Then $Q_{exch}$ can be written as:

$$Q_{exch} = hA\Delta T = hA(T - To) = hA(T - 298) \tag{G}$$

With $V_{BR} = 1{,}000\ \text{L} = 1.00\ \text{m}^3$, 1-m diameter, and with both a flat bottom base and flat top cover, we have:

$$A = \pi DL + 2 * \frac{\pi}{4}D^2 = \pi D\left(\frac{V_{BR}}{\frac{\pi}{4}D^2}\right) + 2\times\frac{\pi}{4}D^2 = \left(\frac{4V_{BR}}{D}\right) + \frac{\pi}{2}D^2 = \left(\frac{4*1}{1}\right) + \frac{\pi}{2}1^2$$
$$= 5.57\ \text{m}^2 \tag{H}$$

$$Q_{exch} = hA(T - 298) = 10 * 5.57 * (T - 298) = 55.7 * (T - 298) \tag{I}$$

So far, we have written down algebraic equations needed for part of the process description. We will write down the differential equations:

**Component Molar Balance for *A***

$$-1.4\times 10^{-5}\frac{1}{s}e^{-\frac{75{,}000}{8.314}\left(\frac{1}{T}-\frac{1}{298}\right)}C_A\frac{\text{mol}}{\text{L}} = \frac{dC_A}{dt} \tag{J}$$

**Energy Balance**

$$-1.4\times 10^{-5}e^{-\frac{75{,}000}{8.314}\left(\frac{1}{T}-\frac{1}{298}\right)}C_A V_{BR}(\Delta H_r(T)) - 55.7 * (T - 298) = n\overline{C_V^{mix}}\frac{d(T)}{dt} \tag{K}$$

Let us define the average temperature, $T_{avg}$, as:

$$T_{avg} = mean\,(T) \tag{L}$$

If we neglect the $O_2$ presence in the liquid water (i.e., $O_2$ gas will be released from the liquid surface), then we will be left with *A* and *B*.

$$\overline{C_V^{mix}} = x_A C_V^A(T_{avg}) + (1 - x_A)C_V^B(T_{avg}) \tag{M}$$

In this way, we neglect the gas-phase dynamics. The gas temperature will be equivalent to that of the liquid. Moreover, the gas pressure will be held at atmospheric pressure, unless otherwise specified.

### 5.8.1 MATLAB Code for Nonadiabatic BR for $H_2O_2$ Decomposition

Figure 5.8 shows MATLAB code (**BR_T_CA5.m**), which is written to simulate the following set of algebraic and differential equations:

| | |
|---|---|
| $\Delta H_r(T) = -98{,}000 + \int_{298}^{T} \left(-C_{P,A} + C_{P,B} + \frac{1}{2}C_{P,C}\right) dT$ | (1) |
| $CpA = (66.86 + 14.988 * x_A) + 0.0229 * T$ | (2) |
| CpB= 276.370 -2.0901*T+0.00812*T^2-0.000014*T^3+9.37e-9^4 | (3) |
| CpC=29.10+10.04*((2526.5/T)/sinh(2526.5/T))^2+9.36*((1153.8/T)/cosh(1153.8/T))^2 | (4) |
| $Q_{exch} = 55.7 * (T - 298)$ | (5) |
| $-1.4 \times 10^{-5} * e^{-\frac{75000}{8.314}\left(\frac{1}{T}-\frac{1}{298}\right)} C_A = \frac{dC_A}{dt}$ | (6) |
| $-1.4 \times 10^{-5} e^{-\frac{75000}{8.314}\left(\frac{1}{T}-\frac{1}{298}\right)} C_A V_{BR}(\Delta H_r(T)) - Qexch = n\overline{C_V^{mix}}\frac{d(T)}{dt}$ | (7) |
| Tavg=mean (T) | (8) |
| $\overline{C_V^{mix}} = x_A C_V^A(Tavg) + (1 - x_A)C_V^B(Tavg)$ | (9) |
| Basis: Initially, 800 L (or 800 kg) of 90 wt. % $H_2O_2$ peroxide. | |
| CA0=26.455 mol/L | (10) |
| CB0=5.5494 mol/L | (11) |
| $C_{Total} = CA0 + CB0$ | (12) |
| $x_A = \frac{C_A}{C_{Total}}$ | (13) |
| $V_{BR} = 800\ L$ | (14) |
| $n = C_{Total} * V_{BR}$ | (15) |
| XA=(1-CA/CA0) | (16) |

In addition, we need to integrate both $C_A$ and $T$ from:

**Initial Condition:** $t = t_0 = 0$; $C_A = C_{A0} = 26.455$ mol/L, $T = T_0 = 298$ K
**To: Final Condition:** $t = t_f = 800$ s

A plot for $C_A$, $X_A$, and $T$ as $f(t)$ will be presented.
User's input parameters are shown below:

```
>> BR_T_CA5
Enter the BR Volume, (L): 800
Enter initial temperature (T0) in K: 298
Enter the final time (tf) in unit of time: 800
```

Figure 5.9 shows the generated plot of the simulation, where $C_A$, $T$, and $X_A$ are drawn versus BR time.

```
clear variables;
% Parameters
R = 8.314; % Universal gas constant, J/(mol·K)
V_BR = input('Enter the BR Volume, (L): ');
%V_BR = 800; % Reactor volume, L
t0 = 0; % Initial time, min
% Prompt user for inputs
T0 = input('Enter initial temperature (T0) in K: '); % Initial temperature
tf = input('Enter the final time (tf) in unit of time: '); % Final time
% Initial concentration of A
%Basis: Initially, 800 L (or 800 kg) of 90 wt% H2O2 peroxide
CA0=26.455; %mol/L
CB0=5.5494; %mol/L
C_Total=CA0+CB0;
n = C_Total * V_BR; % Total moles
% Initial conditions
y0 = [CA0; T0]; % Initial [C_A, T]
% Define the system of ODEs
odefun = @(t, y) reactor_odes(t, y, R, C_Total, V_BR, n);
% Solve ODEs
[t, y] = ode15s(odefun, [t0 tf], y0);
% Extract results
C_A = y(:, 1); % Concentration of A
T = y(:, 2); % Temperature
X_A = 1 - C_A / CA0; % Conversion of A
% Create subplots
figure;
% Subplot 1: Concentration of A
subplot(3, 1, 1);
plot(t, C_A, 'b', 'LineWidth', 2.5);
xlabel('Time (s)');
ylabel('C_A (mol/L)');
title('Concentration of A vs. Time');
grid on;
```

**Figure 5.8** MATLAB code (**BR_T_CA5.m**) to simulate the nonadiabatic BR for liquid-phase water peroxide decomposition with $O_2$ gas relieved off the liquid phase.

```
ax = gca;
ax.FontSize = 16;
ax.FontWeight = 'bold';
% Subplot 2: Temperature
subplot(3, 1, 2);
plot(t, T, 'r', 'LineWidth', 2.5);
xlabel('Time (s)');
ylabel('T (K)');
title('Temperature vs. Time');
grid on;
ax = gca;
ax.FontSize = 16;
ax.FontWeight = 'bold';
% Subplot 3: Conversion
subplot(3, 1, 3);
plot(t, X_A, 'g', 'LineWidth', 2.5);
xlabel('Time (s)');
ylabel('X_A');
title('Conversion of A vs. Time');
grid on;
ax = gca;
ax.FontSize = 16;
ax.FontWeight = 'bold';
% Adjust subplot spacing
sgtitle('Batch Reactor Simulation Results'); % Add a shared title
function dydt = reactor_odes(~, y, R, C_Total, V_BR, n)
   % Unpack variables
   C_A = y(1); % Concentration of A, mol/L
   T = y(2); % Temperature, K
   % Mole fraction of A
   x_A = C_A / C_Total;
   % Reaction rate constant
   k1=1.4e-5*exp(-75000/R*(1/T-1/298));
   % Heat capacities (ideal gas) at T
```

**Figure 5.8** *(Continued)*

```
    C_P_A = (66.86 + 14.988 * x_A) + 0.0229 * T; % Eq. (2)
    C_P_B = 276.370 - 2.0901 * T + 0.00812 * T^2 - 0.000014 * T^3 + 9.37e-9 * T^4; % Eq. (3)
    C_P_C = 29.103 + 10.04 * ((2526.5 / T) / sinh(2526.5 / T))^2 + 9.356 * ((1153.8 / T) / cosh(1153.8 / T))^2; % Eq. (4)
    % Enthalpy change of reaction
    DeltaH_r = -98000 + integral(@(T) (-C_P_A + C_P_B + 0.5 * C_P_C), 298, T, 'ArrayValued', true);
    % Average temperature for mixture heat capacity
    T_avg = mean(T); % Eq. (8)
    % Heat capacities at T_avg
    C_P_A_avg = (66.86 + 14.988 * x_A) + 0.0229 * T_avg; % Eq. (2)
    C_P_B_avg = 276.370 - 2.0901 * T_avg + 0.00812 * T_avg^2 - 0.000014 * T_avg^3 + 9.37e-9 * T_avg^4; % Eq. (3)
    % Average mixture heat capacity, assuming escape of O2 into gas phase
    % above the liquid
    C_V_A_avg = C_P_A_avg; % Constant volume heat capacity of A at T_avg
    C_V_B_avg = C_P_B_avg; % Constant volume heat capacity of B at T_avg
    C_V_mix_avg = x_A * C_V_A_avg + (1 - x_A) * C_V_B_avg; % Eq. (9)
    % Differential equations
    Qexch=55.7*(T-298);
    dC_A_dt = -k1 * C_A; % Component balance
    dT_dt = ((-k1 * C_A * V_BR * DeltaH_r)-Qexch) / (n * C_V_mix_avg); % Energy balance
    % Output derivatives
    dydt = [dC_A_dt; dT_dt];
end
```

**Figure 5.8** (*Continued*)

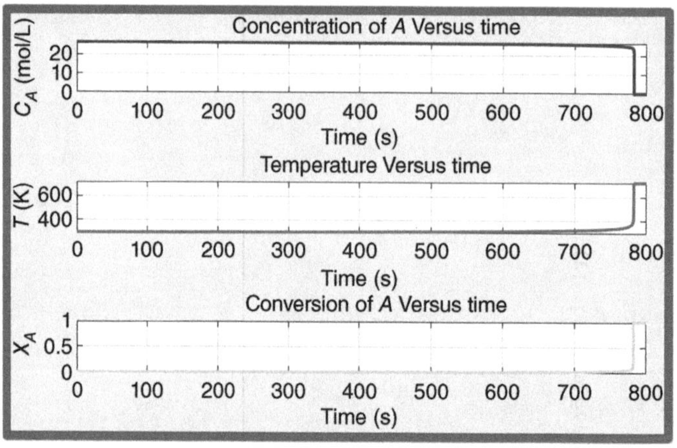


**Figure 5.9** Liquid-phase water peroxide decomposition in BR under a nonadiabatic condition.

## 5.9 Liquid-Phase CSTR Glucose Conversion

In a CSTR, an aqueous glucose feed is continuously supplied while the well-mixed reactor effluent maintains a constant composition at steady state. The isomerization, typically catalyzed by immobilized glucose isomerase, proceeds under controlled temperature (around 55–65 °C) and near-neutral pH to balance reaction rate, equilibrium conversion (≈42–50% fructose), and enzyme stability. Since the reaction is slightly endothermic, mild heat input may be required to maintain the desired operating temperature, and the well-mixed nature of the CSTR ensures uniform conditions throughout the reactor. Reactor design must account for the reversible kinetics, enzyme deactivation over time, and potential side reactions, with strategies such as recycle loops or multistage operation used to improve fructose yield beyond the single-stage equilibrium limit. For simplicity, we will assume it is irreversible.

**Example 5.4** ***Unsteady-state CSTR Liquid-phase First-order Isomerization Reaction***
We will rework the aqueous liquid-phase first-order reaction, which is the isomerization of glucose into fructose in a CSTR. Example 5.1 Eqs. (A) through (F) will be used here, but there is no need to rewrite them.

Write enough number of linearly independent equations, including total material balance, *A*-component molar balance, and energy balance, to describe the adiabatic CSTR system. Notice that both $C_A$ (or $X_A$) and temperature ($T$) do change with CSTR time. The initial concentration of $A$, $C_{A0} = 100$ mol/L, initial volume flow rate of 0.1 L/s, initial temperature of $T_0 = 350$ K, and $V_{CSTR} = 1{,}000$ L.

**Solution**

**Total Molar Balance**

Explained in Example 5.1

$$dn = 0 \quad \rightarrow \quad n = constant = n_A + n_B = constant = n_{A0}$$

It also implies that: $F_{Tot} = FA + FB = constant = FA0$

***A*-Component Material Balance**

$$In - Out + Generation - Consumption = Accumulation = \frac{d(n_A)}{dt}$$

$$F_{A0} - FA - k_1 C_A V_{CSTR} = \frac{C_A d(V_{CSTR})}{dt} + \frac{V_{CSTR} d(C_A)}{dt} = \frac{V_{CSTR} d(C_A)}{dt} \quad \text{(A)}$$

$V_{CSTR}$ is constant; hence, $\frac{C_A d(V_{CSTR})}{dt} = 0$. Equation (A) reduces to:

$$\frac{F_{A0} - FA}{V_{CSTR}} - 1.2 \times 10^{-5} * e^{-\frac{60,000}{8.314}\left(\frac{1}{T} - \frac{1}{298}\right)} C_A = \frac{d(C_A)}{dt} \quad \text{(B)}$$

**Energy Balance**

For simplicity, let us assume that the heat of reaction is $f(T)$. No shaft work is imposed on or out of the system. There is a heat exchange, $Q_{exch,}$ with the surroundings through a heating/cooling jacket.

$$F_{A0} * CpA(T_0 - T_{ref}) - FTot * Cpmix(T - T_{ref}) - k_1 C_A V_{CSTR}(\Delta H_r) + Q_{exch} = \frac{d\left(nC_V^{mix} T\right)}{dt}$$

$$= n\overline{C_V^{mix}} \frac{d(T)}{dt}$$

Select $T_{ref}$ to be zero, and replace $F_{Tot}$ by $F_{A0}$, and combine the two terms on the left side, we have:

$$F_{A0} * (CpA * T_0 - Cpmix * T) - k_1 C_A V_{CSTR}(\Delta H_r) + Q_{exch} = n\overline{C_V^{mix}} \frac{d(T)}{dt} \quad \text{(C)}$$

## 5.9.1 MATLAB Code for Unsteady-State CSTR Liquid-Phase First-Order Glucose Isomerization Reaction

Figure 5.10 shows MATLAB code (**CSTR_T_CA.m**), which is written to simulate the following set of algebraic and differential equations:

$k1 = 1.2 \times 10^{-5} * e^{-\frac{60000}{8.314}\left(\frac{1}{T} - \frac{1}{298}\right)}$

If T < 398 K

$Q_{exch} = UA(T_{Steam} - T) = 500 * 4 * (398 - T);$ *for steam heating*

Else;

$Q_{exch} = UA(T_{Cool.\ Water} - T) = 400 * 4 * (298 - T);$ *for cooling water*

$FA0 * (CpA * T0 - Cpmix * T) - k_1 C_A V_{CSTR}(\Delta H_r) + Q_{exch} = nC_V^{mix} \frac{d(T)}{dt}$

$\frac{FA0-FA}{V_{CSTR}} - k1C_A = \frac{d(C_A)}{dt}$

CA0 = 100; % Initial concentration of A, mol/L

C_Total = CA0; % Total concentration, mol/L

n = C_Total * VCSTR; % Total moles in reactor, mol

VCSTR = 1,000; % Volume of CSTR, L

v0=0.1; % L/s

FA0=CA0*v0; %mol/s

FA=CA*v0; %mol/s

*(Continued)*

**(Continued)**

T0 = 350; % Initial temperature, K
C_P_A = 147.0; % Heat capacity of A, J/(mol*K)
C_P_B = 145.0; % Heat capacity of B, J/(mol*K)
C_V_A = C_P_A; % Cv=Cp for liquids
C_V_B = C_P_B; % Cv=Cp for liquids
C_V_mix = x_A * C_V_A + (1 - x_A) * C_V_B;
$\Delta H_r(T) = 1{,}700\frac{J}{mol} + \int_{298}^{T}(-C\_P\_A + C\_P\_B)dT$
XA=1-(FA/FA0); % Conversion of A
x_A = CA / C_Total;

```
% MATLAB Code to Solve DEs/AEs for CA, T, and XA in a CSTR
clear variables;
R = 8.314; % Universal gas constant, J/(mol*K)
C_P_A = 147.0; % Heat capacity of A, J/(mol*K)
C_P_B = 145.0; % Heat capacity of B, J/(mol*K)
C_V_A = C_P_A; % Cv=Cp for liquids
C_V_B = C_P_B; % Cv=Cp for liquids
VCSTR = input('Enter the CSTR Volume, (L): ');
%VCSTR = 1000; % Volume of CSTR, L
v0 = input('Enter the initial volume flow rate (L/s): ');
%v0 = 0.1; % Volumetric flow rate, L/s
CA0 = input('Enter the initial concentration CA0 (mol/L): ');
%CA0 = 100; % Initial concentration of A, mol/L
C_Total = CA0; % Total concentration, mol/L
n = C_Total * VCSTR; % Total moles in reactor, mol
FA0 = CA0 * v0; % Initial molar flow rate of A, mol/s
% Prompt user for inputs
T0 = input('Enter initial temperature (T0) in K: '); % Initial temperature
tf = input('Enter the final time (tf) in unit of time: '); % Final time
% Initial conditions
t0 = 0; % Initial time, s
y0 = [CA0, T0];
% Define the system of ODEs
odefun = @(t, y) system_of_equations(t, y, R, C_P_A, C_P_B, VCSTR, v0, T0, FA0, n);
% Solve the ODE system
[t, y] = ode45(odefun, [t0 tf], y0);
% Extract results
CA = y(:, 1); % Concentration of A
T = y(:, 2); % Temperature
FA = CA * v0; % Molar flow rate of A
XA = 1 - (FA / FA0); % Conversion of A
% Plot results
figure;
subplot(3, 1, 1);
plot(t, CA, 'b-', 'LineWidth', 1.5);
grid on;
xlabel('Time (s)');
ylabel('C_A (mol/L)');
title('Concentration of A vs. Time');
ax = gca;
```

**Figure 5.10** MATLAB code (**CSTR_T_CA.m**), written to simulate the liquid-phase glucose isomerization reaction in a nonadiabatic CSTR.

```
ax.FontSize = 16;
ax.FontWeight = 'bold';
subplot(3, 1, 2);
plot(t, T, 'r-', 'LineWidth', 1.5);
grid on;
xlabel('Time (s)');
ylabel('T (K)');
title('Temperature vs. Time');
ax = gca;
ax.FontSize = 16;
ax.FontWeight = 'bold';
subplot(3, 1, 3);
plot(t, XA, 'g-', 'LineWidth', 1.5);
grid on;
xlabel('Time (s)');
ylabel('X_A');
title('Conversion of A vs. Time');
ax = gca;
ax.FontSize = 16;
ax.FontWeight = 'bold';
% Function defining the system of ODEs
function dydt = system_of_equations(~, y, R, C_P_A, C_P_B, VCSTR, v0, T0, FA0, n)
   CA = y(1); % Concentration of A, mol/L
   T = y(2); % Temperature, K
   FA = CA * v0; % Molar flow rate of A
   k1 = 1.2e-5 * exp(-60000 / R * (1 / T - 1 / 298));
   x_A = CA / (FA0/v0); % Mole fraction of A
   C_V_A=C_P_A; % for liquids
   C_V_B=C_P_B; % for liquids
   C_V_mix = x_A * C_V_A + (1 - x_A) * C_V_B;
   DeltaH_r = 1700 + integral(@(T) (-C_P_A + C_P_B), 298, T, 'ArrayValued', true);
   if T<398
      Qexch=500*4*(398-T);% U=500 W/m2.K for steam heating medium jacket
   else
      Qexch=400*4*(298-T); % U=400 W/m².K for cooling water medium jacket
   end
   %Or DeltaH_r = 1700 + (-2.0) * (T - 298); % Temperature-dependent reaction enthalpy
   dCA_dt = (FA0 - FA) / VCSTR - k1 * CA;
   dT_dt = (FA0 * (C_P_A * T0 - C_V_mix * T) - k1 * CA * VCSTR * DeltaH_r+Qexch) / (n * C_V_mix);
   dydt = [dCA_dt; dT_dt];
end
```

**Figure 5.10** *(Continued)*

In addition, we need to integrate both $C_A$ and $T$ from:

**Initial Condition:** $t = t_0 = 0$; $C_A = C_{A0} = 100$ mol/L, $T = T_0 = 298$ K
**To: Final Condition:** $t = t_f = 3{,}600$ s

A plot for $C_A$, $X_A$, and $T$ as $f(t)$ will be presented.

The following user's input parameters are used:

**≫ CSTR_T_CA**
**Enter the CSTR Volume, (L): 1000**
**Enter the initial volume flow rate (L/s): 0.1**

*(Continued)*

| (Continued) |
|---|
| **Enter the initial concentration CA0 (mol/L): 100**<br>**Enter initial temperature (T0) in K: 350**<br>**Enter the final time (tf) in unit of time: 3000** |

Notice that the reaction is weakly endothermic, yet the temperature increases with time as a result of steam heating on the jacket side so long as the temperature falls below 398 K. Cooling will not be effective unless $T$ exceeds 398 K. The break temperature can be adjusted by the user such that it does not exceed 60 °C, for example, which keeps a suitable temperature range for the enzymatic catalysis of the glucose isomerization reaction. The high limit of 100 °C is meant to show the effect of steam heating. See Figure 5.11 for the generated profiles of $C_A$, T, and $X_A$.

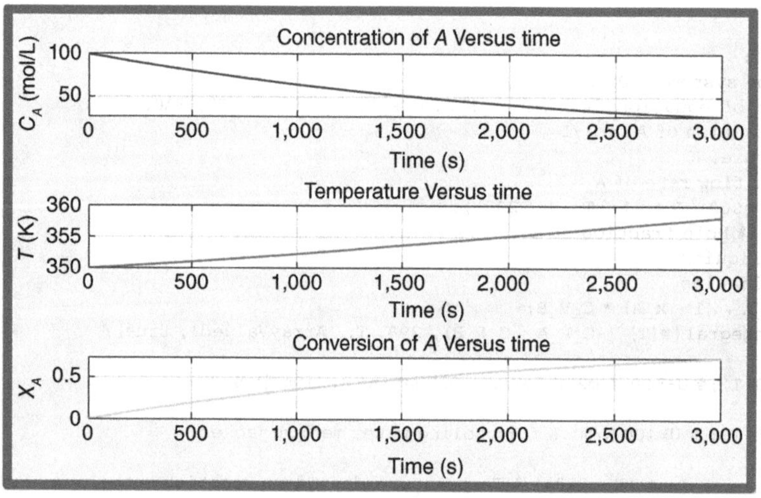


**Figure 5.11** Liquid-phase glucose isomerization in CSTR under a nonadiabatic condition.

**Example 5.5** ***Steady-state CSTR Liquid-phase First-order Isomerization Reaction***

We will rework the aqueous liquid-phase first-order reaction, which is the isomerization of glucose into fructose, in a CSTR. All equations appearing in Example 5.4 will be used or utilized here, except for the unsteady-state differential equations, which will become algebraic equations, representing the steady-state counterpart values of both $C_A$ and $T$ (i.e., $C_{Ass}$ and $T_{ss}$). All other parameters appearing in Example 5.4 will be borrowed or brought here. Notice that, unlike the steady-state PFR case, $C_A$, $T$, and $X$ will be uniform with volume spatial coordinates. We need to find only the steady-state values for a given set of reactor specs and input parameters.

| |
|---|
| $FA0*(CpA*T0-Cpmix*Tss)-k_1C_{Ass}V_{CSTR}(\Delta H_r)+Q_{exch}=0$ **(Steady-state Energy Balance)**<br>$\frac{FA0-FA}{V_{CSTR}}-k1C_{Ass}=0$ **(Steady-state Comp. Molar Balance)**<br>**All other algebraic equations will be borrowed from Example 5.4.** |

### 5.9.2 MATLAB Code for Steady-State CSTR Liquid-Phase First-Order Glucose Isomerization Reaction

Figure 5.12 shows MATLAB code (**CSTR_T_CA_SS.m**), which stands for the steady-state counterpart of the unsteady-state version used in Example 5.4.

```
% MATLAB Code to Compute the Steady-State (dCA/dt=0, dT/dt=0) of a CSTR
clear variables;
%% Constants and Known Parameters
R = 8.314;           % Universal gas constant, J/(mol*K)
C_P_A = 147.0;        % Heat capacity of A, J/(mol*K)
C_P_B = 145.0;        % Heat capacity of B, J/(mol*K)
C_V_A = C_P_A;        % For liquids, Cv = Cp
C_V_B = C_P_B;        % For liquids, Cv = Cp
VCSTR = input('Enter the CSTR Volume, (L): ');
%VCSTR = 800; % Volume of CSTR, L
v0 = input('Enter the initial volume flow rate (L/s): ');
%v0 = 0.1; % Volumetric flow rate, L/s
CA0 = input('Enter feed concentration of A (CA0) in mol/L: ');
C_Total = CA0;        % Total concentration, mol/L (assumed)
FA0 = CA0 * v0;       % Molar flow rate of A, mol/s
% Prompt user for the feed temperature
T0 = input('Enter the feed temperature (T0) in K: ');
%% Solve for Steady-State using fsolve
% Initial guess for [CA, T] (you may need to adjust these)
y0 = [CA0; T0];
lb = [0; 200]; % Lower Bound for CA and for T to avoid non-realistic values
ub = [CA0; 1000]; % Upper Bound for CA and for T to avoid non-sense values.
options = optimoptions('lsqnonlin', ...
    'Display', 'final-detailed', ...
    'TolFun', 1e-6, ...
    'TolX', 1e-6, ...
    'MaxFunctionEvaluations', 20000, ...
    'MaxIterations', 2000);
% % Create an anonymous function that wraps the steady state equations
steady_state_fun = @(y) steady_state_equations(y, R, C_P_A, C_P_B, VCSTR, v0, T0, FA0);
[y_ss, resnorm, residual, exitflag, output] = lsqnonlin(steady_state_fun, y0, lb, ub, options);
% % Solve the system
% [y_ss, fval, exitflag, output] = fsolve(steady_state_fun, y0, options);
if exitflag > 0
    fprintf('Steady-state solution found:\n');
    fprintf('CAss = %.4f mol/L\n', y_ss(1));
    fprintf('Tss  = %.4f K\n', y_ss(2));
else
    fprintf('fsolve did not converge to a solution.\n');
end

%% (Optional) Calculate additional values based on the steady state
CA_ss = y_ss(1);
T_ss  = y_ss(2);
FA_ss = CA_ss * v0;
XA_ss = 1 - (FA_ss / FA0);
Tau=VCSTR/v0;
fprintf('Conversion, XAss = %.4f\n', XA_ss);
disp(['VCSTR = ', num2str(VCSTR), ' L']);
disp(['tCSTR = ', num2str(Tau), ' Unit of time appearing in v0']);
%% Plotting (Optional)
figure;
```

**Figure 5.12** MATLAB code (**CSTR_T_CA_SS.m**), representing the steady-state counterpart of the unsteady-state version used in Example 5.4.

```
subplot(2,1,1);
plot(1, CA_ss, 'bo', 'MarkerSize',10, 'LineWidth',2);
grid on;
xlabel('Steady State');
ylabel('C_A (mol/L)');
title('Steady-State Concentration of A');
subplot(2,1,2);
plot(1, T_ss, 'ro', 'MarkerSize',10, 'LineWidth',2);
grid on;
xlabel('Steady State');
ylabel('T (K)');
title('Steady-State Temperature');
%% Function: steady_state_equations
function F = steady_state_equations(y, R, C_P_A, C_P_B, VCSTR, v0, T0, FA0)
   % Unpack variables
   CA = y(1); % Concentration of A, mol/L
   T  = y(2); % Temperature, K
   % Compute the molar flow rate of A in the reactor
   FA = CA * v0;
   % Rate constant, with an Arrhenius-type temperature dependency
   % (Note: The exponential term is written as exp(-Ea/R*(1/T - 1/298)) )
   k1 = 1.2e-5 * exp(-60000/R * (1/T - 1/298));
   % Mole fraction of A in the reactor (assuming the feed is the only source)
   x_A = CA / (FA0 / v0);
   C_V_A=C_P_A; %For liquids
   C_V_B=C_P_B; % For liquids
   % Calculate the mixed heat capacity
   C_V_mix = x_A * C_V_A + (1 - x_A) * C_V_B;
   % Reaction enthalpy (an example formulation)
   % Here we compute an integrated form; note that the integral is performed
   % from 298 K to T (this is one way to model temperature-dependent ΔH)
   DeltaH_r = 1700 + integral(@(T_int) (-C_P_A + C_P_B), 298, T, 'ArrayValued', true);
    if T<506
      Qexch=500*4*(506-T);% U=500 W/m2.K for high-P steam heating medium jacket
   else
      Qexch=400*4*(298-T); % U=400 W/m2.K for cooling water medium jacket
   end
   % Steady-state material balance for species A (set equal to zero)
   F(1) = (FA0 - FA) / VCSTR - k1 * CA;
   % Steady-state energy balance (set equal to zero)
   F(2) = (FA0 * (C_P_A * T0 - C_V_mix * T) - k1 * CA * VCSTR * DeltaH_r)+Qexch;
   % Return as column vector
   F = F(:);
end
```

**Figure 5.12** *(Continued)*

The following two cases represent running the code for two different pairs of input $T_0$ and $C_{A0}$. Calculated steady-state values: $C_{Ass}$ (i.e., $X_{Ass}$) and $T_{ss}$ are shown beneath each case as well. Varying input parameters will result in different steady-state values of $C_{Ass}$ (i.e., $X_{Ass}$) and $T_{ss}$.

```
>> CSTR_T_CA_SS
Enter the CSTR Volume, (L): 1000
Enter the initial volume flow rate (L/s): 0.1
Enter feed concentration of A (CA0) in mol/L: 100
Enter the feed temperature (T0) in K: 330
```

```
Steady-state solution found:
CAss = 51.2989 mol/L
Tss  = 380.5212 K
Conversion, XAss = 0.4870
VCSTR = 1000 L
tCSTR = 10000 Unit of time appearing in v0

≫ CSTR_T_CA_SS
Enter the CSTR Volume, (L): 1000
Enter the initial volume flow rate (L/s): 0.1
Enter feed concentration of A (CA0) in mol/L: 105
Enter the feed temperature (T0) in K: 335
Steady-state solution found:
CAss = 50.1561 mol/L
Tss  = 381.2158 K
Conversion, XAss = 0.5223
VCSTR = 1000 L
tCSTR = 10000 Unit of time appearing in v0
```

## 5.10 Process Control and Dynamics

Studying the dynamic behavior of chemical reactors is crucial for several reasons beyond merely knowing the steady state. Here are some key points embodying its importance:

1) **Designing Effective Process Control Systems**
   a) **Open-Loop Dynamic Response:** Before installing any closed-loop control system, engineers must understand the reactor's open-loop dynamic response; that is, how the system naturally reacts to changes in inputs such as feed temperature ($T_0$) and feed concentration ($C_{A0}$). This dynamic behavior influences how quickly and in what manner the reactor responds to disturbances.
   b) **Closed-loop Control:** A well-designed control loop (closed-loop system) relies on accurate models of the reactor's dynamics. This understanding ensures that controllers are tuned to achieve desired performance, maintain product quality, and enhance safety. Without a solid grasp of the open-loop dynamics, any control strategy may perform poorly, leading to oscillations, instability, or inefficient operation.
2) **Transient Periods: Start-Up and Shutdown**
   a) **Start-Up:** When a reactor is started from a nonoperational state, the process variables (e.g., concentration, temperature) do not jump instantly to their steady-state values. Instead, the reactor undergoes a transient period where the dynamics dictate how rapidly and smoothly the steady state is achieved. Understanding these dynamics helps in designing safe and efficient start-up procedures, preventing conditions that could lead to runaway reactions or equipment damage.
   b) **Shutdown:** Similarly, during shutdown, the reactor's variables change over time as the process is gradually taken offline. Predicting and controlling these changes is important to avoid abrupt transitions that could result in safety hazards or product quality issues.

3) **Responding to Disturbances**
   **Disturbance Rejection**
   In real-world operations, load disturbances such as variations in feed composition or temperature are inevitable. Studying the reactor dynamics helps in understanding how these disturbances affect the process and how quickly the reactor can recover or adapt. This insight is essential for designing controllers that can robustly handle unexpected changes.
4) **Overall Process Safety and Efficiency**
   a) **Optimized Operation:** Knowledge of reactor dynamics is not only used for control purposes but also for optimizing the overall process. By understanding transient behavior, engineers can design more energy-efficient and safer processes that avoid conditions likely to stress equipment or compromise product quality.
   b) **Predictive Maintenance:** Dynamic models also allow for the prediction of how the system will behave under off-normal conditions, contributing to better maintenance strategies and improved long-term reliability of the process equipment.

Let us take our lovely CSTR, explained in Example 5.5, to demonstrate the importance of process dynamics or its dynamic response to sources of both load and setpoint disturbances. For any given feed temperature $T_0$ and concentration $C_{A0}$, the CSTR will eventually reach a unique steady state characterized by a specific exit concentration $C_{Ass}$ and exit temperature $T_{ss}$. In process control and dynamics, $T_0$ and $C_{A0}$ are treated as separate disturbance inputs or loads. For example, if the CSTR feed is maintained at $T_0 = 330$ K and $C_{A0} = 100$ M, the reactor will eventually settle at a steady state where the exit concentration $C_{Aexit}$ equals $C_{Ass} = 51.3$ mol/L and the exit temperature $T_{exit}$ equals $T_{ss} = 380.5$ K.

If the feed conditions change, for instance, if $C_{A0}$ increases to 105 mol/L and $T_0$ increases to 335 K, these changes act as external load disturbances to the system. As a result, the CSTR will transition from its current steady state ($C_{Ass} = 51.3$ mol/L and $T_{ss} = 380.5$ K) to a new steady state (in Example 5.5, $C_{Ass} = 50.2$ mol/L and $T_{ss} = 381.2$ K). During the transition between these two distinct steady states, the system operates under dynamic conditions; that is, both $C_A$ and $T$ do change with time. Once the new steady state is reached, the system is said to be operating under static conditions, with both $C_A$ and $T$ remaining constant with time. The reactor dynamics determine how fast and smoothly this change occurs. Designing effective control strategies and ensuring safe start-up/shutdown procedures both require an in-depth understanding of these dynamic processes. Notice that in Example 5.5, the CSTR's dynamic response operates under closed-loop control, specifically for temperature regulation. This control system dynamically modulates between cooling and heating modes to sustain the desired setpoint.

It is worth referring to the definition of a flow reactor's runaway condition [5]. The event of "runaway condition" occurs when a small drift in temperature causes the process to erroneously run away down to practically the quenching (i.e., no reaction) zone. The following quote is taken from [6]: "A small minority of processes are neither self-regulating nor integrating and, rather unhelpfully, are unstable under open loop control. That is, if you switch the controller to manual, the PV will continue to rise (or fall), possibly in an exponential fashion, until some physical, possibly disastrous, limit is reached. These are called runaway processes. The most common of these processes is the exothermic reactor with cooling jacket."

### 5.10.1 Multi-Steady States for a Nonadiabatic CSTR

In this section, we will demonstrate how a non-isothermal CSTR with an exothermic first-order reaction can exhibit multiple steady states. The CSTR is equipped with a cooling jacket. Under certain conditions, the system may have more than one steady-state temperature and conversion for the same operating parameters. MATLAB can then be used to find and visualize them using numerical root-finding.

Consider the irreversible, exothermic first-order liquid-phase reaction:

$$A \xrightarrow{k} B$$

**Steady-State Component A-Molar Balance**

$$F_{A0} - F_A - r_A V_{CSTR} = 0$$

Substitute for $C_A$, we have:

$$F_{A0} - F_A - kC_A V_{CSTR} = 0$$

In terms of Conversion, $X$, we have:

$$F_{A0} - F_A - kC_{A0}(1 - X)V_{CSTR} = 0$$

Replace $F_{A0}$ and $F_A$ by their equivalence and $\upsilon = \upsilon_0$, we have:

$$C_{A0}\upsilon_0 - C_A\upsilon_0 - kC_{A0}(1 - X)V_{CSTR} = 0$$

$$C_{A0}\upsilon_0 - C_{A0}(1 - X)\upsilon_0 - kC_{A0}(1 - X)V_{CSTR} = 0$$

$C_{A0}$ will cancel out. Divide through by $v_0$ and given $\tau = \frac{V_{CSTR}}{\upsilon_0}$, we have:

$$1 - (1 - X) - k(1 - X)\tau = 0 \rightarrow X - k(1 - X)\tau = 0$$

$$\tau = \frac{X}{k(1 - X)}$$

**Rate Constant (Arrhenius Law)**

$$k = k_0 e^{-\left(\frac{E}{RT}\right)}$$

**Steady-State Energy Balance**

$$F_{A0} * Cp(T_0 - T_{ref}) - F_{Tot} * Cpmix(T - T_{ref}) - k_1 C_A V_{CSTR}(\Delta H_r) + UA(Tj - T) = 0$$

Select $T_{ref}$ to be zero, and replace $F_{Tot}$ by $F_{A0,}$ and combine the two terms on the left side, we have:

$$F_{A0} * (Cp * T0 - Cp * T) - k_1 C_{A0}(1 - X)V_{CSTR}(\Delta H_r) + UA(Tj - T) = 0$$

Figure 5.13 shows MATLAB code (**MSS_CSTR.m**) that shows a multi-steady state condition for a nonadiabatic CSTR where heat is exchanged with a cooling jacket maintained at a controlled temperature $T_j$.

Figure 5.14 shows the generated plot for conversion, X, as a $f(T_j)$. Obviously, one can see that there exists at least one conversion value for a given $T_j$, maintained at a steady-state condition.

```
% CSTR Multiple Steady States Example
clear; clc;
% Parameters
CA0 = 5.0;           % mol/L
T0 = 300;            % K
tau = 5;             % min
k0 = 7.2e1;          % 1/min
E = 8.00e4;          % J/mol
R = 8.314;           % J/mol-K
dHr = -5e3;          % J/mol (exothermic)
FA0 = 10;            % mol/s
Cp = 72;             % J/mol-K
UA = 5e5;            % J/min-K
VCSTR = 1000;        % L
% Function to solve mass & energy balances
cstr_bal = @(vars, Tj) [
   % Mole balance
   tau * k0 * exp(-E/(R * vars(2))) * (1 - vars(1)) - vars(1);
   % Energy balance
   FA0 * Cp * (T0 - vars(2)) ...
   - k0 * CA0 * (1 - vars(1)) * VCSTR * (dHr) ...
   + UA * (Tj - vars(2))];  % FIXED: multiplication, not indexing
% Coolant temperature sweep
Tj_range = 280:5:320;
X_sol = [];
Tj_vals = [];
T_sol = [];
for Tj = Tj_range
   for guess = [0.1 0.5 0.9] % Different guesses for multiple roots
      sol = fsolve(@(vars) cstr_bal(vars, Tj), [guess, 350], ...
            optimoptions('fsolve','Display','off'));
      X = sol(1); T = sol(2);
      % Keep only physical, non-duplicate solutions
      if X >= 0 && X <= 1 && T > 0
         if isempty(X_sol) || min(abs(X_sol - X) + abs(T_sol - T)) > 1e-3
            X_sol = [X_sol; X];
            T_sol = [T_sol; T];
            Tj_vals = [Tj_vals; Tj];
         end
      end
   end
end
% Plot results
figure;
plot(Tj_vals, X_sol, 'ro', 'MarkerFaceColor', 'r');
xlabel('Coolant Temperature T_j (K)');
ylabel('Conversion X');
title('CSTR Steady States vs Coolant Temperature');
grid on;
ax = gca; ax.FontSize = 14;ax.FontWeight = 'bold';
```

**Figure 5.13** MATLAB code (**MSS_CSTR.m**) for a multi-steady-state condition of a nonadiabatic CSTR where heat is exchanged with a cooling jacket, maintained at a controlled temperature $T_j$.

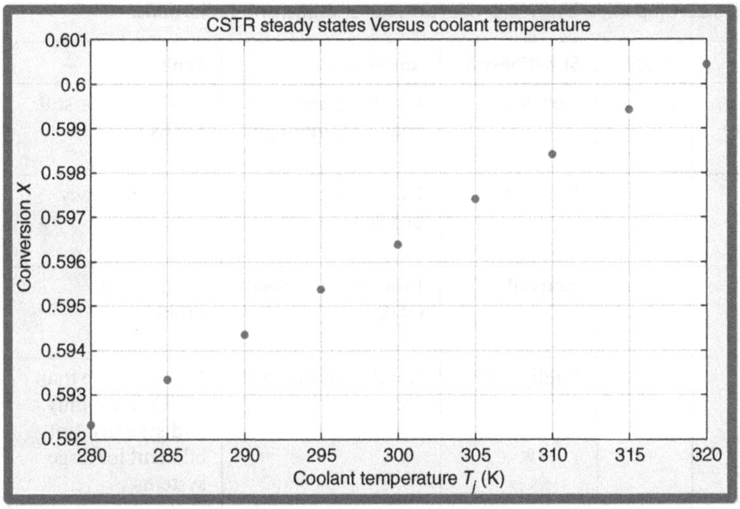


**Figure 5.14** The generated plot for conversion, $X$, as a $f(T_j)$. There exists at least one conversion value for a given $T_j$, maintained at a steady-state condition.

## 5.11 The Stability of Numerical Solutions of Algebraic and Differential Equations

When solving algebraic and differential equations (ADEs and ordinary differential equations [ODEs]), especially in chemical reaction engineering problems such as unsteady-state CSTR and PFR models, numerical stability is a key concern. Stiff systems, characterized by widely varying time scales, pose special challenges because explicit methods require extremely small-time steps to remain stable, leading to impractical computation times. The Jacobian matrix of the system provides insight into stiffness through the magnitude and distribution of its eigenvalues. For nonstiff systems, explicit solvers such as MATLAB's ode45 are efficient, whereas stiff systems demand implicit solvers like ode15s or backward differentiation formula (BDF) methods to maintain stability with larger-time steps.

Stability considerations also vary with system behavior. Oscillatory systems require numerical methods that avoid spurious oscillations or excessive damping, while stiff oscillatory problems may benefit from symplectic or A-stable integrators. Implicit methods tend to be more stable but involve solving nonlinear algebraic equations at each step, creating a trade-off between computational cost and stability. In addition, delay differential equations (DDEs), which depend on past states, introduce extra complexity by affecting feedback mechanisms and often require specialized solvers for accurate, stable integration.

The choice of method, therefore, depends on the stiffness of the system, the presence of oscillations or delays, and the desired balance between accuracy and computational efficiency.

In **Appendix C: The Stability of Numerical Solutions of Algebraic and Differential Equations**, you will find a more detailed treatment, including mathematical definitions of stiffness, eigenvalue analysis, Jacobian formulation for a reactor model, comparisons between implicit and explicit schemes, and specialized techniques for oscillatory and delayed systems, together with solver recommendations and practical MATLAB implementation notes.

Building on the discussion of numerical stability in solving ODEs and ADEs (as expatiated in Appendix C), Table 5.1 summarizes MATLAB ODE solvers along with their properties, advantages,

**Table 5.1** MATLAB ODE solvers accompanied by their properties, advantages, and limitations.

| Solver | Type | Order | Stiff/Nonstiff | Advantages | Limitations |
|---|---|---|---|---|---|
| **ode45** | Explicit RK (Dormand-Prince) | 4(5) | Nonstiff | General-purpose, adaptive, and high accuracy | Inefficient for stiff ODEs |
| **ode23** | Explicit RK (Bogacki-Shampine) | 2(3) | Nonstiff | Fast for simple problems | Lower accuracy |
| **ode113** | Variable Step (Adams-Bashforth-Moulton) | Varies | Nonstiff | Efficient for smooth ODEs | Not for stiff problems |
| **ode23s** | Trapezoidal +Backward Euler (Modified Rosenbrock) | 2(3) | Mildly stiff | Handles mildly stiff ODEs | Less accurate than ode15s for highly stiff systems. Not efficient for large systems |
| **ode15s** | Implicit RK + Backward Differentiation Formula (BDF). Also, called Gear's Method | 1 to 5 | Stiff with large eigenvalues | Best for stiff ODEs and Differential/Algebraic Equations (DAEs). Stiff chemical kinetics | Computationally expensive |
| **ode23t** | Trapezoidal Rule (Implicit RK) | 2 | Stiff with large eigenvalues | Good for oscillations or mildly stiff with oscillations. More accurate than ode23s for the oscillatory cases | Slower than ode15s |
| **ode23tb** | Trapezoidal +BDF (Implicit RK) | 2 | Very stiff | Use when ode15s is too slow. Use for chemical kinetics and reaction modeling | Not as accurate as ode15s for highly stiff systems |
| **dde23** | Delay Differential Equations (Bogacki-Shampine) | 2(3) | Nonstiff | Handles time delays. $dy/dt = f(y(t), y(t\text{-}\tau))$. Delay ODEs in epidemiology, population dynamics, & control systems | Not for stiff delay systems |
| **ddesd** | Delay Differential Equations with state-dependent delays (Explicit) | Varies | Nonstiff | Handles $dy/dt = f(y(t), y(g(y)))$. ODEs with dynamically changing delays | More complex setup than dde23 |
| **ode15i** | Fully Implicit ODEs. Newton Iteration | Varies | Stiff | Solves fully implicit DAEs. $f(t, y, dy/dt) = 0$ | Requires Jacobian. Computationally Expensive |

and limitations. Notably, an ODE solver can also be used alongside algebraic equations, as demonstrated in our examples on the unsteady-state behavior of chemical reactors.

**Example 5.6** ***MATLAB Code for Unsteady-state CSTR Gas-phase Second-order HI Formation***

Let us consider the gas-phase reaction between hydrogen ($H_2$) and iodine ($I_2$) to form hydrogen iodide (HI):

$$H_2(g) + I_2(g) \rightarrow 2HI(g) \qquad \text{Or:} \qquad A(g) + B(g) \rightarrow 2C(g)$$

This reaction is second-order overall, with a first-order dependence on both $H_2$ and $I_2$ concentrations. The rate law is [7]:

$$-r_A = kC_AC_B = 2 \times 10^{11} e^{-\left(\frac{169,900}{8.314*T}\right)} C_AC_B \tag{A}$$

**Heat Capacity Data**

**Source:** NIST Chemistry Webbook, Standard Reference Database Number 69 July 2001 Release https://webbook.nist.gov/chemistry/ (e.g., search for "Hydrogen" then look for gas-phase thermochemistry).

$$C_{P,H2}\left(\frac{\text{J}}{\text{mol.K}}\right) = A + BT + CT^2 + DT^3 + \frac{E}{T^2} \qquad 100\text{ K} \le T \le 1{,}500\text{ K}$$

$$C_{P,H2} = 29.093 - (0.2E-3*T) + (0.4E-6*T^2) - (0.87E-9*T^3) + \left(\frac{0.0055E5}{T^2}\right) \tag{B}$$

$$C_{P,I2}\left(\frac{\text{J}}{\text{mol.K}}\right) = A + BT + CT^2 + DT^3 + \frac{E}{T^2} \qquad 298\text{ K} \le T \le 1{,}000\text{ K}$$

$$C_{P,I2} = 37.4 - (0.0034E-3*T) + (0.18E-6*T^2) - (0.89E-9*T^3) - \left(\frac{0.245E5}{T^2}\right) \tag{C}$$

$$C_{P,HI}\left(\frac{\text{J}}{\text{mol.K}}\right) = A + BT + CT^2 + DT^3 + \frac{E}{T^2} \qquad 200\text{ K} \le \text{T} \le 1{,}500\text{ K}$$

$$C_{P,HI} = 28.14 + (1.88E-3*T) - (6.954E-6*T^2) + (1.08E-9*T^3) - \left(\frac{0.205E5}{T^2}\right) \tag{D}$$

Using Appendix A | A.1 MATLAB (**DelG_Keq.m**) Code for Calculation of $\boldsymbol{\Delta G^0}$, $\boldsymbol{\Delta H^0}$, and $\boldsymbol{K_{eq}}$ for a Chemical Reaction, the following data are extracted:

**Selected Reactants:**
**1 mol of Hydrogen molecule (g) (H2(g))**
**1 mol of Iodine molecule (g) (I2(g))**
**Selected Products:**
**2 mol of Hydrogen iodide (g) (HI(g))**
**Temperature (K): 298**
**Molar Gibbs free energy of reaction (kJ/mol): -15.9**
**Molar heat of reaction @ 298 K, (kJ/mol): -9.4**
**Equilibrium constant (dimensionless): 612.5149**

$$\Delta H_r(T) = -9{,}400\frac{\text{J}}{\text{mol}} + \int_{298}^{T}(-C_{P,A} - C_{P,B} + 2C_{P,C})dT \tag{E}$$

Let us write enough number of differential and algebraic equations to describe the CSTR performance under unsteady-state conditions.

**Component *A*-Molar Balance**

$$F_{A0} - F_A - kC_AC_BV_{CSTR} = \frac{d(V_{CSTR}C_A)}{dt} = V_{CSTR}\frac{dC_A}{dt} + C_A\frac{dV_{CSTR}}{dt}$$

We will assume that the gas mixture will always occupy the entire volume of the CSTR; that is, constant $V_{CSTR}$.

The above equation reduces to:

$$F_{A0} - F_A - kC_AC_BV_{CSTR} = V_{CSTR}\frac{dC_A}{dt}$$

$$C_{A0} = \frac{Y_{A0}P_0}{RT_0} \tag{F}$$

$$C_{B0} = \frac{Y_{B0}P_0}{RT_0} \tag{G}$$

$$\upsilon_0 = \frac{\dot{n}RT_0}{P_0} = \frac{F_0RT_0}{P_0} \tag{H}$$

$$\upsilon = \frac{\dot{n}RT}{P} = \frac{FRT}{P}$$

$$\frac{\upsilon}{\upsilon_0} = \frac{\frac{FRT}{P}}{\frac{F_0RT_0}{P_0}}$$

Since $\Delta n = 2\text{-}1\text{-}1 = 0$, then $F_{Total} = F_0 = F = F_{A0} + F_{B0}$

$$\frac{\upsilon}{\upsilon_0} = \frac{\frac{FRT}{P}}{\frac{F_0RT_0}{P_0}} = \left(\frac{T}{P}\right) * \left(\frac{P_0}{T_0}\right) = \left(\frac{T}{T_0}\right) * \left(\frac{P_0}{P}\right)$$

$$\upsilon = \upsilon_0 * \left(\frac{T}{T_0}\right) * \left(\frac{P_0}{P}\right) \tag{I}$$

$$\frac{\upsilon_0 * C_{A0} - \upsilon * CA}{V_{CSTR}} - 2\times10^{11} * e^{-\left(\frac{169900}{8.314*T}\right)}C_A * C_B = \frac{d(C_A)}{dt} \tag{J}$$

**Component *B*-Molar Balance**

$$\frac{\upsilon_0 * CB_0 - \upsilon * CB}{V_{CSTR}} - 2\times10^{11} * e^{-\left(\frac{169900}{8.314*T}\right)}C_AC_B = \frac{d(C_B)}{dt} \tag{K}$$

With $V_{CSTR}$, lateral area $\pi DL$, and diameter, $D$, we have:

$AH = (4^* V_{CSTR}{}^* 0.001/D)$; %0.001 to convert $L$ to $m^3$ for $D$ in m

(Lateral Area = 4* Volume/Diameter)

if $T < 665$

$Q_{exch} = 500^* AH^* (665 - T)$; $\%U = 500\ \text{W/m}^2\cdot\text{K}$ for steam heating medium jacket (La) else

$$Q_{exch} = 400^{*}\, AH^{*}\, (298 - T); \quad \%U = 400\ \mathrm{W/m^2 \cdot K}\ \text{for cooling water medium jacket} \quad \text{(Lb)}$$

end

**Energy Balance**

$$F_0 * (Cpmix_0 * T_0 - Cpmix * T) - 2 \times 10^{11} * e^{-\left(\frac{169900}{8.314*T}\right)} C_A * C_B * V_{CSTR}(\Delta H_r) + Q_{exch} = \frac{d\left(nC_V^{mix}T\right)}{dt} = n\overline{C_V^{mix}}\frac{d(T)}{dt} \quad \text{(M)}$$

$$CV_A = CP_A - R; \quad \text{(Na)}$$

$$CV_B = CP_B - R; \quad \text{(Nb)}$$

$$CV_C = CP_C - R; \quad \text{(Nc)}$$

$$F_A = \upsilon^{*}\, C_A; \quad \text{(Oa)}$$

$$F_B = \upsilon^{*}\, C_B; \quad \text{(Ob)}$$

$$C_C = 2^{*}\, (C_{A0} - C_A); \quad \text{(Oc)}$$

$$\mathrm{F_C} = \upsilon^{*}\, \mathrm{C_C}; \quad \text{(Od)}$$

$$Y_A = F_A/(F_A + F_B + F_C); \quad \text{(Pa)}$$

$$Y_B = F_B/(F_A + F_B + F_C); \quad \text{(Pb)}$$

$$Y_C = 1 - Y_A - Y_B; \quad \text{(Pc)}$$

$$X_A = (1/[Y_{A0}])^{*}\, (Y_{A0} - Y_A); \quad \text{(Q)}$$

$$T_{avg} = mean\,(T) \quad \text{(R)}$$

$$\overline{C_V^{mix}} = Y_A C_V^A(T_{avg}) + (Y_B) C_V^B(T_{avg}) + (1 - Y_A - Y_B) C_V^C(T_{avg}) \quad \text{(S)}$$

$$Cpmix = Y_A CP_A + (Y_B) * CP_B + (1 - Y_A - Y_B) * CP_C \quad \text{(T)}$$

**CSTR Inside Pressure**

For gas-phase fluid flow, the following considerations apply:

1) The feed stream drives the flow into the reactor, requiring a small pressure drop across the inlet valve or piping system to sustain forward movement.
2) The outlet pressure is typically regulated, either by operating at atmospheric pressure or using a back-pressure control valve to maintain flow continuity.
3) Under continuous operation, the reactor pressure remains quasi-constant and is usually maintained at a value slightly lower than the feed pressure $P_0$.

```
clear variables;
Rgas = 0.08206; % (atm.L)/mol.K
R = 8.314; % J/(mol.K)
% Given Inputs
F0 = input('Enter the feed molar flow rate (F0) in mol/s: ');
%F0 = 0.5; % mol/s
T0 = input('Enter the feed temperature (T0) in K: ');
%T0 = 300; % K
P0 = input('Enter the feed pressure (P0) in atm: ');
%P0 = 40; % atm
v0 = (F0 * Rgas * T0) / P0;
YA0 = input('Enter the feed mole fraction of A (YA0): ');
%YA0 = 0.5;
YB0 = input('Enter the feed mole fraction of B (YB0): ');
%YB0 = 0.5;
CA0 = (YA0 * P0) / (Rgas * T0);
CB0 = (YB0 * P0) / (Rgas * T0);
FA0 = YA0 * F0;
FB0 = YB0 * F0;
VCSTR = input('Enter the CSTR Volume, (L): ');
%VCSTR = 3000; % Volume of CSTR, L
D = input('Enter the BR diameter in m, (D): ');
%D=1.0;
Pzero=P0;% P(0)=1.0 atm at time zero.
Tzero=T0; % T(0)=300 K at time Zero.
% Initial conditions
y0 = [CA0; CB0; Tzero];
% Time span
tspan = [0 800]; % seconds
% Calculate initial heat capacities once
CPA0 = 29.093 - (0.2e-3*T0) + (0.4e-6*T0^2) - (0.87e-9*T0^3) + (0.0055e5/T0^2);
CPB0 = 37.4 - (0.0034e-3*T0) + (0.18e-6*T0^2) - (0.89e-9*T0^3) - (0.245e5/T0^2);
CPC0 = 28.14 + (1.88e-3*T0) - (6.954e-6*T0^2) + (1.08e-9*T0^3) - (0.205e5/T0^2);
Cpmix0 = YA0 * CPA0 + YB0 * CPB0 + (1 - YA0 - YB0) * CPC0; % Assuming C is 0 at feed
% Pass these constants to dCSTR
```

**Figure 5.15** MATLAB code (**CSTR_T_CA_Gas.m**) for the solution of an unsteady-state CSTR using the MATLAB **ode23** solver for HI gas-phase formation. Eqs. A through V are used as DAE's.

```
options = odeset('RelTol', 1e-6, 'AbsTol', 1e-9, 'NonNegative', [1 2 3]);
[t, y] = ode23(@(t, y) dCSTR(t, y, v0, CA0, CB0, P0, R, T0, VCSTR, F0, D, Cpmix0), tspan, y0, options);
%[t, y] = ode78(@(t, y) dCSTR(t, y, v0, CA0, CB0, P0, R, T0, VCSTR, F0, D, Cpmix0), tspan, y0, options);
% Extract results
CA = y(:, 1);
CB = y(:, 2);
T = y(:, 3);
P = 0.98*P0;
CC = 2*(CA0 - CA);
% Molar fractions
YA =CA./(CA+CB+CC);
YB =CB./(CA+CB+CC);
XA=(1/YA0).*(YA0-YA);
% Plot results
figure;
subplot(4,1,1); plot(t, CA, 'b', 'LineWidth', 1.5); xlabel('Time (s)'); ylabel('C_A (mol/L)'); title('Concentration of A');
ax = gca;
ax.FontSize = 14;
ax.FontWeight = 'bold';
subplot(4,1,2); plot(t, CC, 'r', 'LineWidth', 1.5); xlabel('Time (s)'); ylabel('C_C (mol/L)'); title('Concentration of C');
ax = gca;
ax.FontSize = 14;
ax.FontWeight = 'bold';
subplot(4,1,3); plot(t, T, 'g', 'LineWidth', 1.5); xlabel('Time (s)'); ylabel('T (K)'); title('Temperature');
ax = gca;
ax.FontSize = 14;
ax.FontWeight = 'bold';
subplot(4,1,4); plot(t, XA, 'm', 'LineWidth', 1.5); xlabel('Time (s)'); ylabel('X_A'); title('Conversion of A');
ax = gca;
ax.FontSize = 14;
ax.FontWeight = 'bold';
% - - - Inside dCSTR function - - -
function [dydt] = dCSTR(~, y, v0, CA0, CB0, P0, R, T0, VCSTR, F0, D, Cpmix0)
  CA = max(0, min(CA0, y(1)));
  CB = max(0, min(CB0, y(2)));
  T = max(200, min(700, y(3)));
  CC=2*(CA0-CA);
```

**Figure 5.15** (*Continued*)

```
    P = 0.98 * P0;
    v = v0 .* (T ./ T0) .* (P0 ./ P);
    FA = v .* CA;
    FB = v .* CB;
    Ctot = CA + CB + CC;
    n = VCSTR * Ctot;
    % Calculate current mole fractions inside the CSTR
    YA =CA ./ Ctot;
    YB = CB ./ Ctot;
    YC = CC ./ Ctot;
    % Heat capacities at current T
    CPA = 29.093 - (0.2e-3*T) + (0.4e-6*T^2) - (0.87e-9*T^3) + (0.0055e5/T^2);
    CPB = 37.4 - (0.0034e-3*T) + (0.18e-6*T^2) - (0.89e-9*T^3) - (0.245e5/T^2);
    CPC = 28.14 + (1.88e-3*T) - (6.954e-6*T^2) + (1.08e-9*T^3) - (0.205e5/T^2);
    % Average heat capacities (no need for Tavg, just use current T)
    CVA = CPA - R;
    CVB = CPB - R;
    CVC = CPC - R;
    CVavg = YA * CVA + YB * CVB + YC * CVC; % Use current mole fractions
    Cpmix = YA * CPA + YB * CPB + YC * CPC; % Use current mole fractions
    % Reaction rate constant
    k = 2e11 * exp(-169900 / (R * T));
    r = k * CA * CB;
     AH=(4*VCSTR*0.001/D);
    T_switch = 685;
     Qh = 500 * AH * (685 - T);
     Qc = 400 * AH * (298 - T);
    transition = 1 / (1 + exp((T - T_switch))); % smooth switch
    Qexch = transition * Qh + (1 - transition) * Qc;
      dHr = -9400 + integral(@(T) (-CPA - CPB + 2*CPC), 298, T, 'ArrayValued', true);
    % ODEs
    dCA_dt = (v0 * CA0 - FA) / VCSTR - r;
    dCB_dt = (v0 * CB0 - FB) / VCSTR - r;
    dT_dt = (F0 * (Cpmix0 * T0 - Cpmix * T) - r * VCSTR * dHr + Qexch) / (n * CVavg);
    % Return ODEs
    dydt = [dCA_dt; dCB_dt; dT_dt];
end
```

**Figure 5.15** (*Continued*)

Accordingly, we assume the reactor operates at a pressure of:

$$P = 0.98^{*}\ P_0$$

which reflects the small pressure drop necessary to sustain steady-state gas flow through the system.

$V_{CSTR} = 2000;$ %volume of CSTR in L (V)

I.C.'s @t = 0:

$T(0) = Tzero; \quad CA(0) = CA0; \quad CB(0) = CB0; \quad P(0) = Pzero$ (Wa-d)

Figure 5.15 shows MATLAB code (**CSTR_T_CA_Gas.m**) based on translating Eqs. A through W into MATLAB code, which describes HI gas-phase formation in an unsteady-state, nonadiabatic CSTR. The MATLAB **ode23** solver can be used for control systems (see Table 5.1), as we have temperature control.

The following set of input parameters is used to generate the results shown in Figure 5.16 for the profiles of $C_A$, $C_C$, $T$, and $X_A$.

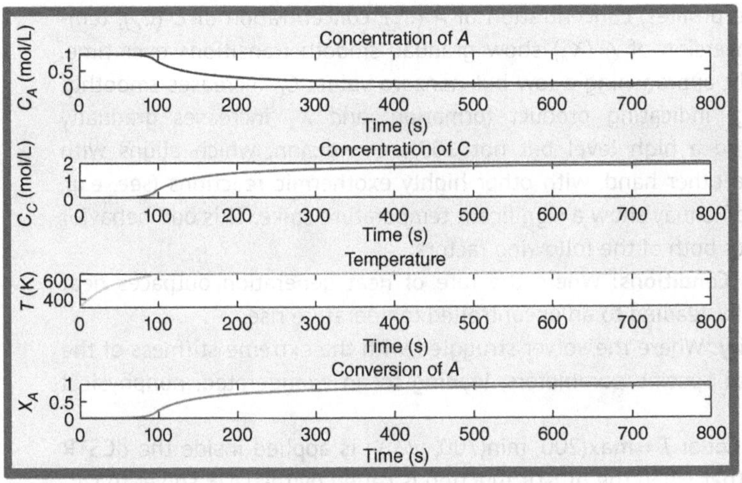


**Figure 5.16** $C_A$, $C_C$, $T$, and $X_A$ profiles for an unsteady-state CSTR using the MATLAB **ode23** solver for HI gas-phase formation, with a plateau conversion, $X_A$, of 0.92.

```
≫ CSTR_T_CA_Gas
Enter the feed molar flow rate (F0) in mol/s: 0.5
Enter the feed temperature (T0) in K: 300
Enter the feed pressure (P0) in atm: 50
Enter the feed mole fraction of A (YA0): 0.5
Enter the feed mole fraction of B (YB0): 0.5
Enter the CSTR Volume, (L): 3000
Enter the CSTR diameter in m, (D): 1.5
```

**NOTE #7**

1) There is a set of adjustable feed input parameters that can be edited by the user, which are shown below:
   a) $\boldsymbol{F_0}$: Feed molar flow rate, mol/s.
   b) $\boldsymbol{T_0}$: Feed temperature, K.
   c) $\boldsymbol{P_0}$: Feed pressure, atm.
   d) $\boldsymbol{Y_{A0}}$: Feed molar composition of *A*.
   e) $\boldsymbol{Y_{B0}}$: Feed molar composition of *B*.
   f) $\boldsymbol{V_{CSTR}}$: Volume of the reactor, L.
2) The MATLAB **ode23** solver is used in Figure 5.15. However, as Table 5.1 shows, there are several ode solvers that can be used to solve a set of DAE's, depending on the condition of the set of DAE's from a numerical stability point of view. The code written in Figure 5.15 can be reused as well for ode solvers; namely: **ode15s, ode23, ode23s, ode23t, ode23tb, and ode113**. Notice that, as far as CPU time is concerned, each ode solver type will have a different CPU duration.
3) In Figure 5.16, all the profiles, concentration of *A* ($C_A$), concentration of *C* ($C_C$), temperature (*T*), and conversion of *A* ($X_A$) show gradual, smooth transitions over time. $C_A$ decreases smoothly, approaching a low but nonzero value; $C_C$ increases smoothly, eventually plateauing, indicating product formation; and $X_A$ increases gradually and smoothly, reaching a high level but not 100% conversion, which aligns with the $C_A$ profile. On the other hand, with other highly exothermic reactions (see, e.g., Figure 5.23), the *T* profile may show a significant temperature spike. This odd behavior may represent either or both of the following factors:
   a) **Runaway Reaction Conditions:** Where the rate of heat generation outpaces heat removal dramatically, leading to an uncontrolled temperature rise.
   b) **Numerical Instability:** Where the solver struggled with the extreme stiffness of the system under those specific parameters, leading to an exaggerated, nonphysical spike.
4) **Notice That the Line Code:** *T* = max(200, min(700, *y*(3))); is applied inside the dCSTR function. This means that when the dCSTR function is called by the ODE solver to calculate the derivatives (*dCA/dt*, *dCB/dt*, and *dT/dt*), the *T* value used for those calculations is capped at 700 K. However, the ODE solver's prediction for the next *y*(3) (which is *T*) is based on these calculated derivatives. If the calculated *dT/dt* is very large and positive, the solver might take a step that results in a *y*(3) value significantly higher than 700 K before it calls dCSTR again for the next iteration. In brief, the max and min functions inside dCSTR affect only the input *T* that the derivative calculation sees, not the output *y*(3) that the solver itself is propagating. The y variable returned by the ode23 solver (and then plotted) contains the raw, unconstrained values that the solver computed. *T* = 700 K is an adjustable parameter, selected to be slightly above the upper *T* control limit, which is 685 K. If the heating medium *T* is changed, then this *T* can be adjusted as well to end up with a smooth *T* profile.

5) Appendix B | B.5 "MATLAB Code (**CSTR_T_CA_Gas_ode15i.m**) for Describing a Gas-phase HI Formation in a nonadiabatic CSTR with Cp = f(T), Using **ode15i** Solver" implements the MATLAB **ode15i** solver, which requires special coding compared with that written for the above listed ode solvers.
6) Appendix B | B.6 "The Volume Expansion Effect due to a Change in Total Number of Moles as a Result of a Gas-Phase Chemical Reaction" shows how to express the volume expansion in terms of the mole fraction of the limiting reactant, $P$, and $T$ for the general case where $\Delta n \neq 0$. Notice that $\Delta n = 0$ for HI formation; that is, Example 5.6.
7) Component **B** profile is essentially the same as that of the component **A** profile. No need to show it.

**Example 5.7** ***MATLAB Code for Steady-state CSTR Gas-phase Second-order HI Formation***

Figure 5.17 shows MATLAB code (**CSTR_T_CA_Gas_SS.m**) for the steady-state counterpart of the unsteady-state, depicted by Figure 5.15. Four different initial guesses are made for steady-state final variables: $C_A$, $C_B$, $T$, and $P$. The sum of the absolute values of the residual vector will be minimized. Ideally, it will reach zero under steady-state conditions.

Below are the results of running the **CSTR_T_CA_Gas_SS.m** file:

```
>> CSTR_T_CA_Gas_SS
Enter the feed molar flow rate (F0) in mol/s: 2
Enter the feed temperature (T0) in K: 300
Enter the feed pressure (P0) in atm: 50
Enter the feed mole fraction of A (YA0): 0.5
Enter the feed mole fraction of B (YB0): 0.5
Enter the CSTR Volume, (L): 2000
Enter the CSTR Diameter, (m): 1
VCSTR = 2000 L
tCSTR = 2031.0342 Unit of time appearing in F0
Steady-state values:
CAss = 0.0826 mol/L
CBss = 0.2214 mol/L
CCss = 1.8658 mol/L
Tss = 680.0585 K
YAss = 0.0381
XAss = 0.9239
```

```
clear; close all;
Rgas = 0.08206; % (atm.L)/mol.K
R = 8.314; % J/(mol.K)
F0 = input('Enter the feed molar flow rate (F0) in mol/s: ');
%F0 = 2; % mol/s
T0 = input('Enter the feed temperature (T0) in K: ');
%T0 = 600; % K
P0 = input('Enter the feed pressure (P0) in atm: ');
%P0 = 50; % atm
YA0 = input('Enter the feed mole fraction of A (YA0): ');
%YA0 = 0.5;
YB0 = input('Enter the feed mole fraction of B (YB0): ');
%YB0 = 0.5;
v0 = (F0 * Rgas * T0) / P0;
CA0 = (YA0 * P0) / (Rgas * T0);
CB0 = (YB0 * P0) / (Rgas * T0);
%MW0 = 2*YA0 + 253.8*YB0;
FA0 = YA0 * F0;
FB0 = YB0 * F0;
VCSTR = input('Enter the CSTR Volume, (L): ');
%VCSTR = 2000; % Volume of CSTR, L
D = input('Enter the CSTR Diameter, (m): ');
%D=1.0;
% Define the steady-state residual function
function residuals = steadyStateCSTR(y, v0, CA0, CB0, P0, R, T0, VCSTR, F0, D)
   CA = y(1);
   CB = y(2);
   T = y(3);
   CA = max(0, min(CA0, CA)); % Ensure CA is within [0, CA0]
   CB = max(0, min(CB0, CB)); % Ensure CB is within [0, CB0]
   T = max(200, min(1000, T)); % Ensure T is within [200, 1000] K
   P = 0.98 * P0;
   CC=2*(CA0-CA);
   v = v0 * (T / T0) * (P0 / P);
   FA = v * CA;
   FB = v * CB;
   FC = v * CC;
   Ctot=CA+CB+CC;
   n=VCSTR*Ctot;
   YA = FA / (FA + FB + FC);
```

**Figure 5.17** MATLAB code (**CSTR_T_CA_Gas_SS.m**) for the steady-state counterpart of the unsteady-state, depicted by Figure 5.15.

```
    YB = FB / (FA + FB + FC);
    YC = 1 - YA - YB;
    % Heat capacities and other calculations (same as before)
    CPA = 29.093 - (0.2e-3*T) + (0.4e-6*T^2) - (0.87e-9*T^3) + (0.0055e5/T^2);
    CPB = 37.4 - (0.0034e-3*T) + (0.18e-6*T^2) - (0.89e-9*T^3) - (0.245e5/T^2);
    CPC = 28.14 + (1.88e-3*T) - (6.954e-6*T^2) + (1.08e-9*T^3) - (0.205e5/T^2);
    Tavg=mean(T);
    CPAavg = 29.093 - (0.2e-3*Tavg) + (0.4e-6*Tavg^2) - (0.87e-9*Tavg^3) + (0.0055e5/Tavg^2);
    CPBavg = 37.4 - (0.0034e-3*Tavg) + (0.18e-6*Tavg^2) - (0.89e-9*Tavg^3) - (0.245e5/Tavg^2);
    CPCavg = 28.14 + (1.88e-3*Tavg) - (6.954e-6*Tavg^2) + (1.08e-9*Tavg^3) - (0.205e5/Tavg^2);
    CVAavg = CPAavg - R;
    CVBavg = CPBavg - R;
    CVCavg = CPCavg - R;
    CVavg = YA * CVAavg + YB * CVBavg + (YC) * CVCavg;
    Cpmix = YA * CPA + YB * CPB + (YC) * CPC;
    CPA0 = 29.093 - (0.2e-3*T0) + (0.4e-6*T0^2) - (0.87e-9*T0^3) + (0.0055e5/T0^2);
    CPB0 = 37.4 - (0.0034e-3*T0) + (0.18e-6*T0^2) - (0.89e-9*T0^3) - (0.245e5/T0^2);
    CPC0 = 28.14 + (1.88e-3*T0) - (6.954e-6*T0^2) + (1.08e-9*T0^3) - (0.205e5/T0^2);
    Cpmix0 = YA * CPA0 + YB * CPB0 + (YC) * CPC0;
    k = 2e11 * exp(-169900 / (R * T));
    r = k * CA * CB;
    % Heat exchange term
    AH = (4 * VCSTR * 0.001 / D); % 0.001 to convert L to m3 for D in m (Area = 4*Volume/Diameter)
    T_switch = 685;
    Qh = 500 * AH * (685 - T);
    Qc = 400 * AH * (298 - T);
    transition = 1 / (1 + exp((T - T_switch))); % smooth switch
    Qexch = transition * Qh + (1 - transition) * Qc;
    dHr = -9400 + integral(@(T) (-CPA - CPB + 2*CPC), 298, T, 'ArrayValued', true);
    % Steady-state residuals
    dCA_dt = (v0 * CA0 - v * CA) / VCSTR - r;
    dCB_dt = (v0 * CB0 - v * CB) / VCSTR - r;
    dT_dt = (F0 * (Cpmix0 * T0 - Cpmix * T) - r * VCSTR * dHr + Qexch) / (n * CVavg);
    residuals = [dCA_dt; dCB_dt; dT_dt];
end
% Multiple Initial Guesses
initialGuesses = [
    CA0, CB0, T0;
    CA0/2, CB0/2, T0+100;
```

**Figure 5.17** *(Continued)*

```
    CA0*0.8, CB0*0.8, T0-100;
    CA0*0.3, CB0*0.7, T0+50;
];
bestSolution = [];
bestResidual = Inf;
options = optimoptions('lsqnonlin', 'Display', 'off', 'TolFun', 1e-9, 'TolX', 1e-9);
lb = [0, 0, 300]; % Lower bounds: CA >= 0, CB >= 0, T >= 200
ub = [CA0, CB0, 1000]; % Upper bounds: CA <= CA0, CB <= CB0, T <= 1000
for i = 1:size(initialGuesses, 1)
  [y_ss, resnorm, residual, exitflag] = lsqnonlin(@(y) steadyStateCSTR(y, v0, CA0, CB0, P0, R, T0, VCSTR, F0, D), initialGuesses(i, :), lb, ub, options);
    totalResidual = sum(abs(residual));
    if totalResidual < bestResidual
      bestResidual = totalResidual;
      bestSolution = y_ss;
    end
end
% Extract steady-state values from the best solution
CA_ss = bestSolution(1);
CB_ss = bestSolution(2);
T_ss = bestSolution(3);
P_ss=0.98*P0;
CC_ss = 2*(CA0-CA_ss);
v_ss = v0 * (T_ss / T0) * (P0 / P_ss);
FA_ss = v_ss * CA_ss;
FB_ss = v_ss * CB_ss;
FC_ss= v_ss * CC_ss;
YA_ss = FA_ss / (FA_ss + FB_ss+FC_ss);
XA_ss=(1/YA0)*(YA0-YA_ss);
Tau=VCSTR/v0;
disp(['VCSTR = ', num2str(VCSTR), ' L']);
disp(['tCSTR = ', num2str(Tau), ' Unit of time appearing in F0']);
% Display steady-state results
fprintf('Steady-state values:\n');
fprintf('CAss = %.4f mol/L\n', CA_ss);
fprintf('CBss = %.4f mol/L\n', CB_ss);
fprintf('CCss = %.4f mol/L\n', CC_ss);
fprintf('Tss = %.4f K\n', T_ss);
fprintf('YAss = %.4f\n', YA_ss);
fprintf('XAss = %.4f\n', XA_ss);
```

**Figure 5.17** *(Continued)*

**NOTE #8**

The final steady-state values are dictated by the initial feed parameters:

1) $F_0$: Feed molar flow rate, mol/s.
2) $T_0$: Feed temperature, K
3) $P_0$: Feed pressure, atm.
4) $Y_{A0}$: Feed molar composition of $A$
5) $Y_{B0}$: Feed molar composition of $B$.
6) $V_{CSTR}$: Volume of the reactor, L

And also, by the selection of the four different initial guesses expected to represent steady-state final variables: $C_A$, $C_B$, $T$, and $P$.

In real life, it says that CSTR may eventually evolve to different scenarios where each scenario represents a unique steady-state condition. Moreover, from a process dynamics and control point of view, CSTR may move from one steady state to another, depending on the magnitude and direction (i.e., increasing or decreasing) of the source of disturbance. The source of disturbance can be one or more of the feed input parameters, as well as the surrounding properties, like the ambient temperature or cooling/heating jacket temperature.

## 5.12 Unsteady-State Liquid-Phase PFR for Glucose Isomerization Reaction

An unsteady-state liquid-phase Plug Flow Reactor (PFR) for the glucose isomerization reaction models the transient behavior of glucose conversion to fructose along the reactor length and over time before steady-state conditions are reached. In this process, an aqueous glucose feed containing dissolved catalyst enters the tubular reactor, and the reversible, slightly endothermic isomerization proceeds according to first-order or more complex kinetics, depending on catalyst type and operating conditions. The unsteady-state formulation couples the molar balance for glucose with an energy balance that accounts for heat effects, variable heat capacities, and possible external heat exchange, while assuming negligible axial dispersion and constant liquid density. Time dependence is introduced because the concentration and temperature profiles evolve as the reactor approaches steady state after a change in inlet flow rate, feed composition, or temperature. Such a model is essential for startup analysis, process control design, and optimization, particularly since the reaction equilibrium and rate are sensitive to temperature, and achieving high-fructose yield requires balancing residence time, thermal management, and catalyst activity.

**Example 5.8** ***Unsteady-state PFR Liquid-phase Glucose Isomerization Reaction***

We will rework the aqueous liquid-phase first-order reaction, which is the isomerization of glucose into fructose in PFR, as shown in Example 5.1.

Write enough number of linearly independent equations, including total material balance, $A$-component molar balance, and energy balance, to describe the unsteady-state PFR system. Notice that both $C_A$ (or $X_A$) and temperature ($T$) do change with PFR time. The initial concentration of $A$, $C_{A0} = 50\,\text{mol/L}$, initial volume flow rate of 0.1 L/s, feed temperature of $T_0 = 380\,\text{K}$, and $V_{PFR} = 2{,}000\,\text{L}$.

## *A*-Component Molar Balance

$$In - Out + Generation - Consumption = Accumulation$$

$$F_{A0} - F_A + 0 - \left(-\int r_A dV\right) = \frac{dN_A}{dt} = \frac{d}{dt}\left(\int_0^{V_{PFR}} C_A(V,t)dV\right)$$

Since $V_{PFR}$ is constant and the integration limits are fixed, we can move the derivative inside the integral (under appropriate mathematical conditions, which generally hold here):

$$F_{A0} - F_A + 0 - \left(-\int r_A dV\right) = \frac{dN_A}{dt} = \int_0^{V_{PFR}} \frac{\partial C_A}{\partial t} dV$$

$$F_{A0} - F_A + 0 - \left(\int_0^{V_{PFR}} kC_A dV\right) = \int_0^{V_{PFR}} \frac{\partial C_A}{\partial t} dV$$

$$F_{A0} - F_A + 0 - \left(\int_0^{V_{PFR}} 1.2 \times 10^{-5} * e^{-\frac{60000}{8.314}\left(\frac{1}{T}-\frac{1}{298}\right)} C_A dV\right) = \int_0^{V_{PFR}} \frac{\partial C_A}{\partial t} dV \quad \text{(A)}$$

**NOTE #9**

**Discretized PFR Approximation**

In this section or in subsequent sections, handling a PFR case is a little bit tricky; we know ahead that PFR, unlike CSTR, is not uniform or homogeneous in the axial or longitudinal direction. My approach here is simply to evaluate the left-hand side of Eq. (A) instantly at each point along the reaction progress or coordinate, and the right-hand side of the same equation will be equated to the instantly evaluated left-hand side. This is equivalent to saying that if we have a series of an infinite number of small-sized perfectly mixed CSTRs, we will asymptotically approach a PFR case. This is how I got around this nonuniform PFR case in MATLAB by slicing the volume into a relatively large number of sub-volumes, and the time-variable properties are evaluated instantly at each sub-volume along the axial direction of PFR.

This is often called a "discretized PFR" or "PFR approximated by a series of CSTRs" model. It converts the partial differential equations (PDEs) that strictly govern a transient PFR into a system of ODEs, which can then be solved by MATLAB's ODE solvers. Two different MATLAB codes are written to describe the "discretized PFR" method. See Example 5.10: Unsteady-state PFR gas-phase second-order HI formation. Both MATLAB codes give the same results, quantitatively and qualitatively.

With a volume $V_{PFR}$, height $L$, and diameter, $D$, we have:

$$A_H = \pi DL = \pi D\left(\frac{V_{PFR}}{\frac{\pi}{4}D^2}\right) = \left(\frac{4V_{PFR}}{D}\right)$$

if $T < 398$

$$Q_{exch} = 500^* AH^* (398 - \text{T}); \quad \%\text{U} = 500\,\text{W/m}^2 \cdot \text{K for steam heating medium jacket} \quad \text{(Ba)}$$

else

$$Q_{exch} = 400^* AH^* (298 - \text{T}); \quad \%U = 400\,\text{W/m}^2 \cdot \text{K for cooling water medium jacket} \quad \text{(Bb)}$$

end

$$\text{Glucose: } C_{PA} = C_{VA} = 147.0 \frac{\text{J}}{\text{mol} \cdot \text{K}} \tag{C}$$

$$\text{Fructose: } C_{PB} = C_{VB} = 145.0 \frac{\text{J}}{\text{mol} \cdot \text{K}} \tag{D}$$

$$\Delta H_r = 1{,}700 \frac{\text{J}}{\text{mol}} + \int_{298}^{T} (-C_{PA} + C_{PB})dT \tag{E}$$

**Energy Balance**

$$F_{A0}C_{PA0}T_0 - FC_{Pmix}T - \left(\int_0^{V_{PFR}} 1.2 \times 10^{-5} * e^{-\frac{60000}{8.314}\left(\frac{1}{T}-\frac{1}{298}\right)} C_A dV\right) \Delta Hr + Q_{exch} = n_{PFR}\overline{C_{Vmix}}\frac{dT}{dt}$$

Since $\Delta n = 1 - 1 = 0$, then $F_{A0} = F_0 = F_A + F_B = F$.

$$F_{A0}C_{PA0}T_0 - F_{A0}C_{Pmix}T - \left(\int_0^{V_{PFR}} 1.2 \times 10^{-5} * e^{-\frac{60000}{8.314}\left(\frac{1}{T}-\frac{1}{298}\right)} C_A dV\right) \Delta Hr + Q_{exch} = n_{PFR}\overline{C_{Vmix}}\frac{dT}{dt} \tag{F}$$

$$n_{PFR} = F_0{}^* (V_{PFR}/v_0) = (C_{A0}{}^* v_0)^* V_{PFR}/v_0 = C_{A0}{}^* V_{PFR} \tag{G}$$

$$x_A = C_A/C_{Total} = C_A/C_{A0}; \tag{H}$$

$$CP_{mix} = x_A{}^* CP_A + (1 - x_A)^* CP_B; \tag{I}$$

$$CV_{mix} = x_A{}^* CV_A + (1 - x_A)^* CV_B; \tag{J}$$

### 5.12.1 MATLAB Code for Unsteady-State PFR Liquid-Phase Glucose Isomerization Reaction

Figure 5.18 shows MATLAB code (**PFR_T_CA.m**), which is written to simulate the following set of algebraic and differential equations:

```
CA0 = 100; % Initial concentration of A, mol/L
CTotal = CA0; % Total concentration, mol/L
VPFR = 2000; % Volume of batch reactor, L
n = CTotal * VPFR; % Total moles in PFR reactor, mol
v0=0.1; % L/s
FA0=CA0*v0; %mol/s
FA=CA*v0; %mol/s
T0 = 380; % Feed temperature, K
```

*(Continued)*

**(Continued)**

CPA = 147.0; % Heat capacity of A, J/(mol*K)

CPB = 145.0; % Heat capacity of B, J/(mol*K)

CVA = CPA; % Cv=Cp for liquids

CVB = CPB; % Cv=Cp for liquids

xA = CA / CTotal; % Mole-fraction of A at any point

xA0=1.0; %Feed mole-fraction of A

CVmix = xA * CVA + (1 - xA) * CVB;

XA=(1/xA0)*(xA0-xA); % Conversion of A

AH=4*VPFR*0.001/D

$$\Delta H_r = 1{,}700\frac{J}{mol} + \int_{298}^{T}(-CPA + CPB)dT$$

$$k1 = 1.2\times 10^{-5} * e^{-\frac{60000}{8.314}\left(\frac{1}{T}-\frac{1}{298}\right)}$$

if T<398

Qexch=500*AH*(398-T); % U=500 W/m2.K for steam heating medium jacket

else

Qexch=400*AH*(298-T); % U=400 W/m2.K for cooling water medium jacket

end

$$F_{Ao} - F_A - \left(\int_0^{V_{PFR}} k1 * C_A dV\right) = \int_0^{V_{PFR}} \frac{\partial C_A}{\partial t} dV$$

$$F_{Ao}C_{PA}T_0 - F_{Ao}C_{Pmix}T - \left(\int_0^{V_{PFR}} k1 * C_A dV\right)\Delta Hr + Q_{exch} = n_{PFR}\overline{C_{Vmix}}\frac{dT}{dt}$$

In addition, we need to integrate both $C_A$ and $T$ from:

**Initial Condition:** $t = t_0 = 0$; $C_A = C_{A0} = 50$ mol/L, $T = T(0) = 298$ K
**To: Final Condition:** $t = t_f = 5{,}400$ s $= 1.5$ hour

A plot for $C_A$, $X_A$, and $T$ as $f(t_{PFR})$ will be presented.

```
clear variables;
R = 8.314; % J/(mol*K)
% User inputs
CA0 = input('Enter the feed molar concentration of A (CA0) in mol/L: ');
VPFR = input('Enter PFR volume (VPFR) in L: ');
D=input('Enter PFR Diameter (D) in m: ');
v0 = input('Enter initial volume flow rate (v0) in L/s: ');
xA0 = input('Enter feed mole fraction of A (xA0): ');
T0 = input('Enter feed temperature (T0) in K: ');
tf = input('Enter final time (tf) in seconds: ');
% Fixed parameters
CTotal = CA0; % mol/L
n = CTotal * VPFR; % mol
FA0 = CA0 * v0; % mol/s
```

**Figure 5.18** MATLAB code (**PFR_T_CA.m**) written to simulate the above set of algebraic and differential equations, which describe an unsteady-state, nonadiabatic PFR for liquid-phase glucose isomerization.

```
CPA = 147.0; % J/(mol*K)
CPB = 145.0; % J/(mol*K)
% Initial conditions
t0 = 0;
Tzero = 298.0; % Initial temp
y0 = [CA0, Tzero];
% ODE solver
odefun = @(t, y) system_of_equations(t, y, R, CPA, CPB, VPFR, v0, FA0, T0, n, D);
[t, y] = ode45(odefun, [t0 tf], y0);
% Extract
CA = y(:,1);
T = y(:,2);
xA = CA ./ CTotal;
XA = xA0 - xA;
% Plotting
figure;
subplot(3,1,1);
plot(t, CA, 'b-', 'LineWidth', 1.5);
xlabel('Time (s)'); ylabel('C_A (mol/L)'); title('C_A vs Time');
grid on;
ax = gca;
ax.FontSize = 16;
ax.FontWeight = 'bold';
subplot(3,1,2);
plot(t, T, 'r-', 'LineWidth', 1.5);
xlabel('Time (s)'); ylabel('T (K)'); title('T vs Time');
grid on;
ax = gca;
ax.FontSize = 16;
ax.FontWeight = 'bold';
subplot(3,1,3);
plot(t, XA, 'g-', 'LineWidth', 1.5);
xlabel('Time (s)'); ylabel('X_A'); title('Conversion vs Time');
grid on;
ax = gca;
ax.FontSize = 16;
ax.FontWeight = 'bold';
% - - - SYSTEM OF ODEs - - -
function dydt = system_of_equations(~, y, R, CPA, CPB, VPFR, v0, FA0, T0, n, D)
   CA = y(1);
   T = y(2);
   CTotal = FA0 / v0;
   xA = CA / CTotal;
   CVmix = xA * CPA + (1 - xA) * CPB;
   % Rate constant
   k1 = 1.2e-5 * exp(-60000 / R * (1/T - 1/298));
   % Heat of reaction with T-dependent correction
  DeltaHr = 1700 + integral(@(Tval) (-CPA + CPB) * ones(size(Tval)), 298, T, 'ArrayValued', true);
   AH=4*VPFR*0.001/D;
   % Heat exchange (steam if T<398, else cooling water)
   if T < 398
      Qexch = 500 * AH * (398 - T);
   else
      Qexch = 400 * AH * (298 - T);
   end
   % - - - Numerical integration over volume: ∫k1*CA dV
   Vspan = linspace(0, VPFR, 400);
   integrand_kCA = k1 * CA * ones(size(Vspan));
   integral_kCA = trapz(Vspan, integrand_kCA);
   % === New: Compute ∫ dCA/dt dV using unknown dCA/dt ===
   % Let RHS_integral = ∫(dCA/dt)dV = F_A0 - F_A - ∫k1*CA dV
```

**Figure 5.18** (*Continued*)

```
    FA = CA * v0;
    RHS_integral = FA0 - FA - integral_kCA;
    % From ∫ dCA/dt dV = VPFR * dCA/dt → isolate dCA/dt:
    dCA_dt = RHS_integral / VPFR;
    RHS_integral2= (FA0 * (CPA * T0 - CVmix * T) - integral_kCA * DeltaHr + Qexch);
    % Energy balance
    dT_dt = RHS_integral2 / (n * CVmix);
    dydt = [dCA_dt; dT_dt];
end
```

**Figure 5.18** (*Continued*)

The following input parameters are used, and the results are shown below:

```
≫ PFR_T_CA
Enter the feed molar concentration of A (CA0) in mol/L: 50
Enter PFR volume (VPFR) in L: 2000
Enter PFR Diameter (D) in m: 1
Enter initial volume flow rate (v0) in L/s: 0.1
Enter feed mole fraction of A (xA0): 1.0
Enter feed temperature (T0) in K:380
Enter the final time (tf) in seconds: 5400
```

Figure 5.19 shows the profile for $C_A$, $X_A$, and $T$ as $f(t_{PFR})$ under unsteady-state, nonadiabatic PFR. Notice that the nonsteady state will eventually evolve toward a steady-state condition, as the process variables asymptotically attain a horizontal profile with time.

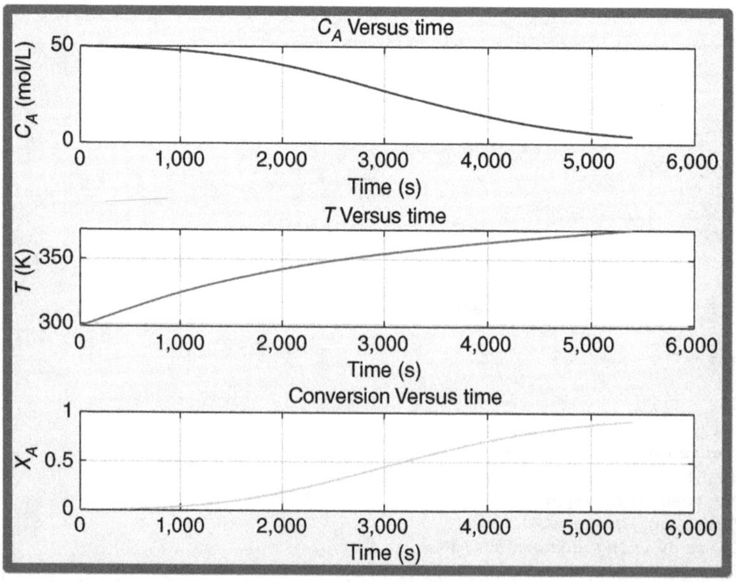


**Figure 5.19** The profile for $C_A$, $X_A$, and $T$ as $f(t_{PFR})$ under unsteady-state, nonadiabatic PFR for liquid-phase glucose isomerization.

**Example 5.9** ***Steady-state PFR Liquid-phase Glucose Isomerization Reaction***
Under steady-state, non-isothermal conditions of a PFR, we have from Eq. (2.21):

$$\frac{dF_A}{dV} = r_A \quad \frac{dF_A}{dV} = \frac{d(vC_A)}{dV} = v\frac{dC_A}{dV} = r_A \quad \rightarrow \quad \frac{dC_A}{dV} = \frac{r_A}{v}$$

Notice that $v$ is assumed constant for liquid-phase reactions.

To save the user from the headache of derivation, please refer to Chapter 12, Eq. (12.6) in Fogler [8] for handling a steady state non-isothermal PFR:

$$\frac{dT}{dV} = \frac{r_A * \Delta Hr - U * \frac{4}{D}(T - Tj)}{F_{A0}\left(\sum_{i=1}^{N}\Theta_i C_{Pi} + \Delta C_P X\right)} = \frac{r_A * \Delta Hr - U * \frac{4}{D}(T - Tj)}{\left(\sum_{i=1}^{N}F_i C_{Pi}\right)}$$

### 5.12.2 MATLAB Code for Steady-State PFR Liquid-Phase Glucose Isomerization Reaction

Figure 5.20 shows MATLAB code (**PFR_T_CA_SS.m**), written to simulate the steady-state counterpart of the unsteady-state, nonadiabatic PFR for liquid-phase glucose isomerization shown in Figure 5.18.

```
clear variables;
R = 8.314; % J/(mol*K)
% User inputs
CA0 = input('Enter the feed molar concentration of A (CA0) in mol/L: ');
VPFR = input('Enter PFR volume (VPFR) in L: ');
D = input('Enter PFR diameter (D) in m: ');
v0 = input('Enter initial volume flow rate (v0) in L/s: ');
xA0 = input('Enter feed mole fraction of A (xA0): ');
T0 = input('Enter feed temperature (T0) in K: ');
% Fixed parameters
CTotal = CA0; % mol/L
FA0 = CA0 * v0; % mol/s
CPA = 147.0; % J/(mol*K)
CPB = 145.0; % J/(mol*K)
% Heat capacity difference (assumed constant)
deltaCP = CPB - CPA;
% Initial conditions
V0 = 0;
CA_init = CA0;
T_init = T0;
y0 = [CA_init, T_init];
% Solve ODEs with ode15s (better for stiff systems)
odefun = @(V, y) steady_state_odes(VPFR, y, R, CPA, deltaCP, v0, FA0,xA0,D);
[Vsol, ysol] = ode15s(odefun, [V0 VPFR], y0); % faster and more stable
% Extract
CA = ysol(:,1);
T = ysol(:,2);
xA = CA ./ CTotal;
XA = xA0 - xA;
% Plotting
figure;
```

**Figure 5.20** MATLAB code (**PFR_T_CA_SS.m**), written to simulate the steady-state counterpart of the unsteady-state, nonadiabatic PFR for liquid-phase glucose isomerization shown in Figure 5.18.

```
subplot(3,1,1);
plot(Vsol, CA, 'b-', 'LineWidth', 1.5);
xlabel('Reactor Volume (L)'); ylabel('C_A (mol/L)'); title('C_A vs V');
grid on;
ax = gca;
ax.FontSize = 14;
ax.FontWeight = 'bold';
subplot(3,1,2);
plot(Vsol, T, 'r-', 'LineWidth', 1.5);
xlabel('Reactor Volume (L)'); ylabel('T (K)'); title('T vs V');
grid on;
ax = gca;
ax.FontSize = 14;
ax.FontWeight = 'bold';
subplot(3,1,3);
plot(Vsol, XA, 'g-', 'LineWidth', 1.5);
xlabel('Reactor Volume (L)'); ylabel('X_A'); title('Conversion vs V');
grid on;
ax = gca;
ax.FontSize = 14;
ax.FontWeight = 'bold';
% - - - Steady-State ODEs - - -
function dydV = steady_state_odes(~, y, R, CPA, deltaCP, v0, FA0, xA0,D)
   CA = y(1);
   T = y(2);
   % Rate constant
   k1 = 1.2e-5 * exp(-60000 / R * (1/T - 1/298));
   rA = -k1 * CA;
   % Mole fraction and heat capacity
   CTotal = FA0 / v0;
   xA = CA / CTotal;
   XA=xA0-xA;
   %CPmix = xA * CPA + (1 - xA) * CPB;
   % Heat of reaction (approximate linear correction)
   DeltaHr = 1700 + deltaCP * (T - 298); % faster than integral
   % % Smooth heat exchange function using a sigmoid transition
   % T_switch = 398;
   % Qh = 500 * 4/D * (398 - T);
   % Qc = 400 * 4/D * (298 - T);
   % transition = 1 / (1 + exp((T - T_switch))); % smooth switch
   % Qexch = transition * Qh + (1 - transition) * Qc;
 % Heat exchange (steam if T<398, else cooling water)
   if T <= 398
      Qexch = 500 * 4/D * (398 - T);
   else
      Qexch = 400 * 4/D * (298 - T);
   end
   % Molar and energy balances
   dCA_dV = rA/v0;
   %FA = CA * v0;
   dT_dV = (Qexch - (DeltaHr) * (rA)) / (FA0*(CPA+deltaCP*XA));
   dydV = [dCA_dV; dT_dV];
end
```

**Figure 5.20** (*Continued*)

The following input parameters are used with and without a smooth heat exchange function using a sigmoid transition. See the commented $Q_{exch}$-related code in Figure 5.20 MATLAB code.

```
>> PFR_T_CA_SS
Enter the initial concentration CA0 (mol/L): 50
Enter PFR volume (VPFR) in L: 1000
Enter PFR diameter (D) in m: 1
Enter the initial volume flow rate (L/s): 0.1
Enter feed mole fraction of A (xA0): 1.0
Enter feed temperature (T0) in K: 320
```

Using MATLAB code (**PFR_T_CA_SS.m**), Figure 5.21 (Top) shows the steady-state but nonuniform profile of $C_A$, $T$, and hence, $X_A$ as a function of PFR axial coordinate, where each of the three properties changes from the feed to the final exit value. Furthermore, the observed oscillations or discontinuity in the temperature ($T$) profile are attributed to the split-range, on–off control loop, which modulates the influent valve for either the cooling or heating medium based on the measured reactor temperature.

On the other hand, using MATLAB code (**PFR_T_CA_SS2.m**), Figure 5.21 (Bottom) shows a smooth behavior of the steady-state but nonuniform profile of $C_A$, $T$, and hence, $X_A$ as a function of PFR axial coordinate. Due to the enforcement of a smooth heat exchange function using a sigmoid transition, the $T$ profile is free of discontinuity or oscillation.

Again, as indicated in NOTE #8, similar to the CSTR steady-state case, the steady-state final solution for PFR can converge to more than one scenario, depending on the feed input parameters and the initial guesses for the steady-state values.

## 5.13 Unsteady-State PFR Gas-Phase Second-Order HI Formation

An unsteady-state PFR for gas-phase second-order hydrogen iodide (HI) formation describes the transient development of reactant and product profiles along the reactor until steady state is achieved. In this system, gaseous hydrogen and iodine enter the tubular reactor and react via an elementary, second-order bimolecular mechanism, with the rate proportional to the product of hydrogen and iodine concentrations. The governing equations couple the transient molar balances of both reactants and the product with the energy balance, accounting for heat release, variable heat capacities, ideal-gas expansion, and possible external heat exchange. The model assumes ideal plug flow with negligible axial and radial dispersion, so concentration and temperature vary only along the axial coordinate and time. Time dependence captures the startup phase or response to inlet condition changes, where both temperature and conversion evolve until the reactor stabilizes. This approach is valuable for analyzing thermal effects, such as temperature rise from exothermicity, and for designing control strategies to maintain safe operation while maximizing HI yield.

**Example 5.10** ***Unsteady-state PFR Gas-phase Second-order HI Formation***
We will rework Example 5.6, but this time for a PFR. We will write enough number of algebraic and differential equations to describe the unsteady-state nonadiabatic PFR for gas-phase HI formation.

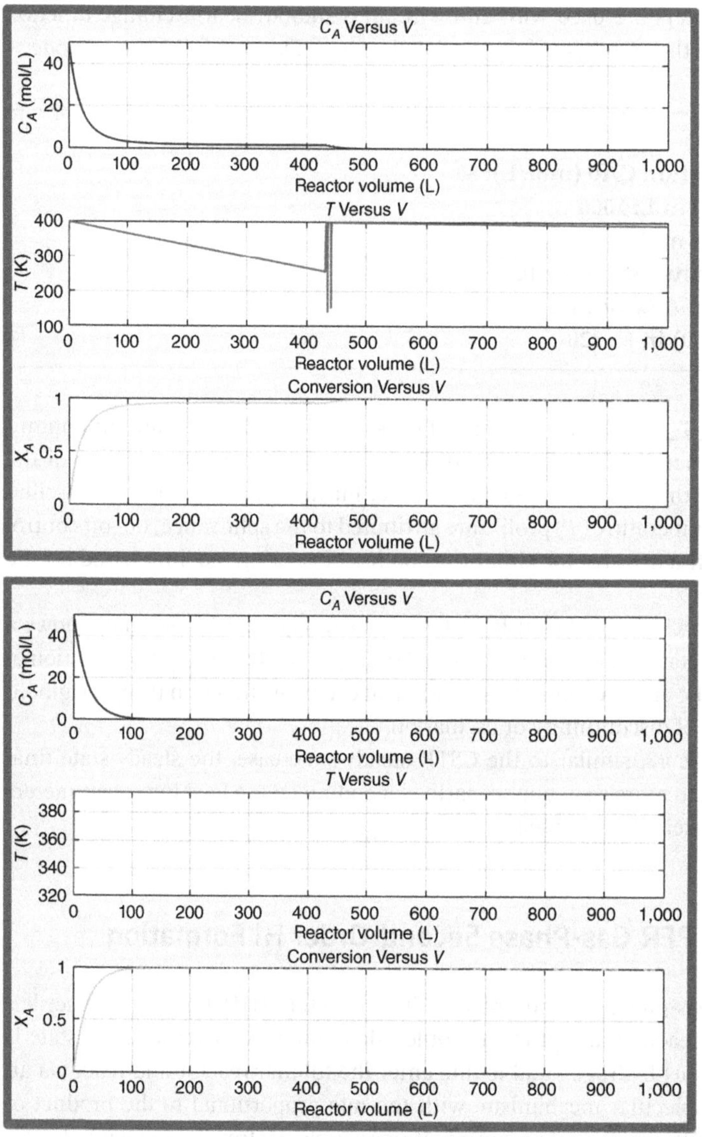


**Figure 5.21** The steady-state but nonuniform profile of $C_A$, $T$, and hence, $X_A$ as a function of PFR axial coordinate, where each of the three properties changes from the feed to the final exit value. (Top) is under the condition of installing a split-range, on–off control loop, which modulates the influent valve for either the cooling or heating medium based on the measured reactor temperature. (Bottom) is under the enforcement of a smooth heat exchange function using a sigmoid transition.

The same feed and reactor conditions of the CSTR will be used here in the PFR.

$$-r_A = kC_AC_B = 2 \times 10^{11} e^{-\left(\frac{169900}{8.314*T}\right)} C_AC_B \tag{A}$$

$$C_{P,H2} = 29.093 - (0.2E-3*T) + (0.4E-6*T^2) - (0.87E-9*T^3) + \left(\frac{0.0055E5}{T^2}\right) \tag{B}$$

$$C_{P,I2} = 37.4 - (0.0034E-3*T) + (0.18E-6*T^2) - (0.89E-9*T^3) - \left(\frac{0.245E5}{T^2}\right) \tag{C}$$

$$C_{P,HI} = 28.14 + (1.88E-3*T) - (6.954E-6*T^2) + (1.08E-9*T^3) - \left(\frac{0.205E5}{T^2}\right) \tag{D}$$

Using Appendix A | A.1 MATLAB (**DelG_Keq.m**) Code for Calculation of $\Delta G^0$, $\Delta H^0$, and $K_{eq}$ for a Chemical Reaction, the molar heat of reaction is found to be: –9.4 kJ/mol.

$$\Delta H_r(T) = -9{,}400\frac{\text{J}}{\text{mol}} + \int_{298}^{T} (-C_{P,A} - C_{P,B} + 2C_{P,C})dT \tag{E}$$

**Component *A*-Molar Balance**

$$F_{A0} - F_A - \int_0^{V_{PFR}} kC_A C_B dV = \int_0^{V_{PFR}} \frac{\partial C_A}{\partial t} dV$$

We will assume that the gas mixture will always occupy the entire volume of the PFR; that is, constant $V_{PFR}$.

The above equation reduces to:

$$v_0 CA0 - vCA - \int_0^{V_{PFR}} kC_A C_B dV = \frac{dN_A}{dt} = \int_0^{V_{PFR}} \frac{\partial C_A}{\partial t} dV \tag{F}$$

$$C_{A0} = \frac{Y_{A0}P_0}{RT_0} \tag{G}$$

$$C_{B0} = \frac{Y_{B0}P_0}{RT_0} \tag{H}$$

$$v0 = \frac{\dot{n}RT_0}{P_0} = \frac{F_0RT_0}{P_0} \tag{I}$$

$$v = \frac{\dot{n}RT}{P} = \frac{FRT}{P}$$

$$\frac{v}{v0} = \frac{\frac{FRT}{P}}{\frac{F_0RT_o}{P_0}}$$

Since $\Delta n = 2\text{-}1\text{-}1 = 0$, then $F_{Total} = F_0 = F = F_{A0} + F_{B0}$

$$\frac{v}{v_0} = \frac{\frac{FRT}{P}}{\frac{F_0RT_0}{P_0}} = \left(\frac{T}{P}\right) * \left(\frac{P_0}{T_0}\right) = \left(\frac{T}{T_0}\right) * \left(\frac{P_0}{P}\right)$$

$$v = v_0 * \left(\frac{T}{T_0}\right) * \left(\frac{P_0}{P}\right) \tag{J}$$

**Component *B*-Molar Balance**

$$v_0 * CB_0 - v * CB - \int_0^{V_{PFR}} kC_A C_B dV = \frac{dN_B}{dt} = \int_0^{V_{PFR}} \frac{\partial C_B}{\partial t} dV \tag{K}$$

Heat transfer area: $A_H = \pi DL = \pi D\left(\frac{V_{PFR}}{\frac{\pi}{4}D^2}\right) = \frac{4V_{PFR}}{D}$

if $T < 685$

$Q_{exch} = 500^* AH^* (685 - \text{T});$ $\quad \%U = 500\ \text{W/m}^2 \cdot \text{K}$ for steam heating medium jacket (La)

else

$Q_{exch} = 400^* AH^* (298 - \text{T});$ $\quad \%U = 400\ \text{W/m}^2 \cdot \text{K}$ for cooling water medium jacket (Lb)

end

**Energy Balance**

$$F0 * (Cpmix0 * (T0) - Cpmix * (T)) - \left(\int_0^{V_{PFR}} kC_A C_B dV\right) * (\Delta H_r) + Q_{exch}$$

$$= \frac{\text{d}\left(nC_V^{mix}T\right)}{dt} = n\overline{C_V^{mix}}\frac{d(T)}{dt} \tag{M}$$

The pressure inside PFR will be assumed constant, as we will assume no pressure drop. Chapter 6 will handle both FBR and PBR, which are known to have a pressure drop in the fluid flow direction. So, $P$ will be assumed as:

$$P = 0.98^{*} P_0 \tag{N}$$

$$CV_A = CP_{A-R}; \tag{Oa}$$

$$CV_B = CP_{B-R}; \tag{Ob}$$

$$CV_C = CP_{C-R}; \tag{Oc}$$

$$F_A = v^{*}\ C_A; \tag{Pa}$$

$$F_B = v^{*}\ C_B; \tag{Pb}$$

$$C_C = 2^{*}\ (C_{A0} - C_A); \tag{Pc}$$

$$F_C = v^{*}\ C_C; \tag{Pd}$$

$$Y_A = F_A/(F_A + F_B + F_C); \tag{Qa}$$

$$Y_B = F_B/(F_A + F_B + F_C); \tag{Qb}$$

$$Y_C = 1 - Y_A - Y_B; \tag{Qc}$$

$$X_A = 2^{*}\ (Y_{A0} - Y_A); \tag{R}$$

$$\%\ \text{Or: } X_A = (C_{A0} - C_A)/C_{A0};$$

$$T_{avg} = mean\,(T) \tag{S}$$

$$\overline{C_V^{mix}} = Y_A C_V^A(T_{avg}) + (Y_B) C_V^B(T_{avg}) + (1 - Y_A - Y_B) C_V^C(T_{avg}) \tag{T}$$

$$Cpmix = Y_A CPA + (Y_B) * CPB + (1 - Y_A - Y_B) * CPC \tag{U}$$

$$V_{PFR} = 3{,}000;\ \%\text{volume of PFR in L} \tag{V}$$

I.C.'s @$t = 0$:

$$T(0) = \text{Tzero};\quad C_A(0) = CA0;\quad CB(0) = CB0;\quad P(0) = Pzero \tag{Wa-d}$$

### 5.13.1 MATLAB Code for Unsteady-State PFR Gas-Phase Second-Order HI Formation

Figure 5.22 shows MATLAB code (**PFR_T_CA_Gas.m**) based on translating Eqs. A through W into MATLAB code, which describes HI gas-phase formation in an unsteady-state, nonadiabatic PFR.

```
clear; close all;
Rgas = 0.08206; % (atm.L)/mol.K
R = 8.314; % J/(mol.K)
% Given Inputs
F0 = input('Enter the feed molar flow rate (F0) in mol/s: ');
%F0 = 1; % mol/s
T0 = input('Enter the feed temperature (T0) in K: ');
%T0 = 300; % K
P0 = input('Enter the feed pressure (P0) in atm: ');
%P0 = 50; % atm
YA0 = input('Enter the feed mole fraction of A (YA0): ');
%YA0 = 0.5;
YB0 = input('Enter the feed mole fraction of B (YB0): ');
%YB0 = 0.5;
VPFR = input('Enter the PFR Volume, VPFR (L): ');
%VPFR = 3000; % Volume of PFR, L
D = input('Enter the PFR Diameter, D (m): ');
%D=1.0;
v0 = (F0 * Rgas * T0) / P0;
CA0 = (YA0 * P0) / (Rgas * T0);
CB0 = (YB0 * P0) / (Rgas * T0);
FA0 = YA0 * F0;
FB0 = YB0 * F0;
Pzero=1.0;% P(0)=1.0 atm at time zero.
Tzero=300; % T(0)=300 K at time Zero.
% Initial conditions
y0 = [CA0; CB0; Tzero];
% Time span
tf = input('Enter the final time tf, (unit of time): ');
%tf=1300;
tspan = [0 tf]; % seconds
% Solve ODEs using ode15s
options = odeset('RelTol', 1e-6, 'AbsTol', 1e-9, 'NonNegative', [1 2 3]);
[t, y] = ode45(@(t, y) dPFR(t, y, v0, CA0, CB0, P0, R, T0, VPFR, D, F0), tspan, y0, options);
%[t, y] = ode113(@(t, y) dPFR(t, y, v0, CA0, CB0, P0, R, T0, VPFR, D, F0), tspan, y0, options);
%[t, y] = ode23(@(t, y) dPFR(t, y, v0, CA0, CB0, P0, R, T0, VPFR, D, F0), tspan, y0, options);
```

**Figure 5.22** MATLAB code (**PFR_T_CA_Gas.m**) based on translating Eqs. A through W into MATLAB code, which describes HI gas-phase formation in an unsteady-state, nonadiabatic PFR.

```
%[t, y] = ode23s(@(t, y) dPFR(t, y, v0, CA0, CB0, P0, R, T0, VPFR, D, F0), tspan, y0, options);
%[t, y] = ode23t(@(t, y) dPFR(t, y, v0, CA0, CB0, P0, R, T0, VPFR, D, F0), tspan, y0, options);
%[t, y] = ode23tb(@(t, y) dPFR(t, y, v0, CA0, CB0, P0, R, T0, VPFR, D, F0), tspan, y0, options);
%[t, y] = ode15s(@(t, y) dPFR(t, y, v0, CA0, CB0, P0, R, T0, VPFR, D, F0), tspan, y0, options);
% Extract results
CA = y(:, 1);
CB = y(:, 2);
CC = 2*(CA0 - CA);
Ctot = CA + CB + CC;
T = y(:, 3);
P = 0.98*P0;
v = v0 * (T / T0) .* (P0 ./ P);
FA = v .* CA;
FB = v .* CB;
FC=v.*CC;
% Molar fractions
YA =FA./(FA+FB+FC);
YB =FB./(FA+FB+FC);
XA=2*(YA0-YA);
%Or: XA=(CA0-CA)/CA0;
Tau=VPFR/v0;
disp(['VPFR = ', num2str(VPFR), ' L']);
disp(['tPFR = ', num2str(Tau), ' Unit of time appearing in F0']);
% Plot results
figure;
subplot(4,1,1); plot(t, CA, 'b', 'LineWidth', 2.5); xlabel('Time (s)'); ylabel('C_A (mol/L)'); title('Concentration of A');
ax = gca;
ax.FontSize = 14;
ax.FontWeight = 'bold';
subplot(4,1,2); plot(t, CC, 'r', 'LineWidth', 2.5); xlabel('Time (s)'); ylabel('C_C (mol/L)'); title('Concentration of C');
ax = gca;
ax.FontSize = 14;
ax.FontWeight = 'bold';
subplot(4,1,3); plot(t, T, 'g', 'LineWidth', 2.5); xlabel('Time (s)'); ylabel('T (K)'); title('Temperature');
ax = gca;
```

**Figure 5.22** (*Continued*)

```
ax.FontSize = 14;
ax.FontWeight = 'bold';
subplot(4,1,4); plot(t, XA, 'k', 'LineWidth', 2.5); xlabel('Time (s)'); ylabel('X_A'); title('Conversion of A');
ax = gca;
ax.FontSize = 14;
ax.FontWeight = 'bold';
% ==================================================================
% Function to Define PFR ODEs
% ==================================================================
function [dydt] = dPFR(~, y, v0, CA0, CB0, P0, R, T0, VPFR, D, F0)
   P = 0.98*P0;
   CA = max(0, min(CA0, y(1))); % Ensure CA is within [0, CA0]
   CB = max(0, min(CB0, y(2))); % Ensure CB is within [0, CB0]
   T = max(200, min(700, y(3))); % Ensure T is within [200, 700] K
   CC = 2*(CA0 - CA);
   v = v0 * (T / T0) * (P0 / P);
   FA = v * CA;
   FB = v * CB;
   FC = v*CC;
   YA =FA/(FA+FB+FC);
   YB =FB/(FA+FB+FC);
   YC = 1 - YA - YB;
   Ctot = CA + CB + CC;
   n = VPFR * Ctot;
   % Heat capacities (J/mol.K)
   CPA = 29.093 - (0.2e-3*T) + (0.4e-6*T^2) - (0.87e-9*T^3) + (0.0055e5/T^2);
   CPB = 37.4 - (0.0034e-3*T) + (0.18e-6*T^2) - (0.89e-9*T^3) - (0.245e5/T^2);
   CPC = 28.14 + (1.88e-3*T) - (6.954e-6*T^2) + (1.08e-9*T^3) - (0.205e5/T^2);
% Average heat capacities
   Tavg=mean(T);
   CPAavg = 29.093 - (0.2e-3*Tavg) + (0.4e-6*Tavg^2) - (0.87e-9*Tavg^3) + (0.0055e5/Tavg^2);
   CPBavg = 37.4 - (0.0034e-3*Tavg) + (0.18e-6*Tavg^2) - (0.89e-9*Tavg^3) - (0.245e5/Tavg^2);
   CPCavg = 28.14 + (1.88e-3*Tavg) - (6.954e-6*Tavg^2) + (1.08e-9*Tavg^3) - (0.205e5/Tavg^2);
   CVAavg = CPAavg - R;
   CVBavg = CPBavg - R;
   CVCavg = CPCavg - R;
```

**Figure 5.22** *(Continued)*

```
    CVavg = YA * CVAavg + YB * CVBavg + (YC) * CVCavg;
    Cpmix = YA * CPA + YB * CPB + (YC) * CPC;
    CPA0 = 29.093 - (0.2e-3*T0) + (0.4e-6*T0^2) - (0.87e-9*T0^3) + (0.0055e5/T0^2);
    CPB0 = 37.4 - (0.0034e-3*T0) + (0.18e-6*T0^2) - (0.89e-9*T0^3) - (0.245e5/T0^2);
    CPC0 = 28.14 + (1.88e-3*T0) - (6.954e-6*T0^2) + (1.08e-9*T0^3) - (0.205e5/T0^2);
    Cpmix0 = YA * CPA0 + YB * CPB0 + (YC) * CPC0;
    % Reaction rate constant
    k = 2e11 * exp(-169900 / (R * T));
    %r = k * CA * CB;
    % Heat exchange term
    AH=(4*VPFR*0.001/D);
% % Smooth heat exchange function using a sigmoid transition
     T_switch = 685;
     Qh = 500 * AH * (685 - T);
     Qc = 400 * AH * (298 - T);
    transition = 1 / (1 + exp((T - T_switch))); % smooth switch
    Qexch = transition * Qh + (1 - transition) * Qc;
    % Enthalpy change of reaction (J/mol)
    dHr = -94000 + integral(@(T) (-CPA - CPB + 2*CPC), 298, T, 'ArrayValued', true);
    % ODEs
    Vspan = linspace(0, VPFR, 400);
    integral_kCACB = trapz(Vspan, k * CA * CB* ones(size(Vspan)));
    % Let RHS_integral = ∫(dCA/dt)dV = F_A0 - F_A - ∫k1*CA*CB dV
    RHS_integral = (v0 * CA0 - v * CA-integral_kCACB);
    % From ∫ dCA/dt dV = VPFR * dCA/dt → isolate dCA/dt:
    dCA_dt = RHS_integral / VPFR;
    % Let RHS_integral = ∫(dCB/dt)dV = F_B0 - F_B - ∫k1*CA*CB dV
    RHS_integral2 = (v0 * CB0 - v * CB-integral_kCACB);
    % From ∫ dCA/dt dV = VPFR * dCA/dt → isolate dCA/dt:
    dCB_dt = RHS_integral2 / VPFR;
    RHS_integral3=(F0 * (Cpmix0*T0 - Cpmix*T) - integral_kCACB * dHr +Qexch);
    dT_dt= RHS_integral3/ (n * CVavg);
    % Return ODEs
    dydt = [dCA_dt; dCB_dt; dT_dt];
end
```

**Figure 5.22** (*Continued*)

The following input parameters are used for Figure 5.22 code:

```
≫ PFR_T_CA_Gas
Enter the feed molar flow rate (F0) in mol/s: 1
Enter the feed temperature (T0) in K: 300
Enter the feed pressure (P0) in atm: 50
Enter the feed mole fraction of A (YA0): 0.5
Enter the feed mole fraction of B (YB0): 0.5
Enter the PFR Volume, (L): 3000
Enter the PFR Diameter, D (m): 1
Enter the final time tf, (unit of time): 1300
VPFR = 3000 L
tPFR = 6093.1026 Unit of time appearing in F0
```

Figure 5.23 shows the generated plot of $C_A$, $C_C$, $T$, and $X_A$ profiles for an unsteady-state PFR using MATLAB **ode45** solver for HI gas-phase formation.

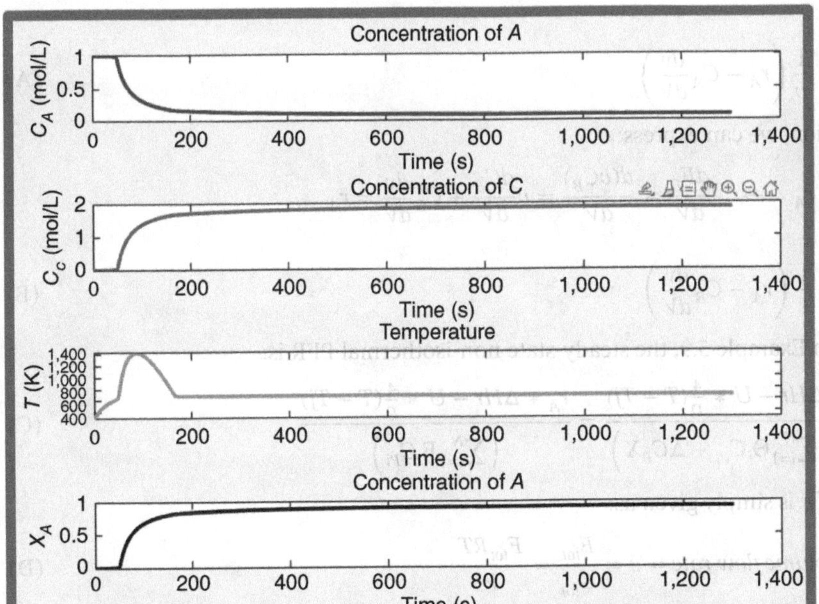


**Figure 5.23** $C_A$, $C_C$, $T$, and $X_A$ profile for an unsteady-state PFR using MATLAB **ode45** solver for HI gas-phase formation.

On the other hand, Appendix B | B.8: The Discretized PFR Approximation Method shows MATLAB code (**DPFRAM.m**) for approximating a nonuniform PFR by a series of perfectly mixed CSTRs, which converts the PDEs that strictly govern a transient PFR into a system of ODEs, which can then be solved by MATLAB's ODE solvers. If one runs the code for the same input parameters, Figure 5.23 will be exactly generated. That simply says both **PFR_T_CA_Gas.m** and **DPFRAM.m** are equivalent in terms of predicting the behavior of transient properties of PFR, like: $C_A$, $C_C$, T, and $X_A$. Either one can be used.

Notice that the line code: $T = \max(200, \min(700, y(3)))$; is applied inside the dPFR function. This means that when the dPFR function is called by the ODE solver to calculate the derivatives ($dCA/dt$, $dCB/dt$, and $dT/dt$), the $T$ value used for those calculations is capped at 700 K. However, the ODE solver's prediction for the next $y(3)$ (which is $T$) is based on these calculated derivatives. If the calculated $dT/dt$ is very large and positive, the solver might take a step that results in a $y(3)$ value significantly higher than 700 K before it calls dPFR again for the next iteration. In brief, the max and min functions inside dPFR affect only the input $T$ that the derivative calculation sees, not the output $y(3)$ that the solver itself is propagating. The $y$ variable returned by the ode113/ode45 solver (and then plotted) contains the raw, unconstrained values that the solver computed. $T = 700$ K is an adjustable parameter, selected to be slightly above the upper $T$ control limit, which is 685 K. If the heating medium $T$ is changed, then this $T$ can be adjusted as well to end up with a smooth $T$ profile. The peak $T$ reaches 1,407 K, which reflects the highly exothermic nature of gas-phase HI formation.

**Example 5.11** ***Steady-state PFR gas-phase second-order HI formation***

Under steady-state, non-isothermal conditions of a PFR, we have from Eq. (2.21):

$$\frac{dF_A}{dV} = r_A \quad \rightarrow \quad \frac{dF_A}{dV} = \frac{d(vC_A)}{dV} = v\frac{dC_A}{dV} + C_A\frac{dv}{dV} = r_A$$

Rearrange to get:

$$\rightarrow \quad \frac{dC_A}{dV} = \frac{1}{v}\left(r_A - C_A\frac{dv}{dV}\right) \tag{A}$$

In a similar fashion, we can express:

$$\frac{dF_B}{dV} = r_B = r_A \quad \rightarrow \quad \frac{dF_B}{dV} = \frac{d(vC_B)}{dV} = v\frac{dC_B}{dV} + C_B\frac{dv}{dV} = r_A$$

$$\rightarrow \quad \frac{dC_B}{dV} = \frac{1}{v}\left(r_A - C_B\frac{dv}{dV}\right) \tag{B}$$

As pointed out in Example 5.9, the steady-state non-isothermal PFR is:

$$\frac{dT}{dV} = \frac{r_A * \Delta Hr - U * \frac{4}{D}(T - Tj)}{F_{A0}\left(\sum_{i=1}^{N}\Theta_i C_{Pi} + \Delta C_P X\right)} = \frac{r_A * \Delta Hr - U * \frac{4}{D}(T - Tj)}{\left(\sum_{i=1}^{N} F_i C_{Pi}\right)} \tag{C}$$

The definition of $v$ is simply given as:

$$\left(\frac{dV}{dt}\right) = \textit{Volume flow rate} = v = \frac{F_{tot}}{C_{tot}} = \frac{F_{tot}RT}{P} \tag{D}$$

**NOTE #10**

P is assumed to linearly decrease with the axial direction, which partially accounts for pressure drop due to the presence of skin friction or drag friction as a result of flow obstruction by a packed/fluidized filling material, as in PBR/FBR. The axial pressure drop in PBR and FBR will be covered in Chapter 6. We will show below the derivation of the volume expansion effect for the case $\varepsilon = 0$.

Please, refer to Appendix B | B.7 The Volume Expansion Effect in a PFR as a Result of a Gas-Phase Chemical Reaction on how to express the volumetric flow rate change with the extent of reaction, pressure, and temperature, for $\varepsilon \neq 0$. See also CHW 5.12 problem.

Taking the natural logarithm of both sides of Eq. (D) and applying differentiation, we have:

$$d\ln(v) = d\ln(F_{tot}) + d\ln(R) + d\ln(T) - d\ln(P) \tag{E}$$

Since both $F_{tot}$ and R terms are constant, then Eq. (E) reduces to:

$$d\ln(v) = d\ln(T) - d\ln(P) \tag{F}$$

$$\frac{1}{v}\frac{dv}{dV} = \frac{1}{T}\frac{dT}{dV} - \frac{1}{P}\frac{dP}{dV} \tag{G}$$

Multiply both sides of Eq. (G) by $v$, we have:

$$\frac{dv}{dV} = \frac{v}{T}\frac{dT}{dV} - \frac{v}{P}\frac{dP}{dV} = v\left(\frac{1}{T}\frac{dT}{dV} - \frac{1}{P}\frac{dP}{dV}\right) \tag{H}$$

If we assume a 10% pressure drop along the tube, then

$$P = P_0\left(1 - 0.1\frac{z}{L}\right) \qquad \frac{dP}{dz} = -0.1\frac{P_0}{L} \tag{I}$$

Divide both sides by the cross-sectional area, $A$, of the tube, we have:

$$\frac{dP}{Adz} = -0.1\frac{P_0}{A.L} \quad \rightarrow \quad \frac{dP}{dV} = -0.1\frac{P_0}{V_{PFR}} \tag{J}$$

Substitute in Eq. (H), we have:

$$\frac{dv}{dV} = v\left(\frac{1}{T}\frac{dT}{dV} - \frac{1}{P}\frac{dP}{dV}\right) = v\left(\frac{1}{T}\frac{dT}{dV} + 0.1 * \frac{P_0}{PV_{PFR}}\right) \tag{K}$$

### 5.13.2 MATLAB Code for Steady-State PFR Gas-Phase Second-Order HI Formation

Figure 5.24 shows MATLAB code (**PFR_T_CA_Gas_SS.m**), written to simulate the steady-state counterpart of the unsteady-state, nonadiabatic PFR for gas-phase HI formation, shown in Figure 5.22.

```
clear; close all;
Rgas = 0.08206; % (atm.L)/mol.K
R = 8.314; % J/(mol.K)
% Given Inputs
VPFR = input('Enter the PFR Volume (L): ');
%VPFR = 1000; % Volume of CSTR, L
D = input('Enter PFR diameter (D) in m: ');
% D=1.0;
F0 = input('Enter the initial molar flow rate, (mol/s), F0: ');
%F0=1.0;
T0 = input('Enter the feed temperature, (K), T0: ');
%T0=300;
P0 = input('Enter the feed pressure, (atm), P0: ');
%P0=50;
v0 = (F0 * Rgas * T0) / P0;
YA0 = input('Enter the feed mole fraction of A (YA0): ');
%YA0 = 0.5;
YB0 = input('Enter the feed mole fraction of B (YB0): ');
```

**Figure 5.24** MATLAB code (**PFR_T_CA_Gas_SS.m**), written to simulate the steady-state, nonadiabatic PFR for gas-phase HI formation.

```
%YB0 = 0.5;
FA0=F0*YA0;
CA0 = (YA0 * P0) / (Rgas * T0);
CB0 = (YB0 * P0) / (Rgas * T0);
% Display best solution
Tau=VPFR/v0;
disp(['VPFR = ', num2str(VPFR), ' L']);
disp(['tPFR = ', num2str(Tau), ' Unit of time appearing in F0'])
V0 = 0;
CA_init = CA0;
CB_init= CB0;
T_init = T0;
%P_init = 1.0;
y0 = [CA_init, CB_init, T_init];
odefun = @(V, y) steadyStateEquations(y, v0, P0, R, T0, CA0,D,VPFR);
[Vsol, ysol] = ode23(odefun, [V0 VPFR], y0); % faster and more stable
% Extract
CA = ysol(:,1);
CB=ysol(:,2);
T = ysol(:,3);
P=0.98*P0;
CC=2*(CA0-CA);
v = v0 * (T / T0) .* (P0 ./ P);
FA = v .* CA;
FB = v .* CB;
FC=v.*CC;
% Molar fractions
YA =FA./(FA+FB+FC);
YB =FB./(FA+FB+FC);
XA=(1/YA0)*(YA0-YA);
% Plotting
figure;
subplot(4,1,1);
plot(Vsol, CA, 'b-', 'LineWidth', 1.5);
xlabel('Reactor Volume (L)'); ylabel('C_A (mol/L)'); title('C_A vs V');
grid on;
ax = gca;
ax.FontSize = 14;
ax.FontWeight = 'bold';
subplot(4,1,2);
plot(Vsol, CC, 'b-', 'LineWidth', 1.5);
xlabel('Reactor Volume (L)'); ylabel('C_C (mol/L)'); title('C_C vs V');
grid on;
ax = gca;
ax.FontSize = 14;
ax.FontWeight = 'bold';
subplot(4,1,3);
plot(Vsol, T, 'r-', 'LineWidth', 1.5);
xlabel('Reactor Volume (L)'); ylabel('T (K)'); title('T vs V');
grid on;
ax = gca;
ax.FontSize = 14;
ax.FontWeight = 'bold';
subplot(4,1,4);
plot(Vsol, XA, 'g-', 'LineWidth', 1.5);
```

**Figure 5.24** (*Continued*)

```
xlabel('Reactor Volume (L)'); ylabel('X_A'); title('Conversion vs V');
grid on;
ax = gca;
ax.FontSize = 14;
ax.FontWeight = 'bold';
function dydV = steadyStateEquations(y, v0, P0, R, T0, CA0, D, VPFR)
  CA = y(1);
  CB = y(2);
  T = max(200, min(700, y(3))); % Ensure T is within [200, 700] K
  P = 0.98*P0;
  CC = 2*(CA0-CA);
  v = v0 * (T / T0) .* (P0 ./ P);
  FA = v * CA;
  FB = v * CB;
  FC=v*CC;
  k = 2e11 * exp(-169900 / (R * T));
  % Heat capacities (J/mol.K)
  CPA = 29.093 - (0.2e-3*T) + (0.4e-6*T^2) - (0.87e-9*T^3) + (0.0055e5/T^2);
  CPB = 37.4 - (0.0034e-3*T) + (0.18e-6*T^2) - (0.89e-9*T^3) - (0.245e5/T^2);
  CPC = 28.14 + (1.88e-3*T) - (6.954e-6*T^2) + (1.08e-9*T^3) - (0.205e5/T^2);
  DeltaHr = -94000 + integral(@(T) (-CPA - CPB + 2*CPC), 298, T, 'ArrayValued', true);
  T_switch = 400;
  Qh = 500 * 4/D * (400 - T);
  Qc = 400 * 4/D * (298 - T);
  transition = 1 / (1 + exp((T - T_switch))); % smooth switch
  Qexch = transition * Qh + (1 - transition) * Qc;
  rA=-k*CA*CB;
  %dCA_dV = -k*CA*CB/v;
  denom=FA*CPA+FB*CPB+FC*CPC;
  dT_dV = (Qexch - (DeltaHr) * (k*CA*CB)) / denom;
  dv_dV = v * ((1/T) * dT_dV+0.1*P0/(P*VPFR)); % (L/s)/L = L/s.L
  dCA_dV=(1/v)*(rA-CA*dv_dV);
  dCB_dV=(1/v)*(rA-CB*dv_dV);
  dydV = [dCA_dV; dCB_dV; dT_dV];
end
```

**Figure 5.24** *(Continued)*

The following input parameters are used, and $C_A$, $C_C$, $T$, and $X_A$ profiles are shown in Figure 5.25.

```
>> PFR_T_CA_Gas_SS
Enter the PFR Volume (L): 2000
Enter PFR diameter (D) in m: 1
Enter the initial molar flow rate, (mol/s), F0: 1
Enter the feed temperature, (K), T0: 300
Enter the feed pressure, (atm), P0: 50
Enter the feed mole fraction of A (YA0): 0.5
Enter the feed mole fraction of B (YB0): 0.5
VPFR = 2000 L
tPFR = 40620.6841 Unit of time appearing in F0
```

As indicated earlier in NOTE #8, similar to the CSTR steady-state case, the steady-state final solution for PFR can converge to more than one scenario, depending on the feed input parameters and the initial guesses for the steady-state values.

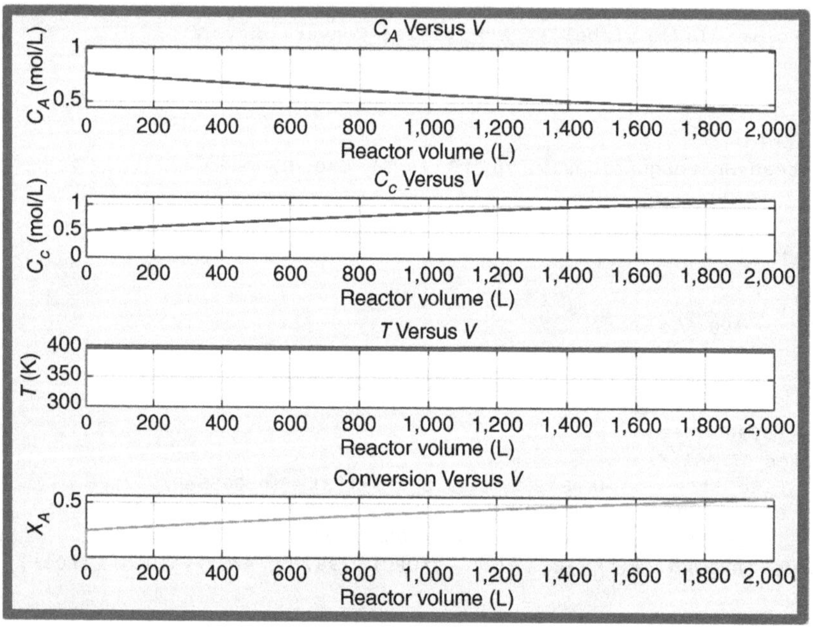


**Figure 5.25** The steady-state version of the gas-phase HI formation with $C_A$, $C_C$, $T$, and $X_A$ profiles as a function of the axial distance of the PFR.

## References


1 Don W. Green (Editor-in-Chief) and Robert H. Perry (Late Editor). *Perry's Chemical Engineers' Handbook*, 8th Edition. Table 2-153: Heat capacities of inorganic and organic liquids [J/(kmol · K)]. Page 2–165 (208/2735). McGraw-Hill Companies, Inc., New York (2008).

2 Don W. Green (Editor-in-Chief) and Robert H. Perry (Late Editor). *Perry's Chemical Engineers' Handbook*, 8th Edition. Table 2-156: Heat capacity at constant pressure of inorganic and organic compounds in the ideal gas state fit to hyperbolic functions. Page 2–176 (219/2735). McGraw-Hill Companies, Inc., New York, USA (2008).

3 Giguére, P.A. and Carmichael, J.L. (1962). Heat capacities for water-hydrogen peroxide systems between 25° and 60° C". *Journal of Chemical and Engineering Data* 7 (4): 526–527. https://doi.org/10.1021/je60015a024.

4 Schumb, W.C., Satterfield, C.N., and Wenworth, R.L. (2025). Report #43. Hydrogen Peroxide Part 2. Ch. 5 Physical Properties, A monograph Prepared with Support from the Office of Naval Research Contract No. N5ovi-07819 NR-092-008. M.I.T., Division of Industrial Cooperation Project 6552, December, 1953. Armed Services Technical Information Agency, AD 0022243. https://apps.dtic.mil/sti/tr/pdf/AD0022243.pdf.

5 Al-Malah, K., "*Aspen Plus: Chemical Engineering Applications*", 2nd Edition, pp. 393–456. Chapter 15: Aspen Plus dynamics. Wiley October, 2022. https://www.wiley.com/en-ae/Aspen+Plus:+Chemical+Engineering+Applications,+2nd+Edition-p-9781119868699.

6 Peacock, F. (2008). PID Tuning Blueprint", V 2.0, Finn Peacock, www.PID-Tuning.com (accessed May 22, 2026).

7 Anderson, J.B. (1996). The hydrogen-iodine reactions: 100 years later. In: *Gas Phase Chemical Reaction Systems*, Springer Series in Chemical Physics, vol. 61 (ed. J. Wolfrum, H.R. Volpp, R.

Rannacher, and J. Warnatz), 167–176. Berlin, Heidelberg. https://doi.org/10.1007/978-3-642-80299-7_12: Springer.
8 H.S. Fogler. "*Elements of Chemical Reaction Engineering*". Chapter 4, Stoichiometry: pp. 270–337. Pearson Education, Inc.; 6th edition (July 24, 2020). ISBN-10: 013548622X. ISBN-13: 978–0135486221.

## End of Chapter Problems

**CHW 5.1 Hydrolysis of Ethyl Acetate with Constant Heat Capacity and Heat of Reaction in an Adiabatic BR**

Consider an irreversible liquid-phase first-order reaction, which is the acid-catalyzed hydrolysis of ethyl acetate:

$$CH_3COOCH_2CH_3(aq) + H_2O(l) \overset{H^+}{\Leftrightarrow} CH_3CH_2OH(aq) + CH_3COOH(aq) \quad (A)$$

Which can be put in the short form:

$$A(aq) + B(l) \rightarrow C(aq) + D(aq) \quad (B)$$

With a reaction rate

$$-r_A = k_1 C_A = k^* e^{-\frac{E_A}{R}\left(\frac{1}{T}-\frac{1}{298}\right)} C_A = 1.5 \times 10^{-5} \frac{1}{s} e^{-\frac{50{,}000}{8.314}\left(\frac{1}{T}-\frac{1}{298}\right)} C_A \frac{\text{mol}}{\text{L}} \quad (C)$$

The heat of reaction is given as:

$$\Delta H_r^0 = 7.6 \frac{\text{kJ}}{\text{mol of } A} \quad (D)$$

The heat capacity data at $T = 350$ K are:

$$\text{Ethyl acetate: } C_P = 187.9 \frac{\text{J}}{\text{mol}\cdot\text{K}} \qquad \text{Water: } C_P = 75.5 \frac{\text{J}}{\text{mol}\cdot\text{K}}$$

$$\text{Ethyl alcohol: } C_P = 137.5 \frac{\text{J}}{\text{mol}\cdot\text{K}} \qquad \text{Acetic acid: } C_P = 137.4 \frac{\text{J}}{\text{mol}\cdot\text{K}}$$

For this question, assume a constant heat of reaction and constant heat capacity, independent of $T$. Moreover, $V_{BR}$ is set to 1,500 L, and the initial temperature is 350 K. The solubility of $A$ is: 83 g per 1,000 g $H_2O$.

$$C_{A0} = \frac{83\ \text{g/L}}{88.11\ \text{g/mol}} = 0.94 \frac{\text{mol}}{\text{L}}$$

Write the total material/M balance, A-component molar balance, $C$- or $D$-component balance, and energy balance to describe the adiabatic batch reactor system. Notice $C_A$ (or $X_A$), $C_C$, and temperature ($T$) do change with batch reactor time. Refer to Example 5.1: Liquid-phase First-order Isomerization Reaction.

**CHW 5.2 Hydrolysis of Ethyl Acetate with Variable Heat Capacity and Heat of Reaction in a Nonadiabatic BR**

Repeat CHW 5.1, but include variable heat capacity and heat of reaction. Use the **CpLiq_T.m** file to find the heat capacity data for reactants and products. A split-range valve system will be used to control the cooling/heating jacket side, such that if $T < 400$ K, a steam heating medium jacket will be enabled; else, a cooling water medium jacket will be enabled. The reactor volume is 2,000 L with a 1-m diameter. The initial temperature is 300 K.

Refer to the "MATLAB Code for Adiabatic BR for Glucose Conversion with $\Delta H_r(T) = f(T)$" section for handling a similar case of a BR with a liquid-phase chemical reaction.

**CHW 5.3 Hydrolysis of Ethyl Acetate with Variable Heat Capacity and Heat of Reaction in a Nonadiabatic CSTR**

Use chemical reaction data in CHW 5.1, with variable heat capacity and heat of reaction. Make use of the **CpLiq_T.m** file to find the heat capacity data for reactants and products. A split-range cooling/heating control loop is installed such that if $T < 400$ K, a steam heating medium jacket will be used; else, a cooling water medium jacket will be used. The reactor volume is 2,000 L with a 1-m diameter. The feed temperature is 300 K. The volumetric flow rate, $v_0$, is 0.1 L/s. The final time is set to 4,000 s. For further help on setting up the required algebraic and differential equations describing a nonadiabatic CSTR, please refer to Example 5.4: Unsteady-state CSTR Liquid-phase First-order Isomerization Reaction.

**CHW 5.4 Hydrolysis of Ethyl Acetate with Variable Heat Capacity and Heat of Reaction in a Nonadiabatic, Steady-state CSTR**

Write down the MATLAB code needed to describe the steady-state counterpart of the CHW 5.3 problem. Report the steady-state values of $C_A$, $C_C$, and $T$ for $V_{CSTR} = 2{,}000$ L, $\upsilon_0 = 0.1$ L/s, and tank diameter, $D = 1.0$ m. $T_0 = 300$ K.

Calculate the steady-state values for a new set of $V_{CSTR} = 2{,}500$ L, $\upsilon_0 = 0.1$ L/s, and tank diameter, $D = 1.2$ m. $T_0 = 300$ K.

Refer to Example 5.5: Steady-state CSTR Liquid-phase First-order Isomerization Reaction.

**CHW 5.5 Hydrolysis of Ethyl Acetate with Variable Heat Capacity and Heat of Reaction in a Nonadiabatic PFR**

Use chemical reaction data in CHW 5.1, with variable heat capacity and heat of reaction. Make use of the **CpLiq_T.m** file to find the heat capacity data for reactants and products. A split-range cooling/heating control loop is installed such that if $T < 400$ K, a steam heating medium jacket will be used; else, a cooling water medium jacket will be used. The PFR volume is 2,000 L with a 1-m diameter. The feed temperature is 300 K. The volumetric flow rate, $v_0$, is 0.1 L/s. The final time is set to 4,000 s. For further help on setting up the required algebraic and differential equations describing a nonadiabatic PFR, please refer to Example 5.8: Unsteady-state PFR liquid-phase glucose isomerization reaction.

**CHW 5.6 Hydrolysis of Ethyl Acetate with Variable Heat Capacity and Heat of Reaction in a Nonadiabatic, Steady-state PFR**

Write down the MATLAB code needed to describe the steady-state counterpart of the CHW 5.5 problem. Show the steady-state profile of $C_A$, $C_C$, $T$, and $X_A$ as a function of the axial distance for $V_{PFR} = 1{,}000$ L, $\upsilon_0 = 0.1$ L/s, and tank diameter, $D = 1.0$ m. The feed temperature is 300 K. Refer to Example 5.9: Steady-state PFR Liquid-phase Glucose Isomerization Reaction.

**CHW 5.7 Thermal Decomposition of Cyclobutane Into Ethylene with Constant Heat Capacity and Heat of Reaction in a Nonadiabatic BR**

Consider the thermal decomposition of cyclobutane into ethylene:

$$C_4H_8(g) \rightarrow 2C_2H_4(g) \qquad A(g) \rightarrow 2B(g)$$

This reaction occurs in the gas phase. It is a unimolecular decomposition reaction, typically exhibiting first-order kinetics. The rate law for this reaction is:

$$-r_A = kC_A = 4 \times 10^{15} \frac{1}{s} e^{-\left(\frac{E_A}{RT}\right)} = 4 \times 10^{15} \frac{1}{s} e^{-\left(\frac{262,000}{8.314*T}\right)}$$

**Reference:** NIST Chemistry WebBook and Laidler, K.J. (1987). Chemical Kinetics (3rd ed.). New York: Harper & Row.

The standard heat of formation data is shown below (https://www.chemeo.com/cid/35-155-1/Cyclobutane, https://www.chemeo.com/cid/56-863-2/Ethylene):

$$\Delta H_f^0(C_4H_8(g)) = 26.67 \frac{kJ}{mol} \qquad \Delta H_f^0(C_2H_4(g)) = 52.37 \frac{kJ}{mol}$$

For this question, assume a constant heat of reaction and constant heat capacity, independent of $T$. Moreover, $V_{BR}$ is set to 1,500 L with a 1.0 m diameter, and the initial temperature is 300 K, and the initial pressure is 10 atm of pure $A$.
The constant heat capacity data at $T = 500$ K are:

$$\text{Cyclobutane: } C_{PA} = 122.2 \frac{J}{mol \cdot K} \qquad \text{Ethylene: } C_{PB} = 62.52 \frac{J}{mol \cdot K}$$

Write an adequate number of algebraic and differential equations, including the total material/M balance, $A$-component molar balance, and energy balance to describe the behavior of a nonadiabatic batch reactor system. A split-range valve system will be used to control the cooling/heating jacket side such that if $T < 900$ K, a steam heating medium jacket will be enabled; else, a cooling water medium jacket will be enabled. Refer to Example 5.2: BR Irreversible Gas-phase First-order Cyclopropane Isomerization Reaction.

**CHW 5.8 Thermal Decomposition of Cyclobutane Into Ethylene with Variable Heat Capacity and Heat of Reaction in a Nonadiabatic BR**

Repeat CHW 5.7, but include variable heat capacity and heat of reaction. Use the **CpGas_T.m** file to find the heat capacity data for reactants and products. A split-range valve system will be used to control the cooling/heating jacket side, such that if $T < 900$ K, a steam heating medium jacket will be enabled; else, a cooling water medium jacket will be enabled. The reactor volume is 1,500 L with a 1-m diameter. Pure $A$ is placed inside the batch reactor with an initial temperature and pressure of 300 K and 10 atm, respectively. Refer to Example 5.2: BR Irreversible Gas-phase First-order Cyclopropane Isomerization Reaction.

**CHW 5.9 Thermal Decomposition of Cyclobutane Into Ethylene with Variable Heat Capacity and Heat of Reaction in a Nonadiabatic, Unsteady-state CSTR**

Use chemical reaction data in CHW 5.7, with variable heat capacity and heat of reaction. A split-range valve system will be used to control the cooling/heating jacket side, such that if $T < 900$ K, a steam heating medium jacket will be enabled; else, a cooling water medium jacket will be enabled. The reactor volume is 3,000 L with a 1-m diameter. The feed molar flow rate is 1 mol/s of pure A with an initial temperature and pressure of 300 K and 50 atm, respectively. Show the profiles of $C_A$, $T$, and $X_A$ as a function of CSTR residence time.

Refer to Example 5.6: Unsteady-state CSTR Gas-phase Second-order HI Formation. Use **Cp_Gas_T.m** to find the heat capacity data for both cyclobutane and ethylene.

**CHW 5.10 Thermal Decomposition of Cyclobutane Into Ethylene with Variable Heat Capacity and Heat of Reaction in a Nonadiabatic Steady-state CSTR**

Use chemical reaction data in CHW 5.7, with variable heat capacity and heat of reaction. A split-range valve system will be used to control the cooling/heating jacket side, such that if $T < 900$ K, a steam heating medium jacket will be enabled; else, a cooling water medium jacket will be enabled. The reactor volume is 1,500 L with a 1-m diameter. The feed flow rate is 1 mol/s of pure $A$ with $T_0 = 300$ K and $P_0 = 50$ atm. Refer to Example 5.7: Steady-state CSTR Gas-phase Second-order HI Formation. Use **Cp_Gas_T.m** to find the heat capacity data for both cyclobutane and ethylene.

**CHW 5.11 Thermal Decomposition of Cyclobutane Into Ethylene with Constant Heat Capacity and Heat of Reaction in a Nonadiabatic, Unsteady-state PFR**

Use chemical reaction data in CHW 5.7, with variable heat capacity and heat of reaction. A split-range valve system will be used to control the cooling/heating jacket side, such that if $T < 900$ K, a steam heating medium jacket will be enabled; else, a cooling water medium jacket will be enabled. The reactor volume is 3,000 L with a 1-m diameter. The feed flow rate is 1 mol/s of pure $A$ with $T_0 = 300$ K and $P_0 = 50$ atm.

Refer to Example 5.10: Unsteady-state PFR Gas-phase Second-order HI Formation. Use **Cp_Gas_T.m** to find the heat capacity data for both cyclobutane and ethylene.

**CHW 5.12 Thermal Decomposition of Cyclobutane Into Ethylene with Constant Heat Capacity and Heat of Reaction in a Nonadiabatic, Steady-state PFR**

Use chemical reaction data in CHW 5.7, with variable heat capacity and heat of reaction. A split-range valve system will be used to control the cooling/heating jacket side, such that if $T < 900$ K, a steam heating medium jacket will be enabled; else, a cooling water medium jacket will be enabled. The reactor volume is 1,000 L with a 0.1-m diameter. The feed pressure and temperature are 40 atm and 300 K, respectively. The feed molar flow rate is 10 mol/s and is pure cyclobutane. Refer to Example 5.11: Steady-state PFR Gas-phase Second-order HI Formation. Use **Cp_Gas_T.m** to find the heat capacity data for both cyclobutane and ethylene.

**CHW 5.13 Dimerization of $NO_2$ Into $N_2O_4$ with Constant Heat Capacity and Heat of Reaction in a Nonadiabatic BR**

Consider the following gas-phase dimerization reaction of nitrogen dioxide into dinitrogen tetroxide:

$$2NO_2(g) \rightarrow N_2O_4(g) \qquad 2A(g) \rightarrow B(g)$$

$$-r_A = kC_A^2 = 1.3 \times 10^9 \left(\frac{\mathrm{L}}{\mathrm{mol \cdot s}}\right) e^{-\frac{50{,}500}{RT}} C_A^2$$

Heat capacity data (298–1,500 K):

$$C_P(NO_2), \quad \frac{\mathrm{J}}{\mathrm{mol \cdot K}} = 16.1086 + 75.895 \times 10^{-3}\,\mathrm{T} - 54.3874 \times 10^{-6}\,\mathrm{T}^2 + 14.3077 \times 10^{-9}\,\mathrm{T}^3$$

$$C_P(N_2O_4), \quad \frac{J}{mol \cdot K} = 19.015 + 115.7235 \times 10^{-3}\,T - 43.8987 \times 10^{-6}\,T^2 + 6.438 \times 10^{-9}\,T^3$$

For this question, assume a constant heat of reaction, $\Delta H_r(298) = -55{,}300$ J/mol, and constant heat capacity, independent of $T$. Moreover, $V_{BR}$ is set to 1,000 L with 0.1-m diameter, the initial temperature is 300 K, and the initial pressure is 1.4 atm for pure $A$.

Use constant heat capacity data evaluated at $T = 300$ K.

Write the total material/M balance, $A$-component molar balance, and energy balance to describe the nonadiabatic batch reactor system. The reaction is highly exothermic and fast; hence, ammonia will be used to cool the reactor and also demonstrate the initial events. A split-range valve system will be used to control the cooling/heating jacket side such that if $T < 300$ K, a steam heating medium jacket will be enabled; else, an ammonia cooling medium jacket will be enabled. Notice $C_A$ (or $X_A$) and temperature ($T$) do change with batch reactor time. Use the **DelG_Keq.m** code to find the standard heat of reaction.

Refer to Example 5.2: BR Irreversible Gas-phase First-order Cyclopropane Isomerization Reaction.

**References:**


1) Laidler, K.J. (1987). *Chemical Kinetics* (3rd ed.). Harper & Row. Page 302 reports.
2) Johnston, H.S., Foering, L., and White, J. (1955). *Journal of the American Chemical Society* 77(2): 420–421. 10.1021/ja01607a034.
3) NIST Chemistry WebBook.
   **$NO_2$:** https://webbook.nist.gov/cgi/cbook.cgi?ID=C10102440
   **$N_2O_4$:** https://webbook.nist.gov/cgi/cbook.cgi?ID=C10544720
*Data are derived from critically evaluated thermochemical tables (NIST-JANAF, TRC).*
4) Atkins, P. and de Paula, J. (2010). *Physical Chemistry* (9th ed.). Oxford University Press. Lists (Table 2.8 in Thermodynamics section).
5) Yaws, C.L. (1999). *Chemical Properties Handbook*. McGraw-Hill. Page 159 ($NO_2$) and Page 176 ($N_2O_4$).


**CHW 5.14 Dimerization of $NO_2$ Into $N_2O_4$ with Variable Heat Capacity and Heat of Reaction in a Nonadiabatic BR**

Repeat CHW 5.13, but include variable heat capacity and heat of reaction. A split-range valve system will be used to control the cooling/heating jacket side such that if $T < 300$ K, a steam heating medium jacket will be enabled; else, an ammonia cooling medium jacket will be enabled. Ammonia is used here to demonstrate the profiles of $C_A$, $T$, and $X_A$. The reaction is so rapid that it reaches the upper limit in a very short time. Keeping $T$ low will enable seeing the events in a time scale of 100 s. The reactor volume is 1,000 L with a 0.5-m diameter. The feed is pure $A$ with an initial pressure of 1.4 atm and an initial temperature of 300 K. Refer to Example 5.2: BR Irreversible Gas-phase First-order Cyclopropane Isomerization Reaction.

**CHW 5.15 Dimerization of $NO_2$ Into $N_2O_4$ with Variable Heat Capacity and Heat of Reaction in a Nonadiabatic Unsteady-state CSTR**

Repeat CHW 5.13, but include variable heat capacity and heat of reaction. A split-range valve system will be used to control the cooling/heating jacket side such that if

$T < 300$ K, a steam heating medium jacket will be enabled; else, an ammonia cooling medium jacket will be enabled. Ammonia is used here to demonstrate the profiles of $C_A$, $T$, and $X_A$. The reaction is so rapid that it reaches the upper limit in a very short time. Keeping $T$ low will enable seeing the events in a time scale of 100 s. The reactor volume is 500 L with a 0.4-m diameter. The feed is 0.01 mol/s of pure A with an initial pressure of 1.4 atm and an initial temperature of 300 K. Refer to Example 5.6: Unsteady-state CSTR Gas-phase Second-order HI Formation.

**CHW 5.16 Dimerization of $NO_2$ Into $N_2O_4$ with Variable Heat Capacity and Heat of Reaction in a Nonadiabatic Steady-state CSTR**

Repeat CHW 5.13, but include variable heat capacity and heat of reaction. A split-range valve system will be used to control the cooling/heating jacket side, such that if $T < 300$ K, a steam heating medium jacket will be enabled; else, an ammonia cooling medium jacket will be enabled. The reactor volume is 1,000 L with a 0.5-m diameter. The feed flow rate is 0.1 mol/s of pure $A$ with an initial temperature of 300 K and an initial pressure of 10 atm. Find the steady-state values. Refer to Example 5.7: Steady-state CSTR Gas-phase Second-order HI Formation.

**CHW 5.17 Dimerization of $NO_2$ Into $N_2O_4$ with Variable Heat Capacity and Heat of Reaction in a Nonadiabatic Unsteady-state PFR**

Repeat CHW 5.13, but include variable heat capacity and heat of reaction. A split-range valve system will be used to control the cooling/heating jacket side, such that if $T < 300$ K, a steam heating medium jacket will be enabled; else, an ammonia cooling medium jacket will be enabled. Ammonia is used here to demonstrate the profiles of $C_A$, $T$, and $X_A$. The reaction is so rapid that it reaches the upper limit in a very short time. Keeping $T$ low will enable seeing the events in a time scale of 100 s. The reactor volume is 500 L with a 0.4-m diameter. The feed is 0.01 mol/s of pure $A$ with an initial pressure of 1.5 atm and an initial temperature of 300 K. Refer to Example 5.10: Unsteady-state PFR Gas-phase Second-order HI Formation.

**CHW 5.18 Dimerization of $NO_2$ Into $N_2O_4$ with Variable Heat Capacity and Heat of Reaction in a Nonadiabatic Steady-state PFR**

Repeat CHW 5.13, but include variable heat capacity and heat of reaction. A split-range valve system will be used to control the cooling/heating jacket side, such that if $T < 300$ K, a steam heating medium jacket will be enabled; else, an ammonia cooling medium jacket will be enabled. The reactor volume is 500 L with a 0.5-m diameter. The feed flow rate is 0.1 mol/s of pure A with an initial temperature of 300 K and an initial pressure of 1.5 atm. Find the steady-state values. Refer to Example 5.11: Steady-state PFR Gas-phase Second-order HI Formation.

# 6
# Catalysis in Packed Bed and Fluidized Bed Reactors

## 6.1 Introduction to Catalysis and Reactor Types

Catalysis, a fundamental process in chemical transformations, involves the acceleration of reaction rates through the introduction of a catalyst, a substance that participates in the reaction mechanism without being consumed in the overall reaction, though it may degrade or deactivate over extended use [1]. The promoting catalyst allows a reaction to proceed at a lower activation energy, often leading to enhanced efficiency and selectivity in chemical processes. Catalysis is broadly classified into two main types: homogeneous and heterogeneous.

Homogeneous catalysis occurs when the catalyst and reactants exist in the same phase, typically a liquid [1]. This phase uniformity ensures intimate molecular contact, facilitating controlled reaction conditions and often leading to high selectivity [2]. The catalytic cycle typically involves reactant activation, transient complex formation, product release, and catalyst regeneration. The solubility of the catalyst in the reaction medium ensures a consistent interaction with the reactants throughout the reaction volume, which can be particularly advantageous for achieving high yields and specific product formation [3].

Heterogeneous catalysis involves a catalyst in a different phase from the reactants, often a solid interacting with gaseous or liquid species [1]. Reactions occur on the catalyst surface, where reactants adsorb onto active sites, undergo transformation, and desorb as products [4]. The effectiveness of heterogeneous catalysis is strongly influenced by surface area, porosity, and the transport of species to and from the catalyst surface [3].

In industrial applications, catalytic reactions are conducted in specialized reactors designed to maximize contact between reactants and the catalyst. Two widely used configurations are packed bed reactors (PBRs) and fluidized bed reactors (FBRs), each suited to specific catalytic processes.

PBRs, also known as fixed bed reactors, consist of a cylindrical vessel filled with a stationary bed of catalyst particles. Reactants in a fluid phase (gas or liquid) flow through the void spaces between the catalyst particles, contacting the catalytic surfaces where the reaction takes place. The fixed nature of the catalyst bed allows for continuous operation and a high conversion rate per weight of catalyst.

While predominantly used for heterogeneous catalysis, PBRs can also accommodate immobilized homogeneous catalysts, where soluble catalysts are anchored onto solid supports to facilitate recovery and reuse [5].

FBRs offer enhanced mixing and heat transfer by suspending solid catalyst particles in an upward fluid flow. This fluidization improves temperature control and catalyst regeneration, making FBRs ideal for highly exothermic reactions and processes requiring frequent catalyst replacement. Thus, fluidization causes the solid material to behave much like a fluid, resulting in excellent mixing and heat-transfer characteristics within the reactor [6]. FBRs are predominantly used for heterogeneous catalytic reactions and are particularly advantageous when processes require efficient heat management or frequent catalyst regeneration or replacement [7].

This chapter will explore both homogeneous and heterogeneous catalysis in the context of packed bed and FBRs, examining their mechanisms, design considerations, and operational principles. Special emphasis will be placed on mass transfer (MT) limitations, which significantly impact reactor performance. Illustrative examples will clarify key concepts, culminating in a comparative analysis to guide reactor selection for specific catalytic applications. Finally, a comparative analysis will offer guidance on selecting the most appropriate reactor type for specific catalytic applications.

## 6.2 Mechanism and Kinetics of Homogeneous Catalysis

In homogeneous catalysis, the catalyst and reactants exist in a single, uniform phase, typically a liquid solution. This arrangement allows for reactions to occur at the molecular level, often involving soluble organometallic complexes. Because every catalyst molecule is a potential active site, this method offers exceptionally high efficiency and control over reaction conditions.

The uniform distribution of the catalyst largely circumvents the MT limitations often inherent in multiphase heterogeneous systems. This leads to several key advantages:

- **High Selectivity:** The well-defined, identical nature of each active site allows for outstanding chemo-, regio-, and enantioselectivity.
- **Mild Reaction Conditions:** Many homogeneous catalysts operate effectively at lower temperatures and pressures.
- **Mechanistic Clarity:** The soluble nature of the catalyst and intermediates makes the system amenable to spectroscopic analysis (e.g., NMR, IR, UV–Vis), facilitating detailed kinetic and mechanistic studies.

The primary operational challenge, however, remains the separation of the catalyst from the product in a post-reaction mixture.

### 6.2.1 The Catalytic Cycle: A Mechanistic Viewpoint

The process is not a simple linear sequence but a catalytic cycle, where the active catalytic species is regenerated after each turnover. While specific cycles vary greatly, a generalized cycle for many cross-coupling or hydrogenation reactions involves several key elementary steps:

1) **Activation and Substrate Coordination:** An inactive pre-catalyst may first be activated. The active catalyst, typically a coordinatively unsaturated metal complex, then binds to one or more reactant molecules (substrates) via ligand substitution.
2) **Oxidative Addition:** A key bond in the substrate breaks, and the two resulting fragments add to the metal center. This process increases the metal's formal oxidation state and its coordination number.
3) **Migratory Insertion (or Other Transformations):** A coordinated ligand inserts into a metal–ligand bond on the same complex. This is often the crucial step where new carbon–carbon or carbon-heteroatom bonds are formed within the coordination sphere of the metal.

4) **Reductive Elimination:** The final step, where the desired product is formed by coupling two ligands. The product is then released (dissociates) from the metal center, which simultaneously reduces the metal's oxidation state and coordination number, thereby regenerating the active catalyst for the next cycle.

A classic example of homogeneous catalysis is acid-catalyzed esterification, in which a proton ($H^+$) activates the carbonyl group of a carboxylic acid, thereby lowering the activation energy for nucleophilic attack by an alcohol. Homogeneous catalysis also plays a vital role in organometallic chemistry; for instance, transition metal complexes, such as those of rhodium and cobalt, are widely used in industrial processes like hydroformylation, a key carbonylation reaction involving the addition of hydrogen and a formyl group (—CHO) across an alkene. These catalysts are also essential in olefin polymerization [8]. For further details and examples, see "6.8 Illustrative Cases of Homogeneous FBRs" section.

The rate laws governing homogeneous catalytic reactions depend on both the reactant and catalyst concentrations. Since the catalyst is uniformly distributed in the reaction medium, its effective concentration can be precisely controlled, making kinetic analysis straightforward. The general rate expression takes the form:

$$-r_A = k[C]^m A^n$$

where

- $r_A$ is the reaction rate,
- $k$ is the rate constant,
- $[C]$ is the catalyst concentration,
- $[A]$ is the reactant (or substrate) concentration,
- $m$ and $n$ are reaction orders, typically determined experimentally.

In many cases, the rate law follows a Michaelis–Menten-type kinetics when the reaction proceeds through an intermediate complex:

$$-r_A = \frac{k_{obs}[A][C]}{K + [A]}$$

where $k_{obs}$ is the observed rate constant, and $K$ is the equilibrium constant for intermediate formation. This expression highlights the saturation effect observed in homogeneous catalytic reactions at high reactant concentrations.

The ability to derive precise kinetic models enables the optimization of reactor conditions, including temperature, pressure, and catalyst concentration. Advanced computational techniques, such as density functional theory (DFT) and molecular dynamics simulations, further aid in elucidating reaction mechanisms and predicting catalytic performance.

Homogeneous catalysis is widely applied in fine chemical synthesis, pharmaceutical manufacturing, and petrochemical processing due to its high efficiency, tunability, and ability to operate under mild conditions. However, catalyst recovery and separation remain key challenges, often necessitating the development of ligand-modified catalysts or biphasic catalytic systems to facilitate catalyst recyclability.

### 6.2.2 Linear Regression of Michaelis–Menten-type Kinetics

**Example 6.1** ***Michaelis–Menten-Type kinetics***

Consider a homogeneous catalytic reaction where the decomposition of a reactant (or substrate) $A$ is catalyzed by a soluble catalyst $C$. The reaction follows a Michaelis–Menten-type kinetics, and

**Table 6.1** A homogeneous catalytic reaction data for the decomposition of reactant $A$, catalyzed by a soluble catalyst $C$.

| [A] (M) | $-r_A$ (mol/L·s) | [A] (M) | $-r_A$ (mol/L·s) |
|---|---|---|---|
| 0.01 | 0.005 | 0.2 | 0.0182 |
| 0.02 | 0.0083 | 0.25 | 0.019 |
| 0.05 | 0.0133 | 0.3 | 0.021 |
| 0.1 | 0.0167 | 0.35 | 0.022 |

experimental data for the reaction rate ($-r_A$) at different reactant concentrations $[A]$ is provided in Table 6.1. The catalyst concentration is kept constant at = 0.01 M.

Determine the values of $k_{obs}$ and $K$.

**Solution**

$$-r_A = \frac{k_{obs}[A][C]}{K + [A]} \tag{A}$$

Let us linearize the model by taking the inverse of both sides:

$$\frac{1}{-r_A} = \frac{K + [A]}{k_{obs}[A][C]} \tag{B}$$

Rearranging, we have:

$$\frac{1}{-r_A} = \frac{K}{k_{obs}[C]} * \frac{1}{[A]} + \frac{1}{k_{obs}[C]} \tag{C}$$

This equation is in the form of a linear equation:

$$y = bx + a \tag{D}$$

Mapping a term by term, we have:

$$y = \frac{1}{-r_A} \tag{E}$$

$$x = \frac{1}{[A]} \tag{F}$$

$$Slope\ b = \frac{K}{k_{obs}[C]} \tag{G}$$

$$Intercept\ a = \frac{1}{k_{obs}[C]} \tag{H}$$

The linear equation will be solved both using hand calculations and by using the MATLAB linear regression solver called **fitlm**.

The least squares method helps us determine the optimal values of $a$ (intercept) and $b$ (slope) that best represent the overall trend in the data. For more information on the method of least squares and on MATLAB's built-in **fitlm** function, the most versatile function for fitting linear models in MATLAB, which provides access to various statistics and diagnostics related to fitness, please refer to my textbook [9].

$$b = \frac{\sum_{i=1}^{N}(X_i - \overline{X})(Y_i - \overline{Y})}{\sum_{i=1}^{N}(X_i - \overline{X})^2} \tag{I.1}$$

Alternatively,

$$b = \frac{N\sum_{i=1}^{N}(X_iY_i) - \left(\sum_{i=1}^{N}Y_i\right)\left(\sum_{i=1}^{N}X_i\right)}{N\sum_{i=1}^{N}X_i^2 - \left(\sum_{i=1}^{N}X_i\right)^2} \tag{I.2}$$

$$a = \overline{Y} - b\overline{X} \tag{J.1}$$

Alternatively,

$$a = \frac{\sum_{i=1}^{N}Y_i}{N} - b\frac{\sum_{i=1}^{N}X_i}{N} \tag{J.2}$$

To analytically find the best values of $a$ and $b$, using Eqs. I and J, we must create new additional columns as shown in Table 6.2.

**Table 6.2** Creation of new columns to facilitate the use of Eqs. I and J and column-wise summation.

| $[A]$ (M) | $-r_A$ (mol/L·s) | $X = 1/[A]$ | $y = 1/(-r_A)$ | $x^2$ | $xy$ |
|---|---|---|---|---|---|
| 0.01 | 0.005 | 100 | 200 | 10,000 | 20,000 |
| 0.02 | 0.0083 | 50 | 120.4819 | 2,500 | 6,024.096 |
| 0.05 | 0.0133 | 20 | 75.18797 | 400 | 1,503.759 |
| 0.1 | 0.0167 | 10 | 59.88024 | 100 | 598.8024 |
| 0.2 | 0.0182 | 5 | 54.94505 | 25 | 274.7253 |
| 0.25 | 0.019 | 4 | 52.63158 | 16 | 210.5263 |
| 0.3 | 0.021 | 3.333 | 47.61905 | 11.11111 | 158.7302 |
| 0.35 | 0.022 | 2.8571 | 45.45455 | 8.163265 | 129.8701 |
| | **Summation** | $\sum_1^8 x_i = \mathbf{195.2}$ | $\sum_1^8 y_i = \mathbf{656.2}$ | $\sum_1^8 x_i^2 = \mathbf{13060.3}$ | $\sum_1^8 x_i y_i = \mathbf{28900.5}$ |

Using Eq. (I.2), we have:

$$b = \frac{N\sum_{i=1}^{N}(X_iY_i) - \left(\sum_{i=1}^{N}Y_i\right)\left(\sum_{i=1}^{N}X_i\right)}{N\sum_{i=1}^{N}X_i^2 - \left(\sum_{i=1}^{N}X_i\right)^2} = \frac{8(28900.5) - 656.2 * 195.2}{8(13060.3) - (195.2)^2} = 1.5534$$

Using Eq. (J.2), we have:

$$a = \frac{\sum_{i=1}^{N}Y_i}{N} - b\frac{\sum_{i=1}^{N}X_i}{N} = \frac{656.2}{8} - 1.5534 * \frac{195.2}{8} = 44.122$$

Using Eq. (H), we have:

$$a = 44.122\,\frac{\text{s}\cdot\text{L}}{\text{mol}} = \frac{1}{k_{obs}[C]} = \frac{1}{k_{obs} * 0.01\frac{\text{mol}}{\text{L}}} \qquad k_{obs} = \frac{1}{44.122\frac{\text{s}\cdot\text{L}}{\text{mol}} * 0.01\frac{\text{mol}}{\text{L}}} = 2.2664\,\text{s}^{-1}$$

Using Eq. (G), we have:

$$b = 1.5534\,s = \frac{K}{k_{obs}[C]} = \frac{1}{k_{obs}[C]} * K = a * K \rightarrow K = \frac{b}{a} = \frac{1.5534\,\text{s}}{44.122\,\frac{\text{s}\cdot\text{L}}{\text{mol}}} = 0.0352\,\frac{\text{mol}}{\text{L}}$$

$$= 0.0352\ \text{M}$$

Back to Eq. (A), we have the final expression for $-r_A$

$$-r_A = \frac{k_{obs}[A][C]}{K + [A]} = \frac{2.2664\,[A][0.01]}{0.0352 + [A]} = \frac{0.022664\,[A]}{0.0352 + [A]} \tag{K}$$

### 6.2.3 MATLAB Code for Predicting Michaelis–Menten-Type Kinetics

Figure 6.1 shows MATLAB code (**Mich_Menten.m**) for carrying out linear regression using MATLAB's built-in **fitlm** function while presenting the regressed model parameters and plotting both the experimental and fitted curves.

```
clear variables;
% Create X data (independent variable)
X = [0.01 0.02 0.05 0.1 0.2 0.25 0.3 0.35];
% Create Y data (dependent variable)
Y = [0.005 0.0083 0.0133 0.0167 0.0182 0.019 0.021 0.022];
% Enzyme Concentration, mol/L
C=0.01;
% Take the inverse of X and Y.
x=1./X;
y=1./Y;
% Fit the model using fitlm
mdl = fitlm(x, y);
%This line of code fits a linear model of the form Y = beta0 + beta1 * X.
% Get the estimated coefficients.
beta0 = mdl.Coefficients.Estimate(1); % Intercept
beta1 = mdl.Coefficients.Estimate(2); % Slope
kobs=1/(beta0*C);
K=beta1/beta0;
% Display the equation
Word1="Equation: y = ";
Word2= num2str(beta0);
Word3= num2str(beta1);
Title=Word1 + Word2 + " + " + Word3 + "x " ;
disp(Title);
Model_Coeffs=mdl.Coefficients;
disp(['Intercept value, a = ', num2str(Model_Coeffs.Estimate(1))]);
disp(['Slope value, b = ', num2str(Model_Coeffs.Estimate(2))]);
disp(['kobs = ', num2str(kobs), ' s-1']);
disp(['K = ', num2str(K), ' mol/L']);
AdjR2=mdl.Rsquared;
disp(['Ordinary R2 = ', num2str(AdjR2.Ordinary)]);
disp(['Adjusted R2 = ', num2str(AdjR2.Adjusted)]);
% Access other statistics like R-squared, p-values, etc.
% See fitlm documentation for details
% Plot the data and the fitted line
plot(x, y, 'o', linewidth =3);
hold on;
plot(x, mdl.Fitted, '-', linewidth =3);
legend('Data', 'Fitted Line','Location','southeast');
xlabel('x', 'FontSize', 16, 'FontWeight', 'bold');
ylabel('y', 'FontSize', 16, 'FontWeight', 'bold');
title(Title, fontsize=16);
ax = gca;
% Set new font size for all axes text (including labels)
ax.FontSize = 16;
grid on;
```

**Figure 6.1** MATLAB code (**Mich_Menten.m**) for carrying out linear regression using MATLAB's built-in **fitlm** function, presenting the regressed model parameters, and plotting both the experimental and fitted curves.

```
figure;
Yfit=kobs.*X*C./(K+X);
plot(X, Y, 'o', linewidth =3);
hold on;
plot(X, Yfit, '-', linewidth =3);
legend('Data', 'Fitted Line','Location','southeast');
grid on;
xlabel('X-Experimental', 'FontSize', 16, 'FontWeight', 'bold');
ylabel('Y_{exp}, Y_{fit}', 'FontSize', 16, 'FontWeight', 'bold');
title('Y_{exp}, Y_{fit} vs. X', fontsize=16);
hold off; ax = gca;
% Set new font size for all axes text (including labels)
ax.FontSize = 16;
```

**Figure 6.1** (*Continued*)

Results are copied from the Command Window. Notice the mapping between the manual calculation results and those obtained via MATLAB regression.

```
≫ Mich_Menten
Equation: y = 44.1236 + 1.5534x
Intercept value, a = 44.1236
Slope value, b = 1.5534
kobs = 2.2664 s-1
K = 0.035206 mol/L
Ordinary R2 = 0.99855
Adjusted R2 = 0.9983
```

Figure 6.2 (left) shows the linearized version of the model, and Figure 6.2 (right) shows the regressed curve for the original data, shown in Table 6.1.

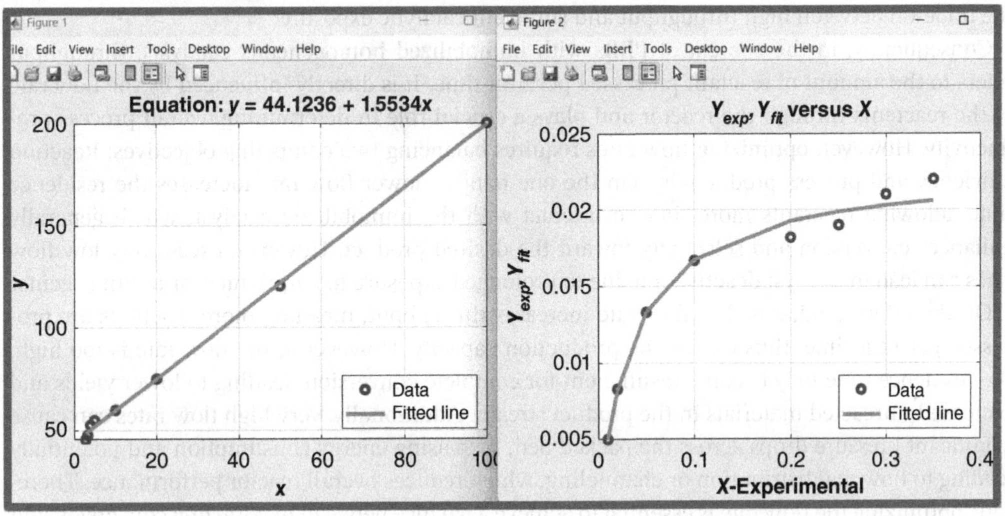


**Figure 6.2** (Left) the linearized version of the model and (right) the regression curve of the original experimental data given in Table 6.1.

## 6.3 Design and Operational Considerations for PBRs with Homogeneous Catalysts

Homogeneous catalysts are inherently soluble in the reaction medium, which poses challenges for their direct use in PBRs, as these reactors typically depend on fixed catalytic sites. To address this, homogeneous catalysts are often immobilized onto solid supports, a process known as *heterogenization*. This strategy combines the high selectivity and efficiency of homogeneous catalysis with the practical advantages of heterogeneous systems, such as ease of catalyst recovery, continuous operation, and enhanced scalability [10].

### 6.3.1 Catalyst Immobilization and Support Selection

The choice of support material is a critical factor in the performance of a PBR utilizing an immobilized homogeneous catalyst. An ideal solid support should exhibit:

✓ Chemical inertness to prevent unwanted interactions with the catalyst or reactants.
✓ High porosity to facilitate the diffusion of reactants and products, minimizing MT limitations.
✓ Large surface area to maximize catalyst loading and improve dispersion, ensuring uniform catalytic activity.

Common support materials include functionalized silica, alumina, zeolites, and polymeric resins, each tailored to specific catalytic applications. Covalent attachment, adsorption, or encapsulation methods are commonly employed to retain the homogeneous catalyst on the support while preserving its activity.

### 6.3.2 Flow and Residence Time Considerations

The residence time ($\tau$) is given by:

$$\tau = \frac{V_{PBR}}{\upsilon_0}$$

where $V_{PBR}$ is the PBR volume, and $\upsilon_0$ is the input volumetric flow rate. This equation highlights the trade-off between high throughput and sufficient catalytic exposure.

Consequently, in the context of PBRs with immobilized homogeneous catalysts, throughput refers to the amount of reactant processed per unit time. It is directly influenced by the flow rate of the reactants through the reactor and plays a crucial role in determining overall process productivity. However, optimizing flow rates requires balancing two competing objectives: Reaction efficiency and process productivity. On the one hand, a lower flow rate increases the residence time, allowing reactants more time to interact with the immobilized catalyst, which generally enhances conversion and selectivity toward the desired product. However, excessively low flow rates can lead to catalyst deactivation due to prolonged exposure to by-products or fouling agents.

On the other hand, a higher flow rate increases throughput, meaning more reactants are processed per unit time, thus enhancing production capacity. However, if the flow rate is too high, the residence time may become insufficient for complete conversion, leading to lower yields and increased unreacted materials in the product stream. Additionally, very high flow rates can cause significant pressure drops across the packed bed, increasing energy consumption and potentially leading to flow maldistribution or channeling, which reduces overall reactor performance. Therefore, optimizing the flow rate is essential to achieve a balance between maximizing conversion and maintaining an economically viable production rate.

### 6.3.3 Thermal Management in PBRs with Immobilized Homogeneous Catalysts

Effective temperature control is vital, particularly for strongly exothermic or endothermic reactions, where thermal gradients can impact reaction performance and catalyst stability [5]. Inadequate heat management can lead to:

⚠ Hot spots, which may cause catalyst deactivation, side reactions, or even structural damage to the reactor.

⚠ Cold zones, which can slow reaction rates and lower conversion efficiency.

To mitigate thermal effects, several strategies can be employed:

✓ **Efficient Heat-transfer Design:** Using high-thermal-conductivity support materials (e.g., metal–organic frameworks or carbon-based supports) to enhance heat dissipation.

✓ **Temperature-controlled Packing Arrangements:** Implementing layered catalyst beds with varying thermal properties to regulate temperature gradients.

✓ **External Heat Exchange Systems:** Employing preheaters, cooling jackets, or embedded heat exchangers to stabilize reactor operating conditions.

By effectively managing heat transfer, maintaining optimal flow dynamics, and carefully selecting support materials, PBRs with immobilized homogeneous catalysts can achieve both high efficiency and long-term durability/stability, making them viable for large-scale industrial scalability.

## 6.4 Advantages and Disadvantages of Immobilized Homogeneous Catalyzed PBRs

The use of immobilized homogeneous catalysts in PBRs presents several advantages, particularly in combining the benefits of homogeneous catalysis with the operational efficiency of heterogeneous systems. Homogeneous catalysts are known for their high selectivity and activity, often leading to improved reaction rates and product yields. By immobilizing these catalysts onto solid supports, their advantages can be harnessed in a continuous flow system, eliminating the need for catalyst recovery and simplifying downstream processing [2]. Unlike traditional homogeneous batch reactors, which require costly separation steps to remove dissolved catalysts from reaction products, PBRs with immobilized catalysts allow for continuous operation, leading to increased productivity and reduced operational costs [10]. Additionally, PBRs typically achieve a high conversion per unit mass of catalyst due to their efficient contact between reactants and catalytic sites, often outperforming other reactor types in terms of catalyst utilization [5].

However, there are significant challenges associated with this approach. One of the most critical concerns is catalyst leaching, where weak interactions between the catalyst and its support can lead to detachment of active species, causing loss of catalytic activity over time and potential contamination of the product stream [11]. This issue is particularly problematic for liquid-phase reactions, where the continuous flow of reactants can gradually wash away loosely bound catalysts, necessitating frequent catalyst replenishment or regeneration.

Another key drawback is MT limitations, which arise when reactants must diffuse through the porous support material to reach the active sites of the immobilized catalyst. If diffusion is too slow compared to the reaction rate, the overall process can become MT-limited, reducing efficiency [12]. Similarly, product molecules may experience diffusion resistance while leaving the catalyst surface, potentially leading to side reactions or deactivation. This issue is especially relevant for reactions involving bulky molecules or viscous media, where molecular transport is inherently slower.

Furthermore, temperature control in PBRs presents a significant challenge. Like heterogeneous catalytic systems, immobilized homogeneous catalysts may experience localized temperature variations due to exothermic or endothermic reactions. Inadequate heat dissipation can result in hotspots, leading to thermal degradation of the catalyst or unwanted side reactions, while excessive cooling can slow reaction rates [5]. Proper reactor design, including the use of heat exchangers, temperature monitoring, and optimized flow conditions, is essential to mitigate these effects.

Despite these limitations, immobilized homogeneous catalysts in PBRs remain an attractive option for many industrial processes, particularly when the advantages of continuous processing, high selectivity, and simplified product separation outweigh the associated challenges. Advances in catalyst design, support materials, and reactor engineering continue to enhance their feasibility and performance in modern chemical manufacturing.

## 6.5 Illustrative Cases of Homogeneous PBRs

### 6.5.1 Hydrolysis of Methyl Acetate in a Packed Bed Reactor

The hydrolysis of methyl acetate (an ester) in an aqueous solution, catalyzed by sulfuric acid, is a classic acid-catalyzed reaction that produces acetic acid and methanol:

$$CH_3COOCH_3 + H_2O \overset{H+}{\Leftrightarrow} CH_3COOH + CH_3OH$$

When sulfuric acid is immobilized on a solid resin with sulfonic acid (—$SO_3H$) functional groups, the reaction can be efficiently conducted in a continuous mode using a PBR. In this setup, the ester solution is pumped through the packed bed containing the immobilized catalyst, ensuring contact between the reactants and the active catalytic sites.

The reaction kinetics follow a pseudo-first-order rate law if water is in large excess:

$$-r_A = kC_{ester}C_{H+} = k_{obs}C_{ester} = k_{obs}C_A$$

where:

- $C_{ester}$ is the concentration of methyl acetate,
- $C_{H+}$ represents the effective concentration of protonic acid sites on the resin,
- $k$ is the intrinsic rate constant, which depends on temperature and catalyst properties,
- $k_{obs}$ is the observed (or apparent) rate constant.

The performance of the reactor depends on several factors:

a) The flow rate of the ester solution, which controls the residence time in the reactor and determines the extent of conversion.
b) The distribution of sulfonic acid sites on the resin affects the local acid strength and reaction rate.
c) The MT effects, since diffusion of reactants into the porous resin and out of the reaction sites, must be considered for optimal performance.
d) The temperature, which influences both the reaction kinetics and the swelling behavior of the resin.

One of the advantages of using an immobilized catalyst is the ease of separation of the catalyst from the reaction mixture, reducing downstream processing costs. This is particularly beneficial in continuous industrial applications where catalyst recovery and reuse are important for cost efficiency.

**Example 6.2** ***Hydrolysis of Methyl Acetate in a PBR***
Calculate the conversion of methyl acetate ($A$) in a PBR with an immobilized sulfuric acid catalyst. The reaction follows pseudo-first-order kinetics, and the reactor operates at steady state with the given parameters.

$$CH_3COOCH_3 + H_2O \overset{H+}{\Leftrightarrow} CH_3COOH + CH_3OH$$

**Given Data**

- **Inlet Concentration of Methyl Acetate:** $C_{A0} = 2\,\text{mol/L}$
- **Volumetric Flow Rate:** $\upsilon_0 = 5\,\text{L/min}$
- **Effective Pseudo-first-order Rate Constant:** $k_{obs} = 0.02\frac{\text{L}}{\text{gcat. min}}$ (see Chapter 2 | NOTE #1)
- **Catalyst Weight:** $W = 300\,\text{g}$

**Assumptions**

- Water exists in a large excess amount (i.e., pseudo-first-order kinetics).
- Liquid-phase reaction with constant density (or constant volume).
- Immobilized $H^+$ concentration is constant ($C_{H+}$ incorporated into $k_{obs}$).

**Solution**
Referring to Eq. (2.30), we have:

$$W_{PBR} = F_{A0}\int_0^X \frac{dX}{-r'_A} \tag{A}$$

$$F_{A0} = C_{A0} * \upsilon_0 = 2\frac{\text{mol}}{\text{L}} * 5\frac{\text{L}}{\text{min}} = 10\frac{\text{mol}}{\text{min}} \tag{B}$$

$$-r'_A = k_{obs} * C_A = k_{obs} * C_{A0}(1 - X_A) \tag{C}$$

Substitute in Eq. (A), we have:

$$W_{PBR} = 300\,\text{g} = F_{A0}\int_0^X \frac{dX}{-r'_A} = F_{A0}\int_0^X \frac{dX}{k_{obs} * C_{A0}(1 - X_A)} = \upsilon_0\int_0^X \frac{dX}{k_{obs} * (1 - X_A)}$$

$$= \frac{5\frac{\text{L}}{\text{min}}}{0.02\frac{\text{L}}{\text{gcat. min}}}\int_0^{XA} \frac{dX}{(1 - X_A)} = 250\,\text{g}\left[\int_0^X \frac{dX}{(1 - X_A)}\right]$$

This implies that:

$$\frac{300\,\text{g}}{250\,\text{g}} = 1.2 = \int_0^{XA} \frac{dX}{(1 - X_A)} = -[\ln(1 - X_A) - \ln(1)] = -\ln(1 - X_A)$$

$$(1 - X_A) = e^{-1.2} = 0.3012 \quad \rightarrow \quad X_A = 0.6988 \cong 0.7$$

### 6.5.2 MATLAB Code for Calculating the Conversion in a PBR for the Acid-Catalyzed Methyl Acetate Hydrolysis

Figure 6.3 shows MATLAB code (**PBR_XA.m**) for calculating the final conversion in a PBR for the acid-catalyzed methyl acetate hydrolysis. Notice that the input parameters: $C_{A0}$, $\upsilon_0$, $k_{obs}$, and $W$ can be modified by the users, and $X_A$ can be calculated accordingly.

```
% MATLAB code to determine the conversion XA in PBR using numerical integration
clear variables;
% Step 1: Prompt user for input values
CA0 = input('Enter the initial concentration of A, CA0 (mol/L): ');
v0 = input('Enter the initial volume flowrate, v0 (L/min): ');
kobs = input('Enter the effective pseudo-first-order rate constant, kobs (L/(gcat.min)): ');
W = input('Enter the catalyst weight, W in (g): ');
% Step 2: Calculate initial molar flow rate
FA0 = CA0 * v0; % mol/min
% Step 3: Define the function to be integrated
fun = @(XA) 1 ./ max(1 - XA, 1e-6); % Avoid division by zero at XA = 1
% Step 4: Solve for XA using numerical integration
options = optimset('TolX', 1e-6); % Set tolerance for fzero
XA_sol = fzero(@(XA) (v0/kobs) * integral(fun, 0, XA, 'RelTol', 1e-6, 'AbsTol', 1e-9) - W, [0, 0.99], options);
% Step 5: Report the conversion XA
fprintf('The final conversion, XA for the given input parameters is: %.4f\n', XA_sol);
```

**Figure 6.3** MATLAB code (**PBR_XA.m**) for $X_A$ calculation in a PBR for acid-catalyzed methyl acetate hydrolysis.

Results are copied from the Command Window. The final conversion $X_A$ is the same as that of the above analytical solution.

```
>> PBR_XA
Enter the initial concentration of A, CA0 (mol/L): 2
Enter the initial volume flowrate, v0 (L/min): 5
Enter the effective pseudo-first-order rate constant, kobs (L/(gcat.min)): 0.02
Enter the catalyst weight, W in (g): 300
The final conversion, XA for the given input parameters is: 0.6988
```

### 6.5.3 Hydrogenation of Hexene in a PBR

Another case of a catalytic reaction in a PBR is the hydrogenation of hexene to hexane using an immobilized iridium complex as the catalyst:

$$C_6H_{12} + H_2 \rightarrow C_6H_{14}$$

The reaction occurs in a continuous gas-phase process, where a mixture of hydrogen gas and hexene vapor (diluted in nitrogen as a carrier gas) is fed into the reactor. The catalyst, an iridium complex immobilized on alumina particles, facilitates the addition of hydrogen across the double bond of hexene, converting it to saturated hexane.

The rate of hydrogenation depends on the following factors:

a) The partial pressure of hydrogen and hexene ($A$), as described by Langmuir-Hinshelwood kinetics:

$$-r_A = \frac{kP_{H2}P_A}{(1 + K_{H2}P_{H2} + K_AP_A)^2}$$

where $k$ is the intrinsic rate constant, and $K_{H2}$ and $K_A$ are adsorption equilibrium constants.

b) The concentration of active iridium sites on the alumina surface determines the overall catalytic activity. A higher density of active sites enhances reaction rates, while catalyst deactivation due to fouling or sintering can reduce efficiency over time.

c) The temperature, which influences reaction kinetics through the Arrhenius relation, affects both the rate constant and the adsorption–desorption equilibrium of hydrogen and hexene on the catalyst surface. An optimal temperature range must be maintained to prevent catalyst deactivation or unwanted side reactions.

d) MT limitations, particularly the diffusion of hydrogen to the catalyst surface in a packed bed. If external MT resistance is significant, the reaction rate may be limited by the rate at which hydrogen reaches the active sites, rather than by the intrinsic reaction kinetics. This issue can be mitigated by optimizing gas flow rates and using catalyst supports with high surface area and porosity.

Using an immobilized organometallic catalyst offers multiple advantages [13]:

✓ Continuous operation without catalyst loss, as the catalyst remains fixed on the alumina support.
✓ Enhanced selectivity, as the catalyst microenvironment can be tailored to avoid side reactions.
✓ Ease of separation, particularly important for expensive catalysts like iridium, where loss must be minimized.

This method of continuous hydrogenation is widely used in fine chemicals and pharmaceuticals, where controlled hydrogenation is essential.

### 6.5.4 MATLAB Code for Hydrogenation of Hexene into Hexane in an Isothermal PBR

**Example 6.3** ***PBR for Hydrogenation of Hexene into Hexane***

Determine the catalyst weight required for 80% conversion in a PBR for the hydrogenation of hexene into hexane. Use the provided Langmuir-Hinshelwood kinetics:

$$-r_A = \frac{kP_{H2}P_A}{(1 + K_{H2}P_{H2} + K_AP_A)^2}$$

$C_6H_{12} + H_2 \rightarrow C_6H_{14}$ In symbols: $A + B \rightarrow C$

**Feed Conditions**

Total pressure $P = P_0 = 10$ atm, initial molar flow rates $F_{A0} = 10$ mol/s, $F_{B0} = 20$ mol/s, $F_{N2} = 70$ mol/s.

**Given Data**

$$k = 0.5\frac{\text{mol}}{\text{kg catalyst} \cdot \text{s}} \times \left(\frac{1}{\text{atm}}\right)^2 \text{ (see Chapter 2 | NOTE\#1)}$$

$$K_{H2} = 0.1 \text{ atm}^{-1}$$

$$K_A = 0.05 \text{ atm}^{-1}$$

$$T_0 = T = \text{Constant} = 600 \text{ K}$$

**Solution**

Referring to Eq. (2.30), we have:

$$W_{PBR} = F_{A0}\int_0^X \frac{dX}{-r'_A} \tag{A}$$

$$F_{A0} = 10\frac{\text{mol}}{\text{s}} \tag{B}$$

For a gas-phase reaction, we need to express partial pressure and/or molar concentration in terms of both volume expansion and conversion, $X$. Please refer to Eqs. (2.74)–(2.77):

$$\varepsilon = \delta \times \frac{F_{A0}}{F_0} = \delta \times y_{A0} = \left(\frac{c}{a} - \frac{b}{a} - \frac{1}{1}\right)y_{A0} = (1 - 1 - 1) * 0.1 = -0.1 \tag{C}$$

$$C_A = C_{A0}\frac{(1 - X)}{(1 + \varepsilon X)}\frac{P}{P_0}\frac{T_0}{T}$$

$P$ and $T$ are held constant; hence, we have:

$$C_A = C_{A0}\frac{(1 - X)}{(1 - 0.1X)}$$

Or

$$P_A = P_{A0}\frac{(1 - X)}{(1 - 0.1X)} = (P * y_{A0}) * \frac{(1 - X)}{(1 - 0.1X)} = (10.0 * 0.1)\text{ atm} * \frac{(1 - X)}{(1 - 0.1X)} \tag{D}$$

$$C_{H2} = C_B = C_{A0}\frac{\left(\theta_{B0} - \frac{b}{a}X\right)}{(1 + \varepsilon X)} = C_{A0}\frac{(2 - X)}{(1 - 0.1X)}$$

Or

$$P_{H2} = P_B = (P * y_{A0})\frac{(2-X)}{(1-0.1X)} = 1.0\,\text{atm} * \frac{(2-X)}{(1-0.1X)} \tag{E}$$

$$W_{PBR} = F_{A0}\int_0^X \frac{dX}{-r'_A} = 10\frac{\text{mol}}{\text{s}}\int_0^X \frac{dX}{\frac{kP_{H2}P_A}{(1+K_{H2}P_{H2}+K_AP_A)^2}} = 10\frac{\text{mol}}{\text{s}}\int_0^X \frac{(1+K_{H2}P_{H2}+K_AP_A)^2dX}{kP_{H2}P_A} \tag{F}$$

Figure 6.4 shows MATLAB code (**PBR_WPBR.m**) for calculating the weight of the catalyst bed, expressed in kg, as a function of input parameters and conversion, $X$, for the limiting reactant, hexene. Equations (A) through (F) are embedded within the code itself. Notice that the MATLAB code can be rerun for different input parameters and final conversion, $X_f$.

```
clear variables;
%% Step 1-4: Prompt for input parameters
FA0 = input('Enter feed molar flow rate of hexene, FA0 (mol/s): ');
FB0 = input('Enter feed molar flow rate of H2, FB0 (mol/s): ');
FN2 = input('Enter feed molar flow rate of N2 (mol/s): ');
P0  = input('Enter feed pressure, P0 (atm): ');
Xf=input('Enter the final conversion for A, Xf: ');
%% Step 5: Calculate change in number of moles
% Given as: deltaN = 1-1-1;
deltaN = 1 - 1 - 1; % This equals -1.
%% Step 6 & 7: Calculate feed mole-fractions of A and B
totalFlow = FA0 + FB0 + FN2;
yA0 = FA0 / totalFlow;
yB0 = FB0 / totalFlow;
%% Step 8: Calculate the volume expansion
eps = deltaN * yA0; % eps = ΔN * yA0
%% Steps 9 & 10: Define conversion-dependent partial pressures
% Here, X is the conversion, which will serve as our integration variable.
% PA = (yA0*P0)*(1-X)/(1+eps*X)
% PB = (yA0*P0)*((FB0/FA0)-X)/(1+eps*X)
% For clarity, we define functions for PA and PB as a function of X:
PA = @(X) (yA0 * P0) .* ((1 - X) ./ (1 + eps * X));
PB = @(X) (yA0 * P0) .* (((FB0/FA0) - X) ./ (1 + eps * X));
%% Step 11: Express rA as a function of X via PA and PB
% The given rate expression is:
% rA = (-0.5 * PB * PA) / (1 + 0.1*PB + 0.05*PA)^2
% We assume here that PH2 = PB.
reactionRate = @(X) ( -0.5 .* PA(X) .* PB(X) ) ./ ( 1 + 0.1 .* PB(X) + 0.05 .* PA(X) ).^2;
%% Step 12: Calculate the catalyst weight, W_PBR.
% The weight is given by:
% W_PBR = FA0 * ∫[0 to Xf] (dX/(-rA(X)))
% We use MATLAB's integral function to perform the integration.
W_PBR = FA0 * integral(@(X) 1 ./ (-reactionRate(X)), 0, Xf);
%% Display the result
fprintf('The weight of catalyst in the PBR is %.3f kg\n', W_PBR);
```

**Figure 6.4** MATLAB code (**PBR_WPBR.m**) for calculating the weight of the catalyst bed, expressed in kg, as a function of input parameters and conversion, $X$, for the limiting reactant, hexene.

Results are copied from the Command Window.

```
>> PBR_WPBR
Enter feed molar flow rate of hexene, FA0 (mol/s): 10
Enter feed molar flow rate of H2, FB0 (mol/s): 20
Enter feed molar flow rate of N2 (mol/s): 70
Enter feed pressure, P0 (atm): 10
Enter the final conversion for A, Xf: 0.8
The weight of catalyst in the PBR is 27.442 kg
```

## 6.6 Adaptation of Homogeneous Catalysis in FBRs

The application of homogeneous catalysis in FBRs is less common compared to PBRs due to the fundamental challenge of retaining soluble catalysts within the fluidized environment [14]. Since homogeneous catalysts dissolve in the reaction medium, they are prone to being carried away by the fluidizing gas or liquid, leading to catalyst loss, decreased catalytic activity, and potential contamination of the product stream. This loss not only affects reactor efficiency but also increases operating costs due to frequent catalyst replenishment and downstream separation requirements. Maintaining a stable and effective catalyst concentration within the reactor is therefore a crucial consideration.

Despite these challenges, several strategies have been developed to adapt homogeneous catalysis for use in FBRs. One approach involves encapsulating the homogeneous catalyst within porous particles [15]. These particles, typically composed of inert materials such as silica, zeolites, or polymeric microcapsules, are fluidized within the reactor. The reactants diffuse into the porous structure to interact with the encapsulated catalyst, while the products diffuse out. This method effectively confines the catalyst within the reactor, reducing its loss due to fluid flow. However, diffusion limitations can impact reaction rates, and long-term catalyst stability may be affected by leaching or deactivation. Advanced encapsulation techniques, such as sol–gel microencapsulation or core-shell structures, have been explored to mitigate these limitations.

Another promising approach is the adsorption of homogeneous catalysts onto the surface of fluidizing particles. By leveraging strong surface interactions, the catalyst can be retained within the reactor while maintaining some of the advantages of homogeneous catalysis. Examples include organometallic complexes adsorbed onto high-surface-area supports such as silica or alumina. The effectiveness of this method depends on adsorption strength, reactor operating conditions, and the tendency of the immobilized catalyst to desorb under operating conditions.

A key challenge with this approach is the risk of catalyst desorption, which can lead to contamination of the product stream. If the catalyst detaches from the support and enters the bulk fluid phase, it may require additional downstream separation, such as membrane filtration, liquid–liquid extraction, or selective precipitation, to remove catalyst residues from the final product. In some cases, catalyst loss can also alter reaction kinetics and selectivity, reducing process efficiency.

To mitigate excessive catalyst loss, continuous regeneration or recycling strategies may be implemented. These include periodic re-adsorption of the catalyst onto fresh or regenerated support particles, in situ catalyst recovery systems, or operating conditions designed to enhance catalyst-support interactions. However, these strategies must be carefully optimized to ensure minimal interference with reactor hydrodynamics and reaction performance.

Additionally, specialized reactor designs have been explored to minimize catalyst carryover. These designs may incorporate internal components such as cyclone separators, dip legs, or baffles to enhance catalyst recovery. Operating conditions, such as lower fluidization velocities or controlled bubble dynamics, can also improve catalyst retention while ensuring adequate mixing and heat transfer. The choice of fluidization regime, whether bubbling, turbulent, or fast fluidization, plays a critical role in determining catalyst stability within the reactor. Industrial applications have demonstrated that modified FBRs can be used effectively for homogeneous catalytic reactions when such design considerations are carefully implemented.

By integrating these adaptation strategies, homogeneous catalysis can be more effectively utilized in FBRs, offering potential advantages in reaction control and selectivity while addressing catalyst retention challenges.

In conclusion, while low-level catalyst loss may be acceptable in some industrial applications, it is typically monitored and controlled within predefined limits. In cases where product contamination is critical, downstream catalyst removal (e.g., filtration, centrifugation, or adsorption) is mandatory. Therefore, industrial designs must balance catalyst retention, product purity, and economic feasibility to determine an acceptable loss threshold.

## 6.7 FRB Design and Operational Considerations

When designing and operating an FBR for homogeneous catalysis, fluidization velocity is a critical parameter. The velocity must be sufficient to suspend the catalyst-containing particles and promote good mixing but not so high as to cause excessive entrainment and catalyst loss [7]. Achieving this delicate balance is essential for effective reactor performance.

The size and density of the encapsulating material or adsorbent particles are also key factors influencing fluidization characteristics [16]. These properties determine the minimum fluidization velocity, bed expansion behavior, and overall hydrodynamics within the reactor. Proper selection of particle properties is necessary to ensure stable and efficient fluidization while minimizing catalyst loss.

One of the key advantages of FBRs is their ability to provide a uniform temperature distribution throughout the reaction bed [7]. This excellent heat-transfer capability is particularly beneficial for homogeneous catalytic reactions, preventing the formation of hot spots and ensuring consistent reaction rates across the entire catalyst bed.

Beyond fluidization control, catalyst retention strategies play a crucial role in maintaining process efficiency. Reactor designs may incorporate internal baffles, cyclone separators, or filtration units to reduce catalyst entrainment. In cases where catalyst loss is unavoidable, recirculating catalyst loops can be employed to recover and reintroduce lost catalyst into the system.

The choice of fluidization regime, whether bubbling, turbulent, or circulating, plays a crucial role in determining the hydrodynamics and overall performance of FBRs, particularly in catalytic processes. In bubbling fluidized beds (BFBs), the gas velocity is only slightly above the minimum fluidization velocity, resulting in the formation of distinct gas bubbles rising through a dense particulate bed. This regime provides moderate mixing and a relatively stable temperature distribution, which is advantageous for reactions sensitive to thermal gradients. Additionally, the relatively low gas velocities help reduce particle attrition and minimize catalyst loss through elutriation, making BFBs suitable for processes where catalyst retention is essential.

On the other hand, circulating fluidized beds (CFBs) operate at significantly higher gas velocities, allowing entrainment of the solid particles throughout the reactor and external riser. This regime offers superior gas–solid contact and enhanced mass and heat transfer due to the vigorous movement of catalyst particles. The circulating nature of the system enables continuous regeneration and recycling of the catalyst, which is particularly valuable for reactions that involve rapid deactivation or coke formation, such as fluid catalytic cracking (FCC). While CFBs can offer higher conversion rates and throughput, they require more complex solid handling systems and cyclone separators to manage the circulation loop and maintain catalyst inventory. Overall, the selected fluidization regime must balance process kinetics, catalyst management, and reactor operability to achieve optimal performance.

Additionally, MT limitations and residence time distribution (RTD) must be carefully evaluated, especially when using encapsulated or immobilized catalysts, where internal diffusion resistance can significantly affect reaction kinetics. Ensuring that gas or liquid reactants have sufficient contact time with the catalyst is critical for achieving high conversion. At the same time, minimizing back-mixing, particularly in fluidized beds operating closer to the bubbling regime, is essential to preserve product selectivity, especially in reactions with competing parallel or consecutive pathways. Reactor configurations must ensure that the gas or liquid residence time is optimized to maximize conversion while minimizing back-mixing, which could reduce selectivity. Reactor design should aim for an optimal trade-off between mixing and plug flow characteristics, depending on the specific reaction and catalyst structure.

Moreover, catalyst deactivation and fouling pose ongoing challenges in fluidized bed systems, particularly in homogeneous or finely dispersed catalyst applications. Loss of catalytic activity may result from several mechanisms, including thermal degradation, chemical leaching, sintering, or the deposition of carbonaceous materials (e.g., coking) and polymeric byproducts. These issues are exacerbated under conditions of high temperature, prolonged residence time, or in the presence of reactive intermediates. To mitigate such effects and sustain long-term reactor efficiency, strategies such as periodic or continuous catalyst regeneration, operating at milder yet effective conditions, and employing multi-stage or segmented fluidization designs can be employed. These approaches help maintain catalyst activity, minimize downtime, and enhance overall process stability.

## 6.8 Illustrative Cases of Homogeneous FBRs

Homogeneous catalysis in FBRs can be adapted through catalyst encapsulation or adsorption onto fluidized particles. The following examples illustrate how these strategies are applied in different catalytic processes, highlighting the unique advantages and challenges of homogeneous FBRs.

**Case #1**

**Biotransformation with Immobilized Enzymes**

Consider a biotransformation process where a homogeneous enzyme catalyzes a reaction in an aqueous solution. To retain the enzyme within the reactor and enable continuous operation, it is immobilized within porous alginate beads. These beads are then fluidized by an upward flow of the reactant solution, allowing the reaction to proceed efficiently [8].

The reaction mechanism involves three key steps:

1) Diffusion of reactant molecules into the porous alginate beads.
2) Catalytic transformation of the reactant by the entrapped enzyme.
3) Diffusion of the product molecules out of the beads into the bulk solution.

This setup provides enhanced mixing and MT, mitigating concentration polarization effects and improving overall reaction efficiency. However, several design and operational factors influence the performance of this system:

a) **Enzyme Loading:** Increasing enzyme concentration within the beads can enhance the intrinsic reaction rate; however, if substrate diffusion into the bead is limiting, excessive enzyme loading may lead to underutilization of enzymes located deeper in the matrix, thereby amplifying internal diffusion resistance.
b) **Porosity of the Alginate Matrix:** Affects both reactant access to the enzyme and product diffusion out of the beads.
c) **Fluidization Velocity:** Must be optimized to prevent bead aggregation while ensuring adequate suspension and mixing.
d) **Enzyme Leaching:** Over time, some enzyme molecules may diffuse out of the beads, leading to potential catalyst loss and reduced activity. Strategies such as cross-linking or co-immobilization with stabilizing agents can mitigate this issue.

A mathematical kinetic model for this system could be formulated based on Michaelis–Menten kinetics, incorporating external MT resistance and internal diffusion limitations inside the porous beads. Such a model would be valuable in optimizing operating conditions and predicting long-term catalyst stability.

**Case #2**

**Olefin Metathesis in a Gas-phase-fluidized Bed**

A more industrially relevant case involves olefin metathesis, a reaction where carbon–carbon double bonds in alkenes are rearranged through the catalytic action of a transition metal complex, such as ruthenium-based catalysts. This reaction is widely used in the production of fine chemicals, polymers, and pharmaceuticals.

**Reaction Overview**

$$R_1R_2C = CR_3R_4 + R_5R_6C = CR_7R_8 \xrightarrow{\text{Cat.}} R_1R_2C = CR_7R_8 + R_5R_6C = CR_3R_4$$

**Reactor Setup**

To facilitate this reaction in an FBR, the ruthenium catalyst is encapsulated within a polymer matrix, forming small, dense particles that are fluidized within the reactor. The olefin reactants are introduced in the gas phase, ensuring intimate gas–solid contact and enhancing reaction efficiency.

(Continued)

**Case #2 (Continued)**

Key advantages of using an FBR for this reaction include:

✓ **Uniform Temperature Distribution:** Prevents hot spots that could lead to catalyst deactivation.
✓ **Efficient Mixing:** Improves MT, particularly for gas-phase reactants.
✓ **Continuous Operation:** Allows for steady catalyst replacement and minimizes batch processing drawbacks.

However, several challenges must be addressed:

⚠ **Catalyst Stability:** The ruthenium complex must remain active within the polymer matrix over extended periods. Exposure to high temperatures or fluidization shear forces may lead to catalyst degradation.
⚠ **Polymer Particle Durability:** The polymer matrix must withstand attrition and mechanical wear in the fluidized bed. Excessive attrition could release homogeneous catalyst fragments, leading to catalyst loss and product contamination.
⚠ **Fluidization Dynamics:** The gas–solid interaction must be optimized to ensure proper catalyst distribution while preventing channeling or dead zones that could reduce conversion efficiency.

To model this reaction system, a simplified kinetic model could be developed based on Langmuir-Hinshelwood kinetics or a pseudo-first-order approach, considering:

a) Concentration of the encapsulated catalyst
b) Diffusion of olefin reactants into the polymer particles
c) Gas–solid MT limitations

Simulation studies incorporating computational fluid dynamics (CFDs) could further optimize reactor design, particle size selection, and operating conditions to maximize olefin conversion efficiency.

**Case #3**

**Hydroformylation in a Liquid–Liquid System with Adsorbed Homogeneous Catalysts**

Hydroformylation, also known as oxo synthesis, is an important reaction in which alkenes react with synthesis gas (CO and $H_2$) in the presence of a homogeneous transition metal catalyst (commonly rhodium or cobalt complexes) to produce aldehydes. To apply this reaction in an FBR, a novel strategy involves adsorbing the homogeneous catalyst onto the surface of hydrophobic fluidizable particles, such as functionalized silica or polymer beads. These particles serve as a pseudo-heterogeneous phase, enabling catalyst retention while maintaining molecular-level activity. The active catalytic species in hydroformylation, for example, $HRh(CO)(PPh_3)_3$ or $HCo(CO)_4$, remains molecularly dispersed and operates in a homogeneous phase, meaning the reaction occurs through molecular coordination complexes rather than through surface-bound active sites typical of heterogeneous catalysts. Notice that $PPh_3$

stands for triphenylphosphine, an organophosphorus ligand commonly used in homogeneous catalysis.

Triphenylphosphine has the formula $P(C_6;H_5)_3$, meaning a central phosphorus atom bonded to three phenyl ($C_6H_5$) rings. Consequently, the catalytic mechanism remains homogeneous because the active metal complex operates as a molecular species in solution, even though it is retained on a solid support for practical processing benefits. Therefore, this is accurately described as pseudo-homogeneous hydroformylation using a supported homogeneous catalyst.

**Reaction Overview**

$$RCH = CH_2 + CO + H_2 \rightarrow RCH_2CH_2CHO \quad \text{(linear aldehyde)}$$

Or

$$RCH = CH_2 + CO + H_2 \rightarrow RCH(CH_3)CHO \quad \text{(branched aldehyde)}$$

**Reactor Setup**

The system operates as a biphasic liquid–liquid reaction: The alkene and syngas are introduced in an organic solvent phase, while the catalyst remains immobilized on fluidized solid particles suspended in a continuous aqueous phase. The fluidized bed configuration enhances phase contact and interfacial area between the organic and aqueous phases, allowing rapid transport of the substrate to the catalyst-active sites. Fluidization also helps prevent phase separation and stagnation zones, improving MT and reducing diffusional limitations across the liquid–liquid interface.

**Advantages and Considerations**

✓ **Catalyst Retention and Reuse:** Adsorption onto solid supports minimizes catalyst leaching and facilitates downstream separation.

✓ **Enhanced Phase Dispersion:** Fluidization maintains effective dispersion of immiscible phases, enhancing interfacial transport.

✓ **Selective Product Formation:** Control of fluidization velocity and phase ratios can influence product selectivity (linear vs. branched aldehydes).

**Challenges Include**

⚠ **Catalyst Desorption:** Prolonged operation or local solvent effects may lead to gradual desorption of the catalyst, requiring stabilizing surface modifications.

⚠ **Support Fouling or Wettability Shift:** Accumulation of reaction byproducts may alter surface properties of the support, affecting catalyst activity and MT efficiency.

⚠ **Multiphase Hydrodynamics:** Balancing densities and viscosities of the biphasic system is critical to achieving stable and uniform fluidization.

A comprehensive reactor model for this system would involve interfacial MT coefficients, partitioning of reactants between phases, and adsorption equilibrium of the catalyst. A kinetic

(Continued)

**Case #3 (Continued)**

scheme based on pseudo-homogeneous assumptions, modified by interfacial transport terms, could be employed. Experimental data combined with multiphase CFD simulations would support the optimization of operating parameters such as phase flow rates, temperature, and particle functionalization strategy.

## 6.9 Mechanism and Steps of Heterogeneous Catalysis

Heterogeneous catalysis is characterized by the presence of reactants and a catalyst in distinct phases, most commonly involving gaseous or liquid reactants interacting with a solid catalyst [1]. The catalytic reaction occurs at the interface, specifically on the active sites located on the surface of the solid. This phase separation allows for easier catalyst recovery and reuse, which is a key advantage in industrial applications. However, because the reaction does not occur uniformly throughout the bulk of the catalyst but rather at the accessible surface sites, surface area, porosity, and catalyst morphology become critical factors in determining overall efficiency. The heterogeneous nature of this process also introduces additional physical phenomena, such as adsorption, diffusion, and desorption, which must be understood and optimized to maximize catalytic performance. Here are significant hurdles that affect the performance of heterogeneous catalysis:

⚠ **External Mass Transfer:** Reactant molecules are first transported from the bulk fluid (gas or liquid) to the external surface of the catalyst particle. This step, known as external MT, is affected by factors such as fluid flow rate, turbulence, and the physical properties (e.g., viscosity, diffusivity) of the fluid medium. A higher flow rate can reduce the thickness of the boundary layer surrounding the catalyst, thereby enhancing the rate of MT [17].

⚠ **Adsorption:** Once at the catalyst surface, reactant molecules are adsorbed onto specific sites known as active sites. Adsorption can be either physical (physisorption) or chemical (chemisorption).

- **Physisorption** involves weak van der Waals forces and is usually reversible, allowing for easy desorption.
- **Chemisorption** involves the formation of chemical bonds between the reactant and the catalyst, typically resulting in stronger interactions and sometimes irreversible binding. The nature and density of active sites, as well as surface morphology, critically determine the adsorption capacity and selectivity of the catalyst.

⚠ **Surface Reaction:** Once adsorbed, the reactant molecules undergo a chemical transformation on the catalyst surface. The surface reaction can involve bond breaking and formation, rearrangement of atoms, or intermediate formation. The rate of this reaction is often influenced by the intrinsic activity of the catalyst and can be enhanced by optimizing the active site structure and electronic properties.

⚠ **Desorption:** Following the surface reaction, the product molecules desorb from the catalyst surface. This step is as important as adsorption since it frees active sites for new reactant molecules. The ease of desorption depends on the binding strength; if products bind too strongly, they may block active sites, reducing the catalyst's effectiveness, a phenomenon known as catalyst poisoning.

⚠ **Internal Mass Transfer (if applicable):** In the case of porous catalysts, an additional MT step may be present: The diffusion of reactants from the external surface into the internal pores and, conversely, the diffusion of products out of the pores into the bulk fluid. This internal MT can become rate-limiting if the pore structure is complex or if the average size of catalyst particles is very large and is coupled with long, irregular pores.

The overall rate of a heterogeneously catalyzed reaction is determined by the slowest of these individual hurdles, which becomes the rate-limiting step [17]. For example, in some systems, external MT may be so rapid that the surface reaction itself limits the rate; in others, strong adsorption or slow internal diffusion might become the bottleneck. Identifying and optimizing the rate-limiting step is crucial for improving catalyst design and reactor performance.

Moreover, the design of the reactor and the operating conditions can be tailored to minimize MT limitations. For instance, increasing turbulence or employing structured catalysts can improve external MT, while optimizing pore size distribution in catalyst particles can enhance internal diffusion.

In summary, heterogeneous catalysis is a complex interplay of transport phenomena and surface reactions. A detailed understanding of each step not only facilitates the rational design of catalysts but also helps in the development of reactors that can operate at optimal efficiency under a variety of conditions.

## 6.10 Design and Operational Principles of PBRs for Heterogeneous Catalysis

PBRs designed for heterogeneous catalysis are widely used in chemical and petrochemical industries due to their high efficiency, scalability, and ability to accommodate solid catalysts. These reactors typically consist of a cylindrical tube filled with catalyst particles, which facilitate chemical reactions by providing active sites for reactant molecules. PBRs are widely used in chemical and petrochemical industries for carrying out heterogeneous catalytic reactions. Their efficiency is governed by multiple factors, including catalyst properties, reactor design, fluid flow patterns, and heat/MT dynamics.

### 6.10.1 PBR Flow Distribution Patterns

The flow pattern within a PBR critically impacts reaction performance, influencing residence time, catalyst utilization, and product yield [18]. Nonuniform flow can lead to inefficiencies, such as incomplete reactant conversion or catalyst underutilization. Here are the factors affecting flow distribution and their implications:

1) **Void Fraction and Channeling**

   The void fraction, or porosity, of the catalyst bed determines how effectively reactants flow through the packed bed. Porosity typically ranges from 0.35 to 0.45 for randomly packed beds, depending on particle size, shape, and packing method. High porosity facilitates fluid permeation but may reduce catalyst loading, while low porosity can increase pressure drops and restrict flow.

   Channeling occurs when fluid follows preferential pathways through regions of higher porosity, bypassing portions of the catalyst bed. This is often caused by:

   a) **Nonuniform Packing:** Uneven particle distribution during reactor loading creates high-porosity zones.

b) **Wall Effects:** Near the reactor walls, porosity is higher due to less dense packing, encouraging flow along the walls.
c) **Fines or Particle Degradation:** Small particles or catalyst fragments can clog interstitial spaces, forcing fluid through larger voids.

Channeling reduces effective catalyst contact, lowering conversion rates and selectivity. To mitigate this, careful packing techniques (e.g., vibratory or sock loading) and uniform particle size distribution are essential. Structured packings, such as monoliths or arranged beds, can further minimize channeling by ensuring consistent porosity.

2) **Radial and Axial Dispersion Effects**

Radial and axial dispersion describe the spreading of fluid in the radial (cross-sectional) and axial (along the flow direction) dimensions of the reactor, respectively. These phenomena are influenced by bed geometry, particle size, and flow velocity:

a) **Radial Dispersion:** In large-diameter reactors, radial variations in velocity and porosity can create temperature and concentration gradients. For example, higher flow near the walls (due to wall effects) may lead to uneven reaction rates. This is particularly problematic in highly exothermic reactions, where poor radial heat transfer can cause hot spots, leading to catalyst deactivation or thermal runaway.
b) **Axial Dispersion:** Axial mixing causes spreading of concentration profiles along the reactor length, reducing the driving force for reaction and deviating from ideal plug flow behavior. Axial dispersion is most pronounced under laminar flow conditions (where molecular diffusion dominates) and in beds with significant channeling. As flow velocity increases and turbulence develops, plug flow behavior generally improves (the Peclet number increases), though the absolute dispersion coefficient may rise. Peclet number measures the ratio between convection and dispersion. The velocity tends to increase faster than the dispersion coefficient, such that the Peclet number increases.

To address dispersion:

a) **Flow Redistributors:** Installing redistributors between catalyst beds (in multi-bed units) promotes uniform flow and resets radial profiles.
b) **Optimized Packing:** Smaller, uniformly sized particles reduce dispersion but may increase pressure drops. A balance must be maintained based on reactor scale and reaction kinetics.
c) **Reactor Aspect Ratio:** A higher length-to-diameter ($L/D$) ratio improves plug flow behavior by reducing the relative influence of inlet/outlet end effects and minimizing the impact of axial dispersion. Low $L/D$ reactors exhibit stronger non-ideal mixing and broader RTDs.

3) **Flow Regimes**

The flow regime in a PBR, determined by the particle Reynolds number ($Re_p = [\rho\, v\, D_p]/\mu$, where $\rho$ is fluid density, $v$ is superficial velocity, $D_p$ is particle diameter, and μ is viscosity), dictates mixing and MT characteristics:

a) **Laminar Flow (Rep < 10):** Common in small-scale, viscous, or low-velocity systems. Laminar flow results in smooth streamlines but poor mixing. This leads to significant external MT limitations, as reactants rely on diffusion to reach catalytic sites.
b) **Turbulent/Inertial Flow ($Re_p$ > 100):** Higher velocities promote inertial flow effects, enhancing mixing and MT to the catalyst surface. This improves reaction rates but increases the pressure drop significantly. If the pressure drop becomes excessive, it risks crushing the catalyst; if the flow is upward, it risks fluidizing the bed and causing attrition.
c) **Transitional Flow (10 < $Re_p$ < 100):** Many industrial PBRs operate in this regime or higher, balancing mixing efficiency with acceptable pressure drops.

To optimize flow regimes:

a) **Adjust Particle Size:** Larger particles reduce pressure drops but may limit internal surface area accessibility, while smaller particles enhance MT at the cost of higher hydraulic resistance.
b) **Control Flow Velocity:** Operating at an optimal Reynolds number ensures sufficient external MT without excessive energy costs (pumping power).
c) **Use Structured Catalysts:** Monoliths can maintain high MT rates with lower pressure drops compared to random packings.

**Ensuring Uniform Flow**
Uniform reactant distribution is essential to maximize catalyst utilization and prevent bypassing. Strategies include:

a) **Pre-distributors:** Inlet designs (e.g., perforated plates, diffuser cones, or spray nozzles) ensure even flow entry across the entire bed cross section.
b) **Regular Bed Maintenance:** Periodic screening, skimming, or repacking prevents fines accumulation and channeling.
c) **Advanced Diagnostics:** Techniques like tracer studies or CFDs modeling can identify and address flow maldistribution.

By optimizing void fraction, minimizing dispersion, and controlling flow regimes, PBRs can achieve uniform flow, maximize catalytic efficiency, and ensure consistent performance for industrial applications.

### 6.10.2 PBR Residence Time

#### 6.10.2.1 Residence Time and Reactor Performance

Residence time ($\tau$) governs how long reactants remain in contact with the catalyst in a PBR, directly affecting reaction efficiency, conversion, and product yield. It is defined as:

$$\tau = \frac{V_{PBR}}{v_0}$$

where $V_{PBR}$ is the PBR volume ($m^3$), and $v_0$ is the input volumetric flow rate ($m^3/s$). Residence time is a critical parameter in PBR design, influencing the trade-off between reaction performance and operational constraints like pressure drop and energy consumption.

#### 6.10.2.2 Factors Influencing Residence Time

a) **Reactor Volume ($V_{PBR}$):** Larger bed volumes increase $\tau$, allowing more time for reactant-catalyst interactions. However, this may require larger reactors via increasing length, increasing capital costs, and potentially exacerbating pressure drops.
b) **Flow Rate ($v_0$):** Lower flow rates extend $\tau$, enhancing conversion by providing sufficient time for reactions to reach completion, especially for slow or kinetically limited reactions. Conversely, higher flow rates reduce $\tau$, increasing throughput but risking incomplete conversion and lower yields.
c) **Catalyst Activity:** Highly active catalysts (e.g., Mn-Ce oxides for VOC oxidation) can achieve high conversion at shorter residence times, while less active catalysts require longer $\tau$ to achieve comparable results.

d) **Bed Geometry:** The $L/D$ ratio strongly influences flow hydrodynamics and RTD, even though the mean residence time $\tau = V/v_0$ is fixed for a given reactor volume and volumetric flow rate. High $L/D$ ratios (long, narrow beds) improve plug flow behavior by reducing the relative impact of inlet and outlet mixing, thus minimizing axial dispersion and promoting more uniform reactant contact with the catalyst. However, increasing $L/D$ also increases pressure drop due to longer flow paths through the packed bed.

#### 6.10.2.3 Impact on Reactor Performance

a) **Conversion and Selectivity:** Longer residence times generally improve conversion by allowing reactants to fully interact with catalytic sites. For example, in the oxidation of volatile organic compounds (VOCs) using Mn-Ce catalysts in PBRs, extended $\tau$ ensures near-complete oxidation to $COH_2$ and $H_2O$, critical for environmental compliance. However, excessively long $\tau$ may lead to side reactions or catalyst deactivation due to prolonged exposure to byproducts (e.g., coke formation).
b) **Pressure Drop:** The pressure drop across a packed bed increases with superficial velocity $v_0$; the relative contribution of viscous and inertial effects is given by the Ergun equation (which combines both effects). See the next section. At low Reynolds numbers, the viscous (linear-in-$v_0$) term dominates, and $\Delta P/L$ is approximately proportional to $v_0$; at high Reynolds numbers, the inertial (quadratic-in-$v_0$) term dominates, and $\Delta P/L$ scales roughly with $v_0^2$. Increasing bed length $L$ or decreasing particle diameter $D_p$ raises total $\Delta P$. High-pressure drops increase pumping power and can cause flow maldistribution or channeling, reducing reactor efficiency.
c) **Thermal Effects:** While smaller reactor diameters reduce the radial heat-transfer path, high L/D packed beds remain susceptible to hot spots because axial heat accumulation and limited radial mixing dominate thermal behavior in exothermic reactions. Thus, long, narrow beds require careful temperature control despite improved radial conduction.

#### 6.10.2.4 Case Study: VOC Oxidation in PBRs

In VOC oxidation, Mn-Ce catalysts are used in PBRs due to their high activity and stability at moderate temperatures (200–400 °C). A high $L/D$ ratio (e.g., ≥10:1) promotes plug flow, ensuring reactants pass through the bed with minimal back-mixing, achieving near-complete VOC destruction (>95% conversion). However, this design increases pressure drop, raising operational costs. To address this:

a) **Segmented Beds:** Dividing the bed into segments with intermediate gas injection or staged heating reduces hot spots and pressure drop. For instance, staged heating maintains optimal temperatures for each segment, enhancing VOC destruction while preventing catalyst overheating.
b) **Catalyst Particle Size Optimization:** Smaller particles increase surface area and catalytic activity but raise pressure drop. A compromise (e.g., 1–3 mm particles) balances MT and flow resistance.
c) **Flow Rate Tuning:** Adjusting $v_0$ to achieve $\tau$ of 0.1–1 s (typical for VOC oxidation) ensures high conversion while maintaining economically viable throughput.

#### 6.10.2.5 Optimization Strategies

To balance residence time, conversion, and operational constraints:

a) **Staged Reactor Designs:** Use multiple shorter beds in series with interstage cooling or heating to maintain optimal $\tau$ and temperature profiles, reducing pressure drop and hot spots.

b) **Variable Flow Control:** Implement dynamic flow adjustments based on real-time monitoring of conversion and pressure drop to optimize $\tau$ for varying feed compositions or reaction conditions.
c) **Structured Catalysts:** Replace random packings with monoliths or foams to reduce pressure drop while maintaining high surface area and with sufficient $\tau$, especially in high-throughput applications.
d) **Computational Modeling:** Use CFDs to simulate flow and reaction kinetics, optimizing $\tau$ and $L/D$ ratios for specific reactions. For VOC oxidation, CFD can predict optimal $\tau$ to achieve >99% destruction efficiency without excessive energy costs.

#### 6.10.2.6 Practical Considerations

In industrial PBRs, residence time must be tailored to the reaction kinetics and process goals. For example, fast reactions (e.g., VOC oxidation) require shorter $\tau$, while slower reactions (e.g., biological processes or selective hydrogenations) demand longer $\tau$. Over-optimizing for excessive $\tau$ can lead to diminishing returns, as conversion plateaus while pressure drop and energy costs rise. Regular monitoring of catalyst performance and bed integrity (e.g., via pressure drop measurements or tracer studies) ensures that $\tau$ remains effective over time, while accounting for catalyst deactivation or fouling.

### 6.10.3 Catalyst Particle Characteristics and Their Role in Reactor Performance

The size, shape, and porosity of catalyst particles affect conversion efficiency, MT, and pressure drop. The physical properties of the catalyst particles, such as size, shape, porosity, and surface area, play a crucial role in determining the reactor's overall efficiency. The choice of catalyst particle configuration is influenced by the following considerations:

⚠ **Surface Area and Reaction Rate:** Smaller catalyst particles provide a higher surface area per unit volume, leading to increased catalytic activity and higher reaction rates. The reaction rate follows a first-order dependence on catalyst surface area for many catalytic reactions. However, excessively small particles can cause excessive pressure drop and hinder fluid flow.

⚠ **Pressure Drop Versus Reaction Efficiency Trade-Off:** A fine catalyst structure enhances reaction rates but increases the resistance to fluid flow, necessitating greater energy input for pumping [19]. Therefore, an optimal particle size is selected to balance the pressure drop and the catalytic efficiency. The pressure drop per unit length in a packed bed is often described by the Ergun equation

$$\frac{\Delta P}{L} = \frac{150(1-\varepsilon)^2}{\varepsilon^3}\frac{\mu v}{d_p^2} + \frac{1.75(1-\varepsilon)}{\varepsilon^3}\frac{\rho v^2}{d_p}$$

where

$\Delta P$ is the pressure drop across the bed (Pa).
$L$ is the length of the bed (m).
$\varepsilon$ is the void fraction (dimensionless).
$\mu$ is the fluid viscosity (Pa·s).
$v$ is the superficial velocity (m/s).
$d_p$ is the catalyst particle diameter (m).
$\rho$ is the fluid density (kg/m$^3$).

⚠ **Particle Shape and Packing Efficiency:** Catalyst particles may be spherical, cylindrical, or irregularly shaped. Spherical pellets tend to offer uniform flow distribution, whereas extruded shapes or Raschig rings may enhance MT by reducing channeling effects. Extruded shapes (like cylinders or trilobes) or Raschig rings are used precisely to increase void fraction near the wall and within the bed, which can reduce wall effects and enhance MT by mitigating channeling (nonuniform flow).

For example, in an ammonia synthesis reactor (Haber-Bosch ammonia synthesis), Fe-based catalysts are used in a PBR. Small catalyst particles (1–5 mm) are used to improve $NH_3$ conversion but increase pressure drop. Using porous catalysts and graded catalyst beds (larger particles at the inlet, smaller particles downstream) balances conversion and pressure drop.

### 6.10.4 Packed Bed Reactor Geometry: Length-to-Diameter (*L*/*D*) Ratio

The reactor's ***L*/*D* ratio** is a critical parameter affecting the residence time, conversion, and pressure drop [20]:

⚠ **Higher L/D Ratios:** Lead to plug flow behavior and longer residence times, allowing for higher reactant conversion. However, they also contribute to greater pressure drops, requiring more energy input.

⚠ **Lower L/D Ratios:** Reduce pressure drop and often facilitate better heat removal (due to a larger surface area to volume ratio) for exothermic reactions, but may result in insufficient residence time for achieving desired conversion levels.

✓ **Optimal L/D Ratio:** An empirical selection for *L*/*D* ratio is based on reaction kinetics and pressure drop constraints. Typical industrial values are:
  $L/D \approx 10–50$ for high-pressure catalytic processes to maximize plug flow.
  $L/D \approx 2–5$ for highly exothermic gas-phase reactions to limit temperature gradients and aid heat transfer.

For example, in the exothermic hydrogenation of benzene to cyclohexane, a nickel-based catalyst is employed in a PBR to facilitate the addition of hydrogen to the aromatic ring, forming cyclohexane, a key intermediate in nylon production. A high *L*/*D* ratio, approximately 20, promotes plug flow behavior, minimizing axial dispersion and ensuring reactants have sufficient residence time to achieve high conversion rates, typically exceeding 95%. The extended path length through the reactor maximizes contact between benzene, hydrogen, and the Ni catalyst, enhancing reaction efficiency. However, the high *L*/*D* ratio increases pressure drop across the bed, as described by the Ergun equation, which can elevate energy costs and cause flow maldistribution if not carefully managed. To mitigate this, uniform catalyst packing and optimized particle sizes (e.g., 1–2 mm) are critical to balance flow resistance with catalytic activity, ensuring consistent performance across the reactor.

The hydrogenation of benzene is highly exothermic ($\Delta H \approx -206$ kJ/mol), and the temperature rise in a high *L*/*D* PBR can lead to thermal runaway and hot spot formation, particularly in adiabatic conditions, which may deactivate the Ni catalyst through sintering or promote byproduct formation, such as methylcyclopentane or cracking products. To address these challenges, a radial-flow PBR design is often adopted, where reactants flow from the reactor's center outward (or vice versa) through a shallow, annular catalyst bed. This configuration shortens the flow path, significantly reducing pressure drop compared to axial-flow designs, and enhances heat transfer by increasing the surface area available for cooling. Incorporating external cooling jackets or internal heat exchange tubes further stabilizes temperatures, maintaining optimal conditions (typically 150–200 °C) to maximize cyclohexane yield while minimizing byproduct formation. For instance, radial-flow PBRs can maintain temperature gradients below 10 °C, compared to 30–50 °C in axial-flow designs, improving catalyst longevity and process selectivity.

### 6.10.5 Heat Management Strategies in PBRs

Temperature control in PBRs is critical because reaction rates are highly sensitive to temperature, and uncontrolled variations can trigger side reactions, reduce selectivity, or deactivate catalysts through sintering or fouling [21]. Effective heat management ensures optimal reaction conditions, enhances catalyst longevity, and maximizes product yield. This is particularly vital for exothermic reactions, like Fischer-Tropsch Synthesis (FTS), where heat release can create hot spots, or endothermic reactions requiring consistent heat input to maintain conversion. Below are some heat management strategies and their application in FTS, focusing on practical implementation and benefits.

#### 6.10.5.1 External Jackets

External jackets involve circulating a heat-transfer fluid (e.g., water, steam, or oil) around the reactor shell to add or remove heat. This method is effective for moderate temperature control in small-to-medium-scale PBRs, where temperature gradients are less severe. For example, in FTS, where iron or cobalt catalysts convert CO and $H_2$ into hydrocarbons ($\Delta H \approx -165$ kJ/mol), jackets maintain reactor temperatures within 200–250 °C to prevent hot spots that could lead to methane formation or catalyst deactivation. Jackets are simple and cost-effective but limited by heat-transfer rates, as heat must conduct through the reactor wall and catalyst bed. In large reactors, this can result in radial temperature gradients (e.g., 20–30 °C), necessitating additional strategies for uniform control.

#### 6.10.5.2 Internal Cooling/Heating Tubes

Embedding cooling or heating tubes within the catalyst bed enhances heat transfer by placing the heat exchange surface closer to the reaction zone. This ensures better temperature uniformity, especially in large-diameter PBRs with exothermic reactions. In FTS, multi-tubular reactors are commonly used, with catalyst packed inside tubes (1–5 cm diameter) and a cooling medium (e.g., boiling water) circulating around them. This design maintains temperature gradients below 5–10 °C, compared to 30–50 °C in jacket-only systems, improving selectivity for desired hydrocarbons (e.g., C5+ chains) while preventing catalyst sintering. However, internal tubes increase reactor complexity and cost, and tube spacing must be optimized to avoid flow maldistribution or reduced catalyst loading.

#### 6.10.5.3 Preheating or Precooling Feed Streams

Adjusting the temperature of inlet feed streams moderates reactor temperature fluctuations and stabilizes reaction conditions. For FTS, preheating the $CO/H_2$ feed to 180–200 °C ensures the reaction initiates at the optimal temperature, reducing the risk of cold zones that slow reaction rates or favor undesired products like $CO_2$. Precooling may be used in highly exothermic systems to prevent initial temperature spikes. This method is energy-efficient and simple to implement using heat exchangers but is less effective for dynamic temperature control during operation, often requiring a combination with jackets or internal tubes for sustained stability.

#### 6.10.5.4 Use of Diluent or Inert Particles

Mixing inert particles (e.g., silica, alumina, or glass beads) with the catalyst dilutes the active bed, reducing heat generation per unit volume in exothermic reactions. In FTS, diluting Fe/Co catalysts with inert particles (e.g., 20–50% by volume) moderates temperature gradients, preventing hot spots that could deactivate catalysts or shift selectivity toward lighter hydrocarbons like methane. Inert particles also improve flow distribution by reducing channeling and enhancing bed porosity. However, dilution reduces catalyst loading, potentially lowering throughput, so the inert-to-catalyst ratio (e.g., 1:1 to 1:3) must be optimized based on reaction kinetics and reactor scale.

#### 6.10.5.5 Additional Considerations

Effective heat management requires tailoring strategies to reaction characteristics and reactor scale. For FTS, CFD modeling can predict temperature profiles and optimize tube spacing or diluent ratios. Regular monitoring (e.g., via thermocouples) detects hot spots early, preventing catalyst damage. For highly exothermic reactions, staged cooling with intermediate heat exchangers between bed segments can further refine temperature control. These strategies, when combined, ensure PBRs operate efficiently, balancing heat transfer, catalyst performance, and process economics for industrial applications like FTS.

## 6.11 Advantages of Heterogeneous PBRs

✓ **High Volumetric Efficiency and Catalyst Utilization**
  a) PBRs enable a high Space–Time Yield (STY), maximizing the conversion rate per unit volume of the reactor.
  b) The high catalyst packing density maximizes the active sites per unit volume, which is ideal for moderately exothermic or thermally stable reactions.

✓ **Simple Construction and Low Maintenance**
  a) PBRs have a relatively simple, static design (compared to FBRs or slurry reactors).
  b) They require minimal complex mechanical components, resulting in lower capital expenditure (CapEx) and reduced maintenance costs.

✓ **Structural Suitability for Extreme Operating Conditions**
  The static, robust nature of the fixed catalyst bed allows PBRs to be built with thick walls, making them highly suitable for withstanding extreme operating conditions:
  PBRs can withstand extreme operating conditions, making them ideal for processes such as:
  a) Ammonia synthesis (Haber-Bosch process) at ~450–500 °C and ~150–300 atm.
  b) Methanol synthesis from syngas using Cu-ZnO/$Al_2O_3$ catalysts at ~200–300 °C and ~50–100 atm.

✓ **Continuous Operation and High Throughput**
  a) Unlike batch reactors, PBRs operate continuously at steady-state, leading to high, consistent production rates.
  b) The steady-state operation allows for consistent product quality and minimal operational intervention.

## 6.12 Disadvantages of Heterogeneous PBRs

Despite their advantages, PBRs have operational limitations that must be addressed to ensure optimal performance.

⚠ **Difficulties in Temperature Control and Hot Spot Formation**

Highly exothermic reactions (e.g., FTS, catalytic oxidation) can create hot spots, leading to nonuniform reaction rates across the bed and catalyst deactivation due to thermal sintering or undesired side reactions.

Remedies include using a multi-tubular reactor design with external cooling and segmented catalyst beds with inert heat-exchange zones.

⚠ **Significant Pressure Drop**

The Ergun equation governs the pressure drop across the packed bed, which increases with smaller catalyst particles (higher surface area but greater resistance to flow) and longer reactor length (higher $L/D$ ratio). A high-pressure drop will increase energy costs for fluid pumping and may limit reactor scalability. Remedies include catalyst grading and shape optimization, where using larger, nonspherical particles (like rings or extrudates) at the inlet and smaller ones downstream helps distribute pressure loss and increase the bed void fraction ($\varepsilon$) to reduce flow resistance. For large-scale industrial processes, adopting a radial-flow PBR is highly effective, as it significantly shortens the fluid flow path, thereby drastically reducing $\Delta P$ compared to standard axial designs. Furthermore, utilizing structured catalyst packing (e.g., monolithic or foam structures) provides high geometrical uniformity and permeability, minimizing frictional pressure loss while maximizing surface area. Finally, for gas-phase systems, operating at a higher overall pressure can be used to reduce the gas velocity for a fixed mass flow, leading to a corresponding reduction in pressure drop.

⚠ **Flow Maldistribution and Channeling**

Uneven flow distribution (maldistribution) is a major disadvantage, especially in wide, shallow reactors (low $L/D$). It causes reactant bypassing, leading to axial dispersion and underutilization of certain catalyst bed sections. Channeling not only reduces overall reaction efficiency but can also create localized hot spots (in exothermic systems) or cold spots (in endothermic systems). Remedies include:

a) **Optimizing $L/D$:** Ensuring a sufficiently high $L/D$ ratio (typically >5) to minimize wall effects.
b) **Structured Packing:** Using monolithic or foam-based catalysts (structured packing) instead of random pellets to enhance flow uniformity.
c) **Flow Redistribution:** Employing radial inlet diffusers, internal baffles, or grading catalyst particle size at the inlet to enforce even flow distribution.

⚠ **Catalyst Regeneration and Replacement Challenges**

Over time, catalysts undergo deactivation due to coking (carbon deposition blocking active sites), poisoning (adsorption of impurities like sulfur, phosphorous, or chlorine compounds), and sintering (loss of active surface area due to high-temperature agglomeration) [22]. On the other hand, catalyst replacement often requires reactor shutdown, leading to operational downtime.

Remedies include regenerative PBR systems, where catalyst beds are switched between reaction and regeneration phases, and the use of bimetallic catalysts that resist deactivation.

⚠ **Limited Scalability for Some Processes**

While PBRs work well at moderate to high flow rates, their efficiency may decline for extremely large-scale processes due to high-pressure drop and thermal gradients.

Alternatively, FBRs offer better scalability and temperature control for very high throughput operations.

## 6.13 Illustrative Cases of Heterogeneous PBRs

**Case #1**

**$SO_3$ Production from $SO_2$**

Consider the catalytic oxidation of sulfur dioxide ($SO_2$) to sulfur trioxide ($SO_3$), a key step in the production of sulfuric acid (contact process). This reaction is typically carried out in a PBR using vanadium pentoxide ($V_2O_5$) mounted at $SiO_2$ support as the heterogeneous catalyst [23]. A mixture of $SO_2$ and air is passed through the packed bed at elevated temperatures.

$$2SO_2 + O_2 \rightarrow 2SO_3$$

The reaction with $\Delta H \approx -99$ kJ/mol, operates at 400–600 °C and is also reversible. The $V_2O_5$ catalyst promotes $SO_2$ adsorption and oxidation on its surface, with the rate governed by Langmuir-Hinshelwood kinetics, accounting for reactant adsorption and surface reaction. Due to equilibrium limitations and heat release, multiple PBRs with intercoolers are employed to cool the gas between stages, maintaining temperatures below 600 °C to prevent catalyst deactivation and achieve conversions >98%. Intercooling shifts the equilibrium toward $SO_3$, and careful bed design minimizes pressure drop while ensuring uniform flow, critical for large-scale plants producing thousands of tons of sulfuric acid daily.

**Case #2**

**Hydrodesulfurization (HDS) of Crude Oil**

The following reaction takes place in PBRs in the presence of a catalyst:

$$\mathrm{R-S+H_2 \rightarrow R-H+H_2S}$$

HDS removes sulfur from crude oil fractions, converting organosulfur compounds (R–S) into hydrocarbons (R–H) and hydrogen sulfide ($H_2S$) using Co-Mo/$Al_2O_3$ or Ni-Mo catalysts in PBRs. Operating at 300–400 °C and 30–100 bar, the catalyst faces challenges, like deactivation due to coke deposition and poisoning by $H_2S$ and nitrogen compounds, reducing its activity over months. Coke forms from heavy hydrocarbons, while H2S and nitrogen compounds adsorb strongly, blocking active sites.

Remedies include: (1) Periodic in situ regeneration (controlled air/oxygen exposure, typically every 6–12 months) to burn off coke deposits without sintering the catalyst structure. (2) Using advanced Co-Mo/$Al_2O_3$ or Ni-Mo catalysts with optimized metal dispersion or promoter additives to enhance sulfur tolerance and extend operating cycles. (3) Operational strategies like high-pressure operation to push the equilibrium to the product side and reduce coke and uniform flow distribution to mitigate localized deactivation. These strategies ensure efficient sulfur removal for cleaner fuels.

**Case #3**

**Catalytic Oxidation of Volatile Organic Compounds (VOCs)**

Catalytic oxidation of VOCs (organic compounds with normal boiling points <250 °C) in PBRs uses Mn-Ce catalysts to convert pollutants into $CO_2$ and $H_2O$, critical for emission control. Operating at 200–400 °C, small catalyst particle sizes (<1 mm) maximize surface area for high conversion (>95%) but increase pressure drop, raising energy costs and risking flow maldistribution.

To mitigate the pressure drop problem, honeycomb monolithic catalysts, made of ceramic or metal, with structured channels (e.g., 100–400 cells per square inch), offer a solution by providing high surface area while reducing pressure drop by 50–70% compared to fine particle beds. These monoliths maintain uniform flow and enhance MT, ensuring effective VOC destruction in applications like industrial exhaust treatment. Regular cleaning or replacement is necessary to prevent performance degradation due to fouling from dust or hydrocarbon residues, sustaining performance over extended operation cycles.

**Case #4**

**Fischer-Tropsch Synthesis (FTS)**

FTS converts CO and $H_2$ into hydrocarbons according to:

$$n\,CO + (2n+1)\,H_2 \rightarrow C_nH_{2n+2} + n\,H_2O \quad \Delta H \approx -165\ \text{kJ/mol}$$

Fe or Co catalysts are typically operated at 200–250 °C in PBRs. The highly exothermic reaction risks hot spots (>300 °C), causing catalyst deactivation via thermal sintering, where metal particles agglomerate, reducing active surface area. Fixed-bed multi-tubular reactors with cooling jackets, using saturated boiling water or oil, maintain temperature gradients below 10 °C, preserving catalyst integrity and selectivity for $C_5$+ hydrocarbons. High-conductivity supports like silicon carbide (SiC) further enhance heat dissipation, reducing hot spot formation by 20–30% compared to alumina supports. Optimizing tube diameter (2–5 cm) and flow rates ensures uniform heat removal, enabling high CO conversion (80–90%) while minimizing methane production (<5%), critical for producing fuels and chemicals in large-scale FTS plants. Alternatively, diluted catalyst beds with inert particles reduce heat release intensity, allowing larger reactors to operate without excessive cooling infrastructure. Combining these strategies, for example, diluted beds in multi-tubular reactors, optimizes heat management, enhances catalyst lifespan (e.g., >1 year), and ensures high selectivity for desired products.

## 6.14 FBR Fluidization Regimes and Reactor Design

In FBRs for heterogeneous catalysis, solid catalyst particles are suspended by an upward flow of a fluid, typically a gas, but may also be a liquid, achieving a state of fluidization. This occurs when the drag force exerted by the fluid stream on the particles balances the gravitational force acting on them [7]. The behavior of the fluidized bed strongly depends on the fluid velocity and particle properties, leading to distinct fluidization regimes [14]. More details on FBRs are shown in Figure 6.11.

a) **Minimum Fluidization:** At the minimum fluidization velocity ($U_{mf}$), the drag force exerted by the fluid just balances the weight of particles. The pressure drop across the bed reaches a constant value, and the packed bed begins to expand and move. This marks the transition from a fixed bed to a fluidized state.
b) **Bubbling (Aggregative) Fluidization:**
For gas fluidization, when the superficial velocity slightly exceeds $U_{mf}$, gas bubbles form and rise through the bed, creating alternating dense and dilute zones. This regime enhances mixing but produces nonuniform gas–solid contact.
For liquid fluidization, increasing velocity beyond $\mathrm{U_{mf}}$ does not generate bubbles. Instead, the bed expands smoothly and uniformly, maintaining a continuous particulate phase. The bed behaves as a homogeneous liquid-like suspension with uniform mixing.
c) **Turbulent Fluidization:**
At higher gas velocities, bubble coalescence and breakup increase, creating a highly mixed turbulent environment in which distinct bubbles are no longer observed. Gas–solid mixing becomes vigorous and more uniform.
In the analogous liquid case, the bed continues to expand smoothly without phase segregation. Mixing intensity increases, but without bubble-driven turbulence typical in gas beds.
d) **Fast Fluidization:**
With further gas velocity increase, a substantial fraction of solids becomes entrained and leaves the dense bed region. External particle recovery equipment (cyclones, dip-legs) becomes necessary.
In liquid–solid systems, the higher drag force of liquids causes entrainment at lower superficial velocities. Solid recovery is typically handled by settlers, hydrocyclones, or filters.
e) **Pneumatic/Hydraulic Transport:**
At very high gas velocities, all particles are fully entrained in the gas stream, transitioning from fluidization to pneumatic conveying.
Similarly, extreme liquid velocities entrain all solids, entering hydraulic conveying or slurry transport mode. This regime is used for solids handling rather than catalytic reaction.

Notice that the most critical differences between gas and liquid fluidization lie in the transition from $U_{mf}$:

- **Gas Fluidization:** The system enters aggregative fluidization (bubbles form), leading to highly nonuniform gas–solid contact or dense/dilute regions. Mixing driven by bubble movement and wakes.
- **Liquid Fluidization:** The system enters particulate fluidization (smooth expansion), leading to highly uniform mixing and contact. Mixing is driven by liquid turbulence and drag forces.

The choice of fluidization regime is critical, as it affects fluid–solid mixing, mass and heat transfer, catalyst attrition, and reactor stability. To ensure efficient operation, FBRs are equipped with [7]:

✓ **Distributor Plates:** Located at the reactor bottom, these plates ensure uniform fluid distribution and prevent dead zones.

✓ **Cyclones and Filters:** These components capture entrained catalyst particles, minimizing losses and maintaining reactor efficiency.
✓ **Bed Expansion Design:** Since fluidization leads to bed expansion, reactors must be designed with sufficient freeboard height to accommodate this expansion.

## 6.15 FBR Operational Principles and Catalyst Behavior

The fundamental principle of an FBR is the suspension of catalyst particles using an upward fluid stream, resulting in [7]:

✓ **Enhanced Mixing:** The continuous movement of particles ensures efficient reactant dispersion and uniform temperature distribution.
✓ **Improved Heat and Mass Transfer:** Compared to fixed beds, FBRs exhibit superior heat and MT rates due to the high contact area between reactants and catalyst.
✓ **Minimized Hot Spots:** The well-mixed nature of the bed prevents localized overheating, crucial for highly exothermic or endothermic reactions.

However, these advantages come with operational challenges [7]:

⚠ **Catalyst Attrition and Loss:** Constant particle motion leads to mechanical wear, reducing particle size and generating fines, which may be lost through entrainment.
⚠ **Catalyst Deactivation:** Exposure to high temperatures, poisoning, and sintering can reduce catalyst activity over time. FBRs often incorporate **in situ or external regeneration** strategies to restore catalyst activity.
⚠ **Particle Agglomeration and De-fluidization:** In reactions involving sticky or high-melting-point materials, catalyst particles may adhere together, leading to poor fluidization. Strategies such as modifying particle size, shape, and surface properties or using inert diluents (e.g., nitrogen or process-compatible gases like hydrogen, where applicable) can mitigate these issues.

## 6.16 Heterogeneous FBRs: Advantages and Limitations

**Advantages:**

✓ **Superior Heat and Mass Transfer:** Efficient mixing prevents temperature gradients and ensures uniform reactant distribution.
✓ **Continuous Catalyst Replacement:** FBRs allow for online catalyst replacement and regeneration, minimizing downtime.
✓ **Scalability:** FBRs handle large feed and catalyst volumes, making them ideal for industrial applications [14].
✓ **Adaptability to Various Feedstocks:** Suitable for processing liquids and gases, including high-viscosity materials [24]. However, stickiness can lead to reactor fouling, particle agglomeration, and poor fluidization. Special antifouling strategies are often required, such as using inert gas diluents (e.g., $N_2$) or modifying catalyst particle size and surface properties.

**Limitations:**

⚠ **Complex Hydrodynamics:** Back-mixing and nonideal flow behavior can deviate from plug flow, affecting selectivity and conversion [15].

⚠ **Pressure Drop and Energy Consumption:** Maintaining fluidization requires higher gas velocities, leading to significant energy consumption.

⚠ **Catalyst and Reactor Wear:** Particle attrition and erosion of reactor internals necessitate robust material selection.

⚠ **Particle Entrainment:** Requires separation units to recover fine catalyst particles, increasing operational complexity.

Despite these challenges, FBRs remain one of the most effective reactor types for catalytic processes requiring efficient heat management, rapid MT, and high operational flexibility.

## 6.17 Illustrative Cases of FBR Heterogeneous Catalysis

FBRs play a critical role in various industrial catalytic processes, particularly where efficient gas–solid contact, continuous catalyst regeneration, and enhanced heat transfer are essential. Below are three prominent examples demonstrating the effectiveness of FBRs in heterogeneous catalysis.

### 6.17.1 Fluid Catalytic Cracking (FCC) in Petroleum Refining

One of the most well-known applications of FBRs in heterogeneous catalysis is FCC, a process widely used in petroleum refineries to convert heavy hydrocarbons into lighter, high-value products such as gasoline, diesel, and olefins [7].

**Process Overview:**

a) A fine-powdered zeolite-based catalyst (typically Faujasite-type Y-zeolite) is fluidized by hydrocarbon vapors in the reactor.
b) The cracking reactions occur at 500–550 °C under moderate pressure (~1–2 atm), where long-chain hydrocarbons undergo catalytic decomposition into smaller molecules.
c) Due to the carbonaceous coke buildup on the catalyst surface, deactivation occurs rapidly. To maintain activity, the catalyst is continuously regenerated in a separate vessel by burning off coke deposits in the presence of air.
d) The exothermic regeneration step provides the heat necessary for the endothermic cracking reactions, ensuring energy efficiency within the system.

**Key Advantages of Using FBR in FCC:**

✓ Superior heat and MT due to fluidized mixing.

✓ Continuous catalyst regeneration, allowing steady-state operation and prolonged catalyst lifetime.

✓ Scalability for high-throughput refinery operations.

**Challenges and Mitigation Strategies:**

⚠ **Catalyst Attrition:** Fine particles may be lost with exhaust gases; cyclones are used for recovery.

⚠ **Coke Deposition:** Requires frequent regeneration to maintain activity.

⚠ **Process Complexity:** Requires precise control of gas flow rates and residence times.

FCC remains an indispensable process for maximizing refinery efficiency and meeting the demand for transportation fuels.

### 6.17.2 Gas-Phase Polymerization of Polyethylene and Polypropylene

FBRs are extensively used in gas-phase polymerization for producing high-density polyethylene (HDPE), linear low-density polyethylene (LLDPE), and polypropylene (PP) using advanced catalyst systems such as Ziegler-Natta, Phillips, and metallocene catalysts [8].

**Process Overview:**

a) Monomer gases (e.g., ethylene, propylene) are introduced into the FBR, where they interact with catalyst particles dispersed in the fluidized bed.
b) The polymerization reaction occurs at moderate temperatures (80–110 °C) and pressures (10–30 bar), leading to the growth of polymer granules directly in the reactor.
c) Due to the highly exothermic nature of polymerization, effective heat removal is crucial, often achieved using recycled gas cooling or liquid condensate injection (condensed-mode operation).
d) The polymer is continuously withdrawn while monomer is replenished, and fresh catalyst is periodically introduced to compensate for the catalyst consumed during polymerization, ensuring continuous production.

In gas-phase polymerization FBRs, notice that catalyst losses primarily occur due to encapsulation within the growing polymer granules, as polymerization proceeds directly on the catalyst surface, incorporating it into the final product. Unlike in FCC, where catalysts are continuously regenerated, polymerization catalysts cannot be recovered for reuse. Additionally, minor losses may arise from entrainment of fine catalyst particles in the gas phase, despite the use of cyclone separators. Over time, these losses reduce the active catalyst concentration in the reactor, necessitating the periodic or continuous addition of fresh catalyst to maintain optimal polymerization rates and ensure stable reactor performance.

**Key Advantages of Using FBR in Polymerization:**

✓ **Excellent Heat Removal:** Prevents overheating and unwanted side reactions.
✓ **High Production Flexibility:** Allows easy switching between polymer grades.
✓ **Uniform Polymer Particle Morphology** due to fluidized mixing.

**Challenges and Mitigation Strategies:**

⚠ **Reactor Fouling and Particle Agglomeration:** Sticky polymer particles may adhere, leading to de-fluidization. Mitigated by precise control of catalyst particle size and anti-fouling additives.

⚠ **Heat Removal Constraints:** Polymerization heat must be efficiently dissipated; advanced cooling strategies are essential.

⚠ **Static Charge Accumulation:** Can cause particle aggregation, necessitating electrostatic control measures.

Gas-phase polymerization in FBRs remains a preferred technology for high-efficiency, large-scale polymer production, offering low operating costs, high reactor throughput, and excellent product uniformity.

### 6.17.3 Methanol Synthesis from Syngas

FBRs are employed in methanol synthesis, converting syngas (a mixture of CO, $CO_2$, and $H_2$) into methanol using copper-based catalysts, such as $Cu/ZnO/Al_2O_3$, for applications in fuel and chemical production.
**Process Overview:**

a) Syngas is fed into the FBR, where it contacts fluidized $Cu/ZnO/Al_2O_3$ catalyst particles at 220–280 °C and 50–100 bar.
b) The reaction ($CO + 2H_2 \rightarrow CH_3OH$, $\Delta H \approx -91$ kJ/mol) is exothermic, requiring efficient heat removal to prevent catalyst deactivation and maintain selectivity.
c) Fluidized beds ensure uniform temperature distribution, with cooling achieved via internal heat-exchange tubes or recycled gas cooling.
d) Unreacted syngas is recycled, and methanol is continuously withdrawn as a liquid product, while fresh catalyst is added to compensate for gradual deactivation.

The fluidized bed enhances mixing, reducing temperature gradients (<5 °C) compared to fixed-bed reactors (10–20 °C), improving methanol yield (up to 90%), and increasing catalyst longevity. Catalyst deactivation occurs due to sintering or poisoning by sulfur impurities in syngas, requiring periodic catalyst replenishment.

**Key Advantages of Using FBR in Methanol Synthesis:**

✓ **Superior Heat and MT:** Minimizes hot spots, enhancing selectivity for methanol over byproducts like dimethyl ether.
✓ **Continuous Operation:** Enables high-throughput production with syngas recycling.
✓ **Uniform Catalyst Activity:** Fluidization ensures consistent catalyst exposure to reactants.

**Challenges and Mitigation Strategies:**

⚠ **Catalyst Deactivation:** Sintering or poisoning by sulfur compounds reduces activity, which is mitigated by using sulfur-resistant catalysts and syngas purification.
⚠ **Pressure Control:** High-pressure operation increases costs; optimized reactor design and gas recycling reduce energy demands.
⚠ **Particle Entrainment:** Fine catalyst particles may be lost; cyclones and filters recover entrained particles.

Methanol synthesis in FBRs offers a robust, scalable solution for producing methanol as a fuel or chemical feedstock, with improved thermal management and operational efficiency.

## 6.18 Introduction to External and Internal Mass Transfer

In catalytic reactors, the overall reaction rate is governed not only by the intrinsic kinetics of the chemical reaction but also by the physical transport of reactants and products to and from the active sites. These transport processes are classified into external and internal MT.

External MT involves the movement of reactants from the bulk fluid phase to the external surface of the catalyst particle (and products in the reverse direction) across a boundary layer surrounding the particle [10]. This process is analogous to *navigating through a crowded hallway*: the rate depends on fluid properties (e.g., viscosity, diffusivity) and flow conditions. For example:

✓ High flow rates reduce boundary layer thickness, enhancing transport.
⚠ Low turbulence or stagnant fluids create thicker boundary layers, slowing the delivery of reactants.

Internal MT, in contrast, refers to the diffusion of reactants from the external surface into the porous catalyst particle to reach internal active sites (and vice versa for products) [19]. This is akin to *traversing a maze*: the rate depends on the catalyst's pore structure (size, tortuosity, and connectivity) and the effective diffusivity of the species within the pores. Key factors include:

✓ Particle size: Larger particles increase diffusion path lengths.
⚠ Pore tortuosity (a measure of pore "twistiness"): Higher tortuosity reduces effective diffusivity.

If internal diffusion is slower than the reaction rate, active sites in the particle's core remain underutilized, lowering the effectiveness factor (see Section 6.21).

## 6.19 Impact of Mass Transfer on PBR Reaction Rates

In a PBR, external MT limitations dominate at low flow rates where the boundary layer thickens, creating a concentration gradient between the bulk fluid and the catalyst surface [12]. For instance, in ammonia synthesis reactors using iron-based catalysts, large particles are often used to minimize pressure drop, but this will risk starving inner active sites of reactants due to slow internal diffusion.

Internal MT limitations become significant with large catalyst particles or low pore diffusivity [19]. For example, reactants may deplete in the outer layers of a particle before reaching the core, effectively wasting the catalyst's interior. This is quantified by the Thiele modulus ($\Phi$). See Section 6.21. In general, a high $\Phi$ ($\Phi > 4$) indicates severe internal diffusion limitations.

⚠ A critical design trade-off arises in PBRs: larger particles reduce pressure drop but increase internal diffusion resistance, while smaller particles improve catalyst utilization at the cost of higher flow resistance [19].

## 6.20 Impact of Mass Transfer on FBR Reaction Rates

FBRs mitigate external MT limitations through intense mixing and turbulence, which thin the boundary layer around catalyst particles. For example, in FCC units, rapid particle-fluid motion ensures efficient reactant delivery to surfaces.

However, internal MT limitations persist and depend on the same factors as in PBRs (particle size, pore structure, etc.). If the catalyst particles are large or the pore diffusivities are low, diffusion within the pores can limit the overall reaction rate 12].

A unique challenge in FBRs is attrition, which gradually breaks particles into smaller fragments [7]. While smaller particles reduce internal diffusion resistance (lower $\Phi$), they also:

⚠ Increase external MT demand (higher surface area requires faster bulk fluid replenishment). Higher surface area (from smaller particles) means more active sites are exposed, which could theoretically boost reaction rates. However, the bulk fluid must now supply reactants to a much larger total surface area. If the fluid-phase transport rate (e.g., flow velocity, turbulence) cannot keep pace with this heightened demand, a steep concentration gradient forms at the particle surface. This leads to external MT limitations, starving the catalyst of reactants and lowering the observed reaction rate.

⚠ Raise entrainment risk, where fine particles escape the reactor, complicating operation and increasing catalyst loss.

Table 6.3 gives a swift summary of MT effects on the catalytic performance of FBRs and PBRs.

**Table 6.3** External, internal mass transfer (MT), and particle size effect on FBR and PBR catalytic performance.

| Factor | FBR | PBR |
|---|---|---|
| **External MT** | Low limitations (high turbulence) | High limitations (laminar flow) |
| **Internal MT** | Governed by particle/pore properties | Governed by particle/pore properties |
| **Particle Size Control** | Dynamic (attrition reduces size) | Fixed (particles retain original size) |

## 6.21 Thiele Modulus and Effectiveness Factor

Since both $C_A$ and $T$ may vary from point to point within a catalyst particle, the rate of reaction also varies accordingly. This may be translated to say that the effectiveness of the catalyst varies within the particle, and this must be taken into account in the rate law.

The overall effectiveness factor ($\eta$) of a spherical catalyst particle, considering both concentration and temperature variations in the radial direction, can be predicted using a coupled reaction–diffusion and heat-transfer model.

For this purpose, we introduce the particle effectiveness factor $\eta$, the ratio of the observed rate of reaction for the particle as a whole to the intrinsic rate at the surface conditions, $C_{As}$ and $T_s$. In terms of a reactant $A$. The effectiveness factor is defined as:

$$\eta = \frac{\textit{Actual reaction rate with diffusion (pore) effect}}{\textit{Intrinsic rate evaluated at surface conditions}} \tag{A}$$

Since this textbook serves as an introduction to chemical reaction engineering, we will not cover nonspherical particles as well as non-first-order reactions.

For a first-order reaction occurring inside a spherical catalyst particle, the steady-state material and energy balances, respectively, lead to the Thiele modulus ($\Phi$), which accounts for internal mass transport limitations, and to the thermal effectiveness factor ($\Phi_T$), which accounts for heat-transfer limitations. We will show the derivation for a spherical catalyst particle, noticing that in real industry, there are other nonspherical particles as well.

1) **Diffusion–Reaction Model (Concentration Variation)**

Consider a first-order irreversible reaction occurring within a spherical catalyst pellet. Under a steady-state condition, the mass balance for a spherical catalyst of radius $R$ and effective diffusivity $D_{eff}$:

$$\frac{1}{r^2}\frac{d}{dr}\left(r^2 D_{eff}\frac{dC}{dr}\right) - kC = 0 \tag{B}$$

where

- $C$ is the reactant concentration,
- $k$ is the reaction rate constant,
- $D_{eff}$ is the effective diffusivity inside the porous catalyst,
- $r$ is the radial position inside the catalyst.

The dimensionless Thiele modulus ($\Phi$) for this scenario is given by:

$$\Phi = R\left(\frac{k}{D_{eff}}\right)^{\frac{1}{2}}, \tag{C}$$

where $R$ is the radius of the pellet, $k$ is the first-order rate constant, and $D_{eff}$ is the effective diffusivity of the reactant within the pellet.

Let us define a dimensionless radius:

$$\xi = \frac{r}{R} \tag{D}$$

Replace $r$ by $R\xi$ and $dr$ by $Rd\xi$, we have:

$$\frac{1}{R^2\xi^2}\frac{d}{Rd\xi}\left(R^2\xi^2 D_{eff}\frac{dC}{Rd\xi}\right) - kC = 0 \tag{E}$$

Assume constant $D_{eff}$, the equation simplifies to:

$$\frac{D_{eff}}{R^2\xi^2}\frac{d}{d\xi}\left(\xi^2\frac{dC}{d\xi}\right) - kC = 0 \tag{F}$$

Rearrangement yields to:

$$\frac{1}{\xi^2}\frac{d}{d\xi}\left(\xi^2\frac{dC}{d\xi}\right) - \frac{kR^2}{D_{eff}}C = 0 \quad \rightarrow \quad \frac{1}{\xi^2}\frac{d}{d\xi}\left(\xi^2\frac{dC}{d\xi}\right) - \Phi^2 C = 0 \tag{G}$$

Carrying out a product differentiation for the bracketed term, we end up with:

$$\frac{1}{\xi^2}\left[\xi^2\frac{d^2C}{d\xi^2} + 2\xi\frac{dC}{d\xi}\right] - \Phi^2 C = 0 \quad \rightarrow \quad \frac{d^2C}{d\xi^2} + \frac{2}{\xi}\frac{dC}{d\xi} - \Phi^2 C = 0 \tag{H}$$

This is a standard form of Bessel's differential equation for a spherical geometry. The general solution is expressed in terms of spherical Bessel functions:

$$C(\xi) = A\frac{\sin h(\Phi\xi)}{\xi} + B\frac{\cos h(\Phi\xi)}{\xi} \tag{J}$$

For a catalyst particle with a finite radius $R$, we impose two boundary conditions at $\xi = 0$ and $\xi = 1$.

The concentration $C$ is finite at the center ($\xi = 0$), as a result of mass flux symmetry. This eliminates the second term, as $\left(\frac{\cos h(\Phi\xi)}{\xi}\right)$ diverges ($\rightarrow\infty$) at $\xi = 0$. This requires $B = 0$.

The equation then reduces to:

$$C(\xi) = A\frac{\sin h(\Phi\xi)}{\xi}$$

For the second boundary condition, the surface concentration, $C = Cs$ at $\xi = 1$ (dimensionless radius), we have:

$$C(1) = Cs = A\frac{\sin h(\Phi)}{1} \quad \rightarrow \quad A = \frac{Cs}{\sin h(\Phi)}$$

Substitute $A$, we have:

$$C(\xi) = A\frac{\sin h(\Phi\xi)}{\xi} = \frac{Cs}{\sin h(\Phi)}\frac{\sin h(\Phi\xi)}{\xi}$$

Define a dimensionless concentration, $\Theta$, we have:

$$\Theta(\xi) = \frac{C(\xi)}{Cs} = \frac{\sin h(\Phi\xi)}{\xi * \sin h(\Phi)} \tag{K}$$

### 6.21.1 MATLAB Code for Pore Concentration Distribution as a Function of Mass Transfer Thiele Modulus

Utilizing Eq. (K), Figure 6.5 shows MATLAB code (***Theta_vs_phi.m***) to generate a plot of $\Theta$ vs. $\xi$ for different parametric curves, each representing a given Thiele modulus $\Phi$ value.

Figure 6.6 shows the generated plot of $\Theta$ vs. $\xi$ for different parametric curves, labeled by a given Thiele modulus $\Phi$ value.

The effectiveness factor ($\eta_C$) for a **sphere** is related to the Thiele modulus by [25]:

$$\eta_C = \frac{3}{\Phi}\left[\frac{1}{\tan h(\Phi)} - \frac{1}{\Phi}\right] \tag{L}$$

The effectiveness factor ($\eta_C$) for a **flat plate** is related to the Thiele modulus by [25]:

$$\eta_C = \frac{\tan h(\Phi)}{\Phi} \tag{M}$$

For $\eta_C \approx 1$, it implies negligible MT limitations.

2) **Heat-Transfer Model (Temperature Variation)**

For exothermic reactions, temperature gradients inside the particle affect the reaction rate. The energy balance is:

$$\frac{1}{r^2}\frac{d}{dr}\left(r^2 k_T \frac{dT}{dr}\right) + (-\Delta H_r)kC = 0 \tag{N}$$

where

- $k_T$ is the thermal conductivity of the catalyst,
- $\Delta H_r$ is the heat of reaction.

The dimensionless, thermal Thiele modulus for a first-order reaction is:

$$\Phi_T = R\left(\frac{k(-\Delta H_r)Cs}{k_T(Tc - Ts)}\right)^{\frac{1}{2}} \tag{O}$$

Introducing the dimensionless temperature:

$$\Theta_T = \left(\frac{T - Tc}{Ts - Tc}\right) \tag{P}$$

where $T_s$ is the surface temperature, and $T_C$ is the center temperature.

The dimensionless radial coordinate

$$\xi = \frac{r}{R} \tag{Q}$$

Similar to what we carried out in the mass balance case, the energy balance equation finally reduces to:

$$\frac{1}{\xi^2}\left[\xi^2\frac{d^2\Theta_T}{d\xi^2} + 2\xi\frac{d\Theta_T}{d\xi}\right] - \Phi_T^2\Theta_T \quad \rightarrow \quad \frac{d^2\Theta_T}{d\xi^2} + \frac{2}{\xi}\frac{d\Theta_T}{d\xi} - \Phi_T^2\Theta_T = 0$$

```
clc; clear; close all;
% Define the range of xi (ξ), avoiding zero to prevent division issues
xi = linspace(0.01, 1, 100);
% Define different values of Phi (Φ)
Phi_values = [0.5, 1, 2, 3, 5];
% Initialize figure
figure;
hold on;
colors = lines(length(Phi_values)); % Generate distinct colors
% Loop over different Phi values
for i = 1:length(Phi_values)
   Phi = Phi_values(i);
   Theta = sinh(Phi * xi) ./ (xi .* sinh(Phi));
   plot(xi, Theta, 'LineWidth', 2, 'Color', colors(i, :), 'DisplayName', ['$\Phi = ' num2str(Phi, '%.1f') '$']);
end
% Labels and title with LaTeX interpreter
xlabel('$\xi$', 'Interpreter', 'latex');
ylabel('$\Theta(\xi)$', 'Interpreter', 'latex');
title('Parametric Curves of $\Theta(\xi)$ for Different $\Phi$ Values', 'Interpreter', 'latex');
legend('Interpreter', 'latex', 'Location', 'best');
grid on;
hold off;
```

**Figure 6.5** MATLAB code (***Theta_vs_phi.m***) to generate a plot of $\Theta$ vs. $\xi$ for different parametric curves, each labeled with a given Thiele modulus $\Phi$ value.

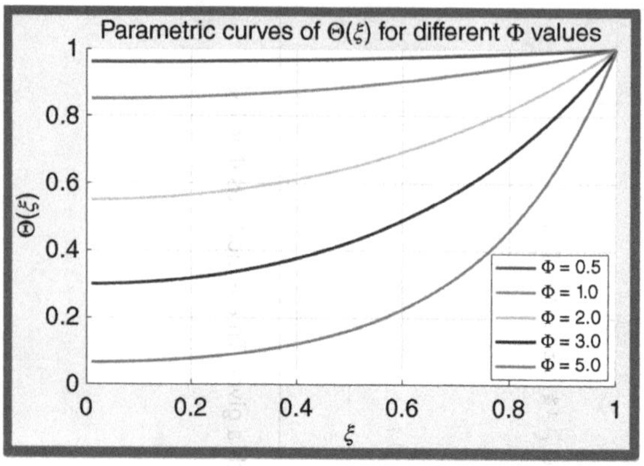


**Figure 6.6** The generated plot of $\Theta$ vs. $\xi$, showing different parametric curves, tagged by a given Thiele modulus $\Phi$ value.

This is a standard form of Bessel's differential equation for a spherical geometry. The general solution is expressed in terms of spherical Bessel functions:

$$\Theta_T(\xi) = A\frac{\sin h(\Phi_T\xi)}{\xi} + B\frac{\cos h(\Phi_T\xi)}{\xi} \tag{R}$$

For a catalyst particle with a finite radius $R$, we impose two boundary conditions at $\xi = 0$ and $\xi = 1$.

The temperature $T$ is finite at the center ($\xi = 0$), as a result of heat flux symmetry. This eliminates the second term, as $\left(\frac{\cos h(\Phi_T\xi)}{\xi}\right)$ diverges ($\rightarrow\infty$) at $\xi = 0$. This requires $B = 0$.

The equation then reduces to:

$$\Theta_T(\xi) = A\frac{\sin h(\Phi_T\xi)}{\xi}$$

For the second boundary condition, the surface temperature, $T = Ts$ (or $\Theta_T = 1$) at $\xi = 1$, we have:

$$\Theta_T(\xi) = 1 = A\frac{\sin h(\Phi_T)}{1} \quad \rightarrow \quad A = \frac{1}{\sin h(\Phi_T)}$$

Substitute $A$, we have:

$$\Theta_T(\xi) = A\frac{\sin h(\Phi_T\xi)}{\xi} = \frac{1}{\sin h(\Phi_T)}\frac{\sin h(\Phi_T\xi)}{\xi}$$

$$\Theta_T = \left(\frac{T - Tc}{Ts - Tc}\right) = \frac{\sin h(\Phi_T\xi)}{\xi * \sin h(\Phi_T)} \tag{S}$$

The effectiveness factor ($\eta_T$) for a **sphere** is related to the thermal Thiele modulus by:

$$\eta_T = \frac{3}{\Phi_T}\left[\frac{1}{\tan h(\Phi_T)} - \frac{1}{\Phi_T}\right] \tag{T}$$

The effectiveness factor ($\eta_C$) for a **flat plate** is related to the Thiele modulus by:

$$\eta_T = \frac{\tan h(\Phi_T)}{\Phi_T} \tag{U}$$

For strong temperature effects, the combined effectiveness factor ($\eta$) is corrected by introducing a thermal correction factor $\eta_T$.

$$\eta = \eta_C * \eta_T \tag{V}$$

**NOTE**

Refer to CHW 6.63 problem. In our initial analysis, we assume a center-to-surface temperature difference of approximately 20 K to estimate the thermal Thiele modulus and assess the extent of internal temperature gradients within the catalyst pellet. However, upon solving the dimensionless temperature profile, we found that the center temperature ($T_c$) was, in fact, very close to the surface temperature ($T_s$), indicating a nearly uniform temperature distribution. This result implies that the thermal Thiele modulus ($\Phi_T$) is much less than unity, confirming that heat generated by the exothermic reaction is rapidly conducted to the pellet surface and is dissipated. Consequently, the radial temperature variation is negligible under the given operating conditions, and the catalyst pellet may be accurately approximated as isothermal. This finding not only simplifies thermal modeling but also supports the validity of using isothermal effectiveness factors in the design and analysis of the catalytic system.

### 6.21.2 MATLAB Code for Evaluating Pore Effectiveness Factor as a Function of Concentration- and Temperature-Dependent Thiele Moduli

Figure 6.7 shows MATLAB code (**Eta_vs_Phi.m**) for calculating both the mass and thermal Thiele modulus factors, the mass, thermal, and overall effectiveness factors for a given fluid, catalyst, and reaction properties. The plot of concentration effectiveness, $\eta_c$, as $f(\Phi)$ is also shown.

```
% MATLAB Code for Effectiveness Factor in a Spherical Catalyst Particle
clc; clear; close all;
% Given parameters (user can modify these values)
R = 0.002; % Catalyst particle radius (m)
Deff = 5e-5; % Effective diffusivity (m^2/s) of a gas @500 K
k = 2; % Reaction rate constant (1/s)
kT = 2.0; % Thermal conductivity of catalyst (W/m.K)
Hrxn = -50000; % Heat of reaction (J/mol)
C_surf = 10; % Surface concentration of reactant (mol/m^3)
T_surf = 500; % Surface temperature (K)
T_cntr = 520; % Center temperature (K)
% Thiele modulus for concentration and temperature
phi_c = R * sqrt(k / Deff);
phi_T = R * sqrt((k * C_surf*(-Hrxn)) / (kT*(T_cntr-T_surf)));
% Define effectiveness factor function for concentration
eta_c = (3/phi_c) * (1/tanh(phi_c) - 1/phi_c);
% Define effectiveness factor function for temperature
eta_T = (3/phi_T) * (1/tanh(phi_T) - 1/phi_T);
% Overall effectiveness factor (combining concentration and temperature effects)
eta = eta_c * eta_T;
% Display results
disp('Effectiveness Factor Results:');
disp(['Thiele Modulus (Concentration): ', num2str(phi_c)]);
disp(['Thiele Modulus (Temperature): ', num2str(phi_T)]);
disp(['Effectiveness Factor (Concentration): ', num2str(eta_c)]);
disp(['Effectiveness Factor (Temperature): ', num2str(eta_T)]);
```

**Figure 6.7** MATLAB code (**Eta_vs_Phi.m**) for calculating both the mass and thermal Thiele modulus factors, the mass, thermal, and overall effectiveness factor for a given fluid, catalyst, and reaction properties.

```
disp(['Overall Effectiveness Factor: ', num2str(eta)]);
% Plot effectiveness factor behavior
phi_range = linspace(0.1, 10, 100);
eta_c_plot = (3./phi_range) .* (1./tanh(phi_range) - 1./phi_range);
figure;
plot(phi_range, eta_c_plot, 'b', 'LineWidth', 2);
xlabel('Thiels Modulus, \phi');
ylabel('Effectiveness Factor, \eta_c');
title('\eta_c vs Thiele Modulus, \phi');
ax = gca;
% Set new font size for all axes text (including labels)
ax.FontSize = 16;
grid on;
```

**Figure 6.7** (*Continued*)

Results are copied from the Command Window:

```
>> Eta_vs_Phi
Effectiveness Factor Results:
Thiele Modulus (Concentration): 0.4
Thiele Modulus (Temperature): 0.31623
Effectiveness Factor (Concentration): 0.98949
Effectiveness Factor (Temperature): 0.9934
Overall Effectiveness Factor: 0.98296
```

Figure 6.8 shows the generated plot of $\eta_C$ versus $\Phi$. Notice that:

- When $\Phi < 1$ (small catalyst particles or fast diffusion), $\eta_C \approx 1$ (reaction rate is not diffusion-limited).
- When $\Phi > 4$ (large catalyst particles or slow diffusion), $\eta_C < 0.5$ (strong internal diffusion limitations).

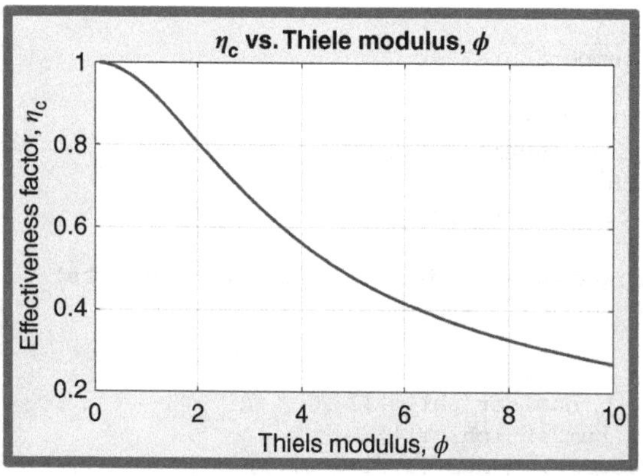


**Figure 6.8** The generated plot of $\eta_C$ versus $\Phi$.

**Example 6.4** Given that the Thiele modulus $\Phi = 3.0$, calculate the effectiveness factor.

$$\eta = \frac{3}{\Phi}\left[\frac{1}{\tan h(\Phi)} - \frac{1}{\Phi}\right] = \frac{3}{3}\left[\frac{1}{\tan h(3)} - \frac{1}{3}\right] = 1 * \left[\frac{1}{0.995} - 0.333\right] \cong 0.672$$

This indicates that the observed (or apparent) reaction rate is only about 67% of what it would be if there were no internal diffusion limitations.

### 6.21.3 MATLAB Code for Thiele Modulus, Effectiveness Factor, and the Concentration Profile Inside the Catalyst Particle

**Example 6.5** ***Catalytic Liquid-phase Reaction***

Consider a first-order liquid-phase reaction $A \rightarrow B$ that occurs in a spherical catalyst particle of radius $R = 0.005$ m with an effective diffusivity of $D_{eff} = 1.5 \times 10^{-9}$ m$^2$/s and a rate constant $k = 0.01$ s$^{-1}$. The reactant concentration at the catalyst surface is $C_s = 200$ mol/m$^3$. Calculate the Thiele modulus, effectiveness factor, and the concentration profile of $A$ inside the catalyst particle as a $f(r)$.

**Solution**

Figure 6.9 shows MATLAB code (**Example6_5.m**) to calculate the Thiele modulus, effectiveness factor, and the concentration profile of $A$ inside the catalyst particle as a function of $r$.

```
% MATLAB code for calculating calculate Thiele modulus, effectiveness factor, and the
% concentration profile of A inside the catalyst particle as a function of r.
clear variables;
% Given parameters
R = 0.005;              % Catalyst particle radius (m)
D_eff = 1.5e-9;         % Effective diffusivity (m²/s)
k = 0.01;               % Reaction rate constant (1/s)
Cs = 200;               % Surface concentration (mol/m³)
% Compute Thiele modulus
Phi = R * sqrt(k / D_eff);
% Compute effectiveness factor
eta_C = (3/Phi) * (1/tanh(Phi) - 1/Phi);
% Compute concentration profile for selected xi values
xi = linspace(0.01, 1, 20); % Avoiding division by zero at xi = 0
Theta_xi = sinh(Phi * xi) ./ (xi .* sinh(Phi));
% Compute actual concentration inside the catalyst
C_xi = Cs * Theta_xi;
% Display results
fprintf('Thiele modulus (Φ): %.4f\n', Phi);
fprintf('Effectiveness factor (η_C): %.4f\n\n', eta_C);
fprintf('ξ\tΘ(ξ)\tC(ξ) (mol/m³)\n');
fprintf('———————————————————————-\n');
for i = 1:length(xi)
    fprintf('%.2f\t%.4f\t%.4f\n', xi(i), Theta_xi(i), C_xi(i));
end
% Plot concentration profile
figure;
plot(xi, C_xi, 'bo-', 'LineWidth', 2, 'MarkerSize', 6);
xlabel('\xi = r/R');
ylabel('Concentration C(\xi) (mol/m³)');
title('Concentration Profile Inside the Catalyst Particle');
ax = gca;ax.FontSize = 16;
grid on;
```

**Figure 6.9** MATLAB code (**Example6_5.m**) to calculate Thiele modulus, effectiveness factor, and the concentration profile of $A$ inside the catalyst particle as a function of $r$.

Results are copied from the Command Window:

```
>> Example6_5
Thiele modulus (Φ): 12.9099
Effectiveness factor (η_C): 0.2144
ξ     Θ(ξ)          C(ξ) (mol/m³)   ξ      Θ(ξ)        C(ξ) (mol/m³)
-------------------------------------------------------------------------
0.01  0.0001        0.0128          0.53   0.0044      0.8844
0.06  0.0001        0.0142          0.58   0.0079      1.5781
0.11  0.0001        0.0179          0.64   0.0142      2.8386
0.17  0.0001        0.0251          0.69   0.0257      5.1405
0.22  0.0002        0.0379          0.74   0.0468      9.3629
0.27  0.0003        0.0600          0.84   0.1575      31.5088
0.32  0.0005        0.0987          0.79   0.0857      17.1387
0.37  0.0008        0.1666          0.90   0.2907      58.1494
0.43  0.0014        0.2866          0.95   0.5384      107.6788
0.48  0.0025        0.5004          1.00   1.0000      200.0000
```

Figure 6.10 shows the generated plot of concentration profile as $f(r)$ within the catalyst particle.

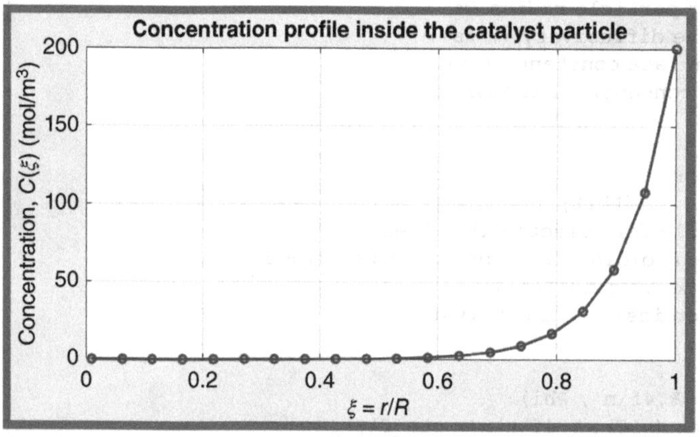


**Figure 6.10** The generated plot of concentration profile as $f(r)$ within the catalyst particle.

## 6.22 Strategies to Identify and Mitigate Mass Transfer Limitations

1) **Diagnostic Methods**
   - **Experimental:**
     a. **Vary Flow Rate:** Increased rate boosting reaction rate implies external limitations.
     b. **Vary Particle Size:** Smaller particles improving rate suggests internal limitations.
   - **Theoretical:**
     a. **Thiele Modulus (Φ):** Values >4 indicate pore diffusion controls the rate.
     b. **Effectiveness Factor:** $\eta c \approx 1$ implies negligible MT limitations.
2) **Mitigation Strategies**
   - **External Mass Transfer:**
     a. Optimize flow rates/turbulence (e.g., baffles in PBRs).
     b. Use smaller catalyst particles to enhance external MT, provided that the resulting pressure drop remains within reasonable operational limits.

- **Internal Mass Transfer:**
  a. **Hierarchical Pore Structures:** Combine macro- (fast transport) and micropores (high surface area). For example, zeolites with mesoporous coatings.
  b. **High Diffusivity Supports:** Metal–organic frameworks (MOFs) or carbon nanotubes enhance $D_{eff}$.
- **Thermal Trade-Offs:** Higher temperatures increase $D_{eff}$ but may deactivate catalysts (e.g., sintering) or reduce selectivity.

## 6.23 Comparison of PBR Versus FBR Catalysis

Table 6.4 summarizes key features, advantages, and disadvantages of homogenous and heterogeneous catalysis in PBRs and FBRs.

**Table 6.4** A comparison of PBR vs. FBR in terms of catalytic features, advantages, and disadvantages.

| Feature/Aspect | PBR (Homogeneous Catalyst) | FBR (Homogeneous Catalyst) | PBR (Heterogeneous Catalyst) | FBR (Heterogeneous Catalyst) |
|---|---|---|---|---|
| Catalyst state | Immobilized on a solid support | Encapsulated within particles or adsorbed on fluidizing particles | Solid particles | Solid particles (fluidized) |
| Flow pattern | Plug flow (ideal) | Complex, can approach mixed flow | Plug flow (ideal) | Complex, turbulent |
| Key advantages | Potential for high selectivity and yield, continuous operation, and easy product separation | Excellent heat transfer, uniform temperature, and continuous operation (with encapsulated catalyst) | High conversion per catalyst weight, simple construction, and continuous operation | Excellent heat transfer, uniform temperature, high mass transfer rates, and easy catalyst regeneration |
| Key disadvantages | Potential catalyst leaching, mass transfer limitations within support, and temperature control challenges | Difficulty in catalyst retention (the catalyst remains encapsulated or adsorbed), particle attrition, complex design, and high pumping costs | Difficult temperature control, high-pressure drop, challenging catalyst replacement, and potential channeling | Increased vessel size, high pumping costs, particle entrainment, catalyst attrition, and potential back-mixing |
| Typical applications | Ester hydrolysis (immobilized acid), hydrogenation (immobilized organometallic), and methanol synthesis, using homogeneous catalysts like rhodium complexes for CO hydrogenation | Enzymatic biotransformation (immobilized enzyme), olefin metathesis (encapsulated catalyst), and biodiesel production, using homogeneous base catalysts like sodium methoxide | Ammonia synthesis, sulfuric acid production, ethene hydrogenation, and hydrodesulfurization (HDS) of crude oil, using Co-Mo/$Al_2O_3$ catalysts | Catalytic cracking of petroleum, polyethylene/polypropylene synthesis, and Fischer-Tropsch synthesis, using Fe/Co catalysts in multi-tubular FBRs, producing synthetic fuels |

## 6.24 Guidelines for Selecting the Appropriate Reactor Type

The selection of the most appropriate reactor type for a catalytic process depends on a multitude of factors related to the catalyst, the reactants, and the desired process characteristics. The phase of the catalyst and reactants is a primary consideration. Heterogeneous catalysts, being in a different phase from the reactants, are commonly employed in both packed beds and FBRs. Packed beds are particularly well-suited for gas–solid or liquid–solid reactions where a stationary bed of catalyst can efficiently contact the flowing reactant stream. Fluidized beds are also used for such systems, especially when excellent mixing and heat transfer are required.

Homogeneous catalysts, which exist in the same phase as the reactants, are less frequently used in their soluble form in these reactor types due to separation challenges. However, homogeneous catalysts can be adapted for use in packed beds by immobilization on a solid support and in fluidized beds through encapsulation within fluidizable particles. Table 6.5

**Table 6.5** Selection factors for homogeneous and heterogeneous catalysis of PBRs and FBRs.

| Selection Factor | PBR (Homogeneous) | FBR (Homogeneous) | PBR (Heterogeneous) | FBR (Heterogeneous) |
|---|---|---|---|---|
| Catalyst phase | Immobilized on a solid support | Encapsulated in fluidizable particles or adsorbed | Solid particles | Solid particles |
| Reactant phase | Gas-phase, liquid-phase | Gas-phase, Liquid-phase | Gas–solid, liquid–solid | Gas–solid, liquid–solid |
| Mixing and contacting | Moderate mixing, immobilized catalyst | Good mixing, encapsulated catalyst | Poor mixing, fixed catalyst bed | Excellent mixing, fluidized catalyst |
| Heat transfer | Moderate | Good | Limited, potential hot spots | Excellent, uniform temperature |
| Temperature control | Moderate control | Good control | Challenging for exother-mic/endothermic reactions | Superior control |
| Continuous operation | Possible, but requires an immobilized catalyst | Possible with specialized designs | Possible, but catalyst replacement can be difficult | Easy catalyst circulation and regeneration |
| Catalyst regeneration | Difficult due to solubility | Moderate, depends on encapsulation | Periodic shutdown required | Continuous or batchwise regeneration is possible |
| Pressure drop | Moderate | Higher than PBR due to particle movement | Lower for large particles | Can be significant, especially for fine particles |
| Particle attrition | Minimal | Moderate | Minimal | High, due to continuous movement |
| Scale-up complexity | Challenging, due to phase interactions | Complex; requires advanced fluidization techniques | Easier, well-understood | More complex due to hydrodynamics |

summarizes the description of the selection factor for both PBR and FBR under homogeneous and heterogeneous cases.

The reaction kinetics and the heat of reaction are also critical factors. Reactions with high heats of reaction, whether exothermic or endothermic, often benefit from the superior heat-transfer capabilities of FBRs, which can maintain a more uniform temperature and prevent hot spots or cold spots. Reactions requiring precise temperature control are also better suited for FBRs.

The need for continuous operation and the frequency of catalyst regeneration or replacement can significantly influence reactor selection. Both packed beds and FBRs can operate continuously. However, fluidized beds offer an advantage when the catalyst deactivates rapidly, as the fluid-like nature of the catalyst allows for easy and continuous circulation to a regeneration unit.

Pressure drop limitations can favor PBRs with larger catalyst particles, as fluidized beds and packed beds with fine particles can exhibit significant pressure drops. Particle size and the potential for attrition are more relevant to FBRs, where the constant movement of particles can lead to wear and tear.

Finally, the scale of operation can play a role. Both reactor types are used in large-scale industrial processes, but specific design and operational considerations may become more pronounced at different scales. For instance, scaling up fluidized beds can be more complex due to the hydrodynamics involved.

## 6.25 FBR Reaction Control: Kinetic Versus Mass Transfer Rate

The design equations for the rate of disappearance of reactant $A$ $(-r_A)$ in an isothermal FBR are derived based on the fundamental principles of chemical reaction engineering. These equations will relate the reaction rate to operational variables and reactor specifications, such as the fluid velocity, particle size, catalyst properties, and temperature.

### 6.25.1 Rate of Disappearance of Reactant $A$ $(-r_A)$

The rate of disappearance of reactant $A$ depends on the reaction kinetics and the MT between the fluid and solid phases. In an isothermal FBR, $(-r_A)$ is typically influenced by two main factors:

- **Intrinsic Reaction Kinetics:** The reaction occurring at the surface of the catalyst particles.
- **Mass Transfer Limitations:** The transport of reactant A from the bulk fluid phase to the catalyst surface.

Thus, the overall rate $(-r_A)$ can be expressed as:

$$-r_A = min(-r_{kinetic}, r_{mass\ transfer}) \tag{A}$$

where

- $r_{kinetic}$ is the intrinsic rate of reaction (reaction kinetics).
- $r_{mass\ transfer}$ is the rate limited by MT between phases.

### 6.25.2 Intrinsic Reaction Rate

The intrinsic rate of reaction for reactant A on the catalyst surface is typically modeled using the following rate law:

$$-r_{kinetic} = kC_A^n \tag{B}$$

### 6.25.3 Mass Transfer Rate

The overall rate of disappearance of reactant A may also be limited by MT from the bulk fluid phase to the surface of the catalyst particles. The MT rate can be modeled using the following equation:

$$r_{mass\ transfer} = k_g a(C_{A,bulk} - C_{A,surface}) \tag{C}$$

where

- $k_g$ is the gas–solid MT coefficient (m/s).
- $a$ is the interfacial area per unit volume of bed ($m^2/m^3$), related to the particle size and reactor geometry.
- $C_{A,bulk}$ is the concentration of reactant A in the bulk phase (mol/L).
- $C_{A,surface}$ is the concentration of reactant A at the catalyst surface (mol/L).

For an FBR, $k_g$ depends on the superficial gas velocity $U_0$, particle diameter $d_p$, and the physical properties of the gas (e.g., diffusivity $D$).

The Sherwood number $N_{Sh}$ relates the MT coefficient $k_g$ to the particle diameter:

$$N_{sh} = \frac{k_g d_p}{D} \tag{D}$$

where $D$ is the diffusivity of reactant $A$ in the fluid phase.

For a fluidized bed, empirical correlations are often used to estimate $N_{Sh}$, which depends on the Reynolds number $N_{Re}$ and Schmidt number $N_{Sc}$:

$$N_{sh} = \frac{k_g d_p}{D} = 2 + 0.6 * N_{Re}^{0.5} N_{sc}^{1/3} \tag{E}$$

where

$$N_{Re} = \frac{\rho_f U_0 d_p}{\mu_f} \tag{F}$$

$$N_{Sc} = \frac{\mu_f}{\rho_f D} \tag{G}$$

$\rho_f$ and $\mu_f$ are the fluid density and viscosity, respectively.

## 6.26 The Minimum Fluidization Velocity and Pressure Drop for FBR

The superficial gas velocity $U_0$ plays a key role in the design of the FBR. The operating fluid velocity must be sufficient to maintain fluidization but not too high to cause excessive entrainment of particles. Let us first explain the concept of fluidization in mass, heat, and chemical reaction engineering applications. The behavior of gas–solid fluidized beds will depend strongly on the superficial gas velocity. As velocity increases, the bed passes through distinct fluidization regimes, as illustrated in Figure 6.11a–f:

a) **Fixed Bed Regime:** At low gas velocities, the gas simply percolates through the void spaces of the packed bed without causing particle movement. This is referred to as a fixed bed, where the solid phase remains immobile, and pressure drop increases linearly with gas velocity. If the fluid flows slowly, it will simply pass through the spaces between the particles without disturbing their arrangement, as shown in Figure 6.11a. This is called a static bed (SB). An SB has a specific porosity (amount of empty space) called $e_{SB}$ and a height called $h_{SB}$.

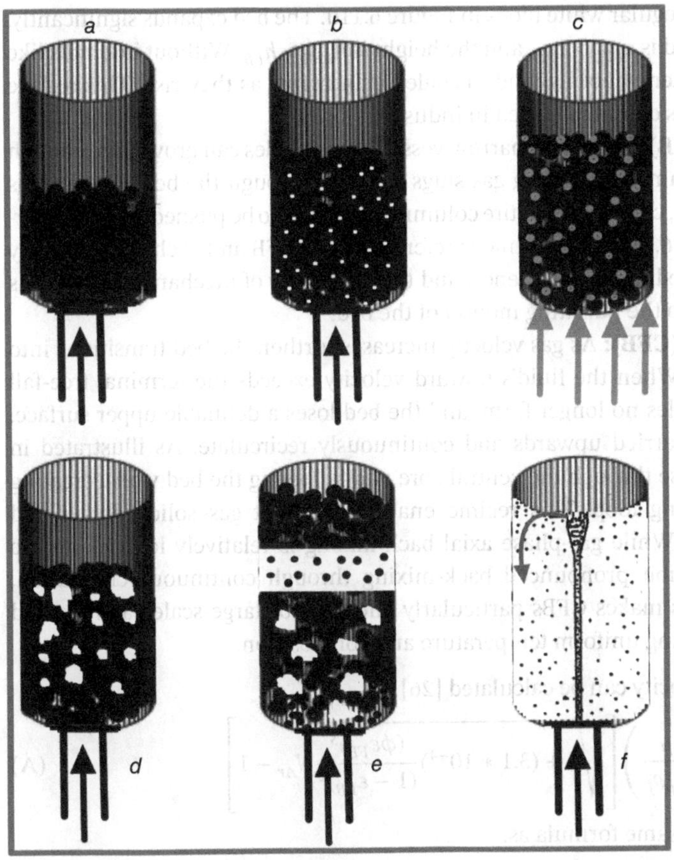


**Figure 6.11** A sketch for FBR. (a) Static bed. (b) Gas-fluidized bed at minimum velocity. (c) Liquid-fluidized bed at minimum velocity. (d) Bubble-fluidized Bed. (e) Slugging-fluidized bed. (f) Stranded solids with circulation at a high superficial gas velocity.

b) **Minimum Fluidization Point:** As the fluid velocity increases, some particles begin to vibrate or shift slightly while most remain in contact, forming a slightly expanded bed. Upon reaching a critical velocity, known as the minimum fluidization velocity ($U_{mf}$), the drag force exerted by the fluid matches the weight of the solid particles per unit area. At this point, the particles become fully suspended, and the bed transitions into a fluid-like state. There is no sustained particle–particle contact, and the pressure drop across the bed equals the weight of the solids and fluid per unit cross-sectional area. This marks the onset of fluidization as illustrated in Figure 6.11b. The white dots represent gas fully dispersed throughout the bed. Compared to the SB, the fluidized bed has a larger void fraction, $\varepsilon_{FB} > \varepsilon_{SB,}$ and the height is $h_{FB} > h_{SB}$.

c) **Particulate (Smooth) Fluidization:** Slightly above the minimum fluidization velocity, the bed enters the particulate fluidization regime, where particles move gently and uniformly throughout the bed. Gas flows uniformly, and mixing is good. This regime is typically observed with fine, uniform particles and is ideal for reactions requiring good gas–solid contact. On the other hand, liquid-fluidized beds respond to an increase in velocity by expanding uniformly. This behavior is referred to as particulate fluidization and is depicted in Figure 6.11c.

d) **Bubbling Fluidized Bed (BFB):** As the gas velocity increases further, excess gas escapes through the bed in the form of relatively large gas bubbles with little to no solid content.

These are clearly shown as irregular white blobs in Figure 6.11d. The bed expands significantly, and the void fraction of the bed is $\varepsilon_{BFB} > \varepsilon_{FB}$, and the height is $h_{BFB} > h_{FB}$. Without internals like baffles or simply flow disrupters, bubbles tend to coalesce and grow as they rise. This regime provides intense mixing and is commonly used in industrial FBRs.

e) **Slugging-Fluidized Bed (SFB):** In tall and narrow vessels, gas bubbles can grow large enough to span the entire column diameter, forming gas slugs that rise through the bed. These slugs displace the solids above them, causing the entire column of particles to be pushed upward intermittently, as shown in Figure 6.11e. This regime is referred to as a SFB and is characterized by poor gas–solid contact, reduced mixing efficiency, and the possibility of mechanical vibrations or pressure fluctuations due to the pulsating motion of the bed.

f) **Circulating Fluidized Bed (CFB):** As gas velocity increases further, the bed transitions into the fast fluidization regime. When the fluid's upward velocity exceeds the terminal free-fall velocity of the particles, bubbles no longer form, and the bed loses a definable upper surface. In this CFB, the solids are carried upwards and continuously recirculate. As illustrated in Figure 6.11f, solid particles rise through the central core and fall along the bed walls, creating a well-established recirculating loop. This regime enables excellent gas–solid contact and high solids circulation rates. While gas-phase axial back-mixing is relatively low due to the fast upward flow, solids exhibit pronounced back-mixing through continuous circulation, enhancing heat and MT. This makes CFBs particularly effective for large-scale catalytic and combustion operations requiring uniform temperature and composition.

The minimum fluidization velocity can be calculated [26] as:

$$U_{mf} = 42.9\left(\frac{1-\varepsilon_{FB}}{\phi}\right)\left(\frac{\mu}{d_p\rho_f}\right)\left[\sqrt{1+(3.1 * 10^{-4})\frac{(\phi\varepsilon_{FB})^3}{(1-\varepsilon_{FB})^2}N_{Ar}} - 1\right] \tag{A}$$

Notice that $N_{Re}$ is given by the same formula as:

$$N_{Re} = 42.9\left(\frac{1-\varepsilon_{FB}}{\phi}\right)\left[\sqrt{1+(3.1 * 10^{-4})\frac{(\phi\varepsilon_{FB})^3}{(1-\varepsilon_{FB})^2}N_{Ar}} - 1\right] \tag{B}$$

where

$\phi$ is the sphericity of the particle, which is defined as: $\phi = \frac{S_V}{S_P}$. $S_V$ is the surface area of a sphere of equal volume, and $S_P$ is the surface area of the actual particle.
$\mu$ is the dynamic viscosity of the fluid.
$\varepsilon_{FB}$ is the void fraction of the fluidized bed.
$d_p$ is the Sauter diameter of the particles.
$N_{Ar}$ is the Archimedes number is defined as:

$$N_{Ar} = \frac{g(d_p)^3(\rho_f)^2}{\mu^2}\left(\frac{\rho_p}{\rho_f} - 1\right) \tag{C}$$

The most widely used formulation for the pressure drop in poly-dispersed beds of irregular particles ≥5 μm in size and uniformly packed at random is the one proposed by Ergun and Orning [27]. This formulation is applicable across the full range of Reynolds numbers for $\varepsilon_{SB} < \varepsilon < 1.0$, which also covers the minimum (onset of) fluidization velocity condition.

$$\frac{\Delta P}{h} = 150\frac{(1-\varepsilon)^2}{\varepsilon^3}\frac{\mu U}{\left(\phi d_p^2\right)^2} + 1.75\frac{1-\varepsilon}{\varepsilon^3}\frac{\rho_f U^2}{\phi d_p} \tag{D}$$

**Table 6.6** The porosity ($\varepsilon_{FB}$) of typical packing materials as a function of the average particle size or the Sauter diameter [28]. $\phi$ is the sphericity of the particle.

| Packing Particles | ($\varepsilon_{FB}$) based on average particle size (μm), Sauter diameter | | | | | | |
|---|---|---|---|---|---|---|---|
| | 20 | 50 | 70 | 100 | 200 | 300 | 400 |
| Angular sand ($\phi$ = 0.67). $\rho_p$ = 2,650 kg/m$^3$ | — | 0.60 | 0.59 | 0.58 | 0.54 | 0.50 | 0.49 |
| Rounded sand ($\phi$ = 0.86). $\rho_p$ = 2,650 kg/m$^3$ | — | 0.56 | 0.52 | 0.48 | 0.44 | 0.42 | — |
| Sand mixture (round particles). $\rho_p$ = 2,650 kg/m$^3$ | — | — | 0.42 | 0.42 | 0.41 | — | — |
| Pulverized coal. $\rho_p$ = 1,400 kg/m$^3$ | 0.72 | 0.67 | 0.64 | 0.62 | 0.57 | 0.56 | — |
| Glass $\rho_p$ = 2,500 kg/m$^3$ | | | | | | | |
| Anthracite ($\phi$ = 0.63). $\rho_p$ = 1,600 kg/m$^3$ | — | 0.62 | 0.61 | 0.60 | 0.56 | 0.53 | 0.51 |
| Activated charcoal. $\rho_p$ = 500 kg/m$^3$ | 0.74 | 0.72 | 0.71 | 0.69 | — | — | — |
| Fischer–Tropsch catalyst ($\phi$ = 0.58). $\rho_p$ = 1,500 kg/m$^3$ | — | — | — | 0.58 | 0.56 | 0.55 | — |
| Carborundum. $\rho_p$ = 3,250 kg/m$^3$ | — | 0.61 | 0.59 | 0.56 | 0.48 | — | — |

Table 6.6 shows the porosity (voidage fraction), $\varepsilon_{FB}$, of typical packing materials as a function of the average particle size or the Sauter diameter.

## 6.27 The Terminal Fluidization Velocity

Terminal fluidization velocity can be calculated from the equation given in the dimensionless form [29]:

$$C_D \times N_{Re}^2 = \frac{4}{3} N_{Ar} \tag{E}$$

where $C_D$ is the drag coefficient, which can be determined experimentally or estimated using various empirical correlations. The value of $C_D$ depends on the Reynolds number.

$$C_D = \frac{24}{N_{Re}} + \frac{4}{\sqrt{N_{Re}}} + 0.4 \tag{F}$$

Substituting Eq. (F) into Eq. (E), we obtain

$$\left( \frac{24}{N_{Re}} + \frac{4}{\sqrt{N_{Re}}} + 0.4 \right) N_{Re}^2 = \frac{4}{3} N_{Ar} \tag{G}$$

Rearrange to find the root (i.e., the correct $N_{Re}$) for such a nonlinear equation:

$$f(N_{Re}) = \left( \frac{24}{N_{Re}} + \frac{4}{\sqrt{N_{Re}}} + 0.4 \right) N_{Re}^2 - \frac{4}{3} N_{Ar} = 0 \tag{H}$$

MATLAB root finding technique will be used to iteratively reach the final solution where $f(N_{Re}) = 0$ is satisfied. Once we manage to find the root of Eq. (H), we can proceed to find the terminal fluidization velocity utilizing the definition of Reynolds number. The superficial velocity will then be a fraction of the terminal velocity and, at the same time, higher than the minimum fluidization velocity. MATLAB code is used to find the root of Eq. (H). See the next section.

## 6.28 Expressing $C_{Asurface}$ as a Function Mass Transfer Resistance

Given or calculated, the MT coefficient, $k_g$, can be used to decide on the surface concentration of reactant species $A$ via equating the MT rate with that of the kinetic rate. For the first-order reaction, we have:

$$r_{mass\ transfer} = k_g a(C_{Abulk} - C_{Asurface}) = -r_{kinetic} = k' C_{Asurface} \tag{A}$$

From which we can express:

$$k_g a(C_{Abulk} - C_{Asurface}) = k' C_{Asurface} \quad \rightarrow \quad k_g a(C_{Abulk}) = (k' + k_g a) C_{Asurface}$$

$$C_{Asurface} = \left( \frac{k_g a}{k' + k_g a} \right) (C_{Abulk}) \tag{B}$$

On the other hand, for an $n^{th}$-order reaction, we have:

$$r_{mass\ transfer} = k_g a(C_{Abulk} - C_{Asurface}) = -r_{kinetic} = k' C^n_{Asurface} \tag{C}$$

From which we can express:

$$k_g a(C_{Abulk} - C_{Asurface}) = k' C^n_{Asurface}$$

$$\rightarrow k_g a(C_{Abulk}) = (k_g a) C_{Asurface} + k' C^n_{Asurface}$$

$$\rightarrow k_g a(C_{Abulk}) = C_{Asurface} \left[ (k_g a) + k' C^{n-1}_{Asurface} \right]$$

$$C_{Asurface} = \left( \frac{k_g a}{(k_g a) + k' C^{n-1}_{Asurface}} \right) C_{Abulk} \tag{D}$$

Or

$$f(C_{Asurface}) = \left[ C_{Asurface} - \left( \frac{k_g a}{(k_g a) + k' C^{n-1}_{Asurface}} \right) C_{Abulk} \right] = 0 \tag{E}$$

This simply says that we account for MT effect on the intrinsic kinetic rate. The surface concentration of A ($C_{Asurface} \le C_{Abulk}$) will be the actual value seen by the catalyst particle surface at which the chemical reaction is supposed to mainly take place, not within the fluid bulk phase. For a given $C_{Abulk}$, $k_g a$, and $k'$, Eq. (B) can be directly solved to find $C_{Asurface}$.

On the contrary, Eq. (E) is implicit in $C_{Asurface}$, which means that it will be iteratively solved for $C_{A,surface}$. Notice that:

$$F_{A0} = U_0 A_{FB} C^0_{Abulk} \neq U_0 A_{FB} C^0_{Asurface} \tag{F}$$

Expressing $C_{Asurface}$ in terms of either Eq. (B) or Eq. (E) is included in the MATLAB code in the next section.

## 6.29 MATLAB Code for FB Flow and Fluid Properties Evaluation

Appendix D shows the MATLAB algorithm for calculations of FB flow and fluid properties. In addition, $C^0_{A,surface}$ is expressed as indicated by Eqs. (B) or (E). The comments are inserted at each step of the algorithm.

Below is a sample run of **Appendix D | FB.m** file showing both the user-entered data and the generated results, copied from the MATLAB Command Window. Notice that I used the first entry or row in Table 6.6, that is, angular sand packing particles with an average particle size (or Sauter particle diameter) equal to 100 μm.

```
≫ FB
Enter the bed porosity (dimensionless. See Table 6.6): 0.58
Enter the particle sphericity (dimensionless). See Table 6.6: 0.67
Enter the Sauter particle diameter (μm). See Table 6.6: 100
Enter the particle density (kg/m^3). See Table 6.6: 2650
Enter FB temperature (Kelvin): 300
Enter FB pressure (atm): 10
Select the fluid type (Enter 'Gas' or 'Liquid'): Gas
Enter the molecular mass (g/mol): 29
Enter the gas viscosity (Pa.s),e.g., mu_air=1.81E-5 Pa.s: 1.81E-5
Schmidt number for gas @T, NSc = 0.71631 (dimensionless)
Gas Diffusivity, D = 2.14496e-06 m^2/s
Gas Diffusivity, D = 2.14496e-02 cm^2/s
Minimum fluidization velocity, Umf = 0.01937 m/s
Equation solved.
fsolve completed because the vector of function values is near zero
as measured by the value of the function tolerance, and
the problem appears regular as measured by the gradient.
<stopping criteria details>
The terminal velocity,Utv is: 0.361016 m/s
The root which is NRetv at the terminal velocity is: 23.496824
Operational superficial velocity, U0 = 0.07581 m/s
Pressure drop per bed length h, evaluated @U0, ΔP/h, = 45.26235 kPa/m
Reynold Number @U0, dimensionless = 4.93433
Sherwood Number @U0, dimensionless = 3.19252
Mass transfer coefficient, kg = 6.84783e-02 m/s
Bed Diameter (m): 0.8
Enter Bed Height (m): 3
FB Superficial Area = 5.02655e-01 m2
Fluid Volume Flow Rate = 3.81080e-02 m3/s
Fluid Mass Flow Rate = 4.48928e-01 kg/s
FB Volume = 1.50796e+00 m3
FB ΔP = 1.35787e+02 kPa
Enter the reaction order (n): 2
Enter the heterogenous rate constant, k (L^n/(kg cat.mol^(n-1).time)): 0.5
Enter the interfacial area per volume, a (m2/m3): 100
Enter FA0 (mol/time): 5
Equation solved.
fsolve completed because the vector of function values is near zero
as measured by the value of the function tolerance, and
the problem appears regular as measured by the gradient.
<stopping criteria details>
CAbulk0 (mol/L) = 1.31206e-01 mol/L
CAsurface0 (mol/L) = 1.29973e-01 mol/L
```

**Notes on MATLAB code of Appendix D | FB.m:**

1) The code basically handles the fluidized bed as far as MT and fluid mechanics aspects are concerned. No chemical reaction is accounted for by the code. Prior to using the fluidized bed as FBR, chemical engineers want to maximize the MT coefficient between the bulk phase and catalyst particle surface via the optimum configuration of the fluidized bed itself and of the fluid/flow properties. In addition, the calculated bed pressure drop between the inlet and exit stream falls within the acceptable limit.
2) Some lump fluid properties, like the Schmidt number, are initially assigned yet allowed to change with system operating temperature; however, those can be modified by the user via writing the proper equation that will describe how each individual physical fluid property changes with the operating conditions, namely, pressure and temperature.
3) FBR conversion: $X = f(H_{FB})$, or vice versa, $H_{FB} = f(X)$ will be considered in the next section.

## 6.30 FBR Design Equations for Height/Conversion

The models developed by Fogler [30], Kunii and Levenspiel 28], and Levenspiel [31] provide a comprehensive framework for analyzing chemical reaction conversion in FBRs, capturing the complex interplay between various operating conditions and reactor specifications. These models intricately incorporate essential factors such as reaction kinetics, which dictate the rate at which reactants are converted to products, alongside MT dynamics that govern the movement of reactants to the catalyst surface. Furthermore, they address heat-transfer processes, ensuring that temperature variations, which can significantly affect reaction rates and product quality, are adequately managed within the reactor. This multifaceted approach allows for a more accurate prediction of conversion rates under different operating scenarios, making the models invaluable for both design and optimization.

In addition to the fundamental reaction and transfer phenomena, these models also consider catalyst deactivation mechanisms, which can arise from factors such as fouling, sintering, or poisoning over time. By incorporating these aspects, the models provide a holistic view of reactor performance, enabling researchers and engineers to anticipate changes in conversion efficiency as the catalyst ages or as operating conditions shift. This level of detail ensures that the models are not only theoretically robust but also practically relevant, allowing for effective scaling up from laboratory experiments to industrial applications. As a result, the models serve as critical tools for optimizing FBR operations, enhancing both efficiency and product yield in various chemical processes. Here, we will present a simplified version of the design equation of a solid catalyzed reaction taking place in an isothermal FBR.

### 6.30.1 Reaction-Rate-Limited FBR Under Isothermal, Steady-State Conditions

$$In - Out + Generation - Consumption = Accumulation$$

Let us assume that $F_A$ changes with the height of the bed. Thus, we have:

$$F_A|_Z - F_A|_{Z+\Delta Z} + 0 - (-r'_A)\,\Delta W = \frac{\Delta N_A}{\Delta t} = 0 \tag{A}$$

The differential weight is expressed in terms of differential length, $\Delta Z$. Thus, we have:

$$\Delta W = \rho_{cat}\Delta V = \rho_{cat}A_{FB}(1 - \varepsilon_{FB})\Delta Z \tag{B}$$

$$F_A|_Z - F_A|_{Z+\Delta Z} + 0 - k' C_{A,surface} \times \rho_{cat}A_{FB}(1 - \varepsilon_{FB})\Delta Z = 0 \tag{C}$$

Divide by $\Delta Z$, and we have:

$$\frac{F_A|_Z - F_A|_{Z+\Delta Z} +}{\Delta Z} - k' C_{A,surface} \rho_{cat} A_{FB}(1 - \varepsilon_{FB}) = 0 \tag{D}$$

Take the limit as $\Delta Z$ approaches zero, we get:

$$\frac{dF_A}{dZ} = -k' C_{A,surface} \rho_{cat} A_{FB}(1 - \varepsilon_{FB}) \tag{E}$$

Given that $F_A = F_{A0}(1 - X)$ and thus $dF_A = -F_{A0}dX$ (F)

Also, $C_{A,surface} = C^0_{A,surface}(1 - X)$ is assumed.

$$\frac{dF_A}{dZ} = -\frac{F_{A0}dX}{dZ} = -k' C^0_{A,surface}(1 - X)\rho_{cat} A_{FB}(1 - \varepsilon_{FB}) \tag{G}$$

Rearrange, we have:

$$\frac{dX}{(1 - X)} = \frac{k' C^0_{A,surface} \times \rho_{cat} A_{FB}(1 - \varepsilon_{FB})}{F_{A0}} dZ \tag{H}$$

Integrate $dX$ from $X = 0$ to $X = X$ and $dZ$ from $Z = 0$ to $Z = H_{FB}$, we have:

$$\int_0^X \frac{dX}{(1 - X)} = \frac{k' C^0_{A,surface} \times \rho_{cat} A_{FB}(1 - \varepsilon_{FB})}{F_{A0}} H_{FB} \tag{I}$$

$$-\ln(1 - X) = \ln\left(\frac{1}{1 - X}\right) = \frac{k' C^0_{A,surface} \times \rho_{cat} A_{FB}(1 - \varepsilon_{FB})}{F_{A0}} H_{FB} \tag{J}$$

Or

$$H_{FB} = \frac{F_{A0}}{k' C^0_{A,surface} \times \rho_{cat} A_{FB}(1 - \varepsilon_{FB})} \ln\left(\frac{1}{1 - X}\right) \tag{K}$$

### 6.30.2 Mass Transfer-Limited FBR Under Isothermal, Steady-State Conditions

The differential weight is expressed in terms of differential length, $\Delta Z$. Thus, we have:

$$\Delta V = A_{FB}(\varepsilon_{FB})\Delta Z \tag{L}$$

$$F_A|_Z - F_A|_{Z+\Delta Z} + 0 - k_g * a * (C_{A,bulk} - C_{A,surface}) \times A_{FB}(\varepsilon_{FB})\Delta Z = 0 \tag{M}$$

Divide by $\Delta Z$, and we have:

$$\frac{F_A|_Z - F_A|_{Z+\Delta Z} +}{\Delta Z} - k_g * a * (C_{A,bulk} - C_{A,surface}) A_{FB} \varepsilon_{FB} = 0 \tag{N}$$

Take the limit as $\Delta Z$ approaches zero, we get:

$$\frac{dF_A}{dZ} = -k_g * a * (C_{A,bulk} - C_{A,surface}) A_{FB} \varepsilon_{FB} \tag{O}$$

Given that $F_A = F_{A0}(1 - X)$ and thus $dF_A = -F_{A0}dX$ (P)

$$\frac{dF_A}{dZ} = -\frac{F_{A0}dX}{dZ} = -k_g * a * (C_{A,bulk} - C_{A,surface}) A_{FB} \varepsilon_{FB} \tag{Q}$$

If both $C_{A,bulk}$ and $C_{A,surface}$ linearly decrease with conversion in the same fashion, we have:

$$\frac{F_{A0}dX}{dZ} = k_g * a * \left(C^0_{A,bulk} - C^0_{A,surface}\right) * (1 - X) * A_{FB}\varepsilon_{FB} \tag{R}$$

Integrate $dX$ from $X = 0$ to $X = X$ and $dZ$ from $Z = 0$ to $Z = H_{FB}$, we have:

$$\int_0^X \frac{dX}{(1-X)} = \frac{k_g * a * \left(C^0_{A,bulk} - C^0_{A,surface}\right) * A_{FB}\varepsilon_{FB}}{F_{A0}} H_{FB} \tag{S}$$

$$-\ln(1-X) = \ln\left(\frac{1}{1-X}\right) = \frac{k_g * a * \left(C^0_{A,bulk} - C^0_{A,surface}\right) * A_{FB}\varepsilon_{FB}}{F_{A0}} H_{FB} \tag{T}$$

Or

$$H_{FB} = \frac{F_{A0}}{k_g * a * \left(C^0_{A,bulk} - C^0_{A,surface}\right) * A_{FB}\varepsilon_{FB}} * \ln\left(\frac{1}{1-X}\right) \tag{U}$$

## 6.31 MATLAB Code for Evaluating Conversion and Bed Height in an Isothermal FBR

Let us manually solve the following example and then use MATLAB code to find the final conversion for the given size of FBR. Given the FBR specifications, we will first see if the reaction is MT limited by comparing the intrinsic kinetic rate and MT rate for reactant A in an isothermal FBR. The diameter of the bed is 0.8 m, and the height is 3 m.

**Example 6.6** Consider a gas-phase first-order reaction: $A \rightarrow B$, which takes place at the surface of the catalyst pellets or particles that form the packing material of an FBR.

**Kinetic Rate Constant:** $k' = 0.05 \frac{\text{L}}{\text{kg catalyst} \cdot \text{s}}$ | **Reaction Order:** $n = 1$
**Bulk Concentration:** $C_{A,bulk} = 0.2$ mol/L | **Surface Concentration:** $C_{\text{A,surface}} = 0.1$ mol/L
**Particle Density:** $\rho_p = 1{,}200$ kg/m³ | **Bed Voidage Fraction:** $\varepsilon_{FB} = 0.56$
**Volume of the Bed:** $V_{FB} = A_{FB} * H_{FB} = \frac{\pi D_{FB}^2}{4} \times H_{FB} = \frac{\pi * 0.8^2}{4} \text{m}^2 \times H_{FB} = 0.503\ \text{m}^2 * 3\ \text{m} = 1.5\ \text{m}^3$
**MT Coefficient:** kg = 0.01 m/s | **Interfacial Area Per Unit Volume:** $a = 200\ \text{m}^2/\text{m}^3$
**Initial Molar Flow Rate of** $A$**:** $F_{A0} = 1$ mol/s
**Find Final Conversion:** X

**Solution**

**Step 1: Calculate the Intrinsic Reaction Rate**

The intrinsic rate of disappearance of reactant $A$ due to reaction kinetics is given by:

$$-r_{kinetic} = k' C_{A,surface} = 0.05 \frac{\text{L}}{\text{kg catalyst} \cdot \text{s}} * 0.1 \frac{\text{mol}}{\text{L}} = 0.005 \frac{\text{mol}}{\text{kg catalyst} \cdot \text{s}}$$

$$-r_{kinetic} W_{cat} = 0.005 \frac{\text{mol}}{\text{kg catalyst} \cdot \text{s}} * \rho_P(1 - \varepsilon_{FB}) V_B$$

$$-r_{kinetic} W_{cat} = 0.005 \frac{\text{mol}}{\text{kg catalyst} \cdot \text{s}} * 1{,}200 \frac{\text{kg}}{\text{m}^3} (1 - 0.56) \times 1.5\ \text{m}^3 = 3.96 \frac{\text{mol}}{\text{s}}$$

**Step 2: Calculate the MT Rate**

The MT rate is calculated using the following equation:

$$r_{mass\ transfer} = k_g a(C_{A,bulk} - C_{A,surface}) = 0.01\ \frac{\text{m}}{\text{s}} * 200\frac{\text{m}^2}{\text{m}^3}\left(0.2\frac{\text{mol}}{\text{L}} - 0.1\frac{\text{mol}}{\text{L}}\right) = 0.2\ \frac{\text{mol}}{\text{L}\cdot\text{s}}$$

$$r_{mass\ transfer}V_{Gas} = 0.2\ \frac{\text{mol}}{\text{L}\cdot\text{s}} * \varepsilon_{FB}V_B = 0.2\ \frac{\text{mol}}{\text{L}\cdot\text{s}} * 0.56 * 1.5\ \text{m}^3 \times \left(\frac{1{,}000\ \text{L}}{\text{m}^3}\right) = 168\ \frac{\text{mol}}{\text{s}}$$

Since $r_{mass\ transfer}V_{Gas} = 168\ \frac{\text{mol}}{\text{s}} > -r_{kinetic}W_{cat} = 3.96\ \frac{\text{mol}}{\text{s}}$, then $-r_{kinetic}W_{cat}$ will be taken into account as the reaction is kinetically limited, but not MT limited.

Notice that if the height of the bed is not initially given, then a comparison between MT and reaction kinetics can be made using one unit length of height, that is, $H_{FB} = 1.0$ m. Alternatively, $V_B$ can be assumed as $V_B = A_{FB}{*}H_{FB} = 1\ \text{m}^3$.

**Step 3: Calculate the Conversion**

From Eq. (K), we have:

$$H_{FB} = \frac{F_{A0}}{k'C_{A,surface} \times \rho_{cat}A_{FB}(1 - \varepsilon_{FB})} \ln\left(\frac{1}{1-X}\right)$$

$$H_{FB} = \frac{1\ \frac{\text{mol}}{\text{s}}}{0.05\frac{\text{L}}{\text{kg cat}\cdot\text{s}} * 0.1\frac{\text{mol}}{\text{L}} \times 1{,}200\ \frac{\text{kg cat}}{\text{m}^3} * 0.503\ \text{m}^2 * (1 - 0.56)} \ln\left(\frac{1}{1-X}\right) = 3\ \text{m}$$

$$0.753\ \text{m} * \ln\left(\frac{1}{1-X}\right) = 3\ \text{m} \cdot \text{Solve for}\ X. \rightarrow X = 0.9813$$

Figure 6.12 shows MATLAB code (**HFB_Or_X.m** file) for either calculating the height of FBR in terms of conversion, or, vice versa, the conversion in terms of the height of FBR.

The following user-entered parameters and generated results are copied from the Command Window, which represent the solution for the manually solved Example 6.6. The code gives the same value for the final conversion.

```
>> HFB_Or_X
Enter FA0 (mol/time): 1
Enter CAsurface0 (mol/L): 0.1
Enter CAbulk0 (mol/L): 0.2
Enter catalyst particle density (kg/m^3): 1200
Enter fluidized bed diameter (m): 0.8
Enter FB void fraction (dimensionless): 0.56
Enter the reaction order (n): 1
Enter the heterogenous rate constant, k (L^n/(kg cat.mol^(n-1).time)): 0.05
Enter the mass transfer coefficient, kg (m/s): 0.01
Enter the interfacial area per volume, a (m2/m3): 200
Do you want to enter (a) Height of the bed or (b) Conversion? Enter "a" or "b": a
Enter the height of the bed, HFB (m): 3
The conversion X is: 0.9813
```

```
clear variables;
% Prompt the user for input values
FA0 = input('Enter FA0 (mol/time): ');
CAsurface0 = input('Enter CAsurface0 (mol/L): ');
CAbulk0 = input('Enter CAbulk0 (mol/L): ');
rowcat = input('Enter catalyst particle density (kg/m^3): ');
dFB = input('Enter fluidized bed diameter (m): ');

%% Calculate the area of the fluidized bed
AFB = pi * (dFB^2) / 4; % m^2

eFB = input('Enter FB void fraction (dimensionless): ');
n = input('Enter the reaction order (n): ');
kprime = input('Enter the heterogenous rate constant, k (L^n/(kg cat.mol^(n-1).time)): ');
kg = input('Enter the mass transfer coefficient, kg (m/s): ');
ia = input('Enter the interfacial area per volume, a (m2/m3): ');

% Calculate intermediate values
rxnrate = kprime * (CAsurface0^n) * rowcat * AFB * (1 - eFB);
Hrxn = FA0 / rxnrate;
massrate = kg *ia* (CAbulk0-CAsurface0) * AFB *eFB*1000;
Hmass = FA0 / massrate;
% Prompt the user to choose between entering bed height or conversion
choice = input('Do you want to enter (a) Height of the bed or (b) Conversion? Enter "a" or "b": ', 's');

if choice == 'a'
  % Option (a): Enter the height of the bed
  HFB = input('Enter the height of the bed, HFB (m): ');
  if rxnrate<=massrate
  rxnterm = HFB / Hrxn;
  X = 1 - 1 / exp(rxnterm);
  else
  massterm = HFB / Hmass;
```

**Figure 6.12** MATLAB code for utilizing Eq. (K) or Eq. (U) for calculating $X$ or $H_{FB}$.

```
  X = 1 - 1 / exp(massterm);
  end
  fprintf('The conversion X is: %.4f\n', X);

elseif choice == 'b'
  % Option (b): Enter the conversion
  X = input('Enter the conversion, X (dimensionless): ');

  if rxnrate<=massrate
  HFB=Hrxn*log(1 / (1 - X));
  else
  HFB=Hmass*log(1 / (1 - X));
  end
  fprintf('The height of the bed HFB is: %.4f m\n', HFB);
else
  fprintf('Invalid choice. Please restart the program and choose "a" or "b".\n');
end
```

**Figure 6.12** (*Continued*)

## References


1 Catalysis of Chemical Reactions – Homogeneous & Heterogeneous. https://byjus.com/chemistry/catalysis/ (accessed 15 March 2025).

2 www.vaia.com. https://www.vaia.com/en-us/explanations/engineering/chemical-engineering/homogeneous-catalysis/#:~:text=Homogeneous%20catalysis%20refers%20to%20chemical,and%20efficiency%20in%20industrial%20processes (accessed 15 March 2025).

3 Examples of Homogeneous Catalysis in Industrial Processes. https://www.solubilityofthings.com/examples-homogeneous-catalysis-industrial-processes (accessed 15 March 2025).

4 library.fiveable.me. https://library.fiveable.me/inorganic-chemistry-ii/unit-10/heterogeneous-catalysis/study-guide/NkI8luVoK0SDgYe0#:~:text=Heterogeneous%20catalysis%20is%20a%20game,and%20then%20unstick%20as%20products (accessed 15 March 2025).

5 Packed Bed (PBR) - Visual Encyclopedia of Chemical Engineering Equipment. https://encyclopedia.che.engin.umich.edu/packed-bed-reactors/ (accessed 15 March 2025).

6 Packed bed reactor | PPT – SlideShare. https://www.slideshare.net/slideshow/packed-bed-reactor/79774792 (accessed 15 March 2025).

7 Fluidized bed reactor – Wikipedia. https://en.wikipedia.org/wiki/Fluidized_bed_reactor (accessed 15 March 2025).

8 Homogeneous catalysis – Wikipedia. https://en.wikipedia.org/wiki/Homogeneous_catalysis (accessed 15 March 2025).

9 Al-Malah, K. (2025). MATLAB-based Computations of Chemical Engineering Principles. In: . Wiley Inpress. https://www.wiley.com/en-us/MATLAB-based+Computations+of+Chemical+Engineering+Principles-p-9781394308828\stquote\stquote.

10 Applications of Heterogeneous Catalysis in Industry. https://www.cademix.org/applications-of-heterogeneous-catalysis-in-industry/ (accessed 15 March 2025).

11 Immobilization and Study of Homogeneous Catalysts in a Continuous Flow Reactor Using Inorganic Particles Coated with Polymer. https://par.nsf.gov/servlets/purl/10359586 (accessed 15 March 2025).

12 Green chemistry – RSC publishing. https://pubs.rsc.org/en/content/getauthorversionpdf/c4gc02160f (accessed 15 March 2025).

13 CHAPTER 7: Fixed-Bed Catalytic Reactors I - UCSB College of Engineering. https://sites.engineering.ucsb.edu/~jbraw/chemreacfun/ch7/slides-masswrxn-2up.pdf (accessed 15 March 2025).

14 Fluidized Bed (FBR) – Visual Encyclopedia of Chemical Engineering Equipment. https://encyclopedia.che.engin.umich.edu/fbr/ (accessed 15 March 2025).

15 Packed bed and fluidised bed bioreactor – YouTube. https://www.youtube.com/watch?v=PBXYl_EY5vc (accessed 15 March 2025).

16 Parr Fluidized Bed Reactors for Solids Processing. https://www.parrinst.com/products/tubular-reactor-systems/fluidized-bed-reactors/parr-fluidized-bed-reactors-for-solids-processing/ (accessed 15 March 2025).

17 10.3 Heterogeneous Catalysis | Inorganic Chemistry II Review, Fiveable. https://library.fiveable.me/inorganic-chemistry-ii/unit-10/heterogeneous-catalysis/study-guide/NkI8luVoK0SDgYe0 (accessed 15 March 2025).

18 Single and Two-phase Flow in a Packed Bed Reactor - JoVE. https://app.jove.com/v/10431/single-and-two-phase-flow-in-a-packed-bed-reactor (accessed 15 March 2025).

19 Really, What Is a Packed Bed Reactor? – HUFOCW. https://www.hufocw.org/Download/file/174 (accessed 15 March 2025).

20 What will be the most appropriate L/D ratio for a cylindrical packed bed reactor. https://www.researchgate.net/post/What-will-be-the-most-appropriate-L-D-ratio-for-a-cylindrical-packed-bed-reactor (accessed 15 March 2025).
21 Packed Bed Reactor | PDF | Chemical Reactor | Catalysis – Scribd. https://fr.scribd.com/document/223520329/Packed-Bed-Reactor (accessed 15 March 2025).
22 Packed Bed Reactors (PBRs). https://websites.umich.edu/~elements/7e/asyLearn/bits/pbr/index.htm (accessed 15 March 2025).
23 3.8: Day 25- Homogeneous and Heterogeneous Catalysis - Chemistry LibreTexts. https://chem.libretexts.org/Bookshelves/General_Chemistry/Interactive_Chemistry_(Moor_Zhou_and_Garand)/03%3A_Unit_Three/3.08%3A_Day_25-_Homogeneous_and_Heterogeneous_Catalysis (accessed 15 March 2025).
24 Fluidized Bed Reactors | SwRI. https://www.swri.org/markets/chemistry-materials/chemistry-chemical-engineering/process-engineering-fuels/fluidized-bed-reactors (accessed 15 March 2025).
25 Missen, R.W., Mims, C.A., and Saville, B.A. (1999). Chapter 8: Catalysis and catalytic reactions. In: *Introduction to Chemical Reaction Engineering and Kinetics*, 206. New York: Wiley.
26 Karl-Ernst Wirth. "Flow patterns and pressure drop in fluidized beds", in "*VDI Heat Atlas*", 2nd Edition. Professor Dr.-Ing. Peter Stephan, Editor-in-Chief, VDI-Gesellschaft Verfahrenstechnik und Chemieingenieurwesen. Springer. 2010. pp 1220–1228. https://doi.org/10.1007/978-3-540-77877-6. https://www.humsterlandenergie.nl/resources/livreHeatAtlas.pdf.
27 Ergun, S. and Orning, A.A. (1949). Fluid flow through randomly packed columns and fluidized beds. *Industrial and Engineering Chemistry* 41: 1179–1184.
28 Kunii, D. and Levenspiel, O. (1969). *Fluidization Engineering*. London: Wiley.
29 Stepien, L. (2015). *Fluidization. Chapter 2: Fluidization Velocities*. Department of Energy and Fuels, AGH University (January 2015). https://home.agh.edu.pl/~lstepien/Fluid/Fluid_2.pdf.
30 H.S. Fogler. "Elements of Chemical Reaction Engineering". Chapter 4: pp. 270–337. Pearson Education, Inc.; 6th edition (July 24, 2020). ISBN-10: 013548622X. ISBN-13: 978-0135486221.
31 Levenspiel, O. (1999). *Chemical Reaction Engineering*, 3e. New York. ISBN 0-471-25424-X: Wiley.

## End of Chapter Problems

**CHW 6.1 What is a key advantage of homogeneous catalysis over heterogeneous catalysis?**

**A** The catalyst is easier to separate from the product mixture.
**B** The uniform distribution of the catalyst minimizes MT limitations.
**C** The catalyst is typically more thermally stable.
**D** The reaction rates are always faster due to the higher surface area.
**E** The catalyst can be easily recycled without loss of activity.
**F** The process requires no solvents.

**CHW 6.2 Which step in a typical homogeneous catalytic cycle involves the formation of new carbon–carbon or carbon–heteroatom bonds?**

**A** Oxidative addition
**B** Reductive elimination
**C** Migratory insertion
**D** Substrate coordination

**E** Catalyst activation
**F** Product dissociation

**CHW 6.3** **Why is spectroscopic analysis (e.g., NMR, IR, and UV–Vis) particularly useful in studying homogeneous catalytic systems?**
**A** The solid nature of the catalyst allows for easy characterization.
**B** The soluble nature of the catalyst and intermediates enables detailed mechanistic studies.
**C** The high temperatures used improve signal resolution.
**D** The heterogeneous nature of the system simplifies sample preparation.
**E** The catalysts are always diamagnetic, simplifying NMR analysis.
**F** The reactions occur too quickly for other analytical methods.

**CHW 6.4** What is the general form of the rate law for a homogeneous catalytic reaction?
**A** $-r_A=k[C]^n$
**B** −rA=k[A]m[C]n
**C** −rA=k[A]/
**D** −rA=k*exp(−Ea/RT)
**E** −rA=kobs[A][C]/(K+[A])
**F** −rA=k[A]m

**CHW 6.5** **What does the Michaelis-Menten-type kinetic expression: $-r_A = k_{obs}[A]/(K + [A])$ describe in homogeneous catalysis?**
**A** A reaction where catalyst decomposition is rate-limiting
**B** A saturation effect at high reactant concentrations due to intermediate complex formation
**C** A zero-order dependence on catalyst concentration
**D** A reaction where product inhibition dominates
**E** A purely diffusion-controlled process
**F** A reaction where the rate is independent of substrate concentration

**CHW 6.6** **Why is homogeneous catalysis particularly useful in pharmaceutical manufacturing?**
**A** The catalysts are always cheaper than heterogeneous alternatives.
**B** The reactions are always faster due to higher activation energies.
**C** It allows for high efficiency, tunability, and operation under mild conditions.
**D** Catalyst separation is never a challenge in pharmaceutical processes.
**E** The reactions do not require any solvents.
**F** The catalysts are always stable at very high temperatures.

**CHW 6.7** **Michaelis-Menten-type Kinetics**
Consider a homogeneous catalytic reaction where the decomposition of a reactant (or substrate) $A$ is catalyzed by a soluble catalyst $C$. The reaction follows Michaelis–Menten-type kinetics, and experimental data for the reaction rate ($-r_A$) at different reactant concentrations $[A]$ are provided in Table 6.7. The catalyst concentration is kept constant at = 0.02 M.

**Table 6.7** A homogeneous catalytic reaction data for the decomposition of reactant A, catalyzed by a soluble catalyst C.

| $[A]$ (M) | $-r_A$ (mol/L·s) | $[A]$ (M) | $-r_A$ (mol/L·s) |
|---|---|---|---|
| 0.01 | 0.009 | 0.2 | 0.037 |
| 0.02 | 0.015 | 0.25 | 0.039 |
| 0.05 | 0.025 | 0.3 | 0.042 |
| 0.1 | 0.034 | 0.35 | 0.045 |

Find Michaelis–Menten parameters: $k_{obs}$ and $K$, using:

1) The least squares method.
2) MATLAB code (**Mich_Menten.m**) for carrying out linear regression using MATLAB's built-in **fitlm** function.

**CHW 6.8** **What are the key characteristics of an ideal support material for immobilizing a homogeneous catalyst in a PBR?**

**A** High electrical conductivity to enhance electron transfer in redox reactions.
**B** Chemical inertness to prevent unwanted interactions with the catalyst or reactants.
**C** High porosity to facilitate reactant and product diffusion, minimizing MT limitations.
**D** Large surface area to maximize catalyst loading and ensure uniform dispersion.
**E** Magnetic properties to simplify catalyst recovery via external fields.
**F** Thermal instability to allow for easy catalyst removal at high temperatures.
**G** $A+B+C$
**H** $B+C+D$

**CHW 6.9** **What effect does a lower flow rate have in a PBR with immobilized homogeneous catalysts?**

**A** Increases pressure drop and energy consumption
**B** Decreases residence time and conversion
**C** Increases residence time and enhances conversion
**D** Causes flow maldistribution and channeling
**E** Reduces throughput and catalyst deactivation
**F** Increases unreacted material in the product stream

**CHW 6.10** **Which of the following best explains the trade-off involved in adjusting the flow rate in a PBR with immobilized homogeneous catalysts?**

**A** Lower flow rates always improve both throughput and conversion.
**B** Higher flow rates reduce catalyst fouling and increase selectivity.
**C** High flow rates improve conversion but reduce throughput.
**D** Low flow rates increase residence time but may lead to catalyst deactivation.
**E** Flow rate adjustments have no significant effect on reactor performance.
**F** Optimal flow rate always corresponds to the maximum pressure drop.

**CHW 6.11** **What is a potential negative effect of increasing the flow rate too much in a PBR?**

**A** Increased conversion and selectivity
**B** Decreased energy consumption
**C** Enhanced residence time
**D** Improved catalyst lifetime
**E** Flow maldistribution and pressure drop
**F** Higher residence time for full conversion

**CHW 6.12** **What is a major consequence of hot spots forming within a PBR?**
**A** Enhanced catalyst lifetime
**B** Reduced energy consumption
**C** Improved conversion efficiency
**D** Catalyst deactivation and side reactions
**E** Uniform temperature distribution
**F** Increased reaction rate without side effects

**CHW 6.13** **Which of the following is not a recommended strategy for mitigating thermal gradients in PBRs?**
**A** Using high-conductivity support materials
**B** Employing cooling jackets or heat exchangers
**C** Operating at very low flow rates
**D** Designing layered catalyst beds with thermal control
**E** Preheating the reactant stream
**F** Embedding heat-exchange systems within the reactor

**CHW 6.14** **What is the primary advantage of using immobilized homogeneous catalysts in a PBR compared to traditional homogeneous batch reactors?**
**A** It completely eliminates any potential for MT limitations.
**B** It guarantees that catalyst leaching will not occur, ensuring product purity.
**C** It allows for continuous operation, simplifying product separation and reducing operational costs.
**D** It is less efficient in terms of conversion per unit mass of catalyst.
**E** It inherently prevents the formation of hotspots, simplifying temperature control.
**F** It reduces the high selectivity and activity typical of homogeneous catalysts.

**CHW 6.15** **Which of the following is identified as a critical challenge that can negatively impact the performance of PBRs with immobilized homogeneous catalysts?**
**A** The inherently low activity of homogeneous catalysts, leading to poor yields
**B** The simplified downstream processing, which adds complexity to the overall plant design
**C** The strong and irreversible bonding of the catalyst to the support, which prevents regeneration
**D** The difficulty in controlling localized temperature variations, which can lead to catalyst degradation
**E** The high cost of the catalyst support material compared to the catalyst itself
**F** The increased productivity, which can overwhelm downstream purification units

**CHW 6.16 What is a primary advantage of using immobilized homogeneous catalysts in PBRs compared to traditional homogeneous batch reactors?**

**A** Elimination of catalyst leaching issues
**B** Higher selectivity due to heterogeneous catalyst properties
**C** Continuous operation without costly catalyst separation steps
**D** Reduced need for porous support materials
**E** Lower sensitivity to temperature variations
**F** Complete avoidance of MT limitations

**CHW 6.17 Which of the following is a key challenge associated with immobilized homogeneous catalysts in PBRs?**

**A** Inability to achieve high conversion rates
**B** Catalyst leaching due to weak support interactions
**C** Excessive heat dissipation leading to overcooling
**D** Lack of selectivity compared to heterogeneous systems
**E** Requirement for batch-wise operation
**F** Incompatibility with liquid-phase reactions

**CHW 6.18 Why might MT limitations occur in PBRs with immobilized homogeneous catalysts?**

**A** Reactants cannot adsorb onto the catalyst surface
**B** Diffusion resistance in porous supports slows reactant access to active sites
**C** The catalyst dissolves into the reaction medium
**D** Homogeneous catalysts inherently react too slowly
**E** Temperature gradients block reactant flow
**F** The reactor design prevents continuous operation

**CHW 6.19 How can temperature control challenges in PBRs with immobilized catalysts be mitigated?**

**A** Using nonporous supports to minimize heat generation
**B** Avoiding liquid-phase reactions entirely
**C** Employing heat exchangers and optimized flow conditions
**D** Reducing catalyst loading to lower reaction rates
**E** Operating only at ambient temperature
**F** Switching to batch reactor designs

**CHW 6.20 Hydrolysis of Methyl Acetate in a PBR**

Calculate the conversion of methyl acetate (A) in a PBR with an immobilized sulfuric acid catalyst. The reaction follows pseudo-first-order kinetics, and the reactor operates at steady state with the given parameters.

$$CH_3COOCH_3 + H_2O \overset{H+}{\Leftrightarrow} CH_3COOH + CH_3OH$$

**Given Data:**

- **Inlet concentration of methyl acetate:** $C_{A0} = 4\ \text{mol/L}$
- **Volumetric flow rate:** $\upsilon_0 = 6\ \text{L/min}$
- **Effective Pseudo-first-order Rate Constant:** $k_{obs} = 0.03 \frac{\text{L}}{\text{gcat} \cdot \text{min}}$
- **Catalyst Weight:** $W = 500\ \text{g}$

Also, use MATLAB code (**PBR_XA.m**) to calculate the final conversion in a PBR for the acid-catalyzed methyl acetate hydrolysis.

**CHW 6.21 Decomposition of Hydrogen Peroxide in a PBR**

Calculate the conversion of hydrogen peroxide ($H_2O_2$) in a PBR with immobilized manganese dioxide ($MnO_2$) catalyst. The reaction follows pseudo-first-order kinetics, and the reactor operates at steady state using the parameters below.

$$\text{Reaction: } H_2O_2(aq) \overset{MnO2}{\Leftrightarrow} H_2O(l) + \frac{1}{2}O_2(g)$$

**Given Data:**

- **Inlet Concentration of $H_2O_2$:** $C_{A0} = 2.5\ \text{mol/L}$
- **Volumetric Flow Rate:** $v_0 = 4\ \text{L/min}$
- **Effective Pseudo-first-order Rate Constant:** $k_{obs} = 0.05\ \text{L/(gcat·min)}$
- **Catalyst Weight:** $W = 300\ \text{g}$

Also, use MATLAB code (**PBR_XA.m**) to calculate the final conversion in a PBR for the water peroxide decomposition.

**CHW 6.22 PBR for Hydrogenation of Hexene Into Hexane**

Determine the catalyst weight required for 90% conversion in a PBR for the hydrogenation of hexene into hexane. Use the provided Langmuir-Hinshelwood kinetics:

$$-r_A = \frac{kP_{H2}P_A}{(1 + K_{H2}P_{H2} + K_AP_A)^2}$$

$C_6H_{12} + H_2 \rightarrow C_6H_{14}$ In symbols: $A + B \rightarrow C$

**Feed Conditions:**

Total pressure $P = P_0 = 12\ \text{atm}$, initial molar flow rates $F_{A0} = 12\ \text{mol/s}$, $F_{B0} = 24\ \text{mol/s}$, $F_{N2} = 64\ \text{mol/s}$.

**Given Data:**

$$k = 0.6 \frac{\text{mol}}{\text{kg catalyst} \cdot \text{s}} \times \left(\frac{1}{\text{atm}}\right)^2$$

$K_{H2} = 0.12\ \text{atm}^{-1}$
$K_A = 0.06\ \text{atm}^{-1}$
$T_0 = T = \text{Constant} = 650\ \text{K}$

Make use of MATLAB code (**PBR_WPBR.m**) for calculating the weight of the catalyst bed, expressed in kg, as a function of input parameters and degree of conversion, $X$, for the limiting reactant, hexene.

**CHW 6.23 PBR for Dehydrogenation of Ethanol to Acetaldehyde**

Determine the catalyst weight required to achieve 90% conversion of ethanol ($A$) in a PBR for the gas-phase dehydrogenation of ethanol over a copper-based catalyst. The reaction follows Langmuir-Hinshelwood kinetics:

$$-r_A = \frac{kP_A}{(1 + K_{H2}P_{H2} + K_AP_A)^2}$$

$C_2H_5OH \rightarrow CH_3CHO + H_2$ In symbols: $A \rightarrow B + C$

**Feed Conditions:**
Total pressure $P = P_0 = 10$ atm, initial molar flow rates $F_{A0} = 8$ mol/s, $F_{N2}$ (inert) = 32 mol/s.

**Given Data:**

$$k = 0.3 \frac{\text{mol}}{\text{kg catalyst} \cdot \text{s}} \times \left(\frac{1}{\text{atm}}\right)$$

$K_{H2} = 0.15\ \text{atm}^{-1}$
$K_A = 0.10\ \text{atm}^{-1}$
$T_0 = T = \text{Constant} = 600$ K

Make use of MATLAB code (**PBR_WPBR.m**) for calculating the weight of the catalyst bed, expressed in kg, as a function of input parameters and degree of conversion, $X$, for the limiting reactant, ethanol.

**CHW 6.24** **What is a primary challenge of using homogeneous catalysis in FBRs?**
**A** High catalyst production costs
**B** Catalyst retention due to solubility in the reaction medium
**C** Inadequate heat transfer in the reactor
**D** Limited availability of fluidizing gases
**E** Poor reactor design for bubbling regimes
**F** Low selectivity of homogeneous catalysts

**CHW 6.25** **Which strategy helps retain homogeneous catalysts in FBRs?**
**A** Increasing fluidization velocity
**B** Encapsulating catalysts within porous particles
**C** Using low-surface-area supports
**D** Eliminating downstream separation processes
**E** Reducing catalyst-support interactions
**F** Operating in fast fluidization regimes only

**CHW 6.26** **What is a potential drawback of adsorbing homogeneous catalysts onto fluidizing particles?**
**A** Increased catalyst production costs
**B** Risk of catalyst desorption and product contamination
**C** Inability to fluidize the reactor
**D** Reduced heat-transfer efficiency
**E** Excessive catalyst stability
**F** Limited reactant diffusion

**CHW 6.27** **Which reactor design feature can enhance catalyst retention in FBRs?**
**A** Higher fluidization velocities
**B** Cyclone separators or baffles
**C** Smaller reactor dimensions
**D** Elimination of internal components
**E** Uniform bubble dynamics
**F** Reduced catalyst loading

**CHW 6.28 What is the primary purpose of controlling fluidization velocity in an FBR?**

**A** To maximize catalyst entrainment for higher reaction rates
**B** To suspend catalyst particles while minimizing excessive loss
**C** To ensure complete settling of the catalyst bed
**D** To reduce gas–solid contact for better selectivity
**E** To increase back-mixing for improved mixing
**F** To eliminate the need for cyclone separators

**CHW 6.29 Why is uniform temperature distribution important in FBRs for homogeneous catalysis?**

**A** It prevents hot spots and ensures consistent reaction rates
**B** It increases catalyst attrition for better mixing
**C** It reduces the need for heat exchangers
**D** It allows for higher gas velocities without entrainment
**E** It promotes faster catalyst deactivation
**F** It minimizes the role of cyclone separators

**CHW 6.30 What mainly determines the minimum fluidization velocity in an FBR?**

**A** Reactor diameter
**B** Catalyst color
**C** Size and density of particles
**D** Feedstock composition
**E** Operating pressure
**F** Reactor length

**CHW 6.31 Which of the following is a key advantage of FBRs in homogeneous catalysis?**

**A** Poor heat-transfer capability
**B** Formation of hot spots
**C** Uniform temperature distribution
**D** High catalyst loss by default
**E** Limited mixing efficiency
**F** Requires no fluidization control

**CHW 6.32 Which fluidization regime operates at gas velocities just above the minimum fluidization velocity, forming distinct gas bubbles?**

**A** Circulating fluidized bed (CFB)
**B** Turbulent fluidized bed
**C** Fast fluidized bed
**D** Bubbling fluidized bed (BFB)
**E** Spouted bed
**F** Fixed bed

**CHW 6.33 What is a key advantage of CFBs over BFBs?**

**A** Lower gas velocities reduce catalyst attrition
**B** Better temperature stability for sensitive reactions
**C** Reduced need for catalyst regeneration
**D** Lower energy consumption

**E** Enhanced mass and heat transfer due to vigorous particle movement
**F** No need for cyclone separators

**CHW 6.34** **What is a major challenge in CFBs compared to BFBs?**
**A** Poor gas–solid contact
**B** Higher catalyst retention
**C** Lower conversion rates
**D** Limited temperature control
**E** Reduced particle entrainment
**F** More complex solid handling systems

**CHW 6.35** **What is a common cause of catalyst deactivation in FBRs?**
**A** Coking or carbon deposition
**B** Excessive plug flow characteristics
**C** Low-temperature operation
**D** Reduced gas–solid contact time
**E** Low catalyst retention
**F** Minimal particle movement

**CHW 6.36** **Which of the following statements most accurately describes the "External Mass MT" step in heterogeneous catalysis?**
**A** It is a process primarily determined by the catalyst's intrinsic activity and active site structure.
**B** It is the release of product molecules from the active sites, making them available for new reactants.
**C** It refers to the chemical transformation of adsorbed reactants into products on the catalyst surface.
**D** It involves the diffusion of reactants from the catalyst's external surface into its internal porous structure.
**E** It is the process of reactant molecules forming strong chemical bonds with the catalyst's active sites.
**F** It is the transportation of reactant molecules from the bulk fluid to the external surface of the catalyst particle.

**CHW 6.37** **Which of the following phenomena are considered significant hurdles that affect the performance of heterogeneous catalysis, as described in the chapter?**
**A** Product degradation and active site regeneration
**B** External mass transfer, adsorption, and internal mass transfer
**C** Catalyst preparation and reactor scaling
**D** Catalyst regeneration and product purification
**E** Heat-transfer limitations and catalyst sintering
**F** Chemical equilibrium and reaction kinetics

**CHW 6.38** **Regarding the adsorption step in heterogeneous catalysis, which statement accurately describes the characteristics of physisorption and chemisorption?**

**A** The nature and density of active sites are critical for physisorption capacity, while surface morphology primarily affects chemisorption selectivity.
**B** Physisorption is usually reversible and involves weak van der Waals forces, whereas chemisorption forms chemical bonds, resulting in stronger and sometimes irreversible binding.
**C** Chemisorption is enhanced by high temperatures and is crucial for freeing active sites, while physisorption occurs at very low temperatures.
**D** Physisorption involves the formation of strong chemical bonds, while chemisorption is characterized by weak van der Waals forces.
**E** Neither physisorption nor chemisorption has a significant impact on the overall rate of a heterogeneously catalyzed reaction.
**F** Both physisorption and chemisorption are typically irreversible processes that lead to catalyst poisoning.

**CHW 6.39** **Which of the following scenarios is an example of the rate-limiting step concept in heterogeneous catalysis?**
**A** A system where external MT is extremely rapid, causing the intrinsic surface reaction to become the slowest and therefore the rate-determining step.
**B** A process where the porosity of the catalyst is so high that internal MT is never a factor.
**C** When the overall reaction rate remains constant despite changes in fluid flow rate or turbulence.
**D** A situation where product molecules bind too weakly to the catalyst surface, leading to rapid desorption and reduced effectiveness.
**E** A reaction where increasing the reactor temperature significantly increases the overall reaction rate, regardless of catalyst type.
**F** A catalyst that requires frequent regeneration due to the accumulation of impurities from the fluid medium.

**CHW 6.40** **In the context of PBR flow regimes, what is a primary drawback of operating in the turbulent flow regime ($Re > 100$)?**
**A** Poor mixing and reliance on diffusion for MT
**B** Significant MT limitations
**C** Reduced driving force for reactions due to blurred concentration profiles
**D** Increased risk of catalyst attrition and higher pressure drops
**E** Potential for hot spots and catalyst deactivation due to radial temperature gradients
**F** Lower reaction rates and potential bypassing of catalyst particles

**CHW 6.41** **Which of the following strategies is mentioned as a method to mitigate radial and/or axial dispersion effects in a PBR?**
**A** Employing inlet designs such as perforated plates or spray nozzles
**B** Increasing the reactor's $L/D$ ratio to minimize axial mixing
**C** Implementing vibratory or slurry packing techniques
**D** Operating in the turbulent flow regime (Reynolds number $Re > 100$)
**E** Installing baffles or redistributors at intervals along the reactor
**F** Adjusting particle size to achieve larger particles and reduced pressure drops

**CHW 6.42 What factor primarily influences residence time ($\tau$) in a PBR?**

A Catalyst particle shape
B Reactor volume and flow rate
C Fluid viscosity
D Catalyst color
E Bed porosity
F Reaction temperature

**CHW 6.43 What is a potential drawback of longer residence times in PBRs for VOC oxidation?**

A Reduced pressure drop
B Increased catalyst activity
C Lower conversion rates
D Decreased energy costs
E Improved radial heat transfer
F Formation of hot spots and side reactions

**CHW 6.44 What is the primary effect of increasing residence time ($\tau$) in a PBR?**

A Decreases pressure drop due to reduced flow resistance
B Enhances conversion by allowing more reactant–catalyst interaction
C Reduces catalyst activity due to shorter exposure times
D Minimizes thermal effects by lowering reactor temperature
E Increases flow rate ($v_0$) to improve throughput
F Leads to higher axial dispersion, reducing plug flow behavior

**CHW 6.45 Which of the following strategies helps mitigate the trade-off between high residence time (i.e., conversion) and pressure drop in a PBR?**

A Using larger catalyst particles to reduce surface area
B Increasing the L/D ratio to maximize axial dispersion
C Implementing segmented beds with intermediate gas injection
D Operating at very high flow rates to minimize $\tau$
E Avoiding structured catalysts to maintain random packing
F Reducing reactor volume to lower capital costs

**CHW 6.46 Which of the following best explains the trade-off when selecting catalyst particle size for a PBR?**

A Larger particles increase surface area and enhance MT
B Smaller particles increase surface area but can cause high-pressure drops
C Smaller particles improve flow uniformity and reduce pressure drop
D Larger particles reduce catalytic activity but improve heat transfer
E All particle sizes offer the same conversion efficiency in PBRs
F Only cylindrical particles are used to minimize flow resistance

**CHW 6.47 In the context of PBRs, which catalyst configuration is most likely to balance pressure drop and catalytic efficiency?**

A Use of very fine catalyst powders throughout the bed
B Use of spherical particles exclusively for all applications

**C** Use of larger particles at the reactor outlet and smaller at the inlet
**D** Use of porous particles and graded beds (large inlet, small outlet)
**E** Packing the bed with high-density metal pellets
**F** Avoiding any variation in catalyst particle size to maintain uniformity

**CHW 6.48 Which of the following best describes the influence of the reactor $L/D$ ratio in PBR design?**
**A** High $L/D$ ratios always minimize pressure drop and maximize throughput
**B** Low $L/D$ ratios ensure complete conversion regardless of reaction kinetics
**C** High $L/D$ ratios promote plug flow behavior and higher conversion but increase pressure drop
**D** $L/D$ ratio has minimal effect on temperature gradients in exothermic reactions
**E** Radial-flow reactors use high $L/D$ ratios to improve residence time
**F** An $L/D$ ratio of 2–5 is typically used to reduce pressure drop in high-pressure liquid-phase reactions

**CHW 6.49 Why is a radial-flow PBR preferred over an axial-flow design in the hydrogenation of benzene to cyclohexane?**
**A** It increases the L/D ratio to enhance plug flow behavior.
**B** It ensures complete conversion without temperature rise.
**C** It reduces flow resistance and improves heat removal to prevent catalyst deactivation.
**D** It maximizes the residence time by increasing reactor volume.
**E** It eliminates the need for external cooling systems.
**F** It enhances mixing by promoting axial dispersion.

**CHW 6.50 What is a key limitation of using external jackets for heat management in PBRs?**
**A** They have limited heat-transfer rates, leading to radial temperature gradients.
**B** They require complex reactor modifications.
**C** They are ineffective for all types of catalytic reactions.
**D** They increase the pressure drop across the reactor bed.
**E** They cannot be used for endothermic reactions.
**F** They significantly reduce the catalyst loading capacity.

**CHW 6.51 Why are internal cooling/heating tubes preferred over external jackets for highly exothermic reactions like FTS?**
**A** They eliminate the need for preheating feed streams.
**B** They allow for higher catalyst loading than external jackets.
**C** They reduce the overall reactor size and cost.
**D** They prevent all side reactions from occurring.
**E** They eliminate the need for inert diluent particles.
**F** They provide better temperature uniformity by placing heat-exchange surfaces closer to the reaction zone.

**CHW 6.52 Which of the following is not a typical advantage of heterogeneous PBRs?**
**A** High catalyst utilization and surface area efficiency

**B** Capability to operate under high-pressure and temperature
**C** Continuous operation with consistent product quality
**D** Excellent temperature control in highly exothermic reactions
**E** Simple construction and reduced mechanical complexity
**F** Suitability for high-throughput industrial processes

**CHW 6.53 Which operational issue in PBRs can be mitigated using structured packing or internal baffles?**
**A** Catalyst deactivation due to coking
**B** Excessive pressure drop in long reactors
**C** Low reaction selectivity at high temperature
**D** Decrease in catalyst surface area over time
**E** Channeling and maldistribution of flow
**F** Formation of unwanted byproducts due to side reactions

**CHW 6.54 What strategy is used in FTS within PBRs to prevent catalyst deactivation by hot spots?**
**A** Operating the reactor at low pressure to reduce heat generation
**B** Using monolithic ceramic supports with large channels
**C** Increasing the catalyst particle size to slow reaction rates
**D** Periodically flushing the bed with nitrogen to remove coke
**E** Operating under adiabatic conditions to maximize conversion
**F** Employing multi-tubular reactors with efficient heat removal systems

**CHW 6.55 Which of the following correctly describes the sequence of fluidization regimes in an FBR as the gas velocity increases?**
**A** Fixed bed → Bubbling fluidization → Minimum fluidization → Turbulent fluidization → Pneumatic transport → Fast fluidization
**B** *Minimum* fluidization → Bubbling fluidization → Turbulent fluidization → Fast fluidization → Pneumatic transport
**C** Bubbling fluidization → Minimum fluidization → Turbulent fluidization → Pneumatic transport → Fixed bed
**D** *Minimum* fluidization → Turbulent fluidization → Bubbling fluidization → Pneumatic transport → Fast fluidization
**E** Fixed bed → Fast fluidization → Minimum fluidization → Turbulent fluidization → Bubbling fluidization
**F** Minimum fluidization → Bubbling fluidization → Pneumatic transport → Fast fluidization → Turbulent fluidization

**CHW 6.56 Which of the following design features or operational strategies is essential for ensuring stable and efficient performance in an FBR?**
**A** Use of back-pressure valves to maintain laminar flow in the bed
**B** Narrow reactor diameter to limit bed expansion
**C** Avoiding cyclones to reduce pressure drop
**D** Operating below the minimum fluidization velocity
**E** Distributor plates for uniform gas distribution
**F** Eliminating freeboard space to reduce reactor volume

**CHW 6.57 Which of the following statements about heterogeneous FBRs is correct?**

**A** FBRs are limited to low-viscosity gas feeds due to poor adaptability to various feedstocks.
**B** Catalyst replacement in FBRs requires a complete shutdown of the reactor.
**C** Complex hydrodynamics in FBRs promote ideal plug flow behavior.
**D** FBRs offer superior heat and MT due to efficient mixing.
**E** Particle entrainment is minimal in FBRs and does not require additional separation systems.
**F** FBRs are unsuitable for large-scale operations due to limited scalability.

**CHW 6.58 In gas-phase polymerization FBRs (e.g., polyethylene production), why is catalyst recovery and regeneration impractical, unlike in FCC?**

**A** Catalyst sintering occurs too rapidly during polymerization.
**B** Catalysts become permanently encapsulated within growing polymer granules.
**C** Static charge accumulation degrades catalyst activity irreversibly.
**D** High-pressure conditions (10–30 bar) prevent safe regeneration.
**E** Coke deposition poisons catalysts faster than in FCC units.
**F** Polymerization catalysts lack thermal stability for regeneration.

**CHW 6.59 What is a key advantage of using FBRs over fixed-bed reactors in methanol synthesis from syngas?**

**A** Lower operating pressures (5–10 bar), reducing equipment costs
**B** Elimination of sulfur-poisoning risks for $Cu/ZnO/Al_2O_3$ catalysts
**C** Superior temperature control, minimizing hot spots and improving yield
**D** Ability to regenerate catalysts continuously without shutdowns
**E** Reduced need for syngas purification steps
**F** Compatibility with non-copper-based catalyst systems

**CHW 6.60 Which of the following statements correctly describes internal or external MT in catalytic reactors?**

**A** External MT refers to the diffusion of reactants through catalyst pores.
**B** Internal MT is unaffected by particle size or pore structure.
**C** High flow rates increase boundary layer thickness, hindering external MT.
**D** High pore tortuosity leads to lower effective diffusivity inside the catalyst.
**E** Internal MT is more important than kinetics in all catalytic reactions.
**F** External MT does not influence the overall reaction rate.

**CHW 6.61 In PBRs, which of the following design or operational choices could increase internal diffusion limitations?**

**A** Decreasing catalyst particle size
**B** Increasing pore diffusivity
**C** Operating at higher flow velocity
**D** Installing static mixers in the reactor bed
**E** Operating at high temperatures to increase diffusivity
**F** Using catalyst particles with higher tortuosity and larger size

**CHW 6.62** **Which of the following best describes a trade-off resulting from smaller catalyst particles in FBRs?**

**A** Smaller particles enhance both external and internal MT without drawbacks.
**B** Finer particles reduce entrainment risk and improve flow control.
**C** Attrition eliminates the need for particle size control.
**D** Larger particles are preferred in FBRs due to better mixing characteristics.
**E** Smaller particles expose more surface area but may cause external MT limitations.
**F** Attrition improves catalyst selectivity by narrowing pore size distribution.

**CHW 6.63** **Calculation of the Reactant Concentration at the Center of a Catalyst Spherical Pellet and the Effectiveness Factor, $\eta_C$**

**Given that:** $R = 0.01$ cm, the rate constant, $k = 0.1\ \mathrm{s}^{-1}$, effective diffusivity, $D_{eff} = 1 \times 10^{-5}\ \mathrm{cm^2/s}$. Calculate the surface concentration at the center of the spherical pellet. The surface concentration, $C_s$, is 2.0 mol/L. Evaluate the concentration-based effectiveness factor ($\eta_C$).

**CHW 6.64** **Calculation of the Temperature at the Center of a Catalyst Spherical Pellet and the Effectiveness Factor, $\eta_T$**

**Given that:** $R = 0.01$ cm, the rate constant, $k = 0.1\ \mathrm{s}^{-1}$, the heat of reaction, $\Delta H_r = -80{,}000$ J/mol, surface concentration, $C_s$, is 2.0 mol/L, surface temperature, $T_s = 500$ K, center temperature, $T_C = 520$ K, and thermal conductivity, $k_T = 0.5\ \mathrm{W/(m \cdot K)}$.

Calculate the Thiele modulus based on the temperature of the spherical pellet. Evaluate T at the center. Evaluate $T$-based effectiveness factor ($\eta_T$).

**CHW 6.65** **Thiele Modulus, Effectiveness Factor, and the Concentration Profile Inside the Catalyst Particle**

Consider a first-order liquid-phase reaction $A \rightarrow B$ occurs in a spherical catalyst particle of radius $R = 0.006$ m with an effective diffusivity of $D_{eff} = 1.8 \times 10^{-9}\ \mathrm{m^2/s}$ and a rate constant $k = 0.02\ \mathrm{s}^{-1}$. The reactant concentration at the catalyst surface is $C_s = 100\ \mathrm{mol/m^3}$. Calculate the Thiele modulus, effectiveness factor, and the concentration profile of $A$ inside the catalyst particle as a $f(r)$, using MATLAB.

**CHW 6.66** **Which of the following is not a correct mitigation strategy for internal MT limitations?**

**A** Use hierarchical pore structures combining micro- and macropores
**B** Apply high diffusivity materials like MOFs and carbon nanotubes
**C** Increase turbulence using baffles
**D** Modify pore architecture to improve intraparticle transport
**E** Choose catalyst supports that enhance effective diffusivity
**F** Apply mesoporous coatings on microporous catalysts like zeolites

**CHW 6.67** **Which of the following statements correctly contrasts heterogeneous catalysis in PBRs and FBRs?**

**A** PBRs have better catalyst regeneration than FBRs.
**B** FBRs typically suffer from high-pressure drop compared to PBRs.
**C** PBRs allow for easier catalyst replacement than FBRs.

**D** FBRs operate under ideal plug flow patterns, unlike PBRs.
**E** Both PBRs and FBRs are unsuitable for large-scale applications.
**F** FBRs offer better heat transfer and MT rates than PBRs.

**CHW 6.68** **Which of the following statements about the minimum fluidization point (Umf) is correct?**
**A** Umf is the velocity at which gas begins to bubble in the bed.
**B** At Umf, gas velocity equals the speed of sound in the bed material.
**C** Umf is reached when the pressure drop equals the weight of the fluid only.
**D** At Umf, the drag force balances the weight of solid particles per unit area.
**E** Umf occurs when particle-particle contact is at its maximum.
**F** Umf defines the point at which the void fraction drops to a minimum.

**CHW 6.69** **Which of the following correctly matches a fluidization regime with its key feature?**
**A** Bubbling Fluidized Bed ↔ Large gas bubbles rise through the bed
**B** Fixed Bed ↔ Gas bubbles form, causing vigorous mixing
**C** Particulate Fluidization ↔ Slugs of gas displace the entire bed
**D** Slugging Bed ↔ Fine particles circulate smoothly and uniformly
**E** Circulating Fluidized Bed ↔ Pressure drop increases linearly with velocity
**F** Static Bed ↔ Gas velocity exceeds the terminal settling velocity of particles

**CHW 6.70** **What happens when the superficial gas velocity increases beyond the bubbling regime in an FBR?**
**A** The bed enters the fixed bed regime with improved porosity.
**B** Slugging occurs, especially in wide-diameter columns.
**C** Particulate fluidization is restored with enhanced mixing.
**D** The bed settles into a static configuration with low void fraction.
**E** CFB forms with high solids recirculation.
**F** Catalyst deactivation accelerates due to increased pressure drop.

**CHW 6.71** **Fluid and Flow Properties of Fluidized Bed Using Water and Anthracite Packing Particles**
Using MATLAB code (Appendix D | **FB.m**), find water flow properties, the minimum fluidization velocity, the terminal velocity, the MT rate based on 50% of the terminal velocity, the pressure drop per unit height of FB ($\Delta P/h$), and the MT coefficient, kg [m/s]. Use anthracite packing particles with an average particle size of 300 μm. Refer to Table 6.6 for anthracite properties. Use an operating temperature of 300 K and a pressure of 10 atm. You are supposed to know the density and viscosity of water at room temperature.

Consider a second-order liquid-phase reaction: $A \rightarrow$ Products, complete the job by entering the reaction order $n = 2$, the reaction rate constant $= 0.5\ \text{L}^2/(\text{kg cat}\cdot\text{mol}\cdot\text{s})$, the fluidize bed diameter $= 0.8$ m, the bed height $= 3$ m, the interfacial area per volume $a = 200\ \text{m}^2/\text{m}^3$, and the inlet molar flow rate of $A$, $F_{A0} = 5$ mol/s.

Report the mass flow rate of $A$, kg/s, the entire column pressure drop, kPa, and both the inlet bulk and inlet surface concentrations of $A$.

**CHW 6.72 Fluid and Flow Properties of Fluidized Bed Using Air and Anthracite Packing Particles**

Using MATLAB code (Appendix D | **FB.m**), find water flow properties, the minimum fluidization velocity, the terminal velocity, the MT rate based on 50% of the terminal velocity, the pressure drop per unit height of FB ($\Delta P/h$), and the MT coefficient, kg [m/s]. Use anthracite packing particles with an average particle size of 300 μm. Refer to Table 6.6 for anthracite properties. Use an operating temperature of 300 K and a pressure of 10 atm.

Consider a second-order gas-phase reaction: $A \rightarrow$ Products, complete the job by entering the reaction order $n = 2$, the reaction rate constant $= 1.5\ \mathrm{L^2/(kg\ cat{\cdot}mol{\cdot}s)}$, the fluidize bed diameter $= 1.5$ m, the bed height $= 3$ m, the interfacial area per volume $a = 200\ \mathrm{m^2/m^3}$, and the inlet molar flow rate of $A$, $F_{A0} = 5$ mol/s.

Report the mass flow rate of $A$, kg/s, the entire column pressure drop, kPa, and both the inlet bulk and inlet surface concentrations of $A$.

**CHW 6.73 Calculation of Conversion in an Isothermal FBR**

Consider a gas-phase second-order reaction: $A \rightarrow B$, which takes place at the surface of the catalyst pellets or particles that form the packing material of FBR. Given the bed volume, area, and void fraction, check to see if the reaction is MT limited by comparing the intrinsic kinetic rate versus the MT rate for reactant A in an isothermal FBR. Calculate the final conversion, X, using both the manual (analytic) and MATLAB-code approaches using the **HFB_Or_X.m** file.

**Kinetic Rate Constant:** $k' = 0.3\frac{\mathrm{L^2}}{\mathrm{kg\ catalyst \cdot mol{\cdot}s}}$ | **Reaction order:** $n = 2$

**Bulk Concentration:** $C_{Abulk} = 0.2$ mol/L | **Surface concentration:** $C_{Asurface} = 0.1$ mol/L

**Particle Density:** $\rho_p = 1{,}200\ \mathrm{kg/m^3}$ | **Bed voidage fraction:** $\varepsilon_{FB} = 0.56$

**Volume of the Bed:** $V_{FB} = \frac{\pi D_{FB}^2}{4} \times H_{FB} = \frac{\pi * 0.8^2}{4}\mathrm{m^2} \times H_{FB} = 0.503\ \mathrm{m^2} * 3\ \mathrm{m} = 1.5\ \mathrm{m^3}$

**MT Coefficient:** kg $= 0.01$ m/s | **Interfacial area per unit volume:** $a = 200\ \mathrm{m^2/m^3}$

**Initial Molar Flow Rate of $A$:** $F_{A0} = 1$ mol/s

**CHW 6.74 Calculation of an Isothermal FBR Bed Height for a Given $X$**

Consider a gas-phase second-order reaction: $A \rightarrow B$, which takes place at the surface of the catalyst pellets or particles that form the packing material of FBR. Using a unit height, tube area, and void fraction, check to see if the reaction is MT limited by comparing the intrinsic kinetic rate versus MT rate for reactant A in an isothermal FBR. Calculate the height of the bed ($H_{FB}$) for a final conversion, $X = 0.95$, using both the manual (analytic) and MATLAB-code approach using the **HFB_Or_X.m** file.

**Kinetic Rate Constant:** $k' = 0.6\frac{\mathrm{L^2}}{\mathrm{kg\ catalyst \cdot mol \cdot s}}$ | **Reaction Order:** $n = 2$

**Bulk Concentration:** $C_{Abulk} = 0.2$ mol/L | **Surface Concentration:** $C_{Asurface} = 0.18$ mol/L

**Particle Density:** $\rho_p = 1{,}200\ \mathrm{kg/m^3}$ | **Bed Voidage Fraction:** $\varepsilon_{FB} = 0.56$

**Volume of the Bed:** $V_{FB} = A_{FB}H_{FB} = \frac{\pi D_{FB}^2}{4} \times H_{FB} = \frac{\pi * 0.8^2}{4}\mathrm{m^2} = 0.503\ H_{FB}\ \mathrm{m^3}$

**MT Coefficient:** kg $= 0.01$ m/s | **Interfacial Area Per Unit Volume:** $a = 80\ \mathrm{m^2 m^3}$

**Initial Molar Flow Rate of $A$:** $F_{A0} = 1$ mol/s

# Appendix A

## $\Delta G^0$, $\Delta H^0$, and $K_{eq}$ Calculations

Here is a human-friendly algorithm of the MATLAB code of Appendix | 1.A Section:

1) **File Handling and Data Loading:**
   a) The program first tries to load thermodynamic data from an Excel file (**Therm.xlsx**).
   b) If the Excel file isn't available, it attempts to load the data from a MATLAB file (**ThermProp.mat**).
   c) If both attempts fail, the program exits with an error message.
2) **Reactants Selection:**
   a) The user is asked to input the number of reactants in the chemical reaction.
   b) For each reactant, the user can search the database using either:
      i) **Compound Name** (e.g., "Water").
      ii) **Chemical Formula** (e.g., "$H_2O$").
   c) Matching entries from the database are displayed, and the user selects the desired compound by its index.
   d) The user also inputs the number of moles for each selected reactant.
3) **Products Selection:**
   a) Similar to the reactants, the user is asked to input the number of products in the reaction.
   b) The search and selection process for products works in the same way as for reactants.
   c) The user specifies the number of moles for each selected product.
4) **Reaction Property Calculations:**
   The program calculates the molar **Gibbs free energy change** ($\Delta G_{rxn}$) and molar **enthalpy change** ($\Delta H_{rxn}$) for the reaction using the following steps:
   a) For reactants and products, their respective thermodynamic properties (molar Gibbs free energy and molar enthalpy of formation) are multiplied by their moles and summed.
   b) The difference between the sums for products and reactants gives the respective reaction properties.
5) **Temperature and Equilibrium Constant:**
   a) The user inputs the temperature in Kelvin.
   b) The program calculates the equilibrium constant ($K_{ex}$) using the relationship:
      i) $$K_{eq} = e^{-\frac{\Delta G^0}{RT}}$$
6) **Results Display:**
   a) The selected reactants and products, along with their moles, are displayed.
   b) The calculated reaction properties $\Delta G^0$, $\Delta H^0$, and $K_{eq}$ are presented.

## A.1 MATLAB (DelG_Keq.m) Code for Calculation of $\Delta G^0$, $\Delta H^0$, and $K_{eq}$ for a Chemical Reaction

```
clear variables;
try
  % Attempt to read the Excel file first
  PureComp = readtable('Therm.xlsx', 'Format', 'auto');
catch ME
  % If reading the Excel file fails, try loading the MAT file
  disp('Failed to read Excel file; it will continue with the saved MATLAB file.');
  try
    load('ThermProp.mat', 'PureComp');
    disp('Loaded PureComp from MAT file.');
  catch ME2
    disp('Failed to read MAT file.');
    return;
  end
end

myvar = PureComp.Name;
myvar2 = PureComp.Formula;


% Prompt user to select reactants and their moles
numReactants = str2double(input('Enter the number of reactants: ', 's'));
muReact = zeros(numReactants, 1);
ReactIdx = zeros(numReactants, 1);

for i = 1:numReactants
  SearchKey = str2double(input('Input "1" to search by Compound Name, "2" to search by Formula: ', 's'));
  if SearchKey == 1
    CompName = input('Input the name of the Compound: ', 's');
    matchStart = regexpi(myvar, CompName);
  elseif SearchKey == 2
    FormName = input('Input the chemical formula: ', 's');
    matchStart = regexpi(myvar2, FormName);
  else
    disp('You did not enter 1 or 2');
    return;
  end

  % Display matching results
  tempindex = 1;
  flag = 0;
  myindx = [];
  for idx = 1:size(matchStart, 1)
    if ~isempty(matchStart{idx})
      myindx(tempindex, 1) = idx;
      tempindex = tempindex + 1;
      flag = 1;
    end
  end

  if flag == 0
    disp('No Entry was Found!; Try Again.');
    return;
  end

  for idex2 = 1:size(myindx, 1)
    value = myindx(idex2, 1);
    disp([num2str(value), ': ', PureComp.Name{value}, ' (', PureComp.Formula{value}, ')']);
```

```
    end

    ReactIdx(i) = str2double(input(['Select the index for reactant ', num2str(i), ': '], 's'));
    muReact(i) = str2double(input(['Enter the number of moles for reactant ', PureComp.Name{ReactIdx(i)}, ': '], 's'));
end

% Prompt user to select products and their moles
numProducts = str2double(input('Enter the number of products: ', 's'));
muProd = zeros(numProducts, 1);
ProdIdx = zeros(numProducts, 1);

for j = 1:numProducts
    SearchKey = str2double(input('Input "1" to search by Compound Name, "2" to search by Formula: ', 's'));
    if SearchKey == 1
        CompName = input('Input the name of the Compound: ', 's');
        matchStart = regexpi(myvar, CompName);
    elseif SearchKey == 2
        FormName = input('Input the chemical formula: ', 's');
        matchStart = regexpi(myvar2, FormName);
    else
        disp('You did not enter 1 or 2');
        return;
    end

    % Display matching results
    tempindex = 1;
    flag = 0;
    myindx = [];
    for idx = 1:size(matchStart, 1)
        if ~isempty(matchStart{idx})
            myindx(tempindex, 1) = idx;
            tempindex = tempindex + 1;
            flag = 1;
        end
    end

    if flag == 0
        disp('No Entry was Found!; Try Again.');
        return;
    end

    for idex2 = 1:size(myindx, 1)
        value = myindx(idex2, 1);
        disp([num2str(value), ': ', PureComp.Name{value}, ' (', PureComp.Formula{value}, ')']);
    end

    ProdIdx(j) = str2double(input(['Select the index for product ', num2str(j), ': '], 's'));
    muProd(j) = str2double(input(['Enter the number of moles for product ', PureComp.Name{ProdIdx(j)}, ': '], 's'));
end

% Calculate reaction properties
DelGProducts = sum(muProd .* PureComp.DelGf_kJ_molinv(ProdIdx));
DelGReactants = sum(muReact .* PureComp.DelGf_kJ_molinv(ReactIdx));
DelGRxn = DelGProducts - DelGReactants;

DelHProducts = sum(muProd .* PureComp.DelHf_kJ_molinv(ProdIdx));
DelHReactants = sum(muReact .* PureComp.DelHf_kJ_molinv(ReactIdx));
DelHRxn = DelHProducts - DelHReactants;

% Prompt user for temperature
T = str2double(input('Enter the temperature in Kelvin (K): ', 's'));
```

(Continued)

**(Continued)**

```
% Calculate equilibrium constant
R = 8.314; % Universal gas constant in J/(mol*K)
Keq = exp(-DelGRxn * 1000 / (R * T));

% Display results
disp('Selected Reactants:');
for i = 1:numReactants
   disp([num2str(muReact(i)), ' mol of ', PureComp.Name{ReactIdx(i)}, ' (', PureComp.Formula{ReactIdx(i)}, ')']);
end

disp('Selected Products:');
for j = 1:numProducts
   disp([num2str(muProd(j)), ' mol of ', PureComp.Name{ProdIdx(j)}, ' (', PureComp.Formula{ProdIdx(j)}, ')']);
end

disp(['Temperature (K): ', num2str(T)]);
disp(['Molar Gibbs free energy of reaction (kJ/mol): ', num2str(DelGRxn)]);
disp(['Molar heat of reaction @ 298 K, (kJ/mol): ', num2str(DelHRxn)]);
disp(['Equilibrium constant (dimensionless): ', num2str(Keq)]);
```

# Appendix B

# Heat Capacity Data, Volume Expansion Effect, and the Discretized PFR Approximation Method

## B.1 MATLAB Code for Liquid Heat Capacity as $f(T)$ for Different Inorganic and Organic Compounds

MATLAB code (**CpLiq_T.m**) for presenting the value of constants appearing in the heat capacity polynomial equation for the selected compound, out of 335 compounds:

$$C_P^{Liq}\left[\frac{\text{J}}{\text{kmol}\cdot\text{K}}\right] = C_1 + C_2T + C_3T^2 + C_4T^3 + C_5T^4$$

Or:

$$C_{Pm}^{Liq}\left[\frac{\text{J}}{\text{kg}\cdot\text{K}}\right] = C_{1m} + C_{2m}T + C_{3m}T^2 + C_{4m}T^3 + C_{5m}T^4$$

In addition, it shows the calculated heat capacity, both per unit mole and per unit mass of the selected compound at a given $T$, such that $T$ lies between $T_{min}$ and $T_{max}$, expressed in $K$. $T_{min}$ and $T_{max}$ are shown ahead of the user.

```
clear variables;
try
  % Attempt to read the Excel file first
  LiqCp = readtable('Cp_Liq_T.xlsx', 'Format', 'auto');
catch ME
  % If reading the Excel file fails, try loading the MAT file
  disp('Failed to read Excel file; it will continue with the saved MATLAB file.');
  try
    load('CpLiqT.mat', 'LiqCp');
    disp('Loaded PureComp from MAT file.');
  catch ME2
    disp('Failed to read MAT file.');
    return;
  end
end
try
  % Display search options to the user
  fprintf('Search the database by: \n');
  fprintf('1. Name \n');
  fprintf('2. Formula \n');
  fprintf('3. CAS_NO \n');
  choice = input('Enter your choice (1, 2, or 3): ');
  % Validate user input
  while ~ismember(choice, [1, 2, 3])
```

*(Continued)*

**(Continued)**

```
    fprintf('Invalid choice. Please enter 1, 2, or 3.');
    choice = input('Enter your choice (1, 2, or 3): ');
  end
  % Determine the field to search based on user choice
  switch choice
    case 1
      field = 'Name';
      searchKeyword = lower(input('Enter a search keyword for Name: ', 's'));
      LiqCp.Name = lower(LiqCp.Name);
    case 2
      field = 'Formula';
      searchKeyword = lower(input('Enter a search keyword for Formula: ', 's'));
      LiqCp.Formula = lower(LiqCp.Formula);
    case 3
      field = 'CAS_NO';
      searchKeyword = lower(input('Enter a search keyword for CAS_NO: ', 's'));
      LiqCp.CAS_NO = lower(LiqCp.CAS_NO);
  end
  % Find rows containing the search keyword in the selected field
  matchRows = find(contains(LiqCp.(field), searchKeyword));
  if isempty(matchRows)
    disp('No entries found matching the search keyword.');
    return;
  else
    % Display matching compounds and row numbers (looping approach)
    disp('Matching entries:');
    for i = 1:length(matchRows)
      disp([' - ', LiqCp.Name{matchRows(i)}, ' (Row ', num2str(matchRows(i)), ')']);
      disp([' - ', upper(LiqCp.Formula{matchRows(i)}), ' (Row ', num2str(matchRows(i)),')']);
    end
  end
  rowNumber = input('Enter the row number for the compound: ');
  % Check for NAN input
  if isnan(rowNumber)
    disp('Terminating program');
    return;
  end
% Get relevant data for the selected compound
  selectedCompound = LiqCp(rowNumber,:);
  Tmin = selectedCompound.Tmin;
  Tmax = selectedCompound.Tmax;
  C1=selectedCompound.C1;
  C1m=C1/selectedCompound.MW;
  C2=selectedCompound.C2;
  C2m=C2/selectedCompound.MW;
  C3=selectedCompound.C3;
  C3m=C3/selectedCompound.MW;
  C4=selectedCompound.C4;
  C4m=C4/selectedCompound.MW;
  C5=selectedCompound.C5;
  C5m=C5/selectedCompound.MW;
  T_min=Tmin-273.15;% Tmin in oC;
  T_max=Tmax-273.15; % Tmax in  oC
  CmpName=selectedCompound.Name;
  CmpFormula=selectedCompound.Formula;
  disp(['Liquid Name: ', CmpName]);
```

```
    disp(['Liquid Formula: ', upper(CmpFormula)]);
    disp(['Tmin: ', num2str(Tmin), ' K']);
    disp(['Tmin: ', num2str(T_min), ' oC']);
    disp(['Tmax: ', num2str(Tmax), ' K']);
    disp(['Tmax: ', num2str(T_max), ' oC']);
    disp(['Cp= ', 'C1+C2*T+C3*T^2+C4*T^3+C5*T^4',' [Molar Basis]']);
    disp(['C1: ', num2str(C1), ' J/(kmol.K)']);
    disp(['C2: ', num2str(C2), ' J/(kmol.K^2)']);
    disp(['C3: ', num2str(C3), ' J/(kmol.K^3)']);
    disp(['C4: ', num2str(C4), ' J/(kmol.K^4)']);
    disp(['C5: ', num2str(C5), ' J/(kmol.K^5)']);
    disp(['Cpm= ', 'C1m+C2m*T+C3m*T^2+C4m*T^3+C5m*T^4',' [Mass Basis]']);
    disp(['C1m: ', num2str(C1m), ' J/(kg.K)']);
    disp(['C2m: ', num2str(C2m), ' J/(kg.K^2)']);
    disp(['C3m: ', num2str(C3m), ' J/(kg.K^3)']);
    disp(['C4m: ', num2str(C4m), ' J/(kg.K^4)']);
    disp(['C5m: ', num2str(C5m), ' J/(kg.K^5)']);
    % User input for T
    T = input('Enter T (must be between Tmin and Tmax, expressed in K): ');
    while T < Tmin
      T = input('Invalid T. Enter a value higher than or equal to Tmin: ');
    end
    while T > Tmax
      T = input('Invalid T. Enter a value lower than or equal to Tmax: ');
    end
    Cp=C1+C2*T+C3*T^2+C4*T^3+C5*T^4;
    Cpm=C1m+C2m*T+C3m*T^2+C4m*T^3+C5m*T^4;
    Cp2=Cp/selectedCompound.MW;
    Cp3=Cp/1000;
    Cp4=Cp2/1000;
    disp(['Liquid Name: ', CmpName]);
    disp(['Liquid Formula: ', upper(CmpFormula)]);
    disp(['At T = ', num2str(T), ' K,', ' Cp=', num2str(Cp), ' J/(kmol.K)']);
    disp("The mass-based heat capacity is the same by curve fitting" + ...
      " and by conversion from mole to mass basis:");
    disp(['At T = ', num2str(T), ' K,', ' Cpm=', num2str(Cpm), ' J/(kg.K)']);
    disp(['At T = ', num2str(T), ' K,', ' Cp2=', num2str(Cp2), ' J/(kg.K)']);
    disp(['At T = ', num2str(T), ' K,', ' Cp3=', num2str(Cp3), ' J/(mol.K)']);
    disp(['At T = ', num2str(T), ' K,', ' Cp4=', num2str(Cp4), ' J/(g.K)']);
catch ME
    disp('An error occurred:');
    disp(ME.message);
end
```

## B.2 MATLAB Code for Ideal Gas Heat Capacity as $f(T)$ for Different Inorganic and Organic Compounds

MATLAB code (**CpGas_T.m**) for presenting the value of constants appearing in the heat capacity hyperbolic equation for the selected compound, out of 341 compounds:

$$C_P^{IG}\left[\frac{\mathrm{J}}{\mathrm{kmol\cdot K}}\right] = C_1 + C_2\left[\frac{\frac{C_3}{T}}{\sin h\left(\frac{C_3}{T}\right)}\right]^2 + C_4\left[\frac{\frac{C_5}{T}}{\cos h\left(\frac{C_5}{T}\right)}\right]^2$$

Or:

$$C_P^{IG}\left[\frac{\text{J}}{\text{kg}\cdot\text{K}}\right] = C_{1m} + C_{2m}\left[\frac{\frac{C_3}{T}}{\sin h\left(\frac{C_3}{T}\right)}\right]^2 + C_{4m}\left[\frac{\frac{C_5}{T}}{\cos h\left(\frac{C_5}{T}\right)}\right]^2$$

In addition, it shows the calculated heat capacity, both per unit mole and per unit mass of the selected compound at a given $T$ such that $T$ lies between $T_{min}$ and $T_{max}$, expressed in $K$. $T_{min}$ and $T_{max}$ are shown ahead to the user.

```
clear variables;
try
  % Attempt to read the Excel file first
  GasCp = readtable('Cp_Gas_T.xlsx', 'Format', 'auto');
catch ME
  % If reading the Excel file fails, try loading the MAT file
  disp('Failed to read Excel file; it will continue with the saved MATLAB file.');
  try
    load('CpGasT.mat', 'GasCp');
    disp('Loaded PureComp from MAT file.');
  catch ME2
    disp('Failed to read MAT file.');
    return;
  end
end

try
  % Display search options to the user
  fprintf('Search the database by: \n');
  fprintf('1. Name \n');
  fprintf('2. Formula \n');
  fprintf('3. CAS_NO \n');
  choice = input('Enter your choice (1, 2, or 3): ');

  % Validate user input
  while ~ismember(choice, [1, 2, 3])
    fprintf('Invalid choice. Please enter 1, 2, or 3.');
    choice = input('Enter your choice (1, 2, or 3): ');
  end

  % Determine the field to search based on user choice
  switch choice
    case 1
      field = 'Name';
      searchKeyword = lower(input('Enter a search keyword for Name: ', 's'));
      GasCp.Name = lower(GasCp.Name);
    case 2
      field = 'Formula';
      searchKeyword = lower(input('Enter a search keyword for Formula: ', 's'));
      GasCp.Formula = lower(GasCp.Formula);
    case 3
      field = 'CAS_NO';
      searchKeyword = lower(input('Enter a search keyword for CAS_NO: ', 's'));
      GasCp.CAS_NO = lower(GasCp.CAS_NO);
  end

  % Find rows containing the search keyword in the selected field
  matchRows = find(contains(GasCp.(field), searchKeyword));
```

```
  if isempty(matchRows)
    disp('No entries found matching the search keyword.');
    return;
  else
    % Display matching compounds and row numbers (looping approach)
    disp('Matching entries:');
    for i = 1:length(matchRows)
      disp([' - ', GasCp.Name{matchRows(i)}, ' (Row ', num2str(matchRows(i)), ')']);
      disp([' - ', upper(GasCp.Formula{matchRows(i)}), ' (Row ', num2str(matchRows(i)),')']);
    end
  end

  rowNumber = input('Enter the row number for the compound: ');

  % Check for NAN input
  if isnan(rowNumber)
    disp('Terminating program');
    return;
  end

% Get relevant data for the selected compound
  selectedCompound = GasCp(rowNumber,:);
  Tmin = selectedCompound.Tmin;
  Tmax = selectedCompound.Tmax;
  C1=1e5*(selectedCompound.C1);
  C1m=C1/selectedCompound.MW;
  C2=1e5*(selectedCompound.C2);
  C2m=C2/selectedCompound.MW;
  C3=1e3*(selectedCompound.C3);
  C3m=C3;
  C4=1e5*(selectedCompound.C4);
  C4m=C4/selectedCompound.MW;
  C5=selectedCompound.C5;
  C5m=C5;
  T_min=Tmin-273.15;% Tmin in oC;
  T_max=Tmax-273.15; % Tmax in  oC
  CmpName=selectedCompound.Name;
  CmpFormula=selectedCompound.Formula;
  disp(['Gas Name: ', CmpName]);
  disp(['Gas Formula: ', upper(CmpFormula)]);
  disp(['Tmin: ', num2str(Tmin), ' K']);
  disp(['Tmin: ', num2str(T_min), ' oC']);
  disp(['Tmax: ', num2str(Tmax), ' K']);
  disp(['Tmax: ', num2str(T_max), ' oC']);
  disp(['Cp= ', 'C1+C2*((C3/T)/sinh(C3/T))^2 + C4*((C5/T)/cosh(C5/T))^2',...
    '[Molar Basis]']);
  disp(['C1= ', num2str(C1), ' % J/(kmol.K)']);
  disp(['C2= ', num2str(C2), ' % J/(kmol.K)']);
  disp(['C3= ', num2str(C3), ' % K']);
  disp(['C4= ', num2str(C4), ' % J/(kmol.K)']);
  disp(['C5= ', num2str(C5), ' % K']);
  % Prompt user for input
choice = input(['Enter your choice (A, B, C, D) for compound to be used in ' ...
  'BR_T_CA.m file: '], 's');

% Convert input to uppercase for case-insensitivity
choice = upper(choice);
```

(Continued)

**(Continued)**

```
% Switch-case to handle the selection
switch choice
  case 'A'
    disp('You selected case A.');
    disp('The following C1A,C2A,C3A,C4A, and C5A parameters can be pasted in BR_T_CA4.m file')
  disp(['C1A= ', num2str(C1/1000.), ' ;% J/(mol.K)']);
  disp(['C2A= ', num2str(C2/1000.), ' ;% J/(mol.K)']);
  disp(['C3A= ', num2str(C3), ' ;% K']);
  disp(['C4A= ', num2str(C4/1000.), ' ;% J/(mol.K)']);
  disp(['C5A= ', num2str(C5), ' ;% K']);
  disp(['CpAm= ', 'C1Am+C2Am*((C3Am/T)/sinh(C3Am/T))^2 + C4Am*((C5Am/T)/cosh(C5Am/T))^2',...
   '[Mass Basis]']);
  disp(['C1Am= ', num2str(C1m), ' ;% J/(kg.K)']);
  disp(['C2Am= ', num2str(C2m), ' ;% J/(kg.K)']);
  disp(['C3Am= ', num2str(C3m), ' ;% K']);
  disp(['C4Am= ', num2str(C4m), ' ;% J/(kg.K)']);
  disp(['C5Am= ', num2str(C5m), ' ;% K)']);

  case 'B'
    disp('You selected case B.');
    disp('The following C1B,C2B,C3B,C4B, and C5B parameters can be pasted in BR_T_CA4.m file')
  disp(['C1B= ', num2str(C1/1000.), ' ;% J/(mol.K)']);
  disp(['C2B= ', num2str(C2/1000.), ' ;% J/(mol.K)']);
  disp(['C3B= ', num2str(C3), ' ;% K']);
  disp(['C4B= ', num2str(C4/1000.), ' ;% J/(mol.K)']);
  disp(['C5B= ', num2str(C5), ' ;% K']);
  disp(['CpBm= ', 'C1Bm+C2Bm*((C3Bm/T)/sinh(C3Bm/T))^2 + C4Bm*((C5Bm/T)/cosh(C5Bm/T))^2',...
   '[Mass Basis]']);
  disp(['C1Bm= ', num2str(C1m), ' ;% J/(kg.K)']);
  disp(['C2Bm= ', num2str(C2m), ' ;% J/(kg.K)']);
  disp(['C3Bm= ', num2str(C3m), ' ;% K']);
  disp(['C4Bm= ', num2str(C4m), ' ;% J/(kg.K)']);
  disp(['C5Bm= ', num2str(C5m), ' ;% K)']);

  case 'C'
    disp('You selected case C.');
    disp('The following C1C,C2C,C3C,C4C, and C5C parameters can be pasted in BR_T_CA4.m file')
  disp(['C1C= ', num2str(C1/1000.), ' ;% J/(mol.K)']);
  disp(['C2C= ', num2str(C2/1000.), ' ;% J/(mol.K)']);
  disp(['C3C= ', num2str(C3), ' ;% K']);
  disp(['C4C= ', num2str(C4/1000.), ' ;% J/(mol.K)']);
  disp(['C5C= ', num2str(C5), ' ;% K']);
  disp(['CpCm= ', 'C1Cm+C2Cm*((C3Cm/T)/sinh(C3Cm/T))^2 + C4Cm*((C5Cm/T)/cosh(C5Cm/T))^2',...
   '[Mass Basis]']);
  disp(['C1Cm= ', num2str(C1m), ' ;% J/(kg.K)']);
  disp(['C2Cm= ', num2str(C2m), ' ;% J/(kg.K)']);
  disp(['C3Cm= ', num2str(C3m), ' ;% K']);
  disp(['C4Cm= ', num2str(C4m), ' ;% J/(kg.K)']);
  disp(['C5Cm= ', num2str(C5m), ' ;% K)']);

  case 'D'
    disp('You selected case D.');
    disp('The following C1D,C2D,C3D,C4D, and C5D parameters can be pasted in BR_T_CA4.m file')
  disp(['C1D= ', num2str(C1/1000.), ' ;% J/(mol.K)']);
  disp(['C2D= ', num2str(C2/1000.), ' ;% J/(mol.K)']);
  disp(['C3D= ', num2str(C3), ' ;% K']);
  disp(['C4D= ', num2str(C4/1000.), ' ;% J/(mol.K)']);
```

```
    disp(['C5D= ', num2str(C5), ' ;% K']);
    disp(['CpDm= ', 'C1Dm+C2Dm*((C3Dm/T)/sinh(C3Dm/T))^2 + C4Dm*((C5Dm/T)/cosh(C5Dm/T))^2',...
     '[Mass Basis]']);
    disp(['C1Dm= ', num2str(C1m), ' ;% J/(kg.K)']);
    disp(['C2Dm= ', num2str(C2m), ' ;% J/(kg.K)']);
    disp(['C3Dm= ', num2str(C3m), ' ;% K']);
    disp(['C4Dm= ', num2str(C4m), ' ;% J/(kg.K)']);
    disp(['C5Dm= ', num2str(C5m), ' ;% K)']);

    otherwise
     disp('Invalid choice. Terminating the program.');
     return; % Terminate the program
  end

    % User input for T
    T = input('Enter T (must be between Tmin and Tmax, expressed in K): ');
    while T < Tmin
      T = input('Invalid T. Enter a value higher than or equal to Tmin: ');
    end

    while T > Tmax
      T = input('Invalid T. Enter a value lower than or equal to Tmax: ');
    end
    Cp=C1+C2*((C3/T)/sinh(C3/T))^2 + C4*((C5/T)/cosh(C5/T))^2;
    Cpm=C1m+C2m*((C3/T)/sinh(C3/T))^2 + C4m*((C5/T)/cosh(C5/T))^2;

    Cp2=Cp/selectedCompound.MW;
    Cp3=Cp/1000;
    Cp4=Cp2/1000;
    disp(['Gas Name: ', CmpName]);
    disp(['Gas Formula: ', upper(CmpFormula)]);
    disp(['At T = ', num2str(T), ' K,', ' Cp=', num2str(Cp), ' J/(kmol.K)']);
    disp("The mass-based heat capacity is the same by curve fitting" + ...
      " and by conversion from mole to mass basis:");
    disp(['At T = ', num2str(T), ' K,', ' Cpm=', num2str(Cpm), ' J/(kg.K)']);
    disp(['At T = ', num2str(T), ' K,', ' Cp2=', num2str(Cp2), ' J/(kg.K)']);
    disp(['At T = ', num2str(T), ' K,', ' Cp3=', num2str(Cp3), ' J/(mol.K)']);
    disp(['At T = ', num2str(T), ' K,', ' Cp4=', num2str(Cp4), ' J/(g.K)']);
  catch ME
    disp('An error occurred:');
    disp(ME.message);
  end
```

## B.3 Algorithm for Solution of Gas-Phase Reaction in an Adiabatic BR with Heat Capacity *f*(*T*)

The following algorithmic steps will be transformed into MATLAB code, which solves a set of algebraic and differential equations describing an adiabatic gas-phase chemical reaction process in a BR with both variable heat capacity and heat of reaction.

The reaction is the irreversible gas-phase first-order isomerization of cyclopropane into propene (or propylene):

$$CH_2CH_2CH_2(g) \rightarrow CH_2 = CH - CH_3(g)$$

Which can be put in the short form:

$$A(g) \rightarrow B(g)$$

$$-r_A = 1.116 \times 10^{17} \frac{1}{\text{min}} * e^{-\frac{273000}{8.314}\left(\frac{1}{T}\right)} C_A$$

The ideal gas heat capacity for $A$ is given by:

$$C_{P,A}^{IG}\left[\frac{\text{J}}{\text{mol}\cdot\text{K}}\right] = C1A + C2A\left[\frac{\frac{C3A}{T}}{\sin h\left(\frac{C3A}{T}\right)}\right]^2 + C4A\left[\frac{\frac{C5A}{T}}{\cos h\left(\frac{C5A}{T}\right)}\right]^2$$

Specifically for cyclopropane, we have:

$$C_{P,A}^{IG}\left[\frac{\text{J}}{\text{mol}\cdot\text{K}}\right] = 33.8 + 168.94\left[\frac{\frac{1613.5}{T}}{\sin h\left(\frac{1613.5}{T}\right)}\right]^2 + 117.68\left[\frac{\frac{722.8}{T}}{\cos h\left(\frac{722.8}{T}\right)}\right]^2$$

On the other hand, the ideal gas heat capacity for B is given by:

$$C_{P,B}^{IG}\left[\frac{\text{J}}{\text{mol}\cdot\text{K}}\right] = C1B + C2B\left[\frac{\frac{C3B}{T}}{\sin h\left(\frac{C3B}{T}\right)}\right]^2 + C4B\left[\frac{\frac{C5B}{T}}{\cos h\left(\frac{C5B}{T}\right)}\right]^2$$

Specifically for propene (propylene), we have:

$$C_{P,B}^{IG}\left[\frac{\text{J}}{\text{mol}\cdot\text{K}}\right] = 43.852 + 150.6\left[\frac{\frac{1398.8}{T}}{\sin h\left(\frac{1398.8}{T}\right)}\right]^2 + 74.754\left[\frac{\frac{616.46}{T}}{\cos h\left(\frac{616.46}{T}\right)}\right]^2$$

Notice that the ideal gas heat capacity data can be retrieved via running the **CpGas_T.m** file and searching first for cyclopropane and second for propene (or propylene). The retrieved constants: C1A, C2A, C3A, C4A, C5A, C1B, C2B, C3B, C4B, and C5B, can be copied from the results and pasted into the **BR_T_CA4.m** file.

$$C_{V,A}^{IG}(T) = C_{P,A}^{IG}(T) - R = C_{P,A}^{IG}(T) - 8.314$$

$$C_{V,B}^{IG}(T) = C_{P,B}^{IG}(T) - R = C_{P,B}^{IG}(T) - 8.314$$

$$\Delta H_r(T) = -33{,}000 + \int_{298}^{T} \left(-C_{V,A}^{IG}(T) + C_{V,B}^{IG}(T)\right) dT$$

$$V_{BR} = 1{,}500$$

$$n = C_{Total} * V_{BR}$$

$$C_{A0} = \frac{\text{P0 atm} * 1}{0.08206 \frac{\text{atm}\cdot\text{L}}{\text{mol}\cdot\text{K}} * T0\ \text{K}}$$

$$C_{Total} = C_{A0}$$

$$x_A = \frac{n_A}{n} = \frac{C_A}{C_{Total}}$$

$$C_V^{mix} = x_A C_V^A + (1 - x_A) C_V^B$$

$$\frac{dx_A}{dt} = \frac{1}{C_{Total}}\frac{dC_A}{dt}$$

$$\frac{dC_V^{mix}}{dt} = x_A\frac{dC_V^A}{dt} + C_V^A\frac{dx_A}{dt} + (1 - x_A)\frac{dC_V^B}{dt} - \frac{dx_A}{dt}C_V^B$$

Use the chain rule differentiation to express:

$$\frac{dC_V^A}{dt} = \left(\frac{dC_V^A}{dT} * \frac{dT}{dt}\right)$$

and

$\frac{dC_V^B}{dt} = \left(\frac{dC_V^B}{dT} * \frac{dT}{dt}\right)$, we have:

$$\frac{dC_V^{mix}}{dt} = x_A\left(\frac{dC_V^A}{dT} * \frac{dT}{dt}\right) + C_V^A\frac{dx_A}{dt} + (1 - x_A)\left(\frac{dC_V^B}{dT} * \frac{dT}{dt}\right) - \frac{dx_A}{dt}C_V^B$$

$$\frac{dx_A}{dt} = \frac{1}{C_{Total}}\frac{dC_A}{dt}$$

$$\frac{dC_A}{dt} = -k_1C_A$$

$$X_A = 1 - \frac{C_A}{C_{A0}}$$

$$nC_V^{mix}\frac{d(T)}{dt} + nT\frac{d\left(C_V^{mix}\right)}{dt} = -k_1C_AV_{BR}\left(\Delta H_r^0\right)$$

We need to integrate both $C_A$ and $T$ from:

**Initial Condition:** $t_0 = 0$; $C_A = C_{A0}$, $T = T_0$
**To:** Final condition: $t = t_f$

One plot for $C_A$, $T$, and $X_A$ as $f(time)$ will be reproduced

Moreover, the temperature derivative of both $\frac{dC_V^A}{dT}$ and $\frac{dC_V^B}{dT}$ are given as separate functions at the end of the main code.

## B.4 MATLAB Code (BR_T_CA4_Gas.m) for Describing a Gas-Phase Reaction in an Adiabatic BR with $C_p = f(T)$

```
clear variables;
T0 = input('Enter initial temperature (T0) in K: '); % Initial temperature
P0 = input('Enter initial pressure (P0) in atm: '); % Initial pressure
tf = input('Enter the final time (tf) in unit of time: '); % Final time
R = 8.314; % Universal gas constant, J/(mol*K)
V_BR = input('Enter the BR Volume, (L): ');
%V_BR = 1500; % Batch reactor volume, L
C_A0 = (P0 * 1) / (0.08206 * T0); % Initial concentration of A, mol/L
C_Total = C_A0; % Total concentration, mol/L
n = C_Total * V_BR; % Total moles in the reactor, mol
% Initial conditions
t0 = 0; % Initial time, unit of time
```

(Continued)

**(Continued)**

```
y0 = [C_A0, T0]; % Initial state vector [C_A, T]
% Define constants for heat capacities
C1A= 33.8 ;% J/(mol.K)
C2A= 168.94 ;% J/(mol.K)
C3A= 1613.5 ;% K
C4A= 117.68 ;% J/(mol.K)
C5A= 722.8 ;% K
C1B= 43.852 ;% J/(mol.K)
C2B= 150.6 ;% J/(mol.K)
C3B= 1398.8 ;% K
C4B= 74.754 ;% J/(mol.K)
C5B= 616.46 ;% K
% Define the system of ODEs
odefun = @(t, y) system_of_equations(t, y, R, C_Total, V_BR, n, C1A, C2A, C3A, C4A, C5A, C1B, C2B, C3B, C4B, C5B);
% Solve the ODE system
options = odeset('RelTol', 1e-8, 'AbsTol', 1e-10);
[t, y] = ode15s(odefun, [t0 tf], y0, options);
% Extract results
C_A = y(:, 1); % Concentration of A
T = y(:, 2); % Temperature
X_A = 1 - C_A / C_A0; % Conversion of A
% Plot results
figure;
subplot(3, 1, 1);
plot(t, C_A, 'b-', 'LineWidth', 1.5);
grid on;
xlabel('Time (min)');
ylabel('C_A (mol/L)');
title('Concentration of A vs. Time');
ax = gca;
ax.FontSize = 16;
ax.FontWeight = 'bold';
subplot(3, 1, 2);
plot(t, T, 'r-', 'LineWidth', 1.5);
grid on;
xlabel('Time (min)');
ylabel('T (K)');
title('Temperature vs. Time');
ax = gca;
ax.FontSize = 16;
ax.FontWeight = 'bold';
subplot(3, 1, 3);
plot(t, X_A, 'g-', 'LineWidth', 1.5);
grid on;
xlabel('Time (min)');
ylabel('X_A');
title('Conversion of A vs. Time');
ax = gca;
ax.FontSize = 16;
ax.FontWeight = 'bold';
% System of equations
function dydt = system_of_equations(~, y, R, C_Total, V_BR, n, C1A, C2A, C3A, C4A, C5A, C1B, C2B, C3B, C4B, C5B)
  % State variables
  C_A = y(1); % Concentration of A, mol/L
  T = y(2); % Temperature, K
  % Reaction rate constant
  k1 = 1.116e17 * exp(-273000 / (R * T));
  % Mole fraction of A
  x_A = C_A / C_Total;
  % Heat capacities
  AsinhTerm = C2A * ((C3A / T) ./ sinh(C3A / T)).^2;
```

```
    AcoshTerm = C4A * ((C5A / T) ./ cosh(C5A / T)).^2;
    C_P_A = C1A + AsinhTerm + AcoshTerm;
    BsinhTerm = C2B * ((C3B / T) ./ sinh(C3B / T)).^2;
    BcoshTerm = C4B * ((C5B / T) ./ cosh(C5B / T)).^2;
    C_P_B = C1B + BsinhTerm + BcoshTerm;
    C_V_A = C_P_A - R; % Constant volume heat capacity of A at T
    C_V_B = C_P_B - R; % Constant volume heat capacity of B at T
    % Mixture heat capacity
    C_V_mix = x_A * C_V_A + (1 - x_A) * C_V_B;

    % Reaction enthalpy
    DeltaH_r = -33000 + integral(@(T) (-C_P_A + C_P_B), 298, T, 'ArrayValued', true);
    % Reaction rate
    dC_A_dt = -k1 * C_A; % Reaction rate for species A
    dx_A_dt = (1 / C_Total) * dC_A_dt; % Conversion rate
    % Derivative of mixture heat capacity
    dC_V_mix_dt = ...
      x_A * derivative_C_V_A(T, C2A, C3A, C4A, C5A) * 0 + ...
      C_V_A * dx_A_dt + ...
      (1 - x_A) * derivative_C_V_B(T, C2B, C3B, C4B, C5B) * 0 – ...
      dx_A_dt * C_V_B;
    % Solve for dT_dt using the energy balance
    dT_dt = (-k1 * C_A * V_BR * DeltaH_r - n * T * dC_V_mix_dt) / (n * C_V_mix);
    dydt = [dC_A_dt; dT_dt];
end

function dC_V_A_dt = derivative_C_V_A(T, C2A, C3A, C4A, C5A)
    % Derivative of C_V^A with respect to T
    dAsinhTerm_dt = -2 * C2A * (C3A^2 / T^3) .* ...
      ((1 ./ sinh(C3A / T)).^2 - (C3A / T) .* cosh(C3A / T) ./ sinh(C3A / T).^3);
    dAcoshTerm_dt = -2 * C4A * (C5A^2 / T^3) .* ...
      ((1 ./ cosh(C5A / T)).^2 + (C5A / T) .* sinh(C5A / T) ./ cosh(C5A / T).^3);
    % Derivative of C_P_A
    dC_P_A_dt = dAsinhTerm_dt + dAcoshTerm_dt;
    % Derivative of C_V_A
    dC_V_A_dt = dC_P_A_dt;
end

function dC_V_B_dt = derivative_C_V_B(T, C2B, C3B, C4B, C5B)
    % Derivative of C_V^B with respect to T
    dBsinhTerm_dt = -2 * C2B * (C3B^2 / T^3) .* ...
      ((1 ./ sinh(C3B / T)).^2 - (C3B / T) .* cosh(C3B / T) ./ sinh(C3B / T).^3);
    dBcoshTerm_dt = -2 * C4B * (C5B^2 / T^3) .* ...
      ((1 ./ cosh(C3B / T)).^2 + (C5B / T) .* sinh(C5B / T) ./ cosh(C5B / T).^3);
    % Derivative of C_P_B
    dC_P_B_dt = dBsinhTerm_dt + dBcoshTerm_dt;
    % Derivative of C_V_B
    dC_V_B_dt = dC_P_B_dt;
end
```

## B.5 MATLAB Code (CSTR_T_CA_Gas_ode15i.m) for Describing a Gas-Phase HI Formation in a Nonadiabatic CSTR with $C_p = f(T)$, Using ode15i Solver

```
clear; close all;
Rgas = 0.08206; % (atm.L)/mol.K
R = 8.314; % J/(mol.K)
% Given Inputs
```

*(Continued)*

**(Continued)**

```
F0 = input('Enter the feed molar flow rate (F0) in mol/s: ');
%F0 = 0.5; % mol/s
T0 = input('Enter the feed temperature (T0) in K: ');
%T0 = 300; % K
P0 = input('Enter the feed pressure (P0) in atm: ');
%P0 = 50; % atm
v0 = (F0 * Rgas * T0) / P0;
YA0 = 0.5;
YB0 = 0.5;
CA0 = (YA0 * P0) / (Rgas * T0);
CB0 = (YB0 * P0) / (Rgas * T0);
FA0 = YA0 * F0;
FB0 = YB0 * F0;
VCSTR = input('Enter the CSTR Volume, (L): ');
%VCSTR = 2000; % L
D = input('Enter the BR diameter in m, (D): ');
%D=1.0;
Pzero = P0; % P(0)=10.0 atm at time zero.
Tzero = T0; % T(0)=300 K at time Zero.
% Initial conditions
y0 = [CA0; CB0; Tzero];
yp0 = [0; 0; 0]; % Initial derivatives set to zero
% Solve for consistent initial conditions
t0 = 0;
[y0, yp0] = decic(@(t, y, yp) dCSTR_implicit(t, y, yp, v0, CA0, CB0, P0, R, T0, VCSTR, F0, D), t0, y0, [], yp0, [], []);
% Time span
tspan = [0 800]; % seconds
options = odeset('RelTol', 1e-6, 'AbsTol', 1e-9);
% Solve DAEs using ode15i
[t, y] = ode15i(@(t, y, yp) dCSTR_implicit(t, y, yp, v0, CA0, CB0, P0, R, T0, VCSTR, F0, D), tspan, y0, yp0, options);
CA = y(:, 1);
CB = y(:, 2);
CC = 2 * (CA0 - CA);
T = y(:, 3);
P = 0.98 * P0;
% Molar fractions
YA =CA./(CA+CB+CC);
YB =CB./(CA+CB+CC);
XA=(1/YA0).*(YA0-YA);
% Plot results
figure;
subplot(4,1,1); plot(t, CA, 'b', 'LineWidth', 1.5); xlabel('Time (s)'); ylabel('C_A (mol/L)'); title('Concentration of A');
subplot(4,1,2); plot(t, CC, 'r', 'LineWidth', 1.5); xlabel('Time (s)'); ylabel('C_C (mol/L)'); title('Concentration of C');
subplot(4,1,3); plot(t, T, 'g', 'LineWidth', 1.5); xlabel('Time (s)'); ylabel('T (K)'); title('Temperature');
subplot(4,1,4); plot(t, XA, 'm', 'LineWidth', 1.5); xlabel('Time (s)'); ylabel('X_A'); title('Conversion of A');
==========================================================
% Function to Define CSTR Implicit DAEs
% ========================================================
function res = dCSTR_implicit(~, y, yp, v0, CA0, CB0, P0, R, T0, VCSTR, F0, D)
   %CA = y(1); CB = y(2); T = y(3); P = y(4);
   dCA_dt = yp(1); dCB_dt = yp(2); dT_dt = yp(3);
   P=0.98*P0;
   CA = max(0, min(CA0, y(1))); % Ensure CA is within [0, CA0]
   CB = max(0, min(CB0, y(2))); % Ensure CB is within [0, CB0]
   CC=2*(CA0-CA);
    T = max(200, min(700, y(3)));
   %T = max(200, min(1000, y(3))); % Ensure T is within [200, 1000] K
   v = v0 .* (T ./ T0) .* (P0 ./ P);
   Ctot = CA + CB + CC;
   n = VCSTR * Ctot;
   % Calculate current mole fractions inside the CSTR
   YA =CA ./ Ctot;
   YB = CB ./ Ctot;
```

```
    YC = CC ./ Ctot;
    % Heat capacities (J/mol.K)
    CPA = 29.093 - (0.2e-3*T) + (0.4e-6*T^2) - (0.87e-9*T^3) + (0.0055e5/T^2);
    CPB = 37.4 - (0.0034e-3*T) + (0.18e-6*T^2) - (0.89e-9*T^3) - (0.245e5/T^2);
    CPC = 28.14 + (1.88e-3*T) - (6.954e-6*T^2) + (1.08e-9*T^3) - (0.205e5/T^2);
  % Average heat capacities
    Tavg=mean(T);
    CPAavg = 29.093 - (0.2e-3*Tavg) + (0.4e-6*Tavg^2) - (0.87e-9*Tavg^3) + (0.0055e5/Tavg^2);
    CPBavg = 37.4 - (0.0034e-3*Tavg) + (0.18e-6*Tavg^2) - (0.89e-9*Tavg^3) - (0.245e5/Tavg^2);
    CPCavg = 28.14 + (1.88e-3*Tavg) - (6.954e-6*Tavg^2) + (1.08e-9*Tavg^3) - (0.205e5/Tavg^2);
    CVAavg = CPAavg - R;
    CVBavg = CPBavg - R;
    CVCavg = CPCavg - R;
    % MWavg = YA * 2 + YB * 253.8 + (YC) * 127.9;
    CVavg = YA * CVAavg + YB * CVBavg + (YC) * CVCavg;
    Cpmix = YA * CPA + YB * CPB + (YC) * CPC;
    CPA0 = 29.093 - (0.2e-3*T0) + (0.4e-6*T0^2) - (0.87e-9*T0^3) + (0.0055e5/T0^2);
    CPB0 = 37.4 - (0.0034e-3*T0) + (0.18e-6*T0^2) - (0.89e-9*T0^3) - (0.245e5/T0^2);
    CPC0 = 28.14 + (1.88e-3*T0) - (6.954e-6*T0^2) + (1.08e-9*T0^3) - (0.205e5/T0^2);
    Cpmix0 = YA * CPA0 + YB * CPB0 + (YC) * CPC0;
    % Reaction rate constant
    k = 2e11 * exp(-169900 / (R * T));
    r = k * CA * CB;
    % Heat exchange term
    AH=(4*VCSTR*0.001/D);
    T_switch = 685;
     Qh = 500 * AH * (685 - T);
     Qc = 400 * AH * (298 - T);
    transition = 1 / (1 + exp((T - T_switch))); % smooth switch
    Qexch = transition * Qh + (1 - transition) * Qc;
    % Enthalpy change of reaction (J/mol)
    dHr = -9400 + integral(@(T) (-CPA - CPB + 2*CPC), 298, T, 'ArrayValued', true);
      res = [
      (v0 * CA0 - v * CA) / VCSTR - r - dCA_dt;
      (v0 * CB0 - v * CB) / VCSTR - r - dCB_dt;
      (F0 * (Cpmix0*T0 - Cpmix*T) - r * VCSTR * dHr + Qexch) / (n * CVavg) - dT_dt;
       ];
end
```

## B.6 The Volume Expansion Effect Due to a Change in Total Number of Moles as a Result of a Gas-Phase Chemical Reaction

For simplicity, consider an ideal gas mixture where a gas-phase reaction takes place in a vessel. Apply the ideal gas law at two points or states: Initial and at any point as the reaction proceeds.

$$V_0 = \frac{n_0 R T_0}{P_0} \tag{A}$$

$$V = \frac{nRT}{P} \tag{B}$$

Divide Equation (B) by (A) side by side, we have:

$$\frac{V}{V_0} = \frac{\upsilon}{\upsilon_0} = \frac{\frac{nRT}{P}}{\frac{n_0 R T_0}{P_0}} = \left(\frac{n_0 + \Delta n}{n_0}\right)\left(\frac{T}{P}\right) * \left(\frac{P_0}{T_0}\right) = \left(\frac{n_0 + \Delta n}{n_0}\right)\left(\frac{T}{T_0}\right) * \left(\frac{P_0}{P}\right) \tag{C}$$

Notice that the volume ratio will be the same whether in absolute or rate form. Thus, we have:

$$\frac{V}{V_0} = \frac{\upsilon}{\upsilon_0} = \left(\frac{n}{n_0}\right)\left(\frac{T}{P}\right) * \left(\frac{P_0}{T_0}\right) = \left(\frac{n_0 + \Delta n}{n_0}\right)\left(\frac{T}{T_0}\right) * \left(\frac{P_0}{P}\right) \tag{D}$$

Let us express the number of moles of the limiting reactant in terms of the molar extent of reaction, $\xi$:

$$\frac{V}{V_0} = \frac{v}{v_0} = \left(\frac{n_0 + \Delta n * \xi}{n_0}\right)\left(\frac{T}{P}\right) * \left(\frac{P_0}{T_0}\right) \tag{E}$$

Simplify, we have:

$$\frac{V}{V_0} = \frac{v}{v_0} = \left(1 + \frac{\Delta n * \xi}{n_0}\right)\left(\frac{T}{P}\right) * \left(\frac{P_0}{T_0}\right) \tag{F}$$

For the limiting reactant, $A$, we have:

$$\xi = \frac{n_A - n_{A0}}{\nu_A} = \frac{N_A - N_{A0}}{\nu_A}\ (Fogler) \tag{G}$$

Replace $\xi$, we have:

$$\frac{V}{V_0} = \frac{v}{v_0} = \left(1 + \frac{\Delta n * (n_A - n_{A0})}{n_0 \nu_A}\right)\left(\frac{T}{P}\right) * \left(\frac{P_0}{T_0}\right) = \left(1 + \frac{\Delta n * (F_A - F_{A0})}{F_0 \nu_A}\right)\left(\frac{T}{T_0}\right) * \left(\frac{P_0}{P}\right) \tag{H}$$

Factor out $n_{A0}$, we have:

$$\frac{V}{V_0} = \frac{v}{v_0} = \left(1 + \frac{\Delta n * n_{A0}\left(\frac{n_A}{n_{A0}} - 1\right)}{n_0 \nu_A}\right)\left(\frac{T}{P}\right) * \left(\frac{P_0}{T_0}\right) = \frac{V}{V_0} = \frac{v}{v_0} \tag{I}$$

Recall that: $Y_{A0} = \frac{n_{A0}}{n_0}$, Eq. (I) becomes:

$$\frac{V}{V_0} = \left(1 + \frac{\Delta n * Y_{A0}\left(\frac{n_A}{n_{A0}} - 1\right)}{\nu_A}\right)\left(\frac{T}{P}\right) * \left(\frac{P_0}{T_0}\right) = \left(1 + \frac{\Delta n * Y_{A0}\left(\frac{F_A}{F_{A0}} - 1\right)}{\nu_A}\right)\left(\frac{T}{T_0}\right) * \left(\frac{P_0}{P}\right) \tag{J}$$

Notice that $\nu_A = -|\nu_A|$, we have:

$$\frac{V}{V_0} = \frac{v}{v_0} = \left(1 + \frac{\Delta n * Y_{A0}\left(\frac{n_A}{n_{A0}} - 1\right)}{-|\nu_A|}\right)\left(\frac{T}{P}\right) * \left(\frac{P_0}{T_0}\right)$$
$$= \left(1 + \frac{\Delta n * Y_{A0}\left(\frac{F_A}{F_{A0}} - 1\right)}{-|\nu_A|}\right)\left(\frac{T}{T_0}\right) * \left(\frac{P_0}{P}\right) \tag{K}$$

To get rid of the minus sign, we multiply the bracketed terms by −1 and rearrange. We have:

$$\frac{V}{V_0} = \frac{v}{v_0} = \left(1 + \frac{\Delta n * Y_{A0}\left(1 - \frac{n_A}{n_{A0}}\right)}{|\nu_A|}\right)\left(\frac{T}{P}\right) * \left(\frac{P_0}{T_0}\right)$$
$$= \left(1 + \frac{\Delta n * Y_{A0}\left(1 - \frac{F_A}{F_{A0}}\right)}{|\nu_A|}\right)\left(\frac{T}{T_0}\right) * \left(\frac{P_0}{P}\right) \tag{L}$$

Equation (L) is synonymous with Fogler's approach 1 where he expressed the volume expansion, due to a chemical reaction, in terms of a parameter ε and conversion $X_A$:

$$\varepsilon X_A = \delta Y_{A0} X_A = \left( \frac{\Delta n * Y_{A0}\left(1 - \frac{F_A}{F_{A0}}\right)}{|\nu_A|} \right) = \left( \frac{\Delta n * Y_{A0}\left(1 - \frac{N_A}{N_{A0}}\right)}{|\nu_A|} \right) \tag{M}$$

where;

$$\delta = \left( \frac{\Delta n}{|\nu_A|} \right) \tag{N}$$

$$\varepsilon = \delta Y_{A0} = \left( \frac{\Delta n * Y_{A0}}{|\nu_A|} \right) \tag{O}$$

Let us take two cases where $\Delta n$ is either positive or negative.

**The First Case:**

$$A + B \Longrightarrow C$$

Let us say that the initial mole fraction of $A$ is $Y_{A0}$. $0 < Y_{A0} \leq 1$. Of course, as the reaction proceeds, $Y_A$ drops from $Y_{A0}$ down to zero. $0 < Y_A \leq Y_{A0}$.

For the above reaction, we can express the conversion in terms of $Y_A$ as:

$$X_A = \frac{1}{Y_{A0}}(Y_{A0} - Y_A) \tag{P}$$

Or:

$$X_A Y_{A0} = (Y_{A0} - Y_A) \tag{Q}$$

Let us check at the two extreme points:

At the inlet of the reactor:

$$X_A = 0 \text{ and } Y_A = Y_{A0}$$

$X_A = \frac{1}{Y_{A0}}(Y_{A0} - Y_A) = \frac{1}{Y_{A0}}(Y_{A0} - Y_{A0}) = 0$. Hence, it satisfies the inlet condition.

At the outlet of the reactor:

$$X_A = 1 \text{ and } Y_A = 0$$

$X_A = \frac{1}{Y_{A0}}(Y_{A0} - Y_A) = \frac{1}{Y_{A0}}(Y_{A0} - 0) = 1$. Hence, it satisfies the outlet condition.

For the above reaction:

$$\Delta n = 1 - 1 - 1 = -1.$$

$$\left( 1 + \frac{\Delta n * Y_{A0}\left(1 - \frac{n_A}{n_{A0}}\right)}{|\nu_A|} \right) = \left( 1 + \frac{-1 * Y_{A0} * X_A}{1} \right) = (1 - Y_{A0} * X_A)$$

Using Eq. (Q), we have:

$$(1 - Y_{A0} * X_A) = 1 - (Y_{A0} - Y_A)$$

Equation (L) becomes:

$$\frac{V}{V_0} = \frac{v}{v_0} = (1 - (Y_{A0} - Y_A))\left(\frac{T}{P}\right) * \left(\frac{P_0}{T_0}\right) \tag{R}$$

Equation (R) picks up the effect of volume expansion via $Y_A$, $T$, and $P$, instead of $X_A$, $T$, and $P$, as in Fogler's approach.

**The Second Case:**

$$A \Longrightarrow B + C$$

For the above reaction:

$$\Delta n = 1 + 1 - 1 = 1.$$

$$\left(1 + \frac{\Delta n * Y_{A0}\left(1 - \frac{n_A}{n_{A0}}\right)}{|v_A|}\right) = \left(1 + \frac{1 * Y_{A0} * X_A}{1}\right) = (1 + Y_{A0} * X_A)$$

Using Eq. (Q), we have:

$$(1 + Y_{A0} * X_A) = 1 + (Y_{A0} - Y_A)$$

Equation (L) becomes:

$$\frac{V}{V_0} = \frac{v}{v_0} = (1 + (Y_{A0} - Y_A))\left(\frac{T}{P}\right) * \left(\frac{P_0}{T_0}\right) \tag{S}$$

Again, Eq. (S) picks up the effect of volume expansion via $Y_A$, $T$, and $P$, instead of $X_A$, $T$, and $P$, as in Fogler's approach.

## B.7 The Volume Expansion Effect in a PFR as a Result of a Gas-Phase Chemical Reaction

The definition of $v$ in a flow reactor is simply given as:

$$\left(\frac{dV}{dt}\right) = Volume\ flow\ rate = v = \frac{F_{tot}}{C_{tot}} = \frac{F_{tot}RT}{P} \tag{A}$$

Taking the natural logarithm of both sides of Eq. (A) and applying differentiation, we have:

$$d\ln(v) = d\ln(F_{tot}) + d\ln(R) + d\ln(T) - d\ln(P) \tag{B}$$

Since $R$ terms are constant, then Eq. (B) reduces to:

$$d\ln(v) = d\ln(F_{tot}) + d\ln(T) - d\ln(P) \tag{C}$$

$$\frac{1}{v}\frac{dv}{dV} = \frac{1}{F_{tot}}\frac{dF_{tot}}{dV} + \frac{1}{T}\frac{dT}{dV} - \frac{1}{P}\frac{dP}{dV} \tag{D}$$

$$F_{tot} = F_{A0} * (1 + \varepsilon * X_A) \qquad dF_{tot} = F_{A0} * (\varepsilon * dX_A) \tag{E}$$

$$\frac{1}{v}\frac{dv}{dV} = \frac{\varepsilon}{(1 + \varepsilon * X_A)}\frac{dX_A}{dV} + \frac{1}{T}\frac{dT}{dV} - \frac{1}{P}\frac{dP}{dV} \tag{F}$$

Multiply both sides of Eq. (D) by $v$, we have:

$$\frac{dv}{dV} = \frac{\varepsilon v}{(1 + \varepsilon * X_A)}\frac{dX_A}{dV} + \frac{v}{T}\frac{dT}{dV} - \frac{v}{P}\frac{dP}{dV} \tag{G}$$

If we assume a 10% pressure drop along the tube, then

$$P = P_0\left(1 - 0.1\frac{z}{L}\right) \quad \rightarrow \quad \frac{dP}{dz} = -0.1\frac{P_0}{L} \tag{H}$$

Divide both sides by the cross-sectional area, $A$, of the tube, we have:

$$\frac{dP}{Adz} = -0.1\frac{P_0}{A.L} \quad \rightarrow \quad \frac{dP}{dV} = -0.1\frac{P_0}{V_{PFR}} \tag{I}$$

Substitute in Eq. (I) in Eq. (G), we have:

$$\frac{dv}{dV} = \frac{\varepsilon v}{(1+\varepsilon * X_A)}\frac{dX_A}{dV} + \frac{v}{T}\frac{dT}{dV} + \frac{v}{P} \times \frac{0.1 * P_0}{V_{PFr}} \tag{J}$$

$$C_{AR} = \frac{C_A}{C_{A0}} \tag{K}$$

$$X_A = \frac{(1 - C_{AR})}{(1 + \varepsilon C_{AR})} \tag{L}$$

$$\frac{dX_A}{dV} = \frac{(1+\varepsilon C_{AR})\left(-\frac{1}{C_{A0}}\frac{dC_A}{dV}\right) - (1 - C_{AR}) * \varepsilon * \frac{1}{C_{A0}}\frac{dC_A}{dV}}{(1+\varepsilon C_{AR})^2}$$

$$= -\frac{(1 + \varepsilon C_{AR} + \varepsilon - \varepsilon C_{AR})}{(1+\varepsilon C_{AR})^2}\frac{1}{C_{A0}}\frac{dC_A}{dV}$$

$$\frac{dX_A}{dV} = -\frac{(1+\varepsilon)}{(1+\varepsilon C_{AR})^2}\frac{1}{C_{A0}}\frac{dC_A}{dV} \tag{M}$$

For $\varepsilon \neq 0$, when we describe the steady-state condition in a PFR, the system of equations we deal with is highly interdependent, or implicitly coupled: Several of the derivatives depend on one another in a circular way, making the system nonlinear and algebraically tangled.

We have this structure:

1) $dT/dV = f(CA)$
2) $dXA/dV = f(dCA/dV)$
3) $dv/dV = f(XA, dXA/dV, dT/dV)$
4) $dCA/dV = f(v, dv/dV)$
5) $dy/dV = [dCA/dV; dT/dV]$

One option is to solve implicitly via iteration (fixed-point loop). In our ODE function, we do an internal iteration (e.g., 5–10 iterations) to break the circular dependence numerically.

## B.8 The Discretized PFR Approximation Method

Similar to my approach of slicing the volume into a relatively large number of sub-volumes and treating them as a series of perfectly mixed CSTRs to approximate a PFR, below is a "discretized PFR" or "PFR approximated by a series of CSTRs" model. It converts the partial differential equations (PDEs) that strictly govern a transient PFR into a system of ordinary differential equations (ODEs), which can then be solved by MATLAB's ODE solvers. **DPFRAM.m** code is shown below, which exploits the concept of discretized PFR approximation.

```
% Main Script: DPFRAM.m
clear; close all;
Rgas = 0.08206; % (atm.L)/mol.K
R = 8.314; % J/(mol.K)
% Given Inputs
F0 = input('Enter the feed molar flow rate (F0) in mol/s: ');
%F0 = 1; % mol/s (Total feed molar flow rate)
T0 = input('Enter the feed temperature (T0) in K: ');
%T0 = 300; % K (Feed temperature)
P0 = input('Enter the feed pressure (P0) in atm: ');
%P0 = 50; % atm (Feed pressure)
YA0 = input('Enter the feed mole fraction of A (YA0): ');
%YA0 = 0.5; % Feed mole fraction of A
YB0 = input('Enter the feed mole fraction of B (YB0): ');
%YB0 = 0.5; % Feed mole fraction of B
VPFR = input('Enter the PFR Volume, VPFR (L): ');
%VPFR = 3000; % Volume of PFR, L
D = input('Enter the PFR Diameter, D (m): ');
%D = 1.0; % PFR Diameter, m
% Calculate initial conditions based on feed
v0 = (F0 * Rgas * T0) / P0; % Initial volumetric flow rate (L/s)
CA0 = (YA0 * P0) / (Rgas * T0); % Initial concentration of A (mol/L)
CB0 = (YB0 * P0) / (Rgas * T0); % Initial concentration of B (mol/L)
% Initial state of the reactor at t=0 (concentrations and temperature)
% For a transient startup, it's common to assume the reactor initially contains
% material at feed conditions, or is empty (which requires different ICs and handling).
y0 = [CA0; CB0; T0]; % [Initial CA in reactor; Initial CB in reactor; Initial T in reactor]
% Time span for simulation
tf = input('Enter the final time tf, (unit of time): ');
%tf = 1300; % seconds
tspan = [0 tf];
% Solve ODEs using ode113
options = odeset('RelTol', 1e-6, 'AbsTol', 1e-9, 'NonNegative', [1 2 3]); % Ensure concentrations/T stay non-negative
[t, y] = ode113(@(t, y) dPFR_ode_function(t, y, v0, CA0, CB0, P0, R, T0, VPFR, D, F0), tspan, y0, options);
% Extract results from the solver output
CA = y(:, 1); % Concentration of A over time
CB = y(:, 2); % Concentration of B over time
CC = 2 * (CA0 - CA); % Concentration of C formed (based on consumption of A)
T = y(:, 3); % Temperature over time
% Calculate other relevant output variables for plotting
P = 0.98 * P0; % Assume constant pressure drop throughout reactor
v = v0 * (T ./ T0) .* (P0 ./ P); % Volumetric flow rate based on current T and P
FA = v .* CA; % Molar flow rate of A (mol/s)
FB = v .* CB; % Molar flow rate of B (mol/s)
FC = v .* CC; % Molar flow rate of C (mol/s)
% Total molar flow rate for mole fraction calculation
F_total = FA + FB + FC;
% Handle potential division by zero if F_total becomes very small
% Initialize with zeros to avoid issues with empty arrays if all F_total are small
YA = zeros(size(F_total));
YB = zeros(size(F_total));
YC = zeros(size(F_total));
valid_indices = F_total > 1e-10; % Only perform division where F_total is significant
YA(valid_indices) = FA(valid_indices) ./ F_total(valid_indices);
YB(valid_indices) = FB(valid_indices) ./ F_total(valid_indices);
YC(valid_indices) = FC(valid_indices) ./ F_total(valid_indices);
% Conversion of A
% XA = (Mols A reacted) / (Mols A fed)
% For a CSTR-like model, if CA0 is feed and CA is outlet concentration:
XA = (CA0 - CA) / CA0;
% Calculate Space Time (Tau)
Tau = VPFR / v0; % This is based on initial volumetric flow rate
```

```
disp(['VPFR = ', num2str(VPFR), ' L']);
disp(['Tau (VPFR/v0) = ', num2str(Tau), ' s']);
% Plot results
figure;
subplot(4,1,1); plot(t, CA, 'b', 'LineWidth', 2.5); xlabel('Time (s)'); ylabel('C_A (mol/L)'); title('Concentration of A');
ax = gca;
ax.FontSize = 14;
ax.FontWeight = 'bold';
subplot(4,1,2); plot(t, CC, 'r', 'LineWidth', 2.5); xlabel('Time (s)'); ylabel('C_C (mol/L)'); title('Concentration of C');
ax = gca;
ax.FontSize = 14;
ax.FontWeight = 'bold';
subplot(4,1,3); plot(t, T, 'g', 'LineWidth', 2.5); xlabel('Time (s)'); ylabel('T (K)'); title('Temperature');
ax = gca;
ax.FontSize = 14;
ax.FontWeight = 'bold';
subplot(4,1,4); plot(t, XA, 'k', 'LineWidth', 2.5); xlabel('Time (s)'); ylabel('X_A'); title('Conversion of A');
ax = gca;
ax.FontSize = 14;
ax.FontWeight = 'bold';
% ===========================================================
% ODE Function: dPFR_ode_function.m (This function must be defined either in its own .m file
% with the same name, or as a local/nested function within the main script.)
% It's generally cleaner to put it in a separate file.
% For simplicity here, I'm providing it as a local function, which works
% if saved in the same .m file as the main script.
% ===========================================================
function dydt = dPFR_ode_function(~, y, v0, CA0_feed, CB0_feed, P0, R, T0_feed, VPFR, D, F0_total_feed)
   % This function calculates the time derivatives for a single, well-mixed reactor volume.
   % Inputs:
   % t: Current time (from ODE solver) - unused, denoted by ~
   % y: Current state vector [CA; CB; T]
   % v0: Initial volumetric feed flow rate (constant parameter)
   % CA0_feed, CB0_feed: Feed concentrations of A and B (constant parameters)
   % P0: Feed pressure (constant parameter)
   % R: Ideal gas constant (constant parameter)
   % T0_feed: Feed temperature (constant parameter)
   % VPFR: Reactor volume (constant parameter)
   % D: Reactor diameter (constant parameter)
   % F0_total_feed: Total molar feed flow rate (constant parameter)
   % Extract current state variables from y
   CA = y(1);
   CB = y(2);
   T = y(3);
   % — Apply input constraints for derivative calculation stability —
   % These clamp the values *used within this function call* to calculate derivatives.
   % They do NOT directly prevent the ODE solver from producing output 'y' values outside these ranges.
   CA = max(0, min(CA0_feed, CA)); % CA between 0 and initial feed CA
   CB = max(0, min(CB0_feed, CB)); % CB between 0 and initial feed CB
   T = max(200, min(700, T));     % T between 200K and 700K (for k and Cp stability)
   % Concentration of product C (based on stoichiometry: 1 A -> 2 C)
   % This assumes CC = 0 initially and is purely formed from CA consumed.
   CC = 2 * (CA0_feed - CA);
   % Pressure in reactor (assuming constant pressure drop)
   P_reactor = 0.98 * P0;
   % Current volumetric flow rate inside the reactor (L/s)
   % Assumes ideal gas behavior and relates flow to temperature and pressure changes.
   v_current = v0 * (T / T0_feed) * (P0 / P_reactor);
   % Heat capacities (J/mol.K) - these depend on the current temperature 'T'
   CPA = 29.093 - (0.2e-3 * T) + (0.4e-6 * T^2) - (0.87e-9 * T^3) + (0.0055e5 / T^2);
   CPB = 37.4 - (0.0034e-3 * T) + (0.18e-6 * T^2) - (0.89e-9 * T^3) - (0.245e5 / T^2);
   CPC = 28.14 + (1.88e-3 * T) - (6.954e-6 * T^2) + (1.08e-9 * T^3) - (0.205e5 / T^2);
   % Molar flow rates of individual species at current conditions
```

(Continued)

**(Continued)**

```
FA_current = v_current * CA;
FB_current = v_current * CB;
FC_current = v_current * CC;
F_total_current = FA_current + FB_current + FC_current; % Current total molar flow rate
% Current mole fractions (for mixture heat capacities)
% Handle potential division by zero
if F_total_current <= 1e-10
  YA_current = 0; YB_current = 0; YC_current = 0;
else
  YA_current = FA_current / F_total_current;
  YB_current = FB_current / F_total_current;
  YC_current = FC_current / F_total_current;
end
% Mixture heat capacity at current conditions (J/mol.K)
Cpmix_current = YA_current * CPA + YB_current * CPB + YC_current * CPC;
% Constant volume heat capacities (J/mol.K)
CVA = CPA - R;
CVB = CPB - R;
CVC = CPC - R;
CVavg_current = YA_current * CVA + YB_current * CVB + YC_current * CVC;
% Reaction rate constant (s^-1 for first order, L/mol.s for second order)
% Here k is for a second order reaction (L/mol.s)
k = 2e11 * exp(-169900 / (R * T));
% Rate of reaction (mol/L.s)
r = k * CA * CB;
% Enthalpy change of reaction (J/mol)
% Delta Cp = (2*CpC) - (CpA + CpB)
Delta_Cp_coeffs = [
  (2 * 28.14 - 29.093 - 37.4), ...
  (2 * 1.88e-3 - (-0.2e-3) - (-0.0034e-3)), ...
  (2 * (-6.954e-6) - (0.4e-6) - (0.18e-6)), ...
  (2 * 1.08e-9 - (-0.87e-9) - (-0.89e-9)), ...
  (2 * (-0.205e5) - (0.0055e5) - (-0.245e5))
];
% Analytical integral of Delta_Cp (from 298 K reference)
integral_Cp_func = @(Temp) (Delta_Cp_coeffs(1) * Temp + ...
                Delta_Cp_coeffs(2) / 2 * Temp.^2 + ...
                Delta_Cp_coeffs(3) / 3 * Temp.^3 + ...
                Delta_Cp_coeffs(4) / 4 * Temp.^4 - ...
                Delta_Cp_coeffs(5) ./ Temp);
dHr = -94000 + (integral_Cp_func(T) - integral_Cp_func(298)); % J/mol (negative for exothermic)
% Heat exchange term (Watts)
AH = (4 * VPFR * 0.001 / D); % Heat transfer area (m^2) - 0.001 converts L to m^3, so AH in m^2
T_switch = 685; % K (your defined switch point for cooling/heating)
Qh = 500 * AH * (685 - T); % Heating duty
Qc = 400 * AH * (298 - T); % Cooling duty
transition = 1 / (1 + exp((T - T_switch))); % Smooth transition function
Qexch = transition * Qh + (1 - transition) * Qc; % Net heat exchange
% Total moles currently in the reactor volume
Ctot_current = CA + CB + CC;
n_total_moles_in_reactor = VPFR * Ctot_current; % Moles
% — ODEs —
% 1. Mass balance for Component A (mol/L.s)
% Accumulation = (Flow In - Flow Out) - Reaction
dCA_dt = (v0 * CA0_feed - v_current * CA) / VPFR - r;
% 2. Mass balance for Component B (mol/L.s)
dCB_dt = (v0 * CB0_feed - v_current * CB) / VPFR - r;
% 3. Energy balance (K/s)
```

```
    % Accumulation = (Flow In Enthalpy - Flow Out Enthalpy) + Heat Generation by Reaction + Heat Exchange
    % (F * Cp * T)_in - (F * Cp * T)_out
    % Need initial mole fractions to calculate Cpmix_feed (at T0_feed)
    YA0_molar = CA0_feed / (CA0_feed + CB0_feed); % Assuming only A and B initially
    YB0_molar = CB0_feed / (CA0_feed + CB0_feed);
    YC0_molar = 0;
    CPA0_feed_calc = 29.093 - (0.2e-3*T0_feed) + (0.4e-6*T0_feed^2) - (0.87e-9*T0_feed^3) + (0.0055e5/T0_feed^2);
    CPB0_feed_calc = 37.4 - (0.0034e-3*T0_feed) + (0.18e-6*T0_feed^2) - (0.89e-9*T0_feed^3) - (0.245e5/T0_feed^2);
    CPC0_feed_calc = 28.14 + (1.88e-3*T0_feed) - (6.954e-6*T0_feed^2) + (1.08e-9*T0_feed^3) - (0.205e5/T0_feed^2);
    Cpmix_feed = YA0_molar * CPA0_feed_calc + YB0_molar * CPB0_feed_calc + YC0_molar * CPC0_feed_calc;
    Heat_Gen_Rxn = -r * dHr * VPFR; % Watts (r in mol/L.s, dHr in J/mol, VPFR in L)
    % The dT/dt equation represents d(U)/dt = ... / (n*Cv_avg)
    dT_dt = (F0_total_feed * Cpmix_feed * T0_feed - F_total_current * Cpmix_current * T + Heat_Gen_Rxn + Qexch) / ...
        (n_total_moles_in_reactor * CVavg_current);
    dydt = [dCA_dt; dCB_dt; dT_dt];
end
```

## Reference


1 H.S. Fogler. "*Elements of Chemical Reaction Engineering*". Chapter 4: "Stoichiometry", pp. 270–337. Pearson Education, Inc.; 6th edition (2020). ISBN-10: 013548622X. ISBN-13: 978-0135486221.

# Appendix C

# The Stability of Numerical Solutions of Algebraic and Differential Equations

Numerical stability is a crucial factor when solving algebraic and differential equations (ADEs and ordinary differential equations [ODEs]), especially in stiff systems and those involving delays or oscillatory behavior. A well-chosen numerical method ensures the accuracy of the solution while maintaining computational efficiency. It appears well in the solution of an unsteady-state gas-phase CSTR or PFR.

## C.1 Defining a Stiff System in Terms of Eigenvalues

A system of ODEs is considered **stiff** if it contains widely varying time scales, leading to rapidly decaying transient components. This stiffness can be understood through the eigenvalues of the Jacobian matrix:

$$J = \frac{\partial f}{\partial y}$$

where $f$ is the vector-valued function defining the system of differential equations.

The system is stiff if:

- The Jacobian matrix has eigenvalues $\lambda_i$ with large negative real parts (e.g., $|\lambda_{max}| \gg |\lambda_{min}|$).
- The stiffness ratio, defined as $\rho = |\lambda_{max}|/|\lambda_{min}|$, is significantly large.

In our 5.4 example, we have:

$$y = \left[\begin{pmatrix} T \\ C_A \end{pmatrix}\right]$$

$$\frac{dy}{dt} = \begin{bmatrix} \dfrac{dT}{dt} \\ \dfrac{dC_A}{dt} \end{bmatrix} = f(y)$$

$$\frac{d(T)}{dt} = \frac{1}{nC_V^{mix}}[FA0 * (CpA * T0 - Cpmix * T) - k_1 C_A V_{CSTR}(\Delta H_r)]$$

$$\frac{d(C_A)}{dt} = \left[\frac{FA0 - FA}{V_{CSTR}} - k1C_A\right] = \left[\frac{FA0 - v_0 C_A}{V_{CSTR}} - k1C_A\right]$$

$$f(y) = \begin{bmatrix} \frac{1}{nC_V^{mix}}[FA0 * (CpA * T0 - Cpmix * T) - k_1 C_A V_{CSTR}(\Delta H_r)] \\ \left[\frac{FA0 - v_0 C_A}{V_{CSTR}} - k1C_A\right] \end{bmatrix} = \begin{bmatrix} f_1 \\ f_2 \end{bmatrix}$$

The Jacobian matrix for the above set of two ODEs is defined as:

$$J = \begin{bmatrix} \frac{\partial f_1}{\partial T} & \frac{\partial f_1}{\partial C_A} \\ \frac{\partial f_2}{\partial T} & \frac{\partial f_2}{\partial C_A} \end{bmatrix}$$

We will not show further details, but to say in words that $J$, the Jacobian matrix, shows how each equation responds to small changes in $T$ and $C_A$.

Explicit numerical methods require extremely small-time steps $\Delta t$ to maintain stability, making them inefficient. Stiff systems commonly arise in chemical kinetics, heat transfer, and control systems, where some processes evolve much faster than others.

## C.2 Stiff Versus Nonstiff Systems on Numerical Stability

Stiff systems require implicit solvers (e.g., backward differentiation formula [BDF] methods, such as MATLAB's **ode15s**), which remain stable even for large time steps.

On the other hand, nonstiff systems can be efficiently solved using explicit solvers like **ode45** (Runge–Kutta methods) that do not require solving nonlinear algebraic equations.

For stiff systems, using explicit solvers can lead to instability unless an extremely small-time step is used, leading to impractical computation times.

## C.3 Trade-Off Between Accuracy and Speed

The choice of a numerical method involves balancing accuracy and computational speed:

a) High-order explicit methods (e.g., Runge–Kutta 4th-order) provide better accuracy but may require small time steps, increasing computation time.
b) Implicit methods (e.g., BDF methods) allow larger time steps while maintaining stability but require solving nonlinear systems at each step, which is computationally expensive.
c) Adaptive time-stepping methods adjust $\Delta t$ based on local error estimates, optimizing both accuracy and efficiency.

The optimal trade-off depends on problem stiffness, available computational resources, and the desired accuracy.

## C.4 Implicit Versus Nonimplicit Solutions

a) **Explicit methods** compute the next step directly from the current values. Example: $y_{n+1} = y_n + hf(y_n, t_n)$. They are efficient for nonstiff ODEs and require small time steps for stability in stiff problems.

b) **Implicit methods** involve solving an equation at each step, typically requiring iterative solvers. Example: $y_{n+1} = y_n + hf(y_{n+1}, t_{n+1})$. They are more stable for stiff problems and require solving nonlinear algebraic equations at each step. Implicit solvers handle stiff problems well, but at the cost of increased computational effort.

## C.5 Oscillatory ODEs/ADEs and Their Effect on Stability

Oscillatory systems contain eigenvalues with significant imaginary parts:

$$\lambda = \alpha \mp i\beta$$

where $\beta$ determines the oscillation frequency. Such systems are sensitive to numerical damping:

Explicit methods may introduce numerical instability (spurious oscillations) if the time step is too large. On the other hand, implicit methods can introduce excessive damping, suppressing real oscillations (undesired behavior in wave problems).

Hence, special **symplectic integrators** and **A-stable methods** are recommended for oscillatory systems, as they preserve the energy of the oscillations. Symplectic integrators are a special class of numerical methods designed to preserve the Hamiltonian structure of dynamical systems, particularly in oscillatory and conservative systems, such as those found in classical mechanics and molecular dynamics simulations. Examples of **symplectic** methods are:

a) **Leapfrog Method (Verlet Integration):** It is used in molecular dynamics and celestial mechanics and preserves energy over long simulations.
b) **Stormer–Verlet Method:** It is a second-order symplectic integrator and is commonly used in N-body problems.
c) **Implicit Midpoint Method:** It is a simple symplectic scheme useful for stiff oscillatory problems.

On the other hand, the **A-stable** method is a numerical method for solving ODEs, which remains stable for all step sizes when applied to the linear test equation: $y' = \lambda y$, with real$(\lambda) \leq 0$. where $\lambda$ is a complex number whose real part is nonpositive.

When solving stiff ODEs (where some components decay rapidly while others evolve slowly), explicit methods require very small step sizes to maintain stability. The A-stable methods allow for large time steps without introducing instability, making them ideal for stiff problems. Examples of **A-stable** methods are:

a) **Backward Euler (BE) Method:**

$$y_{n+1} = y_n + hf(y_{n+1})$$

BE method is fully implicit and unconditionally stable for any step size and is used in stiff problems.

b) **Trapezoidal Rule (Crank-Nicholson Method):**

$$y_{n+1} = y_n + \frac{h}{2}(f(y_n) + f(y_{n+1}))$$

It is a second-order accurate and A-stable. It can exhibit oscillations for very stiff problems.

c) **Backward Differentiation Formulas (BDF Methods):**
BDF1 (first-order Backward Euler) method is both A-stable and L-stable. On the other hand, the BDF2 method (a multistep implicit method, requiring values at two previous steps: $y_n$ and $y_{n-1}$)

is A-stable but not L-stable. On the contrary, the higher-order BDF (≥3) method is neither A-stable nor L-stable. **L-stability** means that as $z \to \infty$ (where $z = \lambda h$, with $\lambda$ being an eigenvalue of the system and $h$ the step size), the method damps out errors completely, preventing numerical oscillations.

Notice that some implicit symplectic integrators are also A-stable. These methods combine both properties: A-stability, which handles stiff problems effectively, and symplecticity, which preserves energy in oscillatory systems. Examples of methods that are both symplectic and A-stable:

a) Implicit Midpoint Rule (A-stable and symplectic)
b) Gauss-Legendre Runge–Kutta Methods (symplectic and A-stable)

## C.6 Delay Differential Equations (DDEs) and Stability

**DDEs Involve Terms Dependent on Past States:**

$$\frac{dy}{dt} = f(y, y(t - \tau), t)$$

where $\tau$ is the delay. The presence of delays impacts stability in two major ways:

a) Delays can introduce oscillations or cause instability by affecting feedback mechanisms.
b) Numerical solvers must account for interpolation of past values to maintain accuracy.

**Methods for DDEs:**

a) Explicit solvers like **dde23** in MATLAB use adaptive time stepping but may struggle with long delays.
b) Implicit solvers like **ddesd** offer better stability but require solving nonlinear algebraic systems.

# Appendix D

# Calculation of Fluidized Bed Fluid and Flow Properties

```
% MATLAB Algorithm for calculating fluid dynamics parameters
clear variables;

% Step 1: Prompt the user for bed porosity, eFB
eFB = input('Enter the bed porosity (dimensionless. See Table 6.6): ');

% Step 2: Prompt the user for particle sphericity, phi
phi = input('Enter the particle sphericity (dimensionless). See Table 6.6: ');

% Step 3: Prompt the user for the Sauter particle diameter, dp (μm)
dp = input('Enter the Sauter particle diameter (μm). See Table 6.6: ');
dp=dp*1.0E-6;

% Step 4: Prompt the user for the particle density, rowp (kg/m^3)
rowp = input('Enter the particle density (kg/m^3). See Table 6.6: ');

% Step 5: Prompt the user for the temperature, T (Kelvin)
T = input('Enter FB temperature (Kelvin): ');
% Step 6: Prompt the user for the pressure, P (atm)
P = input('Enter FB pressure (atm): ');

% Step 7: Prompt the user to select fluid type (Gas or Liquid)
fluid_type = input('Select the fluid type (Enter "Gas" or "Liquid"): ', 's');

if strcmpi(fluid_type, 'Gas')
   % Case A: Gas

   % Step 8: Prompt the user for the molecular mass, MW (g/mol)
   MW = input('Enter the molecular mass (g/mol): ');
   rowf=P*MW/(0.082057*T);

   % Step 9: Prompt the user for gas viscosity, mu (Pa.s)
   mu = input('Enter the gas viscosity (Pa.s),e.g., mu_air=1.81E-5 Pa.s: ');

   % Step 10: Calculate Schmidt number for gas, NSc (dimensionless)
   NSc = 0.70 *(P)^0.01* (T/300)^0.1; % NSc=0.75 is at room temperature for many gases.
   fprintf('Schmidt number for gas @T, NSc = %.5f (dimensionless)\n', NSc);
   % Step 10: Calculate diffusivity, D (m^2/s)
   D = mu / (rowf * NSc);
   D2=D*1.0E4;
   fprintf('Gas Diffusivity, D = %.5e m^2/s\n', D);
   fprintf('Gas Diffusivity, D = %.5e cm^2/s\n', D2);
```

*(Continued)*

**(Continued)**

```
else
  % Case B: Liquid
  % Step 11: Prompt user to select one of seven liquids
  liquid_type = input(['Select one liquid: Water, Glycerol, Ethanol, n-Hexane, ' ...
            'n-Heptane, n-Octane, Toluene: '], 's');
  % Step 12: Prompt the user for liquid viscosity at the given T, mu (Pa.s)
  mu = input('Enter the liquid viscosity (Pa.s) at the given T: ');
  % Step 13: Prompt the user for the liquid density, rowf (kg/m^3)
  rowf = input('Enter the liquid density (kg/m^3): ');
  % Step 14: Assign NSc at 300 K based on the selected liquid
  switch lower(liquid_type)
    case 'water'
      NSc300K = 600;
    case 'glycerol'
      NSc300K = 10000;
    case 'ethanol'
      NSc300K = 1200;
    case 'n-hexane'
      NSc300K = 300;
    case 'n-heptane'
      NSc300K = 400;
    case 'n-octane'
      NSc300K = 500;
    case 'toluene'
      NSc300K = 500;
    otherwise
      error('Invalid liquid selection');
  end

  % Step 15: Calculate Schmidt number for liquid at the given T, NSc (dimensionless)
  NSc = NSc300K * (300 / T)^4.25;
  fprintf('Selected liquid: %s\n', liquid_type);
  fprintf('Schmidt number for the selected liquid @T, NSc = %.5f (dimensionless) \n', NSc);

  % Step 16: Calculate diffusivity, D (m^2/s)
  D = mu / (rowf * NSc);
  D2=D*1.0E4;
  fprintf('Selected Liquid Diffusivity, D = %.5e m^2/s\n', D);
  fprintf('Selected Liquid Diffusivity, D = %.5e cm^2/s\n', D2);
end

% Step 17: Calculate Archimedes number, N_Ar (dimensionless)
g = 9.806; % gravitational acceleration (m/s^2)
N_Ar = ((g * dp^3 * rowf^2) / (mu^2)) * (rowp / rowf - 1);

% Step 18: Calculate minimum fluidization velocity, Umf (m/s)
sqrtTerm= (sqrt(1 + (3.1e-4) * ((phi * eFB)^3 / (1 - eFB)^2) * N_Ar) - 1);
Umf = 42.9 * ((1 - eFB) / phi) * (mu / (dp * rowf)) * sqrtTerm;

% Step 19: Print Umf
fprintf('Minimum fluidization velocity, Umf = %.5f m/s\n', Umf);

NReUmf=(rowf * Umf * dp) / mu;
% Step 20: Find the terminal velocity

% Define the function f(NRe)
f = @(NRetvf) NRetvf^2 * (24/NRetvf + 4/sqrt(NRetvf) + 0.4) - (4/3) * N_Ar;

% Set an initial guess for NRe (a positive number)
initial_guess = 1;
```

```
% Use fsolve to find the root of f(NRe) = 0
options = optimset('Display', 'iter'); % Display iterations (optional)
NRe_root = fsolve(f, initial_guess, options);
Utv=NRe_root*mu/(rowf*dp);
% Display the result
fprintf('The terminal velocity,Utv is: %.6f m/s\n', Utv);
fprintf('The root which is NRetv at the terminal velocity is: %.6f\n', NRe_root);

% Step 20: Calculate superficial velocity, U0 (m/s), assuming 50% of the
% terminal velocity. U0=0.5*U0max=0.5*(1-eFB)*Utv
U0 = 0.5* (1 - eFB)*Utv;

% Step 21: Print U0
fprintf('Operational superficial velocity, U0 = %.5f m/s\n', U0);

%% Step 22: Calculate pressure drop per bed length evaluated at the selected U0.
DelP_over_h=150.0E-3*((1-eFB)^2/eFB^3)*(mu*U0/(phi*dp)^2)+1.75E-3*((1-eFB)/eFB^3)*(rowf*(U0^2))/(phi*dp);

fprintf('Pressure drop per bed length h, evaluated @U0, ΔP/h, = %.5f kPa/m\n',
  DelP_over_h);

% Step 23: Calculate Reynolds number, N_Re (dimensionless)

NReU0 = (rowf * U0 * dp) / mu;
fprintf('Reynold Number @U0, dimensionless = %.5f \n', NReU0);

% Step 24: Calculate Sherwood number, N_Sh (dimensionless)
N_Sh = 2 + 0.6 * NReU0^0.5 * NSc^(1/3);
fprintf('Sherwood Number @U0, dimensionless = %.5f \n', N_Sh)

% Step 25: Calculate mass transfer coefficient, kg (m/s)
kg = N_Sh * D / dp;

% Step 26: Print mass transfer coefficient, kg
fprintf('Mass transfer coefficient, kg = %.5e m/s\n', kg);
% Step #27: Prompt the user to enter the diameter of the bed
DB = input('Bed Diameter (m): ');
% Step #28: Prompt the user to enter the height of the bed
H = input('Enter Bed Height (m): ');
% Step #29. Calculate The FB superficial area, AB, expressed in m2:
AB = (pi/4)*DB^2;
fprintf('FB Superficial Area = %.5e m2\n', AB);
% Step #30. Calculate The fluid volume flow rate, VDot, expressed in m3/s:
VDot = U0*AB;
fprintf('Fluid Volume Flow Rate = %.5e m3/s\n', VDot);
% Step #31. Calculate The fluid mass flow rate, MDot, expressed in kg/s:
MDot = VDot*rowf;
fprintf('Fluid Mass Flow Rate = %.5e kg/s\n', MDot);
% Step #32. Calculate the volume of FB, Vol, expressed in m3:
Vol=AB*H;
fprintf('FB Volume = %.5e m3\n', Vol);
% Step #33. Calculate the pressure drop of FB, ΔP, expressed in kPa:
DPcolumn=DelP_over_h*H;
fprintf('FB ΔP = %.5e kPa\n', DPcolumn);
% Step #34. Enering, reaction order, rate constant, interfacial area of FB, and FA0.
n = input('Enter the reaction order (n): ');
kprime = input('Enter the heterogenous rate constant, k (L^n/(kg cat.mol^(n-1).time)): ');
ia = input('Enter the interfacial area per volume, a (m2/m3): ');
FA0 = input('Enter FA0 (mol/time): ');
% Step #35. CA,bulk0 caculation.
CAbulk0=FA0/(VDot*1000);
```

*(Continued)*

**(Continued)**

```
% Step #36. Based on n, CAsurface0 will be determined.
if n==1
  fprintf('CAbulk0 (mol/L) = %.5e mol/L\n', CAbulk0);

  CAsurface0=(kg*ia)/(kprime+kg*ia)*(CAbulk0);
  fprintf('CAsurface0 (mol/L) = %.5e mol/L\n', CAsurface0);
elseif n ~=1

f2=@(CAsurface0) CAsurface0-(((kg*ia)/((kg*ia )+kprime*CAsurface0^(n-1)))*CAbulk0);

% Set an initial guess for NRe (a positive number)
initial_guess = 0.5* CAbulk0;

% Use fsolve to find the root of f(NRe) = 0
options = optimset('Display', 'iter'); % Display iterations (optional)
CAsurface0 = fsolve(f2, initial_guess, options);
fprintf('CAbulk0 (mol/L) = %.5e mol/L\n', CAbulk0);
fprintf('CAsurface0 (mol/L) = %.5e mol/L\n', CAsurface0);
else
   disp ('You entered a non-realistic reaction order n');
end
```

# Index

## f

## *i*



## *j*



## *k*



## *l*

Printed in the USA/Agawam, MA
September 2, 2026
910009.006